Methods in Enzymology

Volume 166
BRANCHED-CHAIN AMINO ACIDS

METHODS IN ENZYMOLOGY

EDITORS-IN-CHIEF

John N. Abelson   Melvin I. Simon

Methods in Enzymology

Volume 166

Branched-Chain Amino Acids

EDITED BY

Robert A. Harris

DEPARTMENT OF BIOCHEMISTRY
INDIANA UNIVERSITY SCHOOL OF MEDICINE
INDIANAPOLIS, INDIANA

John R. Sokatch

DEPARTMENT OF BIOCHEMISTRY AND MOLECULAR BIOLOGY
THE UNIVERSITY OF OKLAHOMA
OKLAHOMA CITY, OKLAHOMA

ACADEMIC PRESS, INC.
Harcourt Brace Jovanovich, Publishers
San Diego   New York   Berkeley   Boston
London   Sydney   Tokyo   Toronto

ACADEMIC PRESS, INC.
San Diego, California 92101

United Kingdom Edition published by
ACADEMIC PRESS LIMITED
24-28 Oval Road, London NW1 7DX

LIBRARY OF CONGRESS CATALOG CARD NUMBER: 54-9110

ISBN   0-12-182067-X   (alk. paper)

PRINTED IN THE UNITED STATES OF AMERICA
88 89 90 91   9 8 7 6 5 4 3 2 1

Table of Contents

Section I. Analytical and Synthetic Methods

Section II. Enzyme Assay

Section III. Enzymes

Contributors to Volume 166

Article numbers are in parentheses following the names of contributors.
Affiliations listed are current.

D. JOHN ABERHART (3, 4), *Worcester Foundation for Experimental Biology, Shrewsbury, Massachusetts 01545*

R. PAUL AFTRING (24), *Departments of Medicine and Biochemistry, Medical University of South Carolina, Charleston, South Carolina 29425*

ROBERT H. ALLEN (49, 50, 51), *Division of Hematology, University of Colorado Health Sciences Center, Denver, Colorado 80262*

TAMMY K. ANTONUCCI (31), *Department of Immunology and Microbiology, University of Michigan, Medical School, Ann Arbor, Michigan 48109*

ZE'EV BARAK (29, 57), *Department of Biology, Ben Gurion University of the Negev, Beer Sheva 84105, Israel*

KIM BARTLETT (11), *Departments of Child Health and Clinical Biochemistry, The Medical School, University of Newcastle upon Tyne, Newcastle upon Tyne NE2 4HH, England*

MARK BEGGS (21, 22), *Department of Clinical Biochemistry, University of Oxford, John Radcliffe Hospital, Oxford OX3 9DU, England*

L. L. BIEBER (37), *Department of Biochemistry, Michigan State University, East Lansing, Michigan 48824*

KEVIN P. BLOCK (24), *Departments of Medicine and Biochemistry, Medical University of South Carolina, Charleston, South Carolina 29425*

ANDREW P. BRADFORD (2), *Department of Biochemistry, University of Newcastle upon Tyne, Newcastle upon Tyne NE2 4HH, England*

JON BREMER (59), *Institute for Medical Biochemistry, University of Oslo, Oslo 3, Norway*

MARIA GORDON BUSE (24), *Departments of Medicine and Biochemistry, Medical University of South Carolina, Charleston, South Carolina 29425*

DENIS B. BUXTON (60), *Division of Nuclear Medicine and Biophysics, Department of Radiological Sciences, University of California Los Angeles Medical School, Los Angeles, California 90024*

JOSEPH M. CALVO (27, 57), *Department of Biochemistry and Molecular Biology, Cornell University, Ithaca, New York 14850*

ANTHONY G. CAUSEY (11), *Department of Drug Metabolism, Pfizer Central Research, Sandwich, Kent CT13 9NJ, England*

DAVID M. CHIPMAN (29), *Department of Biology, Ben Gurion University of the Negev, Beer Sheva 84 105, Israel*

DAVID T. CHUANG (18, 19), *Department of Medicine, Cleveland Veterans Medical Center, and Department of Medicine and Biochemistry, Case Western Reserve University, Cleveland, Ohio 44106*

KENNETH G. COOK (2, 39), *Department of Biochemistry, University of Newcastle upon Tyne, Newcastle upon Tyne NE2 4HH, England*

BARBARA E. CORKEY (9), *Division of Diabetes and Metabolism, Boston University School of Medicine, Boston, Massachusetts 02118*

NEAL W. CORNELL (10), *Laboratory of Metabolism and Molecular Biology, NIAAA, Rockville, Maryland 20852*

JACQUES-ALAIN COTTING (3), *Worcester Foundation for Experimental Biology, Shrewsbury, Massachusetts 01545*

RODY P. COX (18), *Department of Medicine, Cleveland Veterans Medical Center, and*

xi

Case Western Reserve University, Cleveland, Ohio 44106

PAMELA L. CROWELL (7), Department of Nutritional Sciences, College of Agricultural and Life Sciences, University of Wisconsin-Madison, Madison, Wisconsin 53706

ZAHI DAMUNI (42), Department of Biology, University of South Carolina, Columbia, South Carolina 29208

DEAN J. DANNER (14, 38), Division of Medical Genetics, Department of Pediatrics, Emory University School of Medicine, Atlanta, Georgia 30322

E. JACK DAVIS (59), Indiana University School of Medicine, Department of Biochemistry, Indianapolis, Indiana 46223

MAURILIO DE FELICE (30), International Institute of Genetics and Biophysics, 80125 Naples, Italy, and Istituto di Ingegneria Chimico-Alimentare, Facolté di Ingegneria, Université di Salerno, 84100 Baronissi, Salerno, Italy

WILLIAM T. E. EDWARDS (20), AFRC Institute of Food Research, Norwich Laboratory, Norwich NR4 7UA, England

MARINOS ELIA (5), MRC Dunn Clinical Nutrition Center, Cambridge CB2 1QE, England

LOUIS J. ELSAS (14), Division of Medical Genetics, Department of Pediatrics, Emory University School of Medicine, Atlanta, Georgia 30322

MARK EMPTAGE (12), Central Research Development Department, Experimental Station E328/246A, E. I. du Pont de Nemours & Company, Wilmington, Delaware 19898

LILLIAN EOYANG (55), Department of Molecular Biology, Division of Biological Sciences, Albert Einstein College of Medicine, Bronx, New York 10461

JOSEPH ESPINAL (21, 22), Department of Clinical Biochemistry, University of Oxford, John Radcliffe Hospital, Oxford OX3 9DU, England

ANGELA L. GASKING (5), AFRC Institute of Food Research, Norwich Laboratory, Norwich NR4 7UA, England

KENNETH M. GIBSON (25, 26), Kimberly H. Courtwright and Joseph W. Summers Metabolic Disease Center, Baylor University Medical Center, Dallas, Texas 75246

NATAN GOLLOP (29), Department of Biology, Ben Gurion University of the Negev, Beer Sheva 84 105, Israel

GARY W. GOODWIN (23), Indiana University, School of Medicine, Department of Biochemistry, Indianapolis, Indiana 46223

GIOVANNA GRIFFO (30), International Institute of Genetics and Biophysics, 80125 Naples, Italy

ARNOLD HAMPEL (33), Department of Biological Sciences and Chemistry, Northern Illinois University, DeKalb, Illinois 60115

MARY E. HANDLOGTEN (32), Department of Biochemistry and Molecular Biology, J. Hillis Miller Health Center, University of Florida, Gainesville, Florida 32610

ALFRED E. HARPER (7, 24), Departments of Nutritional Sciences and Biochemistry, University of Wisconsin-Madison, Madison, Wisconsin 53706

ROBERT A. HARRIS (15, 23), Indiana University, School of Medicine, Department of Biochemistry, Indianapolis, Indiana 46223

KENNETH HATTER (48), Department of Biochemistry and Molecular Biology, The University of Oklahoma Health Sciences Center, Oklahoma City, Oklahoma 73190

SUE C. HEFFELFINGER (38), Department of Pathology, Baylor School of Medicine, Texas Children's Hospital, Houston, Texas 77030

ELLEN HILDEBRANDT (60), Department of Biochemistry, Saint Judes Childrens Hospital, Memphis, Tennessee 38101

ANTHONY HOBSON-FROHOCK (5), AFRC Institute of Food Research, Norwich Laboratory, Norwich NR4 7UA, England

YASUYUKI IKEDA (20, 46, 47), National Cardiovascular Center, Research Institute, Suita City 565, Osaka, Japan

JULIUS H. JACKSON (28), *Department of Microbiology, Meharry Medical College, Nashville, Tennessee 37208*

SHEELAGH M. A. JONES (2), *Department of Biochemistry, University of Newcastle upon Tyne, Newcastle upon Tyne NE2 4HH, England*

RYO KIDO (35), *Department of Biochemistry, Wakayama Medical College, Wakayama 640, Japan*

MICHAEL S. KILBERG (32), *Department of Biochemistry and Molecular Biology, J. Hillis Miller Health Center, University of Florida, Gainesville, Florida 32610*

M. TODD KING (10), *Laboratory of Metabolism and Molecular Biology, NIAAA, Rockville, Maryland 20852*

GUNTER B. KOHLHAW (52, 53, 54), *Department of Biochemistry, Purdue University, West Lafayette, Indiana 47907*

J. FRED KOLHOUSE (51), *Division of Hematology, University of Colorado Health Sciences Center, Denver, Colorado 80262*

T. K. KORPELA (34), *Department of Biochemistry, University of Turku, SF-20500 Turku, Finland*

ROBIN J. KOVACHY (49), *Division of Hematology, University of Colorado Health Sciences Center, Denver, Colorado 80262*

MARTHA J. KUNTZ (15), *Department of Biochemistry, Indiana University School of Medicine, Indianapolis, Indiana 46223*

CARMINE T. LAGO (30), *International Institute of Genetics and Biophysics, via G. Marconi 10, 80125 Naples, Italy*

DAVID S. LAPOINTE (60), *Department of Biochemistry, University of Texas Health Science Center, San Antonio, Texas 77284*

ROBERT A. LaROSSA (13), *Central Research and Development Department, E. I. du Pont de Nemours & Company, Wilmington, Delaware 19898*

SUNG-HEE CHO LEE (59), *Hyo-Sung University, Department of Food Science and Nutrition, Taegu, Korea*

DANILA LIMAURO (30), *International Institute of Genetics and Biophysics, via G. Marconi 10, 80125 Naples, Italy*

STUART LITWER (14), *Division of Medical Genetics, Department of Pediatrics, Emory University School of Medicine, Atlanta, Georgia 30322*

GEOFFREY LIVESEY (1, 5, 36), *AFRC Institute of Food Research, Norwich Laboratory, Norwich NR4 7UA, England*

PETER N. LOWE (43), *Department of Molecular Sciences, Wellcome Research Laboratories, Beckenham, Kent BR33BS, England*

PATRICIA LUND (1, 36), *Metabolic Research Laboratory, Nuffield Department of Clinical Medicine, Radcliffe Infirmary, Oxford OX2 6HE, England*

RON MAGOLDA (12), *Central Research Development Department, Experimental Station E328/246A, E. I. du Pont de Nemours & Company, Wilmington, Delaware 19898*

ORVAL A. MAMER (6, 8), *McGill University Biomedical Mass Spectrometry Unit, Montreal, Quebec, Canada H3A 1A3*

ROBERT H. MILLER (7), *Ross Laboratories, Columbus, Ohio 43216*

JANE A. MONTGOMERY (6, 8), *McGill University Biomedical Mass Spectrometry Unit, Montreal, Quebec, Canada H3A 1A3*

GLENN E. MORTIMORE (58), *Department of Physiology, The Milton S. Hershey Medical Center, The Pennsylvania State University, Hershey, Pennsylvania 17033*

MERLE S. OLSON (60), *Department of Biochemistry, University of Texas Health Science Center, San Antonio, Texas 77284*

DALE L. OXENDER (31), *Department of Biological Chemistry, University of Michigan, Ann Arbor, Michigan 48109*

TARUN B. PATEL (60), *Department of Pharmacology, University of Tennessee Center for the Health Sciences, Memphis, Tennessee 38163*

PHILIP A. PATSTON (22), *Department of Clinical Biochemistry, University of Oxford, John Radcliffe Hospital, Oxford OX3 9DU, England*

RALPH PAXTON (23, 41), *Department of Bio-*

logical Sciences, Texas Technical University, Lubbock, Texas 79409

RICHARD N. PERHAM (43), *Department of Biochemistry, University of Cambridge, Cambridge CB2 1QW, England*

FLORA H. PETTIT (40), *Clayton Foundation Biochemical Institute, The University of Texas at Austin, Austin, Texas 78712*

A. REETA PÖSÖ (58), *Department of Biochemistry, College of Veterinary Medicine, Hämeentie 57, SF-00550, Helsinki 55, Finland*

J. MICHAEL POSTON (16, 17), *Laboratory of Biochemistry, National Heart, Lung, and Blood Institute, National Institutes of Health, Bethesda, Maryland 20892*

PHILIP J. RANDLE (21, 22), *Department of Clinical Biochemistry, University of Oxford, John Radcliffe Hospital, Oxford OX3 9DU, England*

LESTER J. REED (40, 42), *Clayton Foundation Biochemical Institute and Department of Chemistry, The University of Texas at Austin, Austin, Texas 78712*

PAUL D. REISS (10), *MilliGen Corporation, Bedford, Massachusetts 01730*

EZIO RICCA (30), *International Institute of Genetics and Biophysics, 80125 Naples, Italy, and IABBAM Consiglio Nazionale della Ricerche, 80147 Ponticelli, Naples, Italy*

P. J. SABOURIN (37), *Department of Biochemistry, Michigan State University, East Lansing, Michigan 48824*

JOHN V. SCHLOSS (12, 56, 57), *Central Research and Development, Experimental Station E328, E. I. du Pont de Nemours & Company, Wilmington, Delaware 19898*

LILLIE L. SEARLES (27), *Department of Biology, University of North Carolina, Chapel Hill, North Carolina 27599-3280*

PHILIP M. SILVERMAN (55), *Department of Molecular Biology, Division of Biological Sciences, Albert Einstein College of Medicine, Bronx, New York 10461*

RONALD SIMPSON (15), *Sandoz Research Institute, Sandoz Pharmaceutical Corporation, East Hanover, New Jersey 07936*

JOHN R. SOKATCH (44, 45, 48), *Department of Biochemistry and Molecular Biology, The University of Oklahoma Health Sciences Center, Oklahoma City, Oklahoma 73190*

ØYSTEIN SPYDEVOLD (59), *Institute for Medical Biochemistry, University of Oslo, Oslo 3, Norway*

SALLY P. STABLER (49, 50, 51), *Division of Hematology, University of Colorado Health Sciences Center, Denver, Colorado 80262*

P. J. SYKES (45), *Department of Haematology, Flinders Medical Centre, Bedford Park, South Australia 5042, Australia*

KAY TANAKA (20, 46, 47), *Yale University School of Medicine, Department of Human Genetics, New Haven, Connecticut 06510*

RICHARD TRITZ (33), *Department of Biological Sciences and Chemistry, Northern Illinois University, DeKalb, Illinois 60115*

DREW E. VAN DYK (56), *Central Research and Development, Experimental Station E328, E. I. du Pont de Nemours & Company, Wilmington, Delaware 19898*

TINA K. VAN DYK (13), *Central Research and Development, Experimental Station E328, E. I. du Pont de Nemours & Company, Wilmington, Delaware 19898*

PARVAN P. WAYMACK (60), *Center for Disease Control, Atlanta, Georgia 30033*

STEPHEN J. YEAMAN (2, 39), *Department of Biochemistry, University of Newcastle upon Tyne, Newcastle upon Tyne NE2 4HH, England*

BEI ZHANG (23), *Indiana University, School of Medicine, Department of Biochemistry, Indianapolis, Indiana 46223*

Preface

Branched-chain amino acids were first discussed in *Methods in Enzymology* in 1970 in a section of Volume XVII: Metabolism of Amino Acids and Amines, Part A. That section dealt primarily with biosynthetic enzymes of branched-chain amino acid pathways in prokaryotes. Although at that time the catabolic pathways for these amino acids had been quite well outlined, few of the catabolic enzymes had been purified. Recently, there has been a renaissance in research on several aspects of branched-chain amino acid metabolism which amply justifies a complete volume on this subject.

The structure and function of branched-chain keto acid dehydrogenase has attracted the attention of several investigators, in part because of the relationship of this enzyme complex to other keto acid dehydrogenase multienzyme complexes, and in part because of the role of this complex in maple syrup urine disease. There are a number of human genetic diseases associated with the metabolism of branched-chain amino acids which has led to the development of sensitive methods, principally high-performance liquid chromatography, for detecting metabolites of these pathways in serum. The pattern of metabolites helps localize the genetic lesion which can then be identified by specific enzyme assays. New enzyme assays have been developed, both for recently described enzymes, such as the branched-chain keto acid dehydrogenase, and well-known enzymes, such as acetolactate synthase. Several enzymes, both catalytic and biosynthetic, have been purified and are described in this volume.

We believe this volume will be useful to investigators of branched-chain amino acid metabolism, to those contemplating entering the field, and to those interested in the genetic diseases affecting branched-chain amino acid metabolism.

ROBERT A. HARRIS
JOHN R. SOKATCH

METHODS IN ENZYMOLOGY

EDITED BY

Sidney P. Colowick and Nathan O. Kaplan

VANDERBILT UNIVERSITY

SCHOOL OF MEDICINE

NASHVILLE, TENNESSEE

DEPARTMENT OF CHEMISTRY

UNIVERSITY OF CALIFORNIA

AT SAN DIEGO

LA JOLLA, CALIFORNIA

METHODS IN ENZYMOLOGY

EDITORS-IN-CHIEF

Sidney P. Colowick and Nathan O. Kaplan

VOLUME LX. Nucleic Acids and Protein Synthesis (Part H)
Edited by KIVIE MOLDAVE AND LAWRENCE GROSSMAN

VOLUME 61. Enzyme Structure (Part H)
Edited by C. H. W. HIRS AND SERGE N. TIMASHEFF

VOLUME 62. Vitamins and Coenzymes (Part D)
Edited by DONALD B. McCORMICK AND LEMUEL D. WRIGHT

VOLUME 63. Enzyme Kinetics and Mechanism (Part A: Initial Rate and Inhibitor Methods)
Edited by DANIEL L. PURICH

VOLUME 64. Enzyme Kinetics and Mechanism (Part B: Isotopic Probes and Complex Enzyme Systems)
Edited by DANIEL L. PURICH

VOLUME 65. Nucleic Acids (Part I)
Edited by LAWRENCE GROSSMAN AND KIVIE MOLDAVE

VOLUME 66. Vitamins and Coenzymes (Part E)
Edited by DONALD B. McCORMICK AND LEMUEL D. WRIGHT

VOLUME 67. Vitamins and Coenzymes (Part F)
Edited by DONALD B. McCORMICK AND LEMUEL D. WRIGHT

VOLUME 68. Recombinant DNA
Edited by RAY WU

VOLUME 69. Photosynthesis and Nitrogen Fixation (Part C)
Edited by ANTHONY SAN PIETRO

VOLUME 70. Immunochemical Techniques (Part A)
Edited by HELEN VAN VUNAKIS AND JOHN J. LANGONE

VOLUME 71. Lipids (Part C)
Edited by JOHN M. LOWENSTEIN

VOLUME 72. Lipids (Part D)
Edited by JOHN M. LOWENSTEIN

Section I

Analytical and Synthetic Methods

[1] Determination of Branched-Chain Amino and Keto Acids with Leucine Dehydrogenase

By GEOFFREY LIVESEY and PATRICIA LUND

Metabolism of branched-chain amino acids in animals is initiated by amino transfer followed by oxidative decarboxylation of the keto acid. This group of amino and keto acids[1,2] is therefore generally considered as a single entity and, for many purposes, it is not important to distinguish between them. They can be assayed relatively cheaply and rapidly with leucine dehydrogenase (L-leucine:NAD$^+$ oxidoreductase, deaminating, EC 1.4.1.9) from microorganisms, especially *Bacillus* species.[3-6] Several other procedures involving ion-exchange chromatography, gas chromatography, or high-performance liquid chromatography have been developed for analysis of the amino[7,8] and keto acids.[9-11] These procedures are more time-consuming but have the advantage of separating the individual amino and keto acids.

Amino Acid Analysis

Principle

The method originates from the work on leucine dehydrogenase (LeuDH) from *Bacillus sphaericus*[5] and *Bacillus subtilis.*[6] The branched-chain amino acids are oxidatively deaminated in the presence of LeuDH.

[1] See also P. L. Crowell, R. H. Miller, and A. E. Harper, this volume [7].

[2] See also A. L. Gasking, W. T. E. Edwards, A. Hobson-Frohock, M. Elia, and G. Livesey, this volume [5].

[3] B. D. Sanwal and M. W. Zink, *Arch. Biochem. Biophys.* **94**, 430 (1961); see also this series, Vol. 17, p. 799.

[4] T. Ohshima, H. Misono, and K. Soda, *J. Biol. Chem.* **253**, 5719 (1978).

[5] T. Ohshima, H. Misono, and K. Soda, *Agric. Biol. Chem.* **42**, 1919 (1978).

[6] G. Livesey and P. Lund, *Biochem. J.* **188**, 705 (1980).

[7] S. Blackburn, "Amino Acid Determinations: Methods and Techniques." Dekker, New York, 1968.

[8] G. J. Hughes, K. H. Winterhaelter, E. Boller, and K. J. Wilson, *J. Chromatogr.* **235**, 417 (1982).

[9] T. C. Cree, S. M. Hutson, and A. E. Harper, *Anal. Biochem.* **92**, 156 (1979).

[10] T. Hayashi, T. Hironori, H. Todoriki, and H. Maruse, *Anal. Biochem.* **122**, 173 (1982).

[11] G. Livesey and W. T. E. Edwards, *J. Chromatogr.* **337**, 98 (1984).

The formation of NADH is measured by an increase in absorption at 340 nm.

$$\text{Leucine} + NAD^+ + H_2O \xrightleftharpoons{\text{LeuDH}} \text{4-methyl-2-ketovalerate}^- \qquad (1)$$
$$+ \; NADH + NH_4^+ + H^+$$

$$\text{Isoleucine} + NAD^+ + H_2O \xrightleftharpoons{\text{LeuDH}} \text{3-methyl-2-ketovalerate}^- \qquad (2)$$
$$+ \; NADH + NH_4^+ + H^+$$

$$\text{Valine} + NAD^+ + H_2O \xrightleftharpoons{\text{LeuDH}} \text{3-methyl-2-ketobutyrate}^- \qquad (3)$$
$$+ \; NADH + NH_4^+ + H^+$$

The equilibrium is dependent on pH. Quantitative oxidation of the branched-chain amino acids (BCAA) occurs when the reaction products are removed from the equilibrium. Protons are trapped by the alkaline reaction medium and the keto acids are converted to the hydrazones.

$$\text{BCAA} + NAD^+ + \text{hydrazine} + H_2O \xrightleftharpoons{\text{LeuDH}} \text{keto acid hydrazone} + NADH + NH_4^+ \quad (4)$$

A large excess of NAD^+ is used and ammonium ions are kept to a minimum. The presence of ethylenediaminetetraacetic acid (EDTA) in the reaction mixture prevents inhibition of the enzyme by heavy metal ions. A relatively high enzyme activity is necessary for a sufficiently rapid reaction.

Reagents

> Hydrazine/Tris buffer pH 9.0: 0.1 M Tris, 0.4 M hydrazine, 2.0 mM EDTA (final concentration) adjusted to pH 9.0 with 5 NHCl. Prepare daily.
> NAD$^+$, 24 mM
> Leucine dehydrogenase stock suspension, LeuDH > 15 U/mg protein from *Bacillus subtilis,* 50 U/ml suspended in 3 M $(NH_4)_2SO_4$.[12]

---

[12] The value of the leucine dehydrogenase from *B. subtilis* for the analysis of the branched-chain substrates in biological samples has been proved.[6] The value of the enzyme from *Bacillus sphaericus* for the analysis of branched-chain substrates in pure solutions[5] and in plasma [J. M. Burrin, J. L. Paterson, and G. M. Hall, *Clin. Chim. Acta* **153**, 37 (1985)] has been demonstrated. The kinetic properties and substrate specificity of leucine dehydrogenase from *Bacillus cereus* and *B. sphaericus* are similar to those for the enzyme from *B. subtilis* (G. Livesey and P. Lund, this volume [36]). Large-scale commercial purification of these enzymes has been described [W. Hummel, H. Schuette, and M. R. Kula, *Eur. J. Appl. Microbiol. Biotechnol.* **12**, 22 (1981); H. Schuette, W. Hummel, H. Tsai, and M.-R. Kula, *Appl. Microbiol. Biotechnol.* **22**, 306 (1985)]. The enzyme from *B. sphaericus* is available from Sigma.

Prepare as described elsewhere.[6,13] The preparation is stable for at least 6 months. The enzyme should not contain more than 0.25% alanine dehydrogenase relative to LeuDH. Less pure preparations require a modified assay procedure, involving alanine dehydrogenase to remove alanine.

Leucine dehydrogenase solution, LeuDH 50 U/ml. Prepare daily by centrifuging 0.5 ml stock suspension, discarding the supernatant, and dissolving the enzyme in 0.5 ml 10 mM K$_2$HPO$_4$/KH$_2$PO$_4$, pH 7.2, containing 2 mM 2-mercaptoethanol.

Assay Method

The oxidative deamination of the branched-chain amino acids is measured at room temperature in 10 mm light path glass cuvettes. The cuvette contains 1.0 ml hydrazine/Tris buffer, 0.1 ml NAD$^+$ solution, and 1.0 ml of either water (reference cuvette), or a solution of leucine, 0.04 to 0.2 mM (standards) or the amino acid extract to be analyzed (sample). After mixing, absorbance is monitored until constant (10–15 min). Leucine dehydrogenase solution (20 μl) is mixed into the reaction mixture and the absorbance is monitored again either after 30, 40, 50, and 60 min for standards or pure amino acid, or after 80, 120, and 150 min for tissue extracts until constant. The increment in absorption, above that found for the reference, gives the quantity of branched-chain amino acid present in tissue or fluid sample.

$$\text{BCAA (m}M) = \frac{\Delta A}{6.22} \times \frac{2.12}{\text{vol. of sample}} \times D$$

ΔA is the increase in absorption produced in the sample cuvette after correction for the change in absorbance in the reference cuvette; 6.22 is the absorption coefficient of NADH (1 mM); 2.12 is the reaction volume; and D is the dilution factor which accounts for the dilution of sample during preparation for assay (see Preparation of Tissue Samples for Analysis).

Comments on the Procedure

The sensitivity of the method is 0.005 A_{340} and the detection limit is 0.01 A_{340}. These values correspond to approximately 7 and 13 μmol/liter of whole blood, respectively. The precision of the method with whole blood containing 0.4 mmol total branched-chain amino acid/liter was found to be ±0.02 mM, which corresponds to a coefficient of variation of 0.05.

---

[13] G. Livesey and P. Lund, this volume [36].

Recovery of amino acids added to perchloric acid extracts of whole blood was $100.1 \pm 1.4\%$ in the concentration range $0.4-2.0$ mmol/liter. The enzymatic method gave a value equal to $101 \pm 4\%$ (mean $\pm$ SD) of that obtained by amino acid analyzer. There are several sources of error. Ammonium ions, too little enzyme, or perchlorate ions slow the reaction. Dialysis of the enzyme to remove NH_4^+ should be against buffer and not against distilled water, which causes complete loss of activity.[14] Tissue extracts prepared with $HClO_4$ should be neutralized with KOH at $0°$ to precipitate $KClO_4$. Samples containing methionine at a concentration equal to or greater than the branched-chain amino acids give a high value because LeuDH will slowly oxidatively deaminate L-methionine. The enzyme also reacts to a limited extent with the straight-chain L-amino acids norvaline, norleucine, and 2-aminobutyrate. When the LeuDH preparation contains alanine dehydrogenase activity in excess of that specified, a high value may be obtained. This is overcome by adding alanine dehydrogenase (alanine:NAD^+ oxidoreductase, deaminating EC 1.4.1.1) to the reaction mixture before monitoring the absorption for the first time. Since L-alanine also reacts quantitatively at pH 9.0, alanine and the branched-chain amino acids can be assayed sequentially. When this is done, it is important to establish that the alanine dehydrogenase is free from LeuDH activity. When the source of alanine dehydrogenase is *B. subtilis* this can be achieved by dialysis of the enzyme against distilled water.[14]

Keto Acid Analysis

Principle

The method originates from work on LeuDH from *Bacillus sphaericus*[5] and from *B. subtilis*.[6] Each branched-chain keto acid is reductively aminated in the presence of LeuDH by reversal of reactions (1)–(3). The oxidation of NADH is measured by decrease in absorption at 340 nm.

Quantitative formation of the amino acids at pH 8.5 occurs in the presence of a high concentration of NH_4^+. The NADH concentration is limited to lower noise, thereby decreasing the detection limit and increasing the sensitivity of the method. Addition of EDTA helps to prevent the inhibition of LeuDH by heavy metal ions. Lactate dehydrogenase (LDH) is added to the reaction mixture to remove pyruvate, 4-methylthio-2-ketobutyrate, and 2-ketobutyrate from the sample.

[14] P. Lund and G. Baverel, *Biochem. J.* **174**, 1079 (1978).

Reagents

Tris buffer, 0.1 M, pH 8.5, containing 2.0 mM EDTA
NADH, 3.6 mM
NH_4Cl, 1 M
Leucine dehydrogenase suspension, LeuDH 50 U/ml. Use stock suspension of enzyme from *B. subtilis,* 25 U, 15 U/mg protein, in 3 M
$(NH_4)SO_4$.[12,13] The suspension is stable for at least 6 months. Prepare as described elsewhere[12,13]
Lactate dehydrogenase suspension, LDH 3000 U/ml; commercial enzyme from pig heart in 3.2 M $(NH_4)_2SO_4$, 5 mg/ml

Measurement Procedure

The reductive amination of keto acids is measured at room temperature (21°) in 10 mm light path glass cuvettes in a reaction volume of
2.52 ml. The cuvette contains 1.5 ml Tris buffer solution, 0.2 ml NH_4Cl,
0.1 ml NADH, 5 μl LDH suspension, 0.7 ml of either a solution of 0–
0.1 mM 4-methyl-2-ketovalerate (standard) or the extract containing the
ketoacids (sample). Two such cuvettes are prepared, one for the assay and
one for the reference (enzyme-free) blank. The difference in absorption in
the cuvettes is monitored for 5–10 min in a high-performance split-beam
spectrophotometer until constant, when the difference in absorption is
recorded. Leucine dehydrogenase, 15 μl, is added to the assay cuvette and
3.2 M ammonium sulfate solution is added to the reference cuvette. After
mixing, the difference in absorption is monitored again for 5–10 min for
standards or pure solutions and for 5–15 min for tissue extracts. By
extrapolation of these values to the time of addition of LeuDH the decrease
in absorption due to reductive amination is obtained. This is proportional
to the concentration of branched-chain 2-keto acid present in the cuvette.
The latter can either be obtained by comparison with a standard curve or,
since the reaction proceeds to completion, the quantity of branched-chain
2-keto acid present in the tissue or fluid sample can be calculated using the
absorption coefficient of NADH.

Branched-chain keto acid content

$$\text{(mmol/kg fresh weight or mmol/liter)} = \frac{\Delta A}{6.22} \times \frac{2.52}{0.7} \times \frac{D}{R}$$

where ΔA is the decrease in absorption due to reductive amination, 6.22 is
the absorption coefficient of NADH (1 mM), 2.52 is the reaction volume,
and 0.7 is the sample volume. D is the dilution factor which accounts for

any dilution of sample during the preparation for assay (see Preparation of Tissue Samples for Analysis). Values are corrected for any incomplete recovery (fractional recovery, R) of the keto acid as occurs when perchloric acid is used to extract the keto acids from tissues.

The procedure is designed to assay the low concentrations of branched-chain keto acids found in normal tissue. Up to 3-fold higher concentrations can be assayed by adding to the reaction cuvette 0.3 ml of NADH in place of the 0.1 ml stated. When this is done, the calculation procedure needs to be amended appropriately and a regular spectrophotometer can be used in place of the high-performance split-beam spectrophotometer. A suitable reference blank may then be the replacement of sample solution with water, LeuDH being added to both assay and reference cuvettes. When assaying tissue extracts, a change in absorption occurs in the absence of enzyme so that a reagent blank containing sample is the more accurate control, any change in absorption due to the leucine dehydrogenase alone being accounted for by the extrapolation procedure.

Comments on the Procedure

The sensitivity of the method using a high-performance split-beam spectrophotometer is $0.0005 \, A_{340}$ and the detection limit is $0.001 \, A_{340}$. These values correspond to approximately 1 and 2 μmol/liter of whole blood, respectively. The precision of the method with plasma containing 70 μmol/liter branched-chain 2-keto acid is $\pm 4 \, \mu M$, which corresponds to a coefficient of variation of 0.057 (5.7%). Recovery of branched-chain keto acids into perchloric acid extracts from plasma is 97% (coefficient of variation 1.1%), from whole blood 77% (coefficient of variation 4.2%), and from solid tissues 80% (coefficient of variation 4.4%). Values for total branched-chain keto acid content of rat tissues and human blood and plasma are similar when determined by the enzymatic method and the gas and high-performance liquid chromatographic methods. There are several sources of error. An imbalance of the addition of sample to assay and reference cuvettes may either overestimate or underestimate the value for branched-chain keto acids. Too low an activity of LeuDH in the cuvette gives low values. Too high an activity of LDH may give low values because 2-keto-3-methylvalerate slowly reacts to give 2-hydroxy-3-methylvalerate. Perchlorate appears to slow the reaction rate so that tissue extracts prepared with perchloric acid need to be adequately neutralized with a basic potassium salt. The LeuDH preparation reacts with 2-keto-4-methylthio-butyrate (ketomethionine), pyruvate, and 2-ketobutyrate; the reduction of these keto acids when present in the sample is brought about by addition of LDH prior to LeuDH, so that these substances do not interfere with the assay.

Preparation of Tissue Samples for Analysis

The enzymatic measurement of the branched-chain amino acids and keto acids has been applied to pure solutions and to extracts of plasma,[6,15,16] whole blood,[6,16] liver, kidney, heart, skeletal muscle, and mammary gland.[6]

Reagents

Perchloric acid, 2.2 M
KOH, 3.6 M
Universal indicator, commercial preparation

Collection and Treatment of Specimens

Blood is taken without stasis from an artery or vein. Additions of heparin (0.2 g/liter) or EDTA (1 g/liter) are without effect on the assay of the branched-chain substrates. One milliliter of blood is mixed with 1 ml of water, to cause hemolysis, then 1 ml of the perchloric acid solution. After thorough mixing, centrifuge for 15 min at approximately 1000 g. A measured volume of supernatant is neutralized with a measured volume of Universal indicator and KOH solution. After 15 min at 0–4°, potassium perchlorate is removed by centrifugation. The supernatant is used in the assay procedures and is approximately 4-fold diluted by comparison with the whole blood. Plasma samples are treated as described for blood. The technique of freeze-clamping is used for solid tissues. Pulverized frozen tissue, 1 g, is homogenized with 4 ml of 1 M perchloric acid. Extraction and neutralization are as described for blood. Acidic supernatants which are turbid (e.g., from mammary gland) are passed through an 8 μm Millipore filter before neutralization. Neutral and acidic tissue extracts can be stored at −15° for several months.

Reference Ranges for Amino Acids and Keto Acids

The content of branched-chain amino acids in tissues of the rat ranges from 300 to 700 μmol/kg fresh weight.[6] Human whole blood normally contains 300–500 μmol/liter, but it is increased by starvation and uncontrolled diabetes[16] and in maple syrup urine disease.[17]

[15] J. M. Burrin, J. L. Paterson, and G. M. Hall, Clin. Chim. Acta 153, 37 (1985).
[16] M. Elia and G. Livesey, Clin. Sci. 64, 517 (1983).
[17] K. Tanaka and L. E. Rosenberg, in "Metabolic Basis of Inherited Disease" (J. B. Stanbury, J. B. Wyngaarden, D. S. Fredrickson, J. L. Goldstein, and M. S. Brown, eds.), pp. 440–470. McGraw-Hill, New York, 1983.

With respect to branched-chain keto acids, rat skeletal muscle normally contains $20-30$ μmol/kg fresh weight; rat blood $30-50$ μmol/liter; rat plasma $30-60$ μmol/liter. The content of branched-chain keto acids in several solid tissues of the rat is below 5 μmol/kg fresh weight. Normal human blood contains $40-60$ μmol/liter, and normal human plasma $60-80$ μmol/liter. The concentration of branched-chain keto acids in plasma is increased in subjects with high circulating levels of branched-chain amino acids.

[2] Analysis of Phosphorylation Sites on Branched-Chain 2-Keto Acid Dehydrogenase Complex

By STEPHEN J. YEAMAN, KENNETH G. COOK, ANDREW P. BRADFORD, and SHEELAGH M. A. JONES

The mammalian branched-chain 2-keto acid dehydrogenase complex, which is located within mitochondria, consists of three enzyme components termed E1, E2, and E3 which catalyze consecutive steps in the overall reaction catalyzed by the complex. The activity of the complex is regulated by reversible phosphorylation of the α-subunit of the E1 component, with phosphorylation causing inactivation of the complex, due to a dramatic decrease in the V_{max} of the E1 component.[1] This phosphorylation is catalyzed by a protein kinase which is apparently tightly bound to the E2 core component of the complex.[2] Dephosphorylation and concomitant reactivation is catalyzed by a specific mitochondrial protein phosphatase.[3]

Buffers and Enzymes

Buffer A: 50 mM imidazole-HCl (pH 7.3), 10% (w/v) glycerol, 0.1 mM ethylenediaminetetraacetic acid (EDTA), 0.1 mM ethylene glycol bis(β-aminoethyl ether)-N,N'-tetraacetic acid (EGTA), 1 mM benzamidine, 1 mM phenylmethylsulfonyl fluoride, 1 mM dithiothreitol (DTT)

[1] P. J. Randle, H. R. Fatania, and K. S. Lau, *Mol. Aspects Cell. Regul.* **3**, 1 (1984).

[2] K. G. Cook, A. P. Bradford, and S. J. Yeaman, *Biochem. J.* **225**, 731 (1985).

[3] Z. Damuni, M. L. Merryfield, and L. J. Reed, *Proc. Natl. Acad. Sci. U.S.A.* **81**, 4335 (1984).

Hexokinase (yeast) is purchased from Boehringer Mannheim, trypsin from Worthington Biochemicals, and $[\gamma\text{-}^{32}P]ATP$ from Amersham International

Procedure

Purified branched-chain 2-keto acid dehydrogenase complex (1– 10 mg/ml in buffer A) is phosphorylated by incubation at 20° in the presence of 10 mM MgCl$_2$, 0.5 mM $[\gamma\text{-}^{32}P]ATP$. Phosphorylation can be monitored conveniently by following inactivation of the complex, using the NADH-production assay.[4] The levels of kinase in the purified complex can vary between preparations, but generally inactivation is complete (>90%) within 20 min. Total phosphorylation proceeds beyond the time necessary for inactivation, due to the presence of a second phosphorylation site,[5,6] and therefore the time allowed for total phosphorylation is routinely three times that for inactivation of the complex.

Phosphorylation can be determined by one of several methods. For correlation of activity with total phosphorylation and phosphorylation of the individual sites, phosphorylation can be terminated by addition of glucose (100 mM) and hexokinase (10 U/ml) to scavenge the ATP. Aliquots can be used for determination of the activity of the complex (using the NADH-production assay) and for total phosphate incorporation (following precipitation of the protein in 10% trichloroacetic acid (TCA)).[7] For analysis of the extent of phosphorylation of the individual sites, the phosphorylated complex can be transferred into 0.2 M NH$_4$HCO$_3$ by desalting using gel filtration on Sephadex G-25 or dialysis overnight. Digestion is achieved by addition of trypsin (1/50 by weight) and incubation for 1 hr at 20°. Ice-cold TCA is then added to a final concentration of 10% (w/v) and, after standing on ice for 10 min, the precipitate is removed by centrifugation at 10,000 g for 5 min. Over 95% of the initial radioactivity should now be in the TCA supernatant, which is then extracted five times with an equal volume of diethyl ether to remove TCA. The extracted supernatant is then lyophilized, and resuspended in a minimal volume of 0.2 M NH$_4$HCO$_3$. Trypsin (corresponding to a 1/1 ratio by weight of trypsin to initial complex) is then added and digestion carried out for 16 hr at 20°. Digestion is again terminated by lyophilization and the phosphopeptides are then redissolved in a minimal volume of 7% formic acid. Insoluble material is

[4] F. H. Pettit, S. J. Yeaman, and L. J. Reed, *Proc. Natl. Acad. Sci. U.S.A.* **75**, 4881 (1978).

[5] R. Lawson, K. G. Cook, and S. J. Yeaman, *FEBS Lett.* **157**, 54 (1983).

[6] K. G. Cook, A. P. Bradford, S. J. Yeaman, A. Aitken, I. M. Fearnley, and J. E. Walker, *Eur. J. Biochem.* **145**, 587 (1984).

[7] J. D. Corbin and E. M. Reimann, this series, Vol. 38, p. 287.

removed by centrifugation at 10,000 g for 5 min prior to analysis of the phosphopeptides. This method is particularly useful for preparative-scale work to allow purification of sufficient amounts of peptides for sequence analysis.

Alternatively, the phosphorylated protein can be precipitated by addition of ice-cold TCA to a final concentration of 10% (w/v). After standing on ice for 10 min, the resulting precipitate is collected by centrifugation at 10,000 g for 5 min. The precipitate is washed twice with ice-cold 10% TCA and then the pellet is extracted by washing once with ice-cold acetone, then twice with ice-cold diethyl ether. Finally the pellet is resuspended in 0.2 M NH_4HCO_3. Phosphorylated complex is then digested by trypsin (approximately 1/50 by weight) for 1 hr at 20°. This digestion solubilizes essentially all the radioactivity from the denatured protein. Insoluble material is removed by centrifugation and the supernatant transferred to another centrifuge tube. The insoluble pellet is washed once with the digestion buffer and the washings added to the supernatant. Trypsin (amount corresponding to a 1/1 ratio by weight of trypsin to initial amount of complex) is added to the supernatant and digestion continued for approximately 16 hr at 20°. Digestion is terminated by lyophilization and the samples processed as described above. This method can be used for preparative-scale work and also for analysis of the phosphate content of each site following the action of protein phosphatase(s) on phosphorylated preparations of the complex. Following incubation of phosphatase with phosphorylated branched-chain 2-keto acid dehydrogenase (prepared by desalting or dialysis into an appropriate buffer) and termination of the reaction by addition of TCA, radioactivity in the supernatant can be determined as a measure of phosphatase activity, and the TCA precipitates processed as described here to determine the activity of the phosphatase against the individual sites.[6]

A third possible method involves resolution of the phosphorylated complex by sodium dodecyl sulfate (SDS)-polyacrylamide gel electrophoresis. Following staining, destaining, drying, and autoradiography, the band corresponding to the α-subunit can be excised from the gel and the protein extracted. The dried gel is chopped, crushed, and then allowed to swell in 50 mM triethanolamine-HCl (pH 8.0), 1% (w/v) SDS, 0.05 mM EDTA, 0.1 mM DTT (bubbled with N_2 immediately prior to use). This suspension is homogenized in a glass/glass homogenizer and the entire contents then transferred to a 1.5-ml plastic centrifuge tube, prior to heating at 65° for 15 min. After centrifugation, the supernatant is removed and the pellet reextracted twice more. The pooled supernatants contain approximately 90% of the radioactivity in the original slice, as estimated by Cerenkov counting. Phosphorylated protein can then be precipitated by addition of ice-cold TCA to a final concentration of 10%. If necessary,

20 μg of sonicated DNA can be used as carrier and the concentration of TCA can be increased to 20% to aid precipitation. The precipitate is then collected, washed, and digested as described above. This method is particularly useful for analysis of sites on the complex phosphorylated within intact mitochondria or in whole cells.[8,9]

These digestion conditions generate two major phosphopeptides, termed T1 and T3. Each peptide contains a single phosphorylation site. T1 contains phosphorylation site 1, which current evidence indicates is responsible for controlling the activity of the complex.[6] Phosphopeptide T3 contains phosphorylation site 2. Incomplete digestion of phosphorylated branched-chain 2-keto acid dehydrogenase may generate two additional phosphopeptides, termed T2 and T1U. T2 is a diphosphorylated peptide, with phosphate equally distributed between phosphorylation sites 1 and 2, whereas T1U is a monophosphorylated peptide, containing phosphate at site 1 or site 2.

The degree of radioactivity at each site can be estimated by high-voltage electrophoresis at pH 1.9 [7% (v/v) formic acid], followed by autoradiography. The electrophoresis is carried out for 40 min at 2.5 kV. Phosphopeptides T1 and T3 have respective mobilities of approximately 0.8 and 0.35 relative to that of L-serine. If digestion is incomplete, then the additional phosphopeptides T2 and T1U are detected, with relative mobilities of 0.6 and 0.75, respectively. Quantitative data can be obtained by cutting out the radioactive spots detected by autoradiography, followed by scintillation counting of the excised spot. If the radioactive peptides are to be recovered subsequently, then the dry paper can be counted by Cerenkov counting.

Alternatively, the tryptic phosphopeptides can be separated by reversed-phase HPLC, using a Beckman Ultrasphere C_{18} ODS or other equivalent column. A linear gradient (0–40%) of acetonitrile in water (containing 0.1% trifluoroacetic acid) is used, with an increase in acetonitrile concentration of 1%/min and a flow rate of 1 ml/min. The effluent can be monitored at 214 nm and by flow-through counting using a Beckman Model 170 Radio-Isotopic Detector. Phosphopeptides T1, T3, and T2 are eluted at approximately 17, 26, and 29% acetonitrile, respectively.

For sequence analysis of the phosphopeptides, further purification by two-dimensional peptide mapping on cellulose sheets is recommended, as described previously.[10] The sequence of phosphopeptides T1 and T3 from the complex from bovine kidney has been determined[6,10] (Table I). Subsequently, the sequence of T1 from rabbit heart has been shown to be

[8] K. G. Cook, R. Lawson, and S. J. Yeaman, *FEBS Lett.* **164**, 85 (1983).
[9] S. M. A. Jones and S. J. Yeaman, *Biochem. J.* **236**, 209 (1986).
[10] K. G. Cook, R. Lawson, S. J. Yeaman, and A. Aitken, *FEBS Lett.* **164**, 47 (1983).

TABLE I
AMINO ACID SEQUENCE AT PHOSPHORYLATION SITES ON BOVINE
KIDNEY BRANCHED-CHAIN 2-KETO ACID DEHYDROGENASE COMPLEX

Peptide	Sequence
T1 (site 1)	Ile-Gly-His-His-Ser(P)-Thr-Ser-Asp-Asp-Ser-Ser-Ala-Tyr-Arg
T3 (site 2)	Ser(P)-Val-Asp-Glu-Val-Asn-Tyr-Trp-Asp-Lys

identical to that from bovine kidney.[11] The sequence of T1, which contains the site primarily responsible for control of the activity of the complex, shows a high degree of homology with the corresponding site from pyruvate dehydrogenase complex,[12,13] although the protein kinases and phosphatases acting on each complex are apparently specific for each complex.

Acknowledgments

Work in the authors' laboratory was supported by grants from the Medical Research Council, U.K. S.J.Y. is a Lister Institute Research Fellow.

[11] R. Paxton, M. Kuntz, and R. A. Harris, *Arch. Biochem. Biophys.* **244**, 187 (1986).
[12] S. J. Yeaman, E. T. Hutcheson, T. E. Roche, F. H. Pettit, J. R. Brown, L. J. Reed, D. C. Watson, and G. H. Dixon, *Biochemistry* **17**, 2364 (1978).
[13] P. H. Sugden, A. L. Kerbey, P. J. Randle, C. A. Waller, and K. B. M. Reid, *Biochem. J.* **181**, 419 (1979).

[3] High-Performance Liquid Chromatographic Separation of (3R)- and (3S)-β-Leucine Using Marfey's Reagent

By D. JOHN ABERHART and JACQUES-ALAIN COTTING

The (3R)-**(1)** and (3S)-**(2)** enantiomers of β-leucine[1] can be cleanly separated by HPLC[2] after first being converted into the corresponding

[1] See also J. M. Poston, this volume [16] and [17]; D. J. Aberhart, this volume [4].
[2] D. J. Aberhart, J.-A. Cotting, and H.-J. Lin, *Anal. Biochem.* **151**, 88 (1985).

diastereomeric derivatives **3** and **4** by treatment with Marfey's reagent[3] **(5)** [N^2-(5-fluoro-2,4-dinitrophenyl)-L-alaninamide].[4,5]

Materials

Marfey's reagent is obtained from Pierce Chemical Co., Rockford, IL 61105

Chromatographic solvent (mobile phase): Triethylamine phosphate (0.05 M) solution is prepared by dissolving 7.0 ml of freshly distilled triethylamine in 1.0 liter of deionized, glass-distilled water and adjusting the pH to 3.0–3.1 with concentrated phosphoric acid. Then 700 ml of this solution is mixed with 300 ml of acetonitrile (HPLC grade). The mixture is filtered through an 0.45 μm nylon filter and degassed by vacuum pumping for a few minutes.

Procedure for Derivatization of DL-β-Leucine

DL-β-Leucine (0.8 mg) is treated with 200 μl of a 1% (w/v) solution of Marfey's reagent in acetone plus 40 μl of 1% NaHCO$_3$. The reaction can be conveniently carried out in a 1-dram vial in an oil or water bath for 1 hr

---

[3] P. Marfey, *Carlsberg Res. Comm.* **49**, 591 (1984).

[4] In earlier publications, this compound was called 1-fluoro-2,4-dinitrophenyl-L-alaninamide. Since this name does not correspond to any accepted system of nomenclature, the name consistent with IUPAC rules of organic chemical nomenclature is used in this article. The current *Chemical Abstracts Index* name of **5** is (S)-2-[(5-fluoro-2,4-dinitrophenyl)-amino]propanamide. We thank Dr. Kurt Loening (Chemical Abstracts Service) for assistance with this nomenclature.

[5] Pierce Chemical Company, "Previews," p. 9. Pierce Chemical Company, Rockford, Illinois, 1985.

with occasional shaking. The mixture is then cooled to room temperature, treated with 20 μl of 2 N HCl, and filtered through an 0.45 μm Acro LC 13 filter (Gelman Ann Arbor, MI 48106). The filtrate is then injected into the HPLC instrument (the solution may be stored at $-78°$ for later injection if desired).

Chromatographic Conditions

> Column: A 3.9-mm i.d. $\times$ 15 cm length stainless steel column packed with Nova-Pak C_{18} (4 μm particle size) (Waters Associates Milford, MA 01757) was used
> Mobile-phase flow rate: 1.0 ml/min (producing $\sim$ 1200 psi back pressure)
> Detector: 340 nm
> Sample size: 1 μl

The chromatogram shown in Fig. 1 was produced under the above conditions. When the chromatography is carried out under isocratic con-

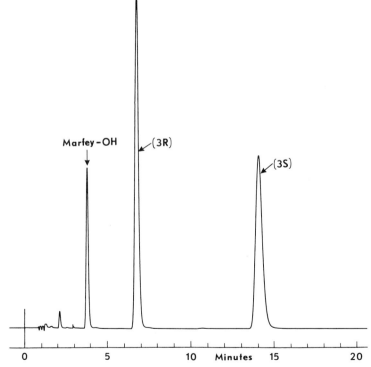

FIG. 1. Chromatogram of (3RS)-β-leucine after derivatization with Marfey's reagent.

ditions as described above, rather than by using a gradient as suggested elsewhere[3] for separations of DL-α-amino acids, the peak (Marfey-OH) resulting from the hydrolysis product of the excess reagent is eluted first rather than between the peaks of interest.

[4] High-Performance Liquid Chromatographic Separation of α- and β-Leucine

By D. JOHN ABERHART

The substrates of leucine 2,3-aminomutase (EC 5.4.3.7),[1,2] α-leucine and β-leucine, can be conveniently separated, without the need for derivatization, by high-performance liquid chromatography (HPLC).[3] The method is an adaptation of a method published by Schuster[4] for separating mixtures of α-amino acids.

Materials

DL-β-Leucine is synthesized as described by Poston,[1,2] except that the reaction of 4-methyl-2-pentenoic acid with concentrated ammonia is carried out in a 450-ml Parr pressure reactor rather than in sealed pressure bottles (which, on several occasions, exploded due to the high pressure generated on heating ammonia to 121°). Solutions of DL-β-leucine (and L-α-leucine) used for injections were filtered through 0.2 μm sterile Acrodisc filters (Gelman) and kept frozen to inhibit bacterial growth.

Chromatography solvent (mobile phase): 900 ml of acetonitrile (HPLC grade): water (100:16 by volume) is mixed with 100 ml of 0.01 M KH$_2$PO$_4$. The mixture is filtered through an 0.45 μm nylon filter and degassed by vacuum pumping before use.

Chromatography column: A 3.9 mm i.d. × 30 cm length stainless steel column packed with μBondapak NH$_2$ (10 μm particle size) (Waters Associates Milford, MA 01757) was used. When used at a mobile-phase flow rate of 2.0 ml/min, this gave a back pressure of ~ 300 psi. A 4.6 mm i.d. × 25 cm length stainless steel Alltech NH$_2$

[1] J. M. Poston, *J. Biol. Chem.* **251**, 1859 (1976).

[2] See also J. M. Poston, this volume [16] and [17]; D. J. Aberhart and J.-A. Cotting, this volume [3].

[3] D. J. Aberhart and H.-J. Lin, *J. Labelled Compd. Radiopharm.* **20**, 611 (1983).

[4] R. Schuster, *Anal. Chem.* **52**, 617 (1980).

column (10 μm particle size) (Alltech Associates Deerfield, IL 60015) was also used with similar results (but with a higher back pressure of ~ 1500 psi).

Precolumn: Connected between the injector and chromatography column was a Waters Associates guard column kit (3.9 mm i.d. × 2.4 cm length) packed with R Sil NH$_2$, 5 μm particle size (Alltech Associates).

Chromatographic conditions: Mobile-phase flow rate: 2.0 ml/min

Detector: 200 nm at 0.04 absorbance units full scale (AUFS)

Column temperature: ambient

Sample size: 1.5 μl of a solution containing 5 mg/ml each of L-α-leucine and DL-β-leucine (7.5 μl of each amino acid)

The chromatogram shown in Fig. 1 was produced under the above conditions. Higher detection sensitivity can be achieved by using, in place of the UV detector, a postcolumn reaction detector system using o-phthalaldehyde (OPA) in conjunction with a fluorescence detector.

For an alternative method of separating α- and β-leucine suitable for assaying leucine 2,3-aminomutase, involving precolumn formation of OPA derivatives, see Reference 5.

Materials and Equipment for Postcolumn Fluorescence Detection of α- and β-Leucine

Postcolumn reactor: A Kratos Model URS 050 reactor having a 1.0 ml reaction chamber was used.

Postcolumn reagent: o-Phthalaldehyde (OPA) (Aldrich Chemical Co.) was recrystallized from CHCl$_3$ – methanol. A solution was prepared of OPA (400 mg) plus 2-mercaptoethanol (0.10 ml) in methanol (5 ml). This was diluted to 500 ml with 0.2 M potassium borate buffer, pH 10.4, prepared by diluting 1 M fluoraldehyde reagent diluent (Pierce Chemical Co.) with deionized, glass-distilled water. The resultant solution was filtered through an 0.45 μm nylon filter and slowly purged with helium while in use. This solution must be prepared fresh daily.

Detector: A Kratos Model FS 950 fluorometer was used, having a 365 nm excitation filter, a 418 nm emission filter, and a Model FSA 111 lamp (365 nm without blue filtering).

Conditions:

Postcolumn reagent flow rate: 1.2 ml/min

Detector settings: sensitivity, 8.0; range, 1.0

Other conditions (column, mobile-phase flow) same as above

[5] D. J. Aberhart, *Anal. Biochem.* **169**, 350 (1988).

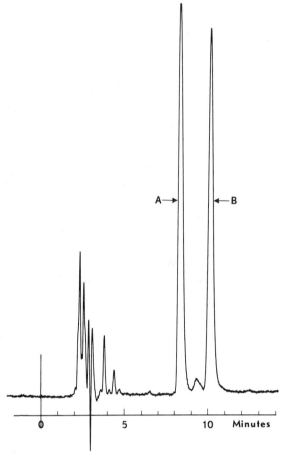

FIG. 1. Chromatogram of a mixture of α- and β-leucine: peak A, α-leucine; peak B, β-leucine.

Under these conditions, an injection of 1 μl of a solution containing L-α-leucine and DL-leucine each at 0.10 mg/ml (100 ng each amino acid) gave a chromatogram similar to that shown in Fig. 1, but with virtually no noise and no extraneous peaks other than those due to α- and β-leucine being visible.

[5] Quantitative High-Performance Liquid Chromatographic Analysis of Branched-Chain 2-Keto Acids in Biological Samples

By ANGELA L. GASKING, WILLIAM T. E. EDWARDS, ANTHONY HOBSON-FROHOCK, MARINOS ELIA, and GEOFFREY LIVESEY

The concentrations of the branched-chain 2-keto acids, 4-methyl-2-ketovaleric acid (ketoleucine), 3-methyl-2-ketovaleric acid (ketoisoleucine), and 3-methyl-2-ketobutyric acid (ketovaline), in mammalian tissues are small,[1-3] making quantification difficult. The major tissue pools of keto acids in the rat are plasma (and whole blood) and muscle.[2,3] In human blood the keto acids predominate in plasma[4] and partly in association with albumin.[5] Human urine normally contains amounts of these keto acids which are too small to quantify with satisfactory accuracy. Analysis of the branched-chain 2-keto acids is important in biochemistry and clinical chemistry with concentrations in blood or plasma being elevated by starvation and diabetes[4] and several disorders of branched-chain amino acid metabolism.[6]

Several analytical methods have been developed to quantify the branched-chain 2-keto acids using gas chromatography,[1,7] high-performance liquid chromatography,[8-10] or enzymatic determination.[2] The enzymatic method is rapid and reliable and sufficiently sensitive for analysis of the keto acids in muscle, plasma, and whole blood, but does not distinguish between the three individual acids. A highly sensitive high-performance liquid chromatographic method is described below involving a single derivatization procedure to the stable quinoxalinols which are detected by fluorescence. The method is based on that of Hayashi and co-workers,[8,9] with a modification to exclude oxygen during derivatization.[10]

---

[1] See also P. L. Crowell, R. H. Miller, and A. E. Harper, this volume [7].

[2] G. Livesey and P. Lund, *Biochem. J.* **188,** 705 (1980); see also G. Livesey and P. Lund, this volume [1].

[3] S. M. Hutson and A. E. Harper, *Am. J. Clin. Nutr.* **34,** 173 (1981).

[4] M. Elia and G. Livesey, *Clin. Sci.* **64,** 517 (1983).

[5] G. Livesey and P. Lund, *Biochem. J.* **212,** 655 (1983).

[6] K. Tanaka and L. E. Rosenberg, *in* "The Metabolic Basis of Inherited Diseases" (J. B. Stanbury, J. B. Wyngaarden, D. S. Fredrickson, J. L. Goldstein, and M. S. Brown, eds.), p. 440. McGraw-Hill, New York, 1983.

[7] T. C. Cree, S. M. Hutson, and A. E. Harper, *Anal. Biochem.* **92,** 156 (1979).

[8] T. Hayashi, T. Hironori, H. Todoriki, and H. Naruse, *Anal. Biochem.* **122,** 173 (1982).

[9] T. Hayashi, H. Tsuchiya, and H. Naruse, *J. Chromatogr.* **273,** 245 (1983).

[10] G. Livesey and W. T. E. Edwards, *J. Chromatogr.* **337,** 98 (1985).

Principles of the Method

The proteins of blood and plasma are precipitated with acid and the solution is purified by retention of the keto acids on a hydrazide gel. The hydrazones are converted anaerobically to the quinoxalinols with o-phenylenediamine which are then separated and quantified by HPLC using a reversed-phase column and a fluorescence detector.

$$Gel-CONH \cdot NH_2 + R \cdot CO \cdot CO_2H \rightarrow Gel-CONHN:CR \cdot CO_2H$$

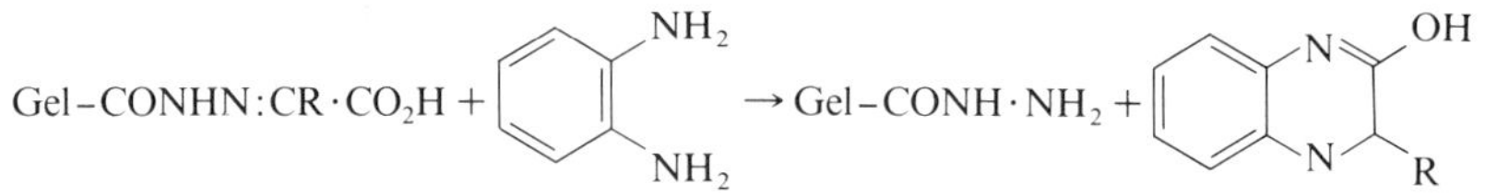

4-Methyl-2-ketovaleric acid [R $= -CH_2CH_2CH(CH_3)_2$

3-Methyl-2-ketovaleric acid [R $= -CH_2CH(CH_3)CH_2CH_3$]

3-Methyl-2-ketobutyric acid [R $= -CH_2CH(CH_3)_2$]

Preparation of the Hydrazide Gel

Reagents

BioGel P-60, 100–200 mesh (Bio-Rad Ltd)

Hydrazine hydrate, 98%

Sodium chloride, 0.1 M

Borate buffer, 0.1 M H_2BO_3 containing 0.2 M NaCl, 0.02 M disodium ethylenediaminetetraacetic acid (Na$_2$ EDTA), 5 μM pentachlorophenol (final concentrations) to pH 7.3 with NaOH

The preparative procedure is based on the method of Hayashi et al.[9] Dry gel (15 g) is added to water (200 ml) in a siliconized flask and left overnight. The gel suspension is heated to 50° and mixed with 80 ml of the hydrazine hydrate also at 50°. The mixture is stirred for 6 hr at 50° then washed with NaCl solution on a Büchner funnel to remove hydrazine. The NaCl solution is displaced with the borate buffer and the hydrazide gel is kept in suspension at 0–4°.

Perchloric Acid Extraction of Keto Acids from Biological Samples

Reagents

Perchloric acid, 2.9 M

Potassium hydroxide, 3.6 M

Universal indicator, commercial preparation

Blood or plasma (0.5 ml) is mixed with water (0.5 ml) and ice-cold perchloric acid solution (0.5 ml) is added. After standing on ice for 15 min, the protein precipitate is removed by centrifugation at 10,000 g for 10 min. The acid extract is neutralized with potassium hydroxide and 25 μl of Universal indicator. Potassium perchlorate is removed by centrifugation. The extract can be stored at $-20°$ or used immediately.

Purification and Derivatization of the 2-Keto Acids

Reagents

Hydrazide gel suspended in borate buffer (see above)
Acetic acid, 0.1 M
Sodium chloride (AR), 0.1 M
o-Phenylenediamine, 2 mg of the dihydrochloride per ml in 2 M HCl containing 0.05% (v/v) 2-mercaptoethanol
N_2 gas (O_2-free)
Sodium dithionite, solid (AR)
Sodium sulfate, saturated aqueous solution
Sodium sulfate, anhydrous (AR)
2-Ketooctanoic acid, 50 nM

The procedure that follows is essentially that of Hayashi et al.[9] as modified by Livesey and Edwards.[10] To the extract from plasma or blood (1.0 ml) is added the 2-ketooctanoic acid (internal standard) (0.5 ml), acetic acid (1.0 ml), and sodium chloride (3 ml). The mixture is transferred to a glass column (150 × 5 mm) containing settled hydrazide gel (0.3 ml) supported on a small pad of glass wool. After the solution has filtered through the gel, it is washed with sodium chloride (5 ml). The gel is transferred to a screw-capped Sovirel tube (160 × 16 mm) and mixed with o-phenylenediamine solution (2.0 ml). The mixture is gassed with N_2 (oxygen-free) and sodium dithionite (1–2 mg) is added to remove residual oxygen 3–5 sec before cessation of gassing. The tube is sealed, heated at 80° for 2.0 hr, then cooled with cold water. Saturated sodium sulfate (4 ml) and ethyl acetate (5 ml) are added and the quinoxalinols extracted into the upper, organic phase by shaking for 5 min. This phase is removed and dried with anhydrous sodium sulfate (100 mg), overnight at 1–4°. This solution is taken to dryness on a rotary or vortex evaporator at 30°. If not used immediately, the residue may be stored for a short time at room temperature in a desiccator.

High-Performance Liquid Chromatography

Conditions

 Column: 5 μm Lichrosorb RP8, 250 mm $\times$ 4.6 mm o.d. fitted with a silica precolumn (HPLC Technology Ltd)

 Liquid chromatograph operating conditions: Flow rate 1.5 ml/min; injection volume 20 μl (Rheodyne valve injector); fluorescence detector set at excitation and emission wavelengths of 322 and 391 nm, respectively; temperature, 50°

 Mobile phase: Solution A, acetonitrile:water (4:1, v/v); solution B, acetonitrile:water:0.1 M Na_2HPO_4 adjusted to pH 7.0 with NaOH (1:12:7, v/v/v)

The data in the present paper were obtained using equipment from Perkin-Elmer Ltd, which comprised a Model 3B pump, LC-100 oven, and a Model 3000 fluorescence detector. On-line degassing of the mobile phase was used to reduce the possibility of quenching of the fluorescence by dissolved oxygen. The silica precolumn was used to saturate the mobile phase with silica. Data were processed by a Pye Unicam Ltd. data control center. The optimum excitation and emission wavelengths given were obtained by scanning the sample in the detector cell under stopped-flow conditions.

Procedure

The dried quinoxalinols are dissolved in dimethylformamide (40 μl) and water (100 μl) and aliquots (20 μl) injected onto the column. A concave gradient (Perkin-Elmer Code 2) is used to change the mobile-phase composition from 20 to 80% solution A over 35 min for plasma or whole blood extracts. The column is equilibrated for 10 min with 20% solution A before injection of the next sample.

Performance of the Method

Variations in retention time for the three keto acids and in their peak area ratios relative to the internal standard are given in Table I and indicate the satisfactory reproducibility of the method.

Plots of the ratio of peak area of keto acid versus keto acid concentration over the range 10 to 60 nmol relative to that of the internal standard (25 nmol) are linear and intercepts on the peak area axis are close to zero ($\pm$ 0.04). Relative standard deviations (RSD) for the slopes of these curves

TABLE I
REPRODUCIBILITY OF RETENTION TIME AND PEAK AREA RATIOS

Keto acid	Retention times (min)		Peak area ratio ± RSD (%)
	Range	Mean	
3-Methyl-2-ketobutyrate	12.9–14.5	13.7	1.19 ± 2.3
4-Methyl-2-ketovalerate	15.5–16.7	15.9	1.25 ± 0.9
3-Methyl-2-ketovalerate	18.0–19.2	18.4	1.19 ± 2.4
2-Ketooctanoate	27.5–28.4	27.8	—

for 3-methyl-2-ketobutyrate, 4-methyl-2-ketovalerate, and 3-methyl-2-ketovalerate are 1.2, 1.0, and 1.3%, respectively.

Typical chromatograms of the keto acids alone and those extracted from plasma and whole blood are shown in Figs. 1, 2, and 3, respectively. The peaks of interest are well resolved and adequate for quantification. The overall performance of the method has been determined by taking blood or plasma spiked with known amounts of each acid through the

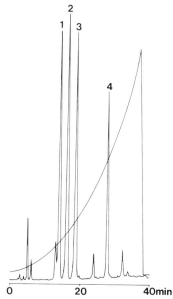

FIG. 1. HPLC chromatogram of quinoxalinols of a standard mixture of keto acids. Peak 1, 3-methyl-2-ketobutyrate; peak 2, 4-methyl-2-ketovalerate; peak 3, 3-methyl-2-ketovalerate; peak 4, 2-ketooctanoate (internal standard).

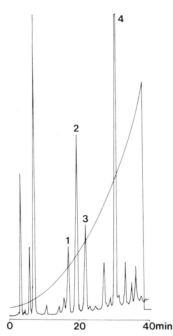

FIG. 2. HPLC chromatogram of quinoxalinols of the branched-chain 2-keto acids from normal human plasma. Peak 1, 3-methyl-2-ketobutyrate; peak 2, 4-methyl-2-ketovalerate; peak 3, 3-methyl-2-ketovalerate; peak 4, 2-ketooctanoate (internal standard).

whole procedure and calculating the amount recovered. Mean recovery values for the three acids were 111% (RSD 2.8%), 105% (RSD 6.3%), and 98% (RSD 2.8%), respectively. Similar experiments with whole blood gave values of 97% (RSD 1.3%), 89% (RSD 2.4%), and 78% (RSD 3.6%), respectively. Recovery values are known to be influenced by the purity of the individual acid,[1] and the analyst is advised to determine these recovery values as frequently as possible as a check on the method. Lower recovery values for 3-methyl-2-ketobutyrate and 4-methyl-2-ketovalerate in whole blood have been noted in the enzymatic method where loss of keto acid with the protein precipitate was held to be responsible.

The reproducibility of the procedures has been determined using samples of plasma and whole blood and has been found to be quite satisfactory. Replicate analyses of plasma samples containing 13.4, 33.9, and 18.7 nmol/ml of the three acids gave RSD values of 1.4, 1.7, and 2.5%, respectively. For whole blood containing 9.5, 18.7, and 11.2 nmol/ml, the values were 2.1, 1.9, and 1.3%.

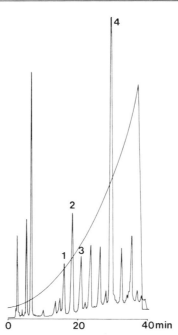

F IG. 3. HPLC chromatogram of quinoxalinols of the branched-chain 2-keto acids from normal human whole blood. Peak 1, 3-methyl-2-ketobutyrate; peak 2, 4-methyl-2-ketovalerate; peak 3, 3-methyl-2-ketovalerate; peak 4, 2-ketooctanoate (internal standard).

Sensitivity, Precision, and Accuracy

This procedure is extremely sensitive and can detect as low as 0.1 nmol of the keto acid. The precision of the assay is within ± 0.6 nmol/ml for 0.5 ml of normal human plasma. The accuracy of the method for plasma is within ± 1 nmol/ml and corresponding values for normal human whole blood are ± 0.3 and ± 0.5 nmol/ml, respectively.

Comments on the Procedure

Standard solutions of the branched-chain 2-keto acids kept at 4° are more stable than that of the internal standard, but all appear to be stable for at least 2 months when frozen at neutral pH. The derivatization procedure is sensitive to oxygen,[10] and results in a marked decrease for the ratio of 3-methyl-2-ketobutyrate relative to the internal standard and increases for 3-methyl-2-ketovalerate and 4-methyl-2-ketovalerate. The absence of oxygen therefore improves the performance of the method and markedly increases the sensitivity of the method for 3-methyl-2-ketobutyrate. The

quinoxalinols are stable for at least 5 days at room temperature; after 7 days there is a decrease in the ratio of the peak area of the internal standard relative to that of each of the three keto acids.

This procedure uses peak areas for quantification. If preferred, it is possible to use peak heights and obtain similar precision and accuracy. Plots of keto acid concentration calculated on peak height compared to those using peak area in 26 analyses of plasma and 26 analyses of whole blood gave linear relationships (passing close to the zero intercept, standard deviation 0.2 nmol/ml) with slopes of 1.03 (RSD 2.0%), 0.99 (2.5%), and 1.01 (0.9%), respectively, for the three acids.

Reference Ranges

One milliliter of human antecubital venous plasma collected after overnight fasting contains 40–80 (mean 63, RSD 25%) nmol of total branched-chain 2-keto acids, 8–16 (mean 13, RSD 22%) nmol of 3-methyl-2-ketobutyrate, 20–40 (mean 31, RSD 25%) nmol of 4-methyl-2-ketovalerate, and 10–30 (mean 19, RSD 32%) nmol of 3-methyl-2-keto-valerate. One milliliter of whole blood contains approximately 60% of the plasma values.

[6] Determination of Branched-Chain 2-Hydroxy and 2-Keto Acids by Mass Spectrometry

By ORVAL A. MAMER and JANE A. MONTGOMERY

Introduction

2-Ketoisovaleric (KIVA), 2-ketoisocaproic (KICA), and 2-keto-3-methyl-valeric (KMVA) acids are the transamination products of valine, leucine, and isoleucine, respectively, and are substrates for the branched-chain keto acid dehydrogenase enzyme complex found principally in liver and kidney. These acids accumulate in the serum and urine of acute patients with inherited errors of this complex in quantities large enough to be found easily by coupled gas chromatography–mass spectrometry (GC-MS). These acids are also of interest, for example, in fasting studies of the chronically obese, where one encounters concentrations of these acids more closely comparable with those for normal individuals. Largely unrecognized is the importance of the concentrations of the corresponding 2-hydroxyisovaleric (HIVA), 2-hydroxyisocaproic (HICA), and 2-hydroxy-

3-methylvaleric (HMVA) acids, which are in an enzyme-mediated equilibrium with the keto acids that is dependent on many factors such as nutritional status, age, and pH of the tissue. In maple syrup urine disease, HIVA is quantitatively more significant than KIVA, both in the acute state and under dietary control.[1,2]

The oxime method[1] was the first method to enable the gas chromatographic measurement of both the keto and the hydroxy acids in a single and relatively simple analysis. The trimethylsilyl (TMS) derivatives of the E and Z isomers of each of the oximes are chromatographically resolvable however, and this leads to uncertainties in peak area measurements. Furthermore, incomplete extraction recoveries, variable chemical yields on oximation, and GC column adsorption add to the difficulty of obtaining reliable results. The quinoxalinol derivatives[3,4] and the pentafluorobenzyl (PFB) esters[5] have also been described as derivatives suitable for quantitative purposes. Although these last methods provide greater specificity and sensitivity than the oxime method, they are sensitive to many or all of the same systematic errors as the oxime method. Finally, the quinoxalinol and the PFB ester methods do not allow a determination of the metabolically significant corresponding hydroxy acids.

The procedure described here is a stable isotope dilution assay for the six acids: HIVA, KIVA, HICA, KICA, HMVA, and KMVA in a single GC-MS analysis that is not sensitive to these errors. It is modified from that of Mamer *et al.*[6]

Assay Principle

Measured volumes of urine and serum are spiked with a known amount of a mixed internal standard composed of 2-hydroxy[2,3-^{2}H$_2$]isovaleric (2-HIVA), 2-hydroxy[2,3,3-^{2}H$_3$]isocaproic (3-HICA), and 2-hydroxy-3-methyl[2,3-^{2}H$_2$]valeric (2-HMVA) acids. A small amount of sodium borodeuteride is then added, sufficient to reduce endogenous KIVA, KICA, and KMVA to the corresponding 2-hydroxy acids bearing one

[1] G. Lancaster, P. Lamm, C. R. Scriver, S. S. Tjoa, and O. A. Mamer, *Clin. Chim. Acta* **48**, 279 (1973).

[2] G. Lancaster, O. A. Mamer, and C. R. Scriver, *Metabolism* **23**, 257 (1974).

[3] U. Langenbeck, U. Wendel, A. Mench-Hoinowski, D. Kuschel, K. Becker, H. Przyrembel, and H. J. Bremer, *Clin. Chim. Acta* **88**, 283 (1978).

[4] U. Langenbeck, H. Luthe, and G. Schaper, *Biomed. Mass Spectrom.* **12**, 507 (1985).

[5] I. Pentilla, A. Huhtikangas, J. Herranen, and O. Moilanen, *J. Chromatogr. Biomed. Appl.* **338**, 265 (1985).

[6] O. A. Mamer, N. S. Laschic, and C. R. Scriver, *Biomed. Environ. Mass Spectrom.* **13**, 553 (1986).

deuterium substitution on the 2-carbon atom (1-HIVA, 1-HICA, and 1-HMVA, respectively). At this point, the sample contains unlabeled endogenous 0-HIVA, 0-HICA, and 0-HMVA, once-labeled 1-HIVA, 1-HICA, and 1-HMVA representing the original endogenous KIVA, KICA, and KMVA, respectively, and last, the added additionally labeled internal standards 2-HIVA, 3-HICA, and 2-HMVA. The diethyl ether extract is made, and the acid mixture is analyzed as the TMS derivatives in selected ion monitoring (SIM) mode on a conventional GC-MS. Fragment ions are selected which include the deuterium labeling, and the areas corresponding to 0-HIVA, 1-HIVA, 2-HIVA, 0-HICA, 1-HICA, 3-HICA, 0-HMVA, 1-HMVA, and 2-HMVA are measured.

These data are entered into a computer program that makes corrections for natural heavy isotope incorporation in the fragments of interest and for isotopic impurity in the standards and sodium borodeuteride reagent. These corrected data are then used to calculate the concentrations of all six of the branched-chain 2-hydroxy and 2-keto acids in the original sample.

Preparation of Standards

Preliminary Remarks

When weighing materials for the purposes of preparing standards, it is strongly suggested that weights given here be reproduced reasonably closely. This is particularly important for the preparation of solution 1 below. This recipe provides enough internal standard for the analysis of approximately 200 samples, and may be scaled up for larger projects as long as all quantities are scaled up in proportion. It is advantageous to have a large enough single lot of solution 1 so that it lasts for all of the samples to be analyzed, and then the determination of the isotopic impurity for this standard needs to be done only once. Synthesis of a new batch of solution 1 will require this analysis to be repeated. In addition, one should purchase a sufficiently large quantity of sodium borodeuteride, all of the same lot number. Changing the lot number will require a redetermination of its isotopic purity.

Solution 1: Labeled Hydroxy Acid Internal Standard

A mixture is made of ethyl 2-ketoisovalerate (0.0847 mmol, 12.2 mg, Sigma Chemical Co., St. Louis, MO), sodium 2-ketoisocaproate (0.0816 mmol, 12.4 mg, Sigma), sodium 2-keto-3-methylvalerate (0.0868 mmol, 13.2 mg, Sigma), sodium deuteroxide (2 ml of a 40% solution in deuterium oxide, MSD Isotopes, Pointe Claire-Dorval, Quebec, Canada), and deute-

rium oxide (10 ml, 99.8% isotopic purity, MSD Isotopes) in a vial closed with a Teflon-lined cap. This mixture is saponified and made to undergo exchange by heating at 50° for 2 hr. The solution is evaporated to near dryness in a dry nitrogen stream, and the residue is taken up in another 10-ml aliquot of deuterium oxide. It is important to exclude from the exchange mixture light water from atmospheric humidity and other sources as completely as possible. This solution is allowed to remain at room temperature overnight sealed as before. Sodium borodeuteride (1.2 mmol, 50 mg, 98% isotopic purity, MSD Isotopes) is added, and the mixture is gently shaken for 4 hr at room temperature. With great care, this solution is acidified to approximately pH 2 by dropwise addition of 5 N hydrochloric acid (caution, vigorous hydrogen evolution). The solution is saturated with sodium chloride (approximately 3 g, analytical grade) and extracted with five 10-ml volumes of ethyl acetate and with five 10-ml volumes of diethyl ether. Solvents must be of microanalytical quality, not pesticide nor spectroscopic grade, as these often have troublesome contaminants that do not impair their usefulness for their declared purposes. Extraction is by intimate mixing with a vortex agitator to form emulsions which can be separated by centrifugation. The organic phases are collected and combined, 5 g of anhydrous sodium sulfate is added to remove water, and the supernatant is concentrated to approximately 8 ml by partial evaporation in a dry nitrogen stream. It is diluted with ethyl acetate to an accurate final volume of 10 ml, and stored in 0.5-ml lots at −40° until needed. An aliquot is assayed by gas chromatography as the TMS derivative against a known amount of capric acid as a quantitating standard; in the present example, it was shown to be composed of 2-HIVA, 3-HICA, and 2-HMVA with concentrations of 0.982, 1.13, and 1.06 mg/ml, respectively.

Solution 2: Nonlabeled Hydroxy Acid Standard

In 2 ml of ethyl acetate are dissolved 0-HIVA (0.0288 mmol, 3.4 mg), 0-HICA (0.0220 mmol, 2.9 mg), and 0-HMVA (0.0197 mmol, 2.6 mg), all from Sigma Chemical Co. This solution is stored at −40° until needed.

Solution 3: Calibrated Mixture

For the purpose of gaining experience with the method, known mixtures of the six acids of interest may be made up and analyzed against the labeled internal standard prepared above. For this purpose, a stock solution of these acids is prepared as follows. To a solution of 5 ml of 5 N aqueous sodium hydroxide are added: 0-HIVA (0.034 mmol, 4.1 mg), 0-HICA (0.0455 mmol, 6.0 mg), sodium 2-hydroxy-3-methylvalerate

(0.0455 mmol, 7.0 mg), ethyl 2-ketoisovalerate (0.0882 mmol, 12.7 mg), sodium 2-ketoisovalerate (0.0967 mmol, 14.7 mg), and sodium 2-keto-3-methylvalerate (0.0803 mmol, 12.2 mg). This solution is saponified at 60° for 2 hr in a sealed vial, cooled, brought to approximately pH 7 with 8 N hydrochloric acid, diluted to 10 ml with water, and stored at $-40°$ in a plastic vial (to prevent breakage on freezing) until required.

A set of six known concentrations may be prepared in 3-ml aliquots of water in centrifuge tubes by the addition of 5, 10, 20, 30, 50, and 80 μl of the stock solution of hydroxy and keto acids just prepared. To each tube is added 60 μl of the labeled internal standard solution, 2 drops of 10% sodium hydroxide, and 2 mg of sodium borodeuteride. The tubes are heated and treated further as described below.

Method

General Extraction Procedure

This extraction procedure is applicable to fluids such as urine that have relatively small protein concentrations that will not result in the precipitation of large amounts of denatured protein during the extraction. It is best carried out with the sample contained in a centrifuge tube having a volume that will accommodate the sample and extracting solvent with sufficient headspace to allow efficient mixing by vortexing. The tube should be closed with a screw cap having a Teflon liner. Into the tube is measured an aliquot of urine (2 ml usually) taken from either a 24-hr collection or a random sample. It is made basic (pH 12–13) with a few drops of 10% sodium hydroxide. The labeled internal standard prepared above is added in the amount of 10 μl per milliliter of urine or similar fluid for samples expected to have near normal concentrations of the branched-chain acids, and 50 μl per milliliter for samples expected to have very high levels of these acids. Sodium borodeuteride (1 mg per milliliter of urine volume) is added, the tube is closed and heated at 60° for 15 min, and then cooled to room temperature. Sufficient sodium chloride is added to saturate the aqueous phase. Residual solid sodium chloride may be left in the tube. A volume of diethyl ether is added that is equal to the aqueous volume, and the contents of the tube are intimately mixed by vortexing. Usually an emulsion is formed that can be made to separate in a small bench-top centrifuge. The ether layer is aspirated away and discarded. This extraction is repeated twice. The sample is acidified by cautious dropwise addition of 5 N hydrochloric acid (hydrogen evolution) to pH 1–2. Vigorous effervescence must be seen with the addition of the first drop of acid; failure indicates exhaustion of the borodeuteride which may be corrected by

addition of an additional 1 mg and repeating the reduction under basic conditions. The sample is again extracted three times with ether, the ether phases are combined and dried with anhydrous sodium sulfate (100 mg approximately).

Modification for Serum and Protein-Containing Fluids

Serum removed from clotted venous blood, CSF, tissue culture media, and other fluids containing high protein concentrations require precipitation and removal of protein that, if left in the sample, will cause difficulty with later separations and handling. A 1-ml aliquot in a centrifuge tube is spiked with 10 to 50 μl of the labeled internal standard as for urine and shaken well. Saturated aqueous sulfosalicylic acid (3 drops per milliliter of serum) is added and the mixture is well shaken and then centrifuged to produce a firm protein pellet and clear supernatant. The supernatant is decanted and made basic with a drop of 10% sodium hydroxide, 1 mg of sodium borodeuteride is added, and the mixture is heated and treated further in the manner described above in the general procedure.

Derivatization

The extract is transferred in aliquots to a 0.5-ml vial suitable for silylation purposes and evaporated to dryness under a nitrogen stream. The extract is derivatized with a volume of *N,O*-bistrimethylsilylacetamide or *N,O*-bistrimethylsilyltrifluoroacetamide (BTMSA and BSTFA, respectively, Pierce Chemical Co. Rockford, IL) approximately equal to five times the volume of the labeled internal standard added at the start of the analysis.

GC-MC Conditions

Trimethylsilylated samples are gas chromatographed on a 4 m × 2 mm i.d. glass column packed with 15% OV-101 on 100–120 mesh Chromosorb W HP (Pierce Chemical Co.) and operated isothermally at 170°. Helium flow rates are optimized for best separator performance (30 ml/ min typically). The injector block and separator are operated at 280°. Successful chromatographic resolution of the HICA and HMVA peaks requires this relatively long and heavily loaded column. Under these conditions, HIVA, HICA, and HMVA have retention times of 6.2, 7.8, and 8.2 min, respectively. Capillary column inlet may also be used to advantage if the GC-MS instrument is capable of a sampling rate sufficient to define very well the narrow peaks that result. Solvents are diverted from the ion source during their elution. To economize on disk storage, it is advantageous to be able to avoid data acquisition until peaks of interest are

about to elute, typically about 5 min after injection. The usual practice of preconditioning the column with multiple injections of the sample at the operating temperature or slightly higher should be followed.

The mass spectrometer is operated in electron impact mode at 70 eV with the ion source temperature between 200° and 240°. It must be capable of measuring ion intensities of a group of at least four ions and be capable of switching to a new group on command or under program control. The fragment ions m/z 247, 248, and 249 corresponding to the $[M-CH_3]^+$ fragments of 0-HIVA, 1-HIVA, and 2-HIVA (Fig. 1) are monitored starting approximately a minute before the expected elution of HIVA and until 15 sec after it has completed elution. Usually a small peak due to the TMS derivative of 2-hydroxy-2-methylbutyric acid which naturally occurs in urine and serum elutes ahead of the HIVA peak without interference and provides an opportunity for fine tuning of the calibration immediately prior to the elution of the HIVA peak. Upon completion of the HIVA elution, the second set of ions is selected for HICA and HMVA; these are m/z 159, 160, 161, and 162. HICA elutes approximately 1.6 min after HIVA. The internal standard for HICA is labeled with three deuterium atoms and produces ion current at m/z 162 (Fig. 1). Responses are noted for HICA principally in the 159, 160, and 162 channels (due to the $[M-COOTMS]^+$ fragments for 0-HICA, 1-HICA, and 3-HICA, respectively), with a diminished response in the 161 channel due to natural abundance heavy isotope substitution in the two lighter fragments. The response in m/z 161 should appear as a shoulder on the very much larger

FRAGMENT STRUCTURE   ASSIGNMENT

$[CH_3-CR(CH_3)-CR'(OTMS)-COOSi(CH_3)_2]^+$

Acid	R	R'	$[M-CH_3]^+$
0-HIVA	H	H	247
1-HIVA	H	D	248
2-HIVA	D	D	249

$[CH_3-CH(CH_3)-CR_2-CR'(OTMS)]^+$

Acid	R	R'	$[M-COOTMS]^+$
0-HICA	H	H	159
1-HICA	H	D	160
3-HICA	D	D	162

$[CH_3-CH_2-CR(CH_3)-CR'(OTMS)]^+$

Acid	R	R'	$[M-COOTMS]^+$
0-HMVA	H	H	159
1-HMVA	H	D	160
2-HMVA	D	D	161

FIG. 1. Structures and labeling in the fragment ions monitored for HIVA, HICA, and HMVA, respectively.

ion current produced during the elution of the HMVA peak 30 seconds after HICA. The ion currents required for HMVA are m/z 159, 160, and 161, the [M-COOTMS]$^+$ fragments respectively for 0-HMVA, 1-HMVA, and 2-HMVA (Fig. 1).

The [M-COOTMS]$^+$ ions for the HIVA peak cannot usefully be monitored because a rearrangement ion at m/z 147 common to most TMS derivatives bearing more than one TMS group interferes with the internal standard. The [M-CH$_3$]$^+$ ions for HICA and HMVA are much less intense than those selected here.

The analysis of the calibrated mixtures should produce correlations that are very good (correlation coefficient > 0.99) when the logarithms of the known and determined concentrations (micromolar) for each of the six acids are plotted (ordinate and abscissa, respectively). The slopes should be between 0.95 and 1.05, and the intercepts on the known axis (i.e., where the determined values are 1 μM, logarithm 1 = 0) should be between -0.05 and 0.05. Poor correlations and significant departures from unit slopes indicate incomplete borodeuteride reduction, and large intercepts ($-$ or $+$) suggest significant impurities in the branched-chain hydroxy and keto acids purchased for the preparations above. It is worth the time to analyze these materials by GC-MS before starting so that their purities may be taken into account in these preparations.

Figure 2 is an example of the data generated with this method using, in this case, a Hewlett-Packard 5980A quadrupole GC-MS. The small peak eluting just before the HIVA peak allows the operator of this particular instrument to choose from several computer switch register selectable mass offsets for optimum response in the channels selected. The computer may then be used to integrate the peak areas for each of the fragmentograms. The example reproduced here was taken from an analysis of serum taken from a member of a group of normal children in the age range 5–10 years cld.

Calculations

The areas obtained for the ion fragments monitored here are related to the partial pressures in the ion source of labeled and unlabeled species, which, in turn, are related to their concentrations in the derivatized extract of the original fluid sample. The relationship between original concentrations and observed peak areas, however, is not one of simple proportionality, but is made complex by the coincidence of several ionic species at the same nominal masses due to isotopic impurities in labeled materials and natural abundance heavy isotope substitution in all materials.

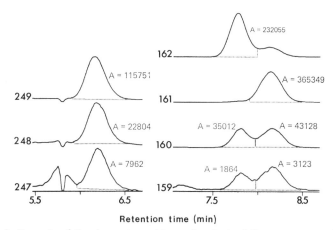

Retention time (min)

Fig. 2. Example of the chromatographic results obtained for a serum sample from a normal child. The ions monitored during the early half of the analysis are m/z 247, 248, and 249, which correspond to the [M-CH₃]⁺ fragments of endogenous HIVA, sodium borodeuteride-reduced endogenous KIVA, and the internal standard 2-HIVA, respectively. The response measured at a retention time of approximately 6.2 min is the HIVA peak. The peak eluting at 5.8 min is due to the TMS derivative of 2-hydroxy-2-methylbutyric acid, which has an ion fragment of similar elemental composition to that of HIVA. This provides an opportunity to manually select a mass offset permitted by the data system for the purpose of optimizing detector response, and this is the cause for the apparent disruption of that peak. After completion of the elution of the HIVA peak, the second group of ions for monitoring is selected for HICA and HMVA eluting with retention times of approximately 7.8 and 8.2 min. The m/z 159–162 cluster represents the [M-COOTMS]⁺ fragment for the TMS derivatives of both acids. Endogenous HICA and HMVA produce principal ion currents in the 159 channel, while the borodeuteride-reduced KICA and KMVA (1-HICA and 1-HMVA, respectively) yield principal responses at m/z 160. The HICA internal standard, 3-HICA, bears three deuterium substitutions and yields its principal ion current at m/z 162. Since 2-HMVA, the internal standard for HMVA and KMVA, bears only two deuterium substitions, its response is at m/z 161. This allows a clearer distinction between the two internal standards and easier assignment of pertinent integrated peak areas.

In the case of the HIVA analysis, the area measured for nominal m/z 247 (A_{247}) will be contributed to by three principal fragments: the fragment due to unlabeled endogenous 0-HIVA; a portion of the borodeuteride-reduced KIVA (1-HIVA) that is unlabeled (i.e., due to isotopic impurity in the sodium borodeuteride); and the completely unlabeled impurity in the internal standard 2-HIVA. If one assumes identical ionization efficiencies for the TMS derivatives of 0-, 1-, and 2-HIVA, and that the substitution of deuterium into the aliphatic chain of HIVA does not produce an appreciable kinetic isotope effect for the formation of this fragment, constants of

proportionality may be collected as K, and

$$A_{247} = K(D_0 + D_1F_1 + D_2F_2) \tag{1}$$

D_0 is the original concentration of 0-HIVA in the sample, D_1 is the concentration of 1-HIVA reduced from KIVA, D_2 is the known concentration of internal standard 2-HIVA in the sample, F_1 is the fraction of 1-HIVA that is unlabeled, and F_2 is the fraction of 2-HIVA that is unlabeled.

In a similar manner,

$$A_{248} = K(D_1 + D_2F_3 + P_1D_0) \tag{2}$$

F_3 is the fraction of internal standard bearing only one deuterium and P_1 is the probability of natural abundance heavy isotope substitution in the endogenous HIVA m/z 247 fragment to yield a current at m/z 248.

Finally,

$$A_{249} = K(D_2 + P_2D_1 + P_3D_0) \tag{3}$$

P_2 is the probability of natural abundance heavy isotope substitution in 1-HIVA to yield ion current at m/z 249 and P_3 is the probability of similar ion current at m/z 249 from 0-HIVA.

F_1 is related to the isotopic purity of the lot number of the sodium borodeuteride in use, and is determined empirically for a single lot only once by reduction of a small amount of KIVA dissolved in distilled water under the conditions of analysis. F_1 is then the ratio of the intensities of the ions m/z 247 : 248 and is typically 0.01 to 0.03. F_2 and F_3 are determined by the analysis of an aliquot of the internal standard and are the intensity ratios of m/z 247 : 249 and 248 : 249, respectively. Probabilities P_1, P_2, and P_3 are, respectively, the intensity ratios: m/z 248 : 247 for unlabeled HIVA, m/z 249 : 248 for 1-HIVA, and m/z 249 : 247 for the internal standard 2-HIVA. These may be calculated for the known elemental compositions of the m/z 247 cluster, or they may be determined empirically as for the fractions F. Since D_2 is the known quantity of internal standard added, and A_{247}, A_{248}, and A_{249} have measured values, the only unknown values are those for D_1 and D_0. If A_1 and A_2 are defined as

$$A_1 = A_{247}/A_{249} \quad \text{and} \quad A_2 = A_{248}/A_{249}$$

the constants of proportionality K cancel, giving:

$$A_1 = \frac{D_0 + F_1D_1 + F_2D_2}{D_2 + P_2D_1 + P_3D_0} \tag{4}$$

and

$$A_2 = \frac{D_1 + F_3D_2 + P_1D_0}{D_2 + P_2D_1 + P_3D_0} \tag{5}$$

Solving these two equations for the unknown concentrations of HIVA and KIVA (D_0 and D_1, respectively):

$$D_1 = D_2 \frac{(A_2 - F_3 D_2)(1 - A_1 P_3) + (A_2 P_3 - P_1)(A_1 - F_2)}{(1 - A_2 P_2)(1 - A_1 P_3) - (A_2 P_3 - P_1)(A_1 P_2 - F_1)} \tag{6}$$

$$D_0 = D_2 \frac{A_1 - F_2}{1 - A_1 P_3} + D_1 \frac{A_1 P_2 - F_1}{1 - A_1 P_3} \tag{7}$$

Analogous equations may be written for the HICA/KICA and HMVA/KMVA pairs using areas for m/z 159, 160, and 162 and for m/z 159, 160, and 161, respectively.

The values of P_1 and P_3 for HIVA are the ratios of the intensities of m/z 248 and 249 relative to 247 and are calculated for the elemental composition $C_{10}H_{23}O_3Si_2$ to be 0.2192 and 0.0942, respectively. P_2 is calculated for $C_{10}H_{22}O_3Si_2$ to be 0.2190. For HICA and HMVA, intensities of m/z 160, 161, and 162 relative to 159 calculated for the two fragments having indentical compositions $C_8H_{19}OSi$ are, respectively, 0.1446, 0.0443, and 0.00375. P_1 and P_3 for HICA are, respectively, 0.1446 and 0.00375, and for HMVA they are 0.1446 and 0.04433. P_2 for HICA is identical to the ratio of m/z 160:158 calculated for $C_8H_{18}OSi$, which is 0.04430. P_2 for HMVA is calculated from this same composition as the intensity ratio of m/z 159:158, and is 0.1444.

The substitution of numerical data into these equations and their solution is most conveniently accomplished with a computer program written for this purpose. Copies of a listing of such a program written in BASIC is available on request from one of us (OAM). It is written to operate with an International Business Machines personal computer, and will accept areas, sample volume, and internal standard concentrations, and calculate concentrations for the six acids.

Suggested Protocol

It is suggested that the procedure be set up in the following order:

1. Prepare the three solutions as described above.
2. A 50-μl aliquot of the nonlabeled hydroxy acid standard (solution 2) may be added to 250 μl of BTMSA or BSTFA and heated at 60° for 10 min to produce final concentrations of approximately 0.2 μg/μl for each acid. This may be used to establish GC conditions and verify proper instrumental operation in SIM. It may also be used to verify independently the calculated intensities of ion fragments bearing natural abundance heavy isotope substitution relative to the unsubstituted species for 0-HIVA, 0-HICA, and 0-HMVA. For this purpose, this silylated sample is analyzed in SIM as indicated above.

3. A 10-μl aliquot of solution 1 is similarly derivatized in a volume of 50 μl and analyzed in SIM to determine the relative abundances of isotopically deficient analogs in the internal standard. F_2 and F_3 for HIVA are, respectively, the area ratios m/z 247:249 and 248:249; for HICA, they are m/z 159:162 and 160:162; and for HMVA, 159:161 and 160:161. This analysis is only required once per batch of internal standard.

4. A solution of KIVA saponified from approximately 5 mg of the ethyl ester (or an equivalent amount of the free acid if available) in 2 ml of aqueous 5 N sodium hydroxide is treated with 2 mg of sodium borodeuteride. The 1-HIVA produced is isolated by solvent extraction as described for solution 1. An aliquot of the product is derivatized and analyzed in SIM. The area ratio m/z 247:248 is a measure of the isotopic purity of the sodium borodeuteride which is F_1 required for the calculations for HIVA, HICA, and HMVA. The ratio for m/z 249:248 may be determined (P_2) and compared to the P_1 value calculated for HIVA above; they should be very similar. This analysis need only be done once for each batch or lot of sodium borodeuteride; it is useful to purchase this material in small units all having the same lot number and storing them sealed in a vacuum desiccator at $-20°$ until required.

5. At this point, all the preparatory steps that only need be done once have been completed, and the samples selected for analysis may be extracted and analyzed. The samples must be analyzed under the same conditions as the standards were. It is useful when analyzing samples at some later date to reestablish instrumental conditions with an aliquot of solution 2 as in step 2 above. Areas for the ions of interest may be determined and factors P_1 and P_3 calculated for comparison with earlier values.

Normal Concentration Values in Serum

The normal micromolar concentrations for HIVA, KIVA, HICA, KICA, HMVA, and KMVA in the sera of normal children ($N = 14$) have been reported (6) to be respectively [mean, (standard deviation)]: 3.84, (1.55); 9.03, (4.55); 0.71, (0.51); 20.85, (10.14); 0.39, (0.27); 14.54, (6.65). Concentrations found for patients with α-ketonuria range from near normal to more than 50 times normal for each of the six acids.

[7] Measurement of Plasma and Tissue Levels of Branched-Chain α-Keto Acids by Gas–Liquid Chromatography

By Pamela L. Crowell, Robert H. Miller, and Alfred E. Harper

Introduction

The branched-chain α-keto acids (BCKA) α-ketoisovalerate (KIV), α-keto-β-methylvalerate (KMV), and α-ketoisocaproate (KIC) are the products of transamination of valine, isoleucine, and leucine, respectively. Quantification of BCKA is of interest in investigations of branched-chain amino acid metabolism,[1-4] maple syrup urine disease, and other organic acidurias,[5] and conditions in which branched-chain amino or keto acids are used as therapeutic agents.[6,7] BCKA have been determined by gas–liquid chromatography (GLC), [8-13] enzymatic assay,[2,14] and high-performance liquid chromatography (HPLC).[1,15-19] The enzymatic assay is rapid but does not distinguish between the three BCKA.[2,14] Physiological concentrations of all three BCKA can be measured by both GLC and HPLC;

---

[1] See also A. L. Gasking, W. T. E. Edwards, A. Hobson-Frohock, M. Elia, and G. Livesey, this volume [5].

[2] See also G. Livesey and P. Lund, this volume [1].

[3] A. E. Harper, R. H. Miller, and K. P. Block, *Annu. Rev. Nutr.* **4,** 409 (1984).

[4] R. H. Miller and A. E. Harper, *Biochem. J.* **224,** 109 (1984).

[5] K. Tanaka and L. E. Rosenberg, *in* "The Metabolic Basis of Inherited Disease" (J. B. Stanbury, J. B. Wyngaarden, D. S. Fredrickson, J. S. Goldstein, and M. S. Brown, eds.), p. 440. McGraw-Hill, New York, 1983.

[6] R. J. Amen and N. N. Yoshimura, *in* "Nutritional Pharmacology" (G. A. Spiller, ed.), p. 73. Liss, New York, 1981.

[7] M. Walser, *Annu. Rev. Nutr.* **3,** 125 (1983).

[8] G. Lancaster, P. Lamm, C. R. Scriver, S. S. Tjoa, and O. A. Mamer, *Clin. Chim. Acta* **48,** 279 (1973).

[9] C. Jakobs, E. Solem, J. Ek, K. Halvorsen, and E. Jellum, *J. Chromatogr.* **143,** 31 (1977).

[10] U. Langenbeck, A. Hoinowski, K. Mantel, and H. U. Mohring, *J. Chromatogr.* **143,** 39 (1977).

[11] T. C. Cree, S. M. Hutson, and A. E. Harper, *Anal. Biochem.* **92,** 156 (1979).

[12] L. I. Woolf, C. Hasinoff, and A. Perry, *J. Chromatogr.* **231,** 237 (1982).

[13] S. Leslie and C. L. Saunderson, *Comp. Biochem. Physiol. B* **80,** 99 (1985).

[14] G. Livesey and P. Lund, *Biochem. J.* **188,** 705 (1980).

[15] T. Hayashi, H. Tsuchiya, H. Todoriki, and H. Naruse, *Anal. Biochem.* **122,** 173 (1982).

[16] T. Hayashi, H. Tsuchiya, and H. Naruse, *J. Chromatogr.* **273,** 245 (1983).

[17] D. J. Kieber and K. Mopper, *J. Chromatogr.* **281,** 135 (1983).

[18] K. Koike and M. Koike, *Anal. Biochem.* **141,** 481 (1984).

[19] G. Livesey and W. T. E. Edwards, *J. Chromatogr.* **337,** 98 (1985).

METHODS IN ENZYMOLOGY, VOL. 166

although HPLC methods can be more rapid and more sensitive than GLC methods, they have been used less extensively for analysis of tissue BCKA concentrations.

The procedure described is essentially that of Cree et al.[11] with minor modifications.[20] The BCKA are measured as their O-trimethylsilated quinoxalinol derivatives in whole blood, plasma, tissues, and urine. The BCKA react with o-phenylenediamine (OPD) in acidic solution to form the respective quinoxalinols, which are further derivatized with N,O-bis(trimethylsilyl)trifluoroacetamide (BSTFA) in pyridine to form the O-trimethylsilyl derivatives.[21,22] The limit of detection is approximately 2 nmol/g tissue or 2 μM in fluids.

Materials and Methods

Reagents and Solutions

o-Phenylenediamine (Aldrich Chemical Co.): OPD must be purified prior to use since it can autoxidize, forming phenazine and diaminophenazine.[22] To recrystallize, 10 g of OPD is dissolved in 200 ml of distilled water and 35 ml of concentrated H_2SO_4; heating may be required to dissolve the OPD. Crystal formation is allowed to take place overnight at 4°. The white crystals are harvested in a Büchner funnel and rinsed several times with cold, distilled water. After drying, the OPD is recrystallized once more. The final crystals are stored in a desiccator at 4°.

OPD stock solution: For each sample, 30 mg OPD is dissolved in 1 ml of 2 N HCl. The solution is made freshly each day.

N,O-Bis(trimethylsilyl)trifluoroacetamide (Pierce Chemical Co.): A fresh 1-ml vial of BSTFA is usually opened each day. Opened vials of BSTFA are stored in a desiccator at −20° throughout the day and can be kept for up to 3 days.

Pyridine, derivatization grade (Regis Chemical Co.)

3% OV-1 on Chromasorb Q (Supelco, Inc.)

α-Ketovaleric acid (KV), α-ketocaproic acid (KC), KIV, KMV, and KIC, sodium salts (Sigma Chemical Co.)

Internal standard solution: A water solution containing both KV and KC, each 1 mM. This solution is distributed in small tubes which are stored at −20°.

[20] S. M. Hutson, C. Zapalowski, T. C. Cree, and A. E. Harper, J. Biol. Chem. **255**, 2418 (1980).

[21] A. Frigerio, P. Martelli, K. M. Baker, and P. A. Biondi, J. Chromatogr. **81**, 139 (1973).

[22] N. L. Edson, Biochem. J. **29**, 2082 (1935).

BCKA standard solution: A water solution containing KIV, KMV, and KIC, each 1 mM. This solution is also distributed in small tubes which are stored at $-20°$.

Chromatographic Conditions

A Packard-Becker model 417 gas chromatograph with a flame-ionization detector is used for analysis. The 12-foot glass column, outer diameter 3 mm and inner diameter 1.5 mm, is packed with 3% OV-1 on Chromasorb Q. The temperature program is from 130° to 270° at 5°/min followed by a 19-min isothermal period. The injection port temperature is 190° and that of the detector is 260°. Gas flows are as follows: prepurified N_2 (carrier), 15 ml/min; H_2, 30 ml/min; and compressed air, 260 ml/min. Peak areas are recorded on a Hewlett-Packard model 3390A integrator or other suitable recorder; a typical chromatogram of plasma to which the internal standards have been added is shown in Fig. 1. The limit of detection of the instrument is approximately 100 pmol per injection.

Sample Preparation

Plasma. Whole blood is collected in heparinized centrifuge tubes and spun at 12,000 g for 20 min in a refrigerated (4°) Sorvall RC2-B centrifuge. The supernatant solution is then collected with a Pasteur pipet. Plasma can be frozen at $-20°$ until the time of analysis. For analysis, 50 μl of the internal standard solution is added to 1 ml of plasma. The sample is deproteinized by the addition of 2.5 volumes of cold ($-20°$) acetone followed by rapid mixing with a Vortex Mixer (American Scientific Products). The protein is sedimented by spinning at 27,000 g for 20 min in a precooled ($-20°$) centrifuge; the supernatant solution is decanted into a

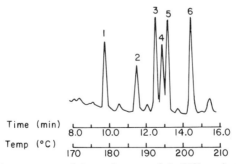

FIG. 1. BCKA chromatogram of plasma from a rat fed a 20% casein diet. The sample was prepared as described in the text. Peaks correspond to (1) pyruvate, (2) KIV, (3) KV, (4) KMV, (5) KIC, (6) KC.

culture tube. For human plasma, it has been helpful to include a second centrifugation step, identical with the first, in order to separate the acetone extract from fat. The extract is then evaporated to dryness under reduced pressure in a Büchler vortex evaporator and derivatized as described below. It is necessary to precool the evaporator sample block and then allow the block temperature to increase slowly in order to prevent bumping of the extract containing acetone. The procedure can be scaled down for as little as 0.3 ml plasma.

Tissues, Whole Blood, and Urine. Tissues are either freeze-clamped between aluminum tongs precooled in liquid nitrogen or simply dropped into liquid nitrogen. The frozen tissue is pulverized between stainless steel blocks which have been precooled in liquid nitrogen. The powdered tissue can be stored at $-80°$ or in liquid nitrogen until the time of analysis. For analysis, 1 g of frozen tissue is weighed in a glass homogenizing tube on dry ice and the tube is then placed in a container of crushed ice. Fifty microliters of the internal standard solution, 3 ml of cold ($-20°$) ethanol, and 1 ml of water are added, and the tissue is homogenized with a Teflon pestle driven by an electric motor. An additional 3 ml of ethanol is added and the sample is again homogenized. The sample is then transferred to a plastic centrifuge tube, placed in a boiling water bath for 2 min, mixed rapidly, and allowed to cool on ice. Following centrifugation at 12,000 g for 15 min at $4°$, the supernatant solution is decanted into a large culture tube. Two milliliters of ethanol/water (6:2, v/v) and 20 ml of chloroform are added, and the sample is mixed and centrifuged at 300 g for 5 min using a swinging bucket rotor in a refrigerated centrifuge. In order to facilitate pipetting of the aqueous layer (main top layer), 1 ml of ethanol/water (6:2, v/v) is gently added and the aqueous layer is removed. The aqueous extract is then evaporated to dryness at room temperature under reduced pressure in a Büchler vortex evaporator. This ethanol extraction procedure is also used for 1 ml of whole blood or urine.

Analysis of Samples

One milliliter of the stock solution of OPD is added to the dried tissue or physiological fluid residue, and the sample is covered and left overnight in the dark. Alternatively, the sample can be placed in a boiling water bath for 30 min in a sealed, screw-capped, culture tube followed by cooling to room temperature. The branched-chain quinoxalinols are then extracted: 1 ml of dichloromethane is added, the sample is mixed vigorously for 1 min, and the lower layer is transferred to a clean culture tube through Whatman #2 filter paper. The filter paper is then washed with an additional 1 ml of dichloromethane. The extraction is repeated twice, and the

combined extracts are then evaporated to dryness. The dried sample may be stored, covered, in a desiccator in the dark at room temperature for up to 2 weeks; however, care should be taken to avoid contact with moisture since water can hydrolyze the quinoxalinols. Just prior to injection onto the column, the sample is trimethylsilated: 10 μl of pyridine and 50 μl of BSTFA are added and the sample is mixed at room temperature. After derivatization for 10 min, 2−4 μl of the sample is injected onto the column. After each injection the syringe is rinsed with chloroform and then with acetone.

Potential Difficulties

The technique described is well established and has given satisfactory results in this laboratory for nearly a decade. The most common problem is from residues accumulating on the column. This is prevented by (1) using the ethanol extraction procedure for samples high in lipid content, (2) routine cleaning of the detector and changing of the septum, and (3) occasional heating of the column overnight at 270° to burn off contaminants.

Calculation and Expression of Results

Internal Standards. KV or KC is used as the internal standard; both chromatograph in the same region as the BCKA but are not normally present in body fluids or tissues. Addition of these internal standards from the beginning of the procedure subjects them to the same losses as the BCKA and, thus, obviates the need for correction for BCKA recovery.

Standard Curves. Standard curves based on the ratios of peak areas are used to estimate BCKA concentrations. The peak area ratio, calculated separately for each BCKA, is the ratio of the BCKA peak area for a physiological sample containing a known amount of BCKA standard to the peak area for the internal standard.

To prepare a standard curve, samples of body fluids or tissue preparations containing both BCKA and internal standards are analyzed as described above. These samples are prepared by adding to 1 ml of body fluid or 1 g of homogenized tissue, 50 μl of the internal standard solution and from 0 to 80 μl of the BCKA standard solution. For each BCKA, the peak area ratio is then plotted against nanomoles of BCKA added to the 1 ml of body fluid or 1 g of homogenized tissue. Separate standard curves are prepared for each physiological fluid or tissue.

Typical standard curves for the three BCKA for plasma are shown in Fig. 2. The peak area ratios for zero nanomoles of keto acid added are the values for the amounts of BCKA in the samples of plasma. The slopes of

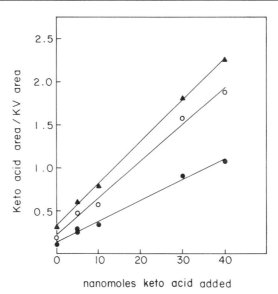

Fig. 2. Plasma BCKA standard curves. Each point represents an individual sample. ●, KIV; ○, KMV; ▲, KIC.

the curves give the relationship between peak area and BCKA concentration per milliliter of fluid or per gram of tissue. If an integrator is not available, peak height rather than peak area can be used as the basis of the standard curves. In our laboratory, KV is used as the internal standard, although KC is equally satisfactory.

Calculations. To determine the concentrations of BCKA in a sample containing unknown amounts of BCKA, the peak area ratio for each BCKA is calculated. Each value is then divided by the slope of the appropriate standard curve to obtain the concentration in nanomoles/milliliter of fluid or nanomoles/gram of tissue.

Representative BCKA Concentrations in Plasma and Tissues

Plasma BCKA concentrations of rats fed diets differing in protein content or branched-chain amino acid content are shown in Table I.[23] For rats fed diets ranging from 0 to 20% casein, BCKA concentrations rose in proportion to the dietary protein content; BCKA concentrations of rats fed 20 or 50% casein were not statistically different. When rats were fed a high leucine diet, the concentration of KIC rose, whereas those of KIV and

[23] P. L. Crowell and A. E. Harper, unpublished observations (1984).

TABLE I

EFFECTS OF DIET ON RAT PLASMA BCKA CONCENTRATIONS

Diet	BCKA concentration (μM) [a]		
	KIV	KMV	KIC
Controlled feeding[b]			
0% casein	9.3 ± 0.9[d]	2.1 ± 0.2[d]	4.4 ± 0.3[d]
5% casein	12.9 ± 2.0[d]	8.8 ± 0.7[d]	13.9 ± 0.8[d]
20% casein (control)	22.7 ± 3.2	19.1 ± 3.2	28.2 ± 3.1
50% casein	20.4 ± 1.7	19.4 ± 3.7	28.1 ± 1.8
Ad libitum feeding[c]			
9% casein (control)	7.7 ± 1.2	15.9 ± 0.8	18.9 ± 1.4
9% casein + 5% Leu	1.9 ± 1.1[d]	8.8 ± 0.8[d]	48.7 ± 5.2[d]
9% casein + 5% LIV	17.0 ± 2.8[d]	23.0 ± 1.3[d]	28.0 ± 2.3[d]

a Values represent means ±SEM with five rats per group. Statistical analysis was done by the students *t*-test (G. W. Snedecor and W. G. Cochran, "Statistical Methods." Iowa State Univ. Press, Ames, Iowa, 1980).

b From 1000–1500 hours each day for 9 days rats were fed diets containing 0, 5, 20, or 50% casein. Rats were killed at 1300 hours on the nineth day.

c For 2 days rats were fed *ad libitum* diets containing 9% casein with the following additions: no addition (control); 5% leucine; or 5% mixture of leucine, isoleucine, and valine (LIV) in the proportions found in casein.

d Significantly different from the control.

KMV fell sharply. Consumption of a diet containing elevated amounts of leucine, isoleucine, and valine resulted in increases in the concentrations of all three BCKA.

Rats fed a standard 20% casein diet have plasma concentrations of 22.7, 19.1, and 28.2 μM for KIV, KMV, and KIC, respectively (Table I). Human plasma BCKA concentrations are the same or higher: 19.7 ± 0.8, 28.7 ± 1.6, and 37.0 ± 2.1 μM for KIV, KMV, and KIC, respectively (mean $\pm$ SEM, $n = 25$).[24] The effects of ingestion of 100 mg valine per kilogram of body weight on human plasma BCKA concentrations are shown in Table II.[24] As in rats fed a high leucine diet, ingestion of a large amount of a single branched-chain amino acid resulted in a rise in the plasma concentration of its corresponding BCKA; however, after valine ingestion by human subjects, changes in the concentrations of the other BCKA were small.

[24] R. H. Miller and A. E. Harper, unpublished observations (1985).

TABLE II
EFFECT OF VALINE INGESTION ON HUMAN
PLASMA BCKA CONCENTRATIONS[a]

	BCKA concentration (μM)		
Time (hr)	KIV	KMV	KIC
0	18.5, 24.5	29.9, 28.5	42.8, 34.7
2	87.4, 81.4	26.1, 20.3	41.3, 29.5
4	58.6, 67.7	26.6, 21.3	43.9, 33.6
6	60.2, 72.3	26.1, 24.6	48.4, 41.3

a Subjects were given 100 mg of valine per kilogram of body weight and blood was drawn for BCKA analysis at 0, 2, 4, and 6 hr. Values represent individual samples from two subjects.

KIC concentrations in heart, liver, kidney, and mammary gland are 2–4 nmol/g, whereas KIV and KMV concentrations are below the limits of detection.[14,25] All three BCKA are present at about 5 nmol/g in skeletal muscle but are undetectable in brain.[23,25] Catabolic conditions such as starvation and diabetes result in increases in most tissue BCKA concentrations.[25]

Acknowledgment

This work was supported in part by USPHS Grant AM10748 and the College of Agricultural and Life Sciences, University of Wisconsin-Madison, Madison, WI. We thank Drs. Stanley Berlow and George Hoganson for the human plasma samples.

[25] S. M. Hutson and A. E. Harper, *Am. J. Clin. Nutr.* **34**, 173 (1981).

[8] Determination of Methylmalonic Acid in Biological Fluids by Mass Spectrometry

By JANE A. MONTGOMERY and ORVAL A. MAMER

Introduction

Methylmalonyl-CoA is a metabolite produced during the catabolism of valine and isoleucine by the carboxylation of propionyl-CoA by propionyl-CoA carboxylase (EC 6.4.1.3). Under normal circumstances, small amounts of methylmalonyl-CoA are hydrolyzed and free methylmalonic acid (MMA) can be measured in urine and in serum.[1] Greatly increased concentrations of MMA may occur when inheritance or other causes lead to faulty or insufficient synthesis of any of several enzymes and factors involved with the metabolism of methylmalonyl-CoA. Some of these are methylmalonyl-CoA mutase (EC 5.4.99.2) which converts R-methylmalonyl-CoA to succinyl-CoA, enzymes responsible for conversion of dietary vitamin B_{12} to adenosylcobalamin required by the mutase, and intrinsic factor and other proteins required for cobalamin absorption and transport.

Gas chromatography–mass spectrometry (GC-MS) procedures for MMA have been reported previously. A stable isotope assay has been developed using 2-[$methyl$-$^2H_3$]methylmalonic acid (MMAD$_3$) as internal standard, methyl ester derivatization, and selected ion monitoring detection (SIM) under chemical ionization conditions.[2] A second employs ethylmalonic acid as internal standard, cyclohexyl ester derivatives, electron impact ionization, and SIM.[3] A third assay has been reported using MMAD$_3$, the trimethylsilyl (TMS) esters, electron impact ionization, and SIM determination of the ratio of ions characteristic of the ratio MMA : MMAD$_3$.[4,5] The first two methods have a variety of difficulties and inherent sources of error that the third method was developed to avoid. TMS derivatives have the advantage over methyl esters in that they are more easily and safely prepared and provide TMS ethers of hydroxylated and enolizable keto acids that are more stable and amenable to GC conditions of analysis. Furthermore, if the MMA concentration is found to be

[1] J. A. Montgomery, O. A. Mamer, and C. R. Scriver, *J. Clin. Invest.* **72,** 1937 (1983).

[2] A. B. Zinn, D. G. Hine, M. J. Mahoney, and K. Tanaka, *Pediatr. Res.* **16,** 740 (1982).

[3] E. J. Norman, H. K. Berry, and M. D. Denton, *Biomed. Mass Spectrom.* **6,** 546 (1979).

[4] F. K. Trefz, H. Schmidt, B. Tauscher, E. Depene, R. Baumgartner, G. Hammersen, and W. Kochen, *Eur. J. Pediatr.* **137,** 261 (1981).

[5] J. A. Montgomery, M.Sc. thesis. McGill University, Montreal, Quebec, Canada, 1982.

METHODS IN ENZYMOLOGY, VOL. 166

normal, the remaining silylate may be used for conventional organic acid profiling for other organic acids. The high temperatures and acidic conditions required for the preparation of the cyclohexyl esters are severe compared to those for silylation, and risk decarboxylative loss of MMA. $MMAD_3$ is better than ethylmalonic acid (EMA) as internal standard because it is chemically much more similar to MMA, and it avoids the problem of the occurrence of endogenous EMA in concentrations greater than normal concentrations for MMA in both urine and serum. Electron impact ionization is used; it is nearly universally available in mass spectrometry laboratories and avoids the problems of reestablishing chemical ionization conditions that reproduce the same ratio of $M^{+\cdot}$ to $[M + H]^+$ required to avoid uncertainties in the measurement of $MMAD_3$ due to natural heavy isotope abundance incorporation in endogenous MMA.

Assay Principle

This is a stable isotope dilution assay which measures MMA against a known amount of $MMAD_3$ as the internal standard. The peak representing the coeluting internal standard and endogenous MMA is examined with GC inlet under SIM conditions. The most intense fragments that display the isotopic labeling are selected, as these intensities affect the ultimate sensitivity of the assay. The relative proportions of the labeled and unlabeled ions are determined and are related to concentrations using a calibration curve.

Preparation of $MMAD_3$ Internal Standard

A standard malonate ester synthesis is employed. Sodium ethoxide is produced by dissolving 230 mg of sodium metal (10 mmol) in 20 ml of absolute ethanol in a 50-ml round-bottom flask. When the sodium is completely dissolved, 500 mg of diethyl malonate (3.1 mmol) is added to the flask with stirring. [$^2H_3$]Methyliodide (500 mg, 3.4 mmol) is taken up in a syringe and added dropwise with stirring to the flask. A reflux condenser is attached and the reaction is heated at reflux for 5 hr. The mixture is transferred to a 100-ml beaker containing 10 ml of water and the ethanol is allowed to evaporate in a gentle airstream overnight. Rotary evaporation of the ethanol is not recommended as the diethyl methylmalonate is volatile. The ester is saponified by adding 20 ml of water and 1 g of sodium hydroxide (25 mmol) and stirring at room temperature for 12 hr. Residual ester is removed by extracting the basic saponificate three times with 20 ml of diethyl ether (extract is discarded). The product mixture will be contaminated with the unreacted malonic acid and the dialkylated product,

2,2-[*dimethyl*-^{2}H$_6$]dimethylmalonic acid (DMMAD$_6$). The latter is differentially extracted from the acidified aqueous phase (pH 2 or less with HCl) with three 10-ml volumes of diethyl ether/hexane (1:3). This extract should be retained until it has been ascertained that the MMAD$_3$ has been satisfactorily recovered. MMAD$_3$ is isolated by extracting the acidified reaction mixture three times with diethyl ether alone (20 ml). The extract is dried over anhydrous MgSO$_4$ and the ether is removed by rotary evaporation. The crude MMAD$_3$ is recrystallized from hot benzene. The yield should be at least 50%. Residual DMMAD$_6$ and malonic acid contamination can be determined by GC-MS by making and analyzing the TMS esters of approximately 20 μg of the product under the conditions of the assay described below. The malonic acid peak will elute well resolved from MMAD$_3$ and DMMAD$_6$, which coelute, and the ratio of total ion current may be used to relate the ratio of malonic to the other two acids. The ratio of MMAD$_3$ to DMMAD$_6$ may in turn be estimated from the ratio of their respective [M-CH$_3$]$^+$ peaks, m/z 250 and 267, which bear similar fractions of the total ion current for each derivative. The degree of purity should be in excess of 80%, and this factor is to be taken into account when weighing this material.

Calibration Curve

A known solution of the internal standard is prepared by dissolving 40 mg of MMAD$_3$ in 10 ml ethyl acetate. Since each analytical sample is prepared using 20 μg internal standard, this stock solution will be sufficient for approximately 2000 analyses.

Two stock solutions of unlabeled MMA having concentrations 0.01 μg/μl and 1.0 μg/μl (I and II, respectively) are prepared by dissolution of 10 mg of MMA in 10 ml of distilled water, and subsequent dilution of 0.10 ml of this solution to 10 ml. The eight mixtures described in Table I should produce a calibration curve suitable for fluids such as urine.

The acids are recovered by acidifying the solutions with 10% HCl and extracting them three times with 1-ml volumes of diethyl ether. The ether extracts are dried over MgSO$_4$ and evaporated under a gentle stream of nitrogen. The TMS derivatives are made and these are analyzed by GC-MS as described below.

The blank sample allows the determination of the amount of unlabeled MMA in the internal standard. If the blank is comparable to the quantity of endogenous MMA present in the sample (i.e., > 2%) it will adversely affect the lower limit of reliable quantitation. The blank is typically of the order of 1%, and must be determined each time a new batch of MMAD$_3$ is synthesized.

TABLE I
CALIBRATION CURVE FOR METHYLMALONIC ACID[a]

| Sample No. | Unlabeled MMA | | |
	Stock Solution	Vol (μl)	Weight (μg)
1	I	20	0.2
2	I	50	0.5
3	I	100	1.0
4	II	5	5.0
5	II	10	10.0
6	II	20	20.0
7	II	50	50.0
8	Blank	0	0

a All samples contain 5 μl of the internal standard solution (20 μg of MMAD$_3$) and varying amounts of the unlabeled stock solutions I and II (concentrations 0.01 and 1.0 μg/μl). All samples are diluted to 1 ml with distilled water.

Figure 1 shows a typical calibration curve prepared by the above method. The logarithms of the observed ratios of the signals of the m/z 247 to 250 ions (ordinate) are plotted against the logarithms of the known weights of unlabeled MMA added for each of the eight mixtures prepared. The lower end of the curve is nonlinear due to the unlabeled content of the internal standard and the ordinate approaches the limiting blank value asymptotically. The upper end of the curve is linear (slope = 1.0211, y-intercept = -1.3513, $r = 0.9989$).

Sample Preparation

General Extraction Procedure

For fluids having low protein concentrations, such as urine, that are expected to have near normal concentrations of MMA, 2-ml aliquots produce ratios of endogenous MMA to MMAD$_3$ suitable for this analysis. If inherited methylmalonic aciduria is suspected, a smaller aliquot (0.5 ml or less) must be used to avoid overwhelming the MMAD$_3$ internal standard. A constant quantity of internal standard (20 μg) is added to each of the samples contained in 5- to 10-ml centrifuge tubes that may be closed tightly with Teflon-lined screw caps. If necessary, these are diluted to 2 ml with distilled water and shaken to ensure sample homogeneity. The sample

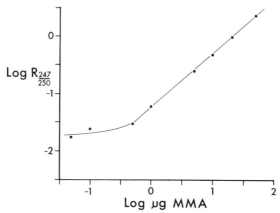

FIG. 1. Sample calibration curve for MMA analysis. The logarithm of observed ratio of the intensities or areas for m/z 247:250 (uncorrected for isotopic impurities) is plotted versus the logarithm of the known amount of unlabeled MMA added to each calibration mixture (see Table I). For subsequent samples, one locates the point on the calibration curve having as ordinate the logarithm of the observed m/z 247:250 area ratio. This point will have as the abscissa the logarithm of the weight of MMA (in micrograms) contained in the sample. Plotting of the ratios and weights on log–log graph paper may be preferred as an alternative.

is saturated with approximately 1 g of sodium chloride and made basic (pH 12–14) by adding five drops of 10% aqueous sodium hydroxide. It must be verified that the pH is greater than 12. The sample is extracted twice with 2 ml of diethyl ether with vortex mixing to remove basic and neutral components that cause interference at low concentrations. A small inexpensive benchtop centrifuge may be used to separate the organic and aqueous phases of the resulting emulsion. The ether layer is discarded. The sample is acidified with 10 drops of 10% HCl and checked to ensure that the pH is less than 2. The sample is extracted three times with 2 ml diethyl ether as above. The ether layers are combined, dried over approximately 500 mg of anhydrous MgSO$_4$, and concentrated to 0.5 ml by evaporation under a stream of dry nitrogen. The extract is transferred to a 0.5-ml vial that may be sealed with a Teflon-lined cap and reduced to dryness under a slow nitrogen stream.

In the case of protein-rich fluids (CSF, serum, amniotic fluid, culture and perfusing media), it is important to avoid the formation of intractable emulsions by precipitating the protein before starting. As many fluids other than urine contain lower concentrations of MMA, larger aliquots may be required. To each aliquot is added 20 μg of MMAD$_3$ and these are shaken to mix thoroughly. The protein is precipitated by adding 5–10 drops of saturated aqueous sulfosalicylic acid, mixing well, and then centrifuging to

form a proteinaceous pellet. The clear supernatant is transferred to a clean centrifuge tube and the extraction is continued as described for urine samples.

Derivatization

To the extract residue is added 80 μl of TRISIL/BSA (Pierce Chemical Co., Rockford, Il.), a commercial trimethylsilylating reagent containing N,O-*bis*(trimethylsilyl)acetamide. The resulting mixtures are heated at 60° for 15 min.

Analysis

Equipment Requirements

A GC-MS instrument is required that is capable of SIM in electron impact mode and data presentation either in an ion current plot integrated by a data system or in an oscillographic recording that allows measurement of the relative intensities of the ions of interest. Analytical conditions are described below.

Gas Chromatography

Separation of methylmalonic acid from interfering endogenous organic acids can be accomplished using a 2 m × 2 mm i.d. glass column packed with 6% OV-17 on Chromosorb W/HP. Dimethylsilicone (e.g., SE-30, OV-1, and OV-101) is unsuitable because the TMS derivative of MMA coelutes on this liquid phase with those for 3-hydroxyisovaleric and 2-methyl-3-hydroxybutyric acids, both of which have fragment ions common to MMA. The carrier gas is helium and flow rates are optimized for best separator operation (20 ml/min approximately). MMA has a retention time of about 5 min at an isothermal column temperature of 110°. Column conditioning is important, and this is easily done by making several injections of a sample under analytical conditions before beginning an analysis.

Mass Spectrometry

The mass spectra of the bis-TMS derivatives of unlabeled MMA and MMAD$_3$ are reproduced in Fig. 2. Moderately intense ions characteristic of the two compounds are the [M-CH$_3$]$^+$ ions at m/z 247 and 250 for endogenous and labeled MMA, respectively. These are the ions selected for the quantitation described here. The ions at m/z 218 and 221 may also be

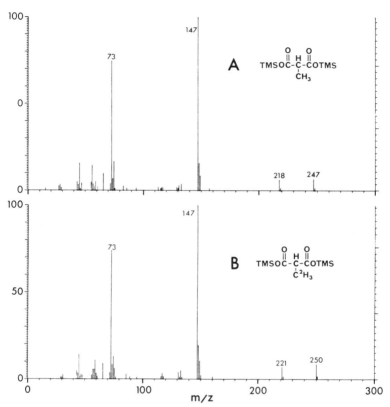

FIG. 2. The 70 eV electron-impact mass spectra of the TMS derivatives of MMA and MMAD$_3$ by GC inlet under the conditions of the assay. The ions selected for quantitative purposes are m/z 247 and 250, and are the fragments formed by the loss of a CH$_3$ radical from one of the TMS groups in the molecular ions of MMA and MMAD$_3$, respectively. The labeled methyl group appears to be refractory to this loss. The molecular ions (m/z 262 and 265, respectively) are present but have intensities too low to be useful. Subsidiary ions for monitoring are m/z 218 and 221, formed by the loss of CO$_2$ from the molecular ions. This last ion pair is used for confirmation of peak identity only, and not for quantitative purposes because column bleed and siloxanes often present a fragment at m/z 221 of highly variable intensity in their spectra.

monitored to confirm the identity of the peak as MMA. A number of peaks will be seen to occur along the m/z 247 trace, but only one will correspond in time to large responses at m/z 218, 221, and 250. Where small amounts of MMA are expected, it is advisable to monitor only m/z 247 and 250, representative of endogenous MMA and MMAD$_3$. This will allow increased sampling time at m/z 247 and therefore more accurate area measurement. The sampling frequency and dwell time per ion should be

chosen so that at least 10 samples are taken in the time interval between the half-height points of the peak of interest. This frequency is approximately 3 Hz with the column and conditions described above. Dwell times per ion should be chosen to permit measurements of their intensities at this rate. Suggested are 0.100 sec for both m/z 247 and 250, and 0.050 sec each for m/z 218 and 221.

Quantitation

Measurement of peak area is more reliable than peak height because of the signal averaging effect that SIM produces. Determinations made using areas will be more accurate than those made with peak heights at low concentrations where signals are small. The logarithm of the ratio of the observed areas for m/z 247:250 is determined and the point on the calibration curve having this value as ordinate will have as abscissa the logarithm of the weight in micrograms of MMA in the extracted volume of the sample. This can be converted to standard concentration values by the following equations:

$$MMA \; (mM) = \frac{\text{Weight MMA } (\mu g)}{\text{Volume extracted (ml)} \times 118} \qquad (1)$$

$$MMA \; (\text{mg/g creatinine}) = \frac{\text{Weight MMA } (\mu g) \times 100}{\text{Volume extracted (ml)} \times \text{creatinine (mg\%)}}$$
$$(2)$$

It is most useful to use the linear portion of the calibration curve. If very low ratios of MMA : MMAD$_3$ are obtained consistently (for example when analyzing CSF or perfusate media), potential errors arising from the use of the lower nonlinear part of the curve may be avoided by increasing the volume of sample extracted, or by establishing another calibration curve using a smaller quantity of MMAD$_3$. Using 4 μg of internal standard, for example, instead of 20 μg will increase the m/z 247:250 intensity ratios fivefold and also reduce the blank MMAD$_3$ contribution to m/z 247. In the case of very low concentrations of MMA, the lower limit of addition of MMAD$_3$ will depend on losses on the gas chromatographic column, which are usually greater for packed columns than for capillary. It will also depend of course on the sensitivity of the mass spectrometer.

Conclusion

Using this method, the concentration of MMA in random samples of urine of normal adults was found to be 3.27 mg/g creatinine (standard deviation, 4.19 mg/g creatinine; $n - 23$).

While newborn screening programs for MMA are well established, there is increased interest in MMA excretion in adult populations where vitamin B_{12} deficiency or malabsorption is suspected to contribute to chronic neurological disorders. This method provides a rapid, specific, and sensitive assay for monitoring or screening patients with these disorders and may also be used to assay MMA in a wide variety of other fluids.

A simple modification of this assay will also allow the measurement of ethylmalonic acid (EMA), which is a metabolite of *2S,3R*-isoleucine [L(+)-alloisoleucine][6] and also the hydrolysis product of the carboxylation of butyryl-CoA by propionyl-CoA carboxylase.[7] The internal standard used is 2-[*ethyl*-$^2H_5$]ethylmalonic acid (EMAD$_5$) which can be synthesized in a manner analogous to that for MMAD$_3$, by substituting [$^2H_5$]ethyl iodide for [$^2H_3$]methyl iodide. The analysis is performed in SIM using ions m/z 261 and 266 for EMA and EMAD$_5$, respectively, and calibrations and measurements analogous to those used for MMA. As most mass spectrometers will allow the monitoring of more than one group of ions, EMA and MMA can be determined simultaneously in one sample simply by adding both internal standards prior to extraction.

[6] O. A. Mamer, S. S. Tjoa, C. R. Scriver, and G. A. Klassen, *Biochem. J.* **160**, 417 (1976).
[7] C. S. Hegre, D. R. Halenz, and M. D. Lane, *J. Am. Chem. Soc.* **81**, 6526 (1959).

[9] Analysis of Acyl-Coenzyme A Esters in Biological Samples

By BARBARA E. CORKEY

Estimates of acylcoenzyme A (CoA)[1] content of tissues have generally been restricted to the enzymatic analysis of CoASH and acetyl-CoA in the acid-soluble fraction of tissue extracts and of total CoASH, after alkaline hydrolysis of the esters, in the acid-soluble and insoluble fractions of tissue extracts.[2-5] Before the development of high-performance liquid chromatographic (HPLC) methods, specific assays had been developed for only a

[1] See also M. T. King, P. D. Reiss, and N. W. Cornell, this volume [10]; K. Bortlett and A. G. Causey, this volume [11].
[2] J. R. Williamson and B. E. Corkey, this series, Vol. 13, p. 434.
[3] P. K. Tubbs and P. B. Garland, this series, Vol. 13, p. 535.
[4] J. R. Williamson and B. E. Corkey, this series, Vol. 55, p. 200.
[5] A. Olbrich, B. Dietl, and F. Lynen, *Anal. Biochem.* **113**, 386 (1981).

small number of compounds including free CoASH, acetyl-CoA, succinyl-CoA, acetoacetyl-CoA, and propionyl-CoA.[2-4,6]

HPLC of acyl-CoA compounds was first described by Baker and Schooley.[7,8] These authors used reversed-phase paired-ion chromatography to separate mixtures of various chain length acyl-CoA thioesters. Their method was not applied to biological samples. Direct reversed-phase HPLC methods have subsequently been developed and applied to the measurement of free CoA,[9] short-chain CoA,[10-13] and long-chain acyl-CoA esters.[14]

The intermediates of branched-chain keto acid (BCKA) oxidation are mainly CoA esters. Evaluation and understanding the regulation of the BCKA pathway requires methods for analyzing pathway metabolites. Studies in the 1950s used column and paper chromatography to identify the intermediates formed from branched-chain amino acids (BCAA). This required hydrolysis of the acyl-CoA derivatives prior to separation, was time consuming, and provided only qualitative data. The HPLC method of Corkey et al.[10] was specifically developed for the purpose of directly measuring acyl-CoA compounds derived from BCKA and has been used extensively to analyze the intermediates of BCKA oxidation in both rat and human liver samples.[15-21] The focus of this article is on high-sensitivity HPLC methods for analysis of acyl-CoA compounds, with particular attention on methods that are useful for measuring metabolites of BCKA metabolism.

[6] W. T. Hron, H. M. Miziorko, and L. A. Menahan, Anal. Biochem. 113, 379 (1981).

[7] F. C. Baker and D. A. Schooley, Anal. Biochem. 94, 417 (1979).

[8] F. C. Baker and D. A. Schooley, this series, Vol. 72, p. 41.

[9] O. C. Ingebretsen and M. Farstad, J. Chromatogr. 202, 439 (1980).

[10] B. E. Corkey, M. Brandt, R. J. Williams, and J. R. Williamson, Anal. Biochem. 118, 30 (1981).

[11] M. S. DeBuysere and M. S. Olson, Anal. Biochem. 133, 373 (1983).

[12] M. T. King and P. D. Reiss, Anal. Biochem. 146, 173 (1985).

[13] Y. Hosokawa, Y. Shimomura, R. A. Harris, and T. Ozawa, Anal. Biochem. 153, 45 (1986).

[14] G. Woldegiorgis, T. Spennetta, B. E. Corkey, J. R. Williamson, and E. Shrago, Anal. Biochem. 150, 8 (1985).

[15] B. E. Corkey, A. Martin-Requero, E. Walajtys-Rode, R. J. Williams, and J. R. Williamson, J. Biol. Chem. 257, 9668 (1982).

[16] A. Martin-Requero, B. E. Corkey, S. Cerdan, E. Walajtys-Rode, R. L. Parrilla, and J. R. Williamson, J. Biol. Chem. 258, 3673 (1983).

[17] J. R. Williamson, B. E. Corkey, A. Martin-Requero, E. Walajtys-Rode, and K. E. Coll, in "Proceedings of an International Symposium on Branched Aminoacids and Ketoacids in Health and Disease" (P. Schauder, ed.), p. 41. Karger, Basel, Switzerland, 1985.

[18] J. R. Williamson, B. E. Corkey, A. Martin-Requero, E. Walajtys-Rode, and K. E. Coll, in "Problems and Potential of Branched Chain Amino Acids in Physiology and Medicine" (R. Odessey, ed.), p. 135. Elsevier/North-Holland, New York, 1986.

Sample Preparation

Acid-Soluble Acyl-CoA Compounds

Samples for analysis of acid-soluble acyl-CoA esters are prepared by minor modifications of standard extraction procedures using perchloric acid [6% (w/v), final concentration][2-4] or trichloroacetic acid [10% (w/v), final concentration].[21,22] Application of large volumes of sample, high salt concentrations, or residual perchloric acid appears to be deleterious to both the quality of the separation and the longevity of the reversed-phase columns. Several techniques have been developed to circumvent these problems. Highly concentrated extracts are obtained by centrifuging cell or mitochondrial suspensions through oil into a small volume of perchloric acid (PCA) as described in Refs. 4 and 10. Attempts to further concentrate samples obtained in this manner by lyophilization or evaporation causes increased residual PCA or salt concentrations, and resulting deterioration of the column with poor separation due to lack of retention of compounds. In order to prevent precipitation of $KClO_4$ on the column, it is essential to adjust the pH of PCA extracts to a value equal to or higher than the pH of the mobile phase. At the same time it is desirable to adjust the pH of acyl-CoA-containing solutions to as low a value as possible for maximum stability of the compounds. We usually adjust the pH of tissue extracts to a value between 5 and 6.

An alternative procedure is described by Hosokawa *et al.*[13] in which large volumes of neutralized PCA extracts are partially prepurified by passing over Sep-Pak C_{18} columns (Waters) and are then concentrated by evaporation. This technique is particularly useful for preparing extracts from whole freeze-clamped or perfused organs.

The procedure currently used in the author's laboratory involves extraction with 10% trichloroacetic acid (TCA)[21,22] followed by TCA removal and sample neutralization by ether extraction (five times with one volume of ether per volume of TCA extract) and sample concentration using a Speed-Vac Centrifugal Evaporator (Savant Instruments Inc., Farmingdale,

[19] B. E. Corkey, A. Martin-Requero, M. Brandt, and J. R. Williamson, *in* "Metabolism and Clinical Implications of Branched Chain Amino and Ketoacids" (M. Walser and J. R. Williamson, eds.), p. 119. Elsevier/North-Holland, New York, 1981.

[20] J. R. Williamson, A. Martin-Requero, B. E. Corkey, M. Brandt, and R. Rothman, *in* "Metabolism and Clinical Implications of Branched Chain Amino and Ketoacids" (M. Walser and J. R. Williamson, eds.), p. 105. Elsevier/North-Holland, New York, 1981.

[21] B. E. Corkey, D. E. Hale, M. C. Glennon, R. I. Kelley, P. M. Coates, L. Kilpatrick, and C. A. Stanley, *J. Clin. Invest.*, in press.

[22] L. C. MacGregor and F. M. Matschinsky, *Anal. Biochem.* **141**, 382 (1984).

NY). Dried samples prepared in this manner are resuspended in a small volume of water (approximately 5 μl/mg dry weight of tissue) just prior to analysis. Typically, 20-μl volumes of sample are analyzed. Recoveries of acyl-CoA esters prepared by TCA extraction are in excess of 80% and column life appears to be at least double that observed with untreated neutralized PCA extracts prepared by centrifuging cell suspensions through oil into PCA. TCA extraction has the advantage of being simple, rapid, and not requiring salt addition for neutralization.

Long-Chain Acyl-CoA Compounds

The method of Mancha et al.[23] as modified by Woldegiorgis et al.[14] is used to extract long-chain acyl-CoA esters from tissues. Powdered tissue is suspended in 2-propanol (20 μl/mg) and 50 mM KH$_2$PO$_4$, pH 7.2 (20 μl/mg) to extract the lipids. The suspension is then acidified by addition of acetic acid (12.5 μl/ml of extract) and extracted with petroleum ether saturated with 50% aqueous 2-propanol to remove free fatty acids and less polar lipids. After addition of saturated (NH$_4$)$_2$SO$_4$ (25 μl/ml of extract) to the washed aqueous phase, the acyl-CoA esters are extracted with chloroform:methanol (1:2 v/v). These extracts are then applied to a neutral alumina AG7 column. The column is washed with chloroform:methanol to remove complex lipids, flushed with nitrogen, and the acyl-CoA esters eluted with 100 mM KH$_2$PO$_4$, pH 7.2, in 50% methanol. The pH of the eluate is adjusted to 5.0 with acetic acid, the methanol removed by evaporation, and the sample lyophilized and stored until analysis. Samples prepared by this procedure do not contain the short-chain acyl-CoA esters which are retained in the first aqueous phase.

Equipment and Operating Conditions

HPLC Equipment

An HPLC system consisting of (1) an injector, (2) a 254 nm detector, (3) two pumps and a gradient programmer or a single pump with a programmable apportioning valve, and (4) an integrator or recorder is required for acyl-CoA ester analysis. The HPLC systems used in the author's laboratory have been either a Beckman Model 324, which includes two pumps, a variable wavelength detector, injector, programmer, and

---

[23] M. Mancha, G. B. Stokes, and P. K. Stumpf, *Anal. Biochem.* **68,** 600 (1975).

integrator, or a modular Waters system, consisting of injector, two Model 6000A pumps, a filter spectrophotometer, a solvent flow programmer, and a data module. Other systems are also suitable.

Reversed-Phase Columns

A variety of reversed-phase columns have been used successfully for acyl-CoA separations, including the μBondapak C_{18} (Waters Inc.),[10] Nova Pak C_{18} Radial Pak Cartridge (Waters Inc.),[14] Lichrosorb RP-8 (Brownlee Laboratories),[7,8] Spherisorb 5 ODS (Custom LC Inc.),[9,11] and Develosil ODS (Nomura Chemical, Seto).[13] The early acyl-CoA separations were performed using reversed-phase C_{18} columns of 10 μm particle size. As technology advanced, available particle sizes have decreased to 5 μm and, more recently, 3 μm, leading to better and faster separations on shorter columns. It is thus preferable to use the smaller particle size reversed-phase columns as they become economically feasible. In the author's laboratory, a 5 μm particle size Novapak C_{18} column (Waters) is currently used for routine acyl-CoA analysis. Theoretically, far greater sensitivity can be achieved using microbore columns of 1–2 mm internal diameter. However, current technology, in the author's experience, has not yet solved the problem of clogging of such narrow bore columns in separations involving high salt mobile phases and biological samples.

Care of Columns and HPLC Equipment

Acyl-CoA separations involve the use of high salt buffers which (a) corrode the metal components of the HPLC, and (b) are not instantly miscible with some of the organic solvents used in these separations (leading to formation of precipitates on the column or in the tubing). Partial protection against these problems is achieved by (1) maintaining the flow of salt-containing solutions through the column at all times, (2) by extensively flushing the column with water before storing in organic solvent solutions, (3) by using the lowest salt concentration consistent with good separation, and (4) by premixing the organic solvent and buffer solutions. The appearance of an irregular baseline or mountainous, broad, asymmetrical peaks at irregular intervals suggests that the retention of uneluted compounds has exceeded the capacity of the column. Thorough purging of the column with a range of hydrophilic to lipophilic solvents generally restores the stable baseline. It is essential to remove all of the buffer solutions from the column with several column volumes of water before introducing organic solvents.

Operating Conditions

Temperature. Separations are carried out at room temperature,[9-11,14,15] 30°,[12] or 35°.[13] The chromatographic profile varies with temperature. This has encouraged some investigators to elevate and regulate the temperature slightly above room temperature to eliminate fluctuations in this important parameter. However, since chromatograms appear to deteriorate with increasing temperature and CoA esters are somewhat labile, it may be preferable to cool the column in order to improve separations, bearing in mind the need to prevent artifacts due to the formation of gas bubbles induced by elevated temperatures in uncooled parts of the system.

Mobile Phase. Solvents are prepared using HPLC grade reagents, prefiltered through Millipore type filters of less than 0.5 μm pore size, degassed under vacuum, and stirred slowly and continuously during chromatography by magnetic stirrers. A two-solvent system is used for multicomponent separations consisting of KH_2PO_4, pH 5.0, or equivalent sodium or ammonium salts at pH values ranging from 4.0 to 5.3 with either acetonitrile or methanol as the organic component. King *et al.*[1,12] have found that prepurification of NaH_2PO_4 minimizes problems associated with baseline shifts during gradient separation. The choice between isocratic and gradient elutions is determined by the nature of the desired separation (see Strategies for Optimizing Assays below). The separations appear to be quite sensitive to variations in pH between 4 and 6. This characteristic can prove useful in separating poorly resolved peaks.

Identification, Standardization, and Quantitation of Acyl-CoA Compounds

Identification

The identity of acyl-CoA esters is determined by comparing the elution times of sample peaks with those of acyl-CoA standards of the highest quality available. The elution profile of each individual standard (0.1 to 0.5 nmol) is determined with a separate chromatogram. The time of elution of free CoASH is determined first, since CoA standards may be contaminated with small amounts of CoASH due to degradation of the ester.

Standarization and Quantitation

The concentration of the standard acyl-CoA ester solution (about 0.1 mM) is determined spectrophotometrically at 260 nm using a millimolar extinction coefficient of 15.4 (see P-L Biochemicals Reference Guide). This concentration is corrected for impurities or degradation of the

acyl-CoA stock solution based on the percentage of absorbing material attributed to the peak of interest. This is determined by calculating the areas of peaks obtained from the chromatogram of the compound alone. Standards containing more than 10% free CoASH or other impurities are replaced. Following standardization and correction for impurities, it is convenient to prepare a complex standard solution containing each of the CoA esters of interest. This stock solution may be stored in small aliquots at $-70°$ and thawed just prior to use. Quantitation of acyl-CoA compounds in samples is determined from comparison of standards of known concentration with sample peak area or height. Excellent linearity is obtained under optimal chromatographic conditions with different aliquots of standard solutions using either peak heights or areas. The linearity of peak height responses over a limited but severalfold range of concentrations permits use of a standard recorder, rather than an integrator, for acyl-CoA analyses.

Validation of Methods Using Biological Samples

Identification

Several methods have been used to validate the identity and quantity of acyl-CoA esters determined by HPLC analysis of biological samples. Initial identification is generally based on similarities of elution times of standard compounds and sample peaks as described above. Verification of the identity of sample and standard peaks is achieved by simultaneous injection of sample and standard as described previously.[10] However, this method cannot differentiate CoA and non-CoA peaks with similar elution times. Since acyl-CoA esters are hydrolized by alkaline treatment,[2,4] further validation of the identity of peaks as CoA esters is based on the demonstration that acyl-CoA peaks disappear and free CoASH appears following alkaline hydrolysis.[10,11,13]

Quantitation

The accuracy of the quantitation of acyl-CoA esters determined by comparison of sample peaks with standard peaks has been validated in four ways. First, the measured content of acyl-CoA has been shown to be linear with increasing sample volume over an approximately 10-fold range.[10] Second, addition of varying amounts of standard to biological samples results in the expected increases in the amount of additional peak height or area obtained with approximately 100% recovery of added standard. Third, repeated analyses of the same sample using the same or different volumes

yield values that do not differ by more than 5%. Finally, enzymatic determination of acetyl-CoA yields values within about 4% of the values obtained on the same samples by HPLC.[10]

Free CoASH

Free CoASH forms a complex with glutathione in the absence of sulfhydryl-protecting agents. This can be prevented by addition of 1 mM dithiothreitol (DTT). However, alterations in the chromatograms are observed in the presence of DTT, possibly due to binding of this compound to the column packing. The problem is expressed as decreasing free CoASH peaks with repeated injections of the same standard or sample. For this reason DTT is not recommended for the extraction procedure. As a result the levels of free CoASH decrease with time after extraction while the levels of glutathione CoA increase. The total of CoASH plus glutathione-CoA is constant in a given sample and agrees within 8% with values obtained by direct enzymatic analysis performed on separate samples extracted in the presence of 1 mM DTT.[2] A similar conclusion has been reached by King and Reiss,[12] who quantitated the appearance of glutathione-CoA at different pH values and found it to be both time and pH dependent. This problem has been addressed in different ways by other investigators. Ingebretsen and Farstad[9] included 0.05% (v/v) thiodiglycol in the mobile phase to avoid oxidation and extracted tissue samples in the presence of 10 mM 2-mercaptoethanol (which elutes in two peaks before CoASH). DeBuysere and Olson[11] used the same mobile phase, but included DTT instead of 2-mercaptoethanol in their extraction procedure without loss of CoASH. The values they obtained from extracts of hearts perfused with α-ketoisovalerate by HPLC (32.3 ± 3.6 nmol/g wet weight) or enzymatic determination[2] (34.3 ± 3.7 nmol/g wet weight) were in good agreement. Hosokawa et al.[13] also extracted tissue in the presence of DTT. Their recovery of CoASH plus succinyl-CoA was rather low (56%). Since their assay did not separate the two compounds it is not clear whether CoASH was in fact poorly recovered. In the author's laboratory, the diminution of the CoASH peak only appears after separation of multiple DTT-containing samples. The basis for apparent discrepancies among various investigators is not known. It is possible that DTT may not interact similarly with all column packing material. Nevertheless, it is clearly essential to determine the recovery of free CoASH under each set of assay conditions.

Recovery

Accurate estimates of acyl-CoA esters require that the percentage recovery of known standards added prior to sample extraction be determined

TABLE I
RECOVERY OF STANDARDS ADDED BEFORE EXTRACTION OF
TISSUE SAMPLES

Acyl-CoA	Recovery (%)		
	Ref. a	Ref. b	Ref. c
Malonyl-CoA		90 ± 5	62
Glutathione-CoA		99 ± 7	
CoASH		83 ± 5	
Methylmalonyl-CoA	100 ± 6	95 ± 6	74
Succinyl-CoA	112 ± 5	92 ± 11	
β-Hydroxy-β-methylglutaryl-CoA	96 ± 6	94 ± 12	74
Acetyl-CoA	96 ± 9	107 ± 5	64
Acetoacetyl-CoA			77
Propionyl-CoA	90 ± 3	101 ± 11	68
Crotonyl-CoA		96 ± 3	
Isobutyryl-CoA	88 ± 3		71
Butyryl-CoA		95 ± 6	
β-Methylcrotonyl-CoA	91 ± 4		
Isovaleryl-CoA	89 ± 3		
Valeryl-CoA		96 ± 9	

a From Ref. 10.
b From Ref. 12.
c CoASH and succinyl CoA were measured as a single peak
with 56% recovery of the combined peak observed.[13]

and that these recovery values be used to correct measured estimates. A summary of recovery values obtained by various investigators is presented in Table I. The extreme lability of some acyl-CoA esters at pH values employed for their determination[2-4,12] requires that recovery values reflect the precise conditions used for sample analysis with regard to pH, elapsed time, and storage conditions.

Separations and Strategies for Optimizing Assays

Variables

The time of elution of a given acyl-CoA peak varies as a consequence of the percentage of organic solvent, lipophilicity of the organic solvent, pH, flow rate, salt concentration of the mobile phase, and choice of reversed-phase column (particle composition, particle size, length, and diameter of column). It is possible to design a gradient to elute 20 or more acyl-CoA esters in a single run, or to design an isocratic system that will rapidly and repeatedly elute a small number of compounds. The choice of program

depends on the goal of the particular separations. The first objective in any separation is to achieve baseline before the first peak of interest elutes. Hence, speed of elution of acyl-CoA compounds is not a meaningful criterion if non-CoA compounds interfere.

Increasing the percentage of organic solvent causes peaks to elute earlier. Changing to a more lipophilic solvent, e.g., substituting acetonitrile for methanol also causes peaks to elute earlier. Very similar separations are achieved with concentrations of acetonitrile that are approximately half those of methanol. Substitution of acetonitrile for methanol is useful in permitting elution of longer-chain-length acyl-CoA esters without precipitating salts contained in the buffers.

The chemical basis for pH effects is not clear. Empirical observation has shown that separation of certain pairs of compounds is very pH sensitive in the range of values from 4 to 6. At pH values appreciably above or below 5.0, in the multicomponent assay described below, baseline resolution is not achieved between CoASH and methylmalonyl-CoA, between succinyl-CoA and HMG-CoA, or between acetyl-CoA and acetoacetyl-CoA.

Programming the flow rate is a very useful technique in optimizing separations. Decreasing flow can markedly improve the separation of two poorly resolved components. Increasing the flow rate is an excellent way to close the gap between two widely separated compounds of interest. Flow programming, unlike gradient programming, does not require column equilibration.

Salt concentration influences chromatographic separations even in the reversed-phase mode. Broad peaks can be sharpened by increasing the ionic strength of the mobile phase. However, high salt concentrations cause deterioration of the pump gaskets and erosion of the stainless steel components. Hence, an ionic strength should be sought that is just sufficient to obtain sharp peaks.

Longer columns or the placement of two or more columns in series enhances the separation of similar compounds. Decreasing particle size also improves the separation between similar compounds or permits the same separation with a much shorter column.

Assay of Short and Medium-Chain Acyl-CoA Esters

Figure 1 illustrates a complex elution profile. Panel A represents a mixture containing between 100 and 200 pmol of standards. Panel B represents a sample of rat liver containing 2 to 3 mg dry weight. Separation is achieved using two buffers: buffer A (0.1 M KH$_2$PO$_4$, pH 5.0) and buffer B (0.1 M KH$_2$PO$_4$, 40% acetonitrile, pH 5.0) with gradient and flow pro-

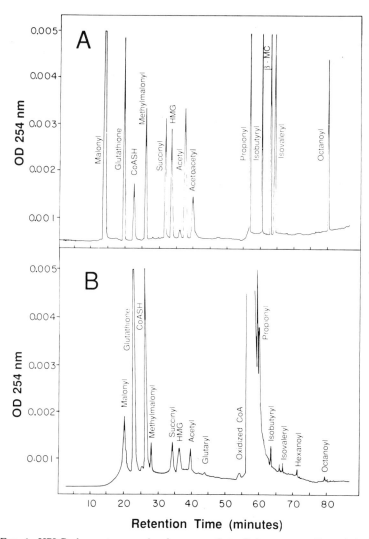

Fig. 1. HPLC chromatogram showing separation of short- to medium-chain-length acyl-CoA esters. (A) A standard mixture of esters containing 100 to 200 pmol of each. (B) The pattern obtained from about 10 mg of liver obtained from a fed, riboflavin-deficient rat. The mobile phases used were (a) 0.1 M KH$_2$PO$_4$, pH 5, and (b) 0.1 M KH$_2$PO$_4$, 40% acetonitrile, pH 5. The initial flow rate was 0.4 ml/min. The gradient program used is described in the text. Abbreviations used are HMG, β-hydroxy-β-methylglutaryl-CoA; β-MC, β-methylcrotonyl-CoA.

gramming beginning with 7.5% buffer B and a flow rate of 0.4 ml/min. At 20 min, the flow is increased to 0.8 ml/min over a period of 1 min. Buffer B is increased to 10% over 2 min beginning at 22 min, and then to 80% for 40 min beginning at 50 min. At 55 min, the flow is increased to 1.2 ml/min for 5 min. At 90 min when the separation is complete, the column is cleansed of retained lipophilic material by increasing buffer B to 100% for 5 min. When a stable baseline has been obtained, buffer B is returned to 7.5% and the flow rate to 0.4 ml/min preparatory to analyzing another sample. Standard runs are interspersed between every 4 to 6 samples. Any portion of the gradient run can be selected when a small number of compounds are to be analyzed. For example, the shorter-chain-length compounds can be determined by following the program described until the peaks of interest have eluted, and then condensing the remainder of the program using a steep gradient to the cleansing step. Alternatively, the medium-chain-length compounds can be analyzed by starting the chromatogram at a higher percentage of buffer B (15–25%).

The assay for long-chain acyl-CoA esters is shown in Fig. 2.[14] Panel A represents a mixture of standards of about 500 pmol each. Panel B represents a liver sample of about 250 mg dry weight. The two mobile phases are (A) 25 mM KH$_2$PO$_4$ pH 5.3, and (B) 100% acetonitrile. The starting conditions were 2 ml/min flow rate and 30% buffer B. Acetonitrile was then increased to 40% over 5 min, increased to 46% over 9 min, increased to 62% over 5 min, and held at 62% for an additional 5 min before returning to the starting conditions.

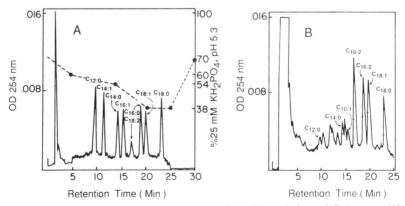

FIG. 2. HPLC chromatogram showing separation of long-chain acyl-CoA esters. (A) A standard mixture containing 500 pmol of the acyl-CoA ester of indicated chain length. (B) Results obtained from an extract of frozen powdered livers from fed and fasted rats. The mobile phases were (A) 25 mM KH$_2$PO$_4$, pH 5.3, and (B) acetonitrile. The flow rate was 2 ml/min and the gradient is indicated in A and described in detail in the text. From Woldegiorgis et al.[14]

Acyl Carnitine Determination. A new method has recently been published that permits analysis of short-chain acyl carnitine esters in urine.[24] First, acyl carnitines are converted to their CoA esters enzymatically using carnitine acetyltransferase (CAT) in the presence of a 25-fold excess of CoASH to acyl carnitine. Next, CAT is removed and the reaction stopped by perchloric acid extraction. Finally, acyl-CoA esters are analyzed by HPLC. The recovery of added acyl carnitines approaches 100% when free carnitine levels are low.

In principle, this technique may prove useful in determining short and branched chain acyl carnitine esters in tissues, provided correction is made for acyl-CoA esters and the decrease in recovery caused by free carnitine. In addition, the long chain acyl carnitine esters may be assayed by a similar technique using carnitine palmitoyl transferase for transesterification coupled to the HPLC assay for long-chain acyl-CoA esters. Application of this technique to tissue carnitine esters has not been reported.

Tissue Acyl-CoA Content

Tissue acyl-CoA contents have been determined in rapidly frozen human liver biopsy samples,[21] freeze-clamped rat livers, isolated hepatocytes, liver mitochondria, and heart using the HPLC techniques described in this chapter. Free CoASH, acetyl-CoA, and succinyl-CoA are the dominant CoA substituents in intact rat liver or isolated hepatocytes (Table II). Starvation causes increases in total CoA, the ratio of acetyl-CoA to CoASH, and the long-chain fatty acyl-CoA esters, myristoyl, linoleoyl, palmitoyl, oleoyl, and stearoyl-CoAs. The influence of the BCKAs on the acyl-CoA profile in isolated hepatocytes is to cause redistribution of CoA esters from free CoASH and succinyl-CoA into intermediates of the BCKA pathway (Table II).

The distribution of acyl-CoA esters in isolated mitochondria is similar to that found in cells. Acyl-CoA levels in isolated rat liver mitochondria incubated in the presence of α-ketoisocaproate (KIC) with and without added succinate or palmitoyl carnitine are shown in Table III. KIC causes appearance of isovaleryl- and β-methylcrotonyl-CoA at the expense of free CoASH and succinyl-CoA. Succinate addition greatly increases the levels of succinyl-CoA while diminishing the levels of all other CoA esters. Palmitoylcarnitine blocks KIC oxidation and isovaleryl-CoA production while decreasing the total acid-soluble CoA pool, presumably as a consequence of insoluble palmitoyl-CoA formation.

Data from experiments performed with perfused rat hearts are shown

[24] R. E. Dugan, M. J. Schmidt, G. E. Hoganson, J. Steele, B. A. Gilles, and A. L. Shug, *Anal. Biochem.* **160**, 275 (1987).

TABLE II
ACYL-CoA LEVELS IN RAT LIVER

| | Acyl-CoA level (nmol/g dry weight) | | | | | |
| | Freeze-clamped liver | | | Isolated hepatocytes | | |
Acyl-CoA	Fed[a]	Fasted[a]	Fasted[b]	Control[c]	α-KIV[c]	α-KIC[c]
CoASH	350[d]	455[d]	313[e]	699	296	258
Malonyl	9.5	3.4	7.5			
Methylmalonyl	5.5	4.1	23	20	530	
Succinyl			68	275	47	44
HMG[f]	8.6	2.9	35			27
Acetyl	180	325	505	369	230	307
Acetoacetyl	5.0	8.2				
Propionyl	19.5	14.9			95	
Isobutyryl					104	
β-Methylglutaryl						91
Isovaleryl						286

	Fed[g]	Fasted[g]				
Lauroyl	2.4	1.4				
Myristoleoyl	4.7	2.8				
Myristoyl	5.8	12				
Palmitoleoyl	6.6	3.3				
Linoleoyl	28	84				
Palmitoyl	20	45				
Oleoyl	22	41				
Stearoyl	4.8	29				

a Values were obtained from freeze-clamped liver from fed or 48 hr-fasted rats. Original data were expressed as nmol/g wet weight and converted to dry weight assuming that 80% of the liver was water.[13]

b Values were obtained from freeze-clamped liver from 48 hr-fasted rats. Original data were expressed as nmol/g wet weight and converted to dry weight assuming that 80% of the liver was water.[12]

c Values were obtained using isolated hepatocytes prepared from rats fasted for 18 hr. Control cells (5 mg/ml) were incubated at 37° for 5 min in Krebs-Ringer bicarbonate buffer without added substrate. α-Ketoisovalerate (α-KIV) and α-ketoisocaproate (α-KIC) were present at 1 mM.[15]

d Values represent the sum of CoASH and succinyl-CoA.

e Reported values for CoASH and glutathione-CoA were added together.

f HMG represents β-hydroxy-β-methylglutaryl CoA.

g Values were obtained from freeze-clamped liver from fed or 48 hr-fasted rats.[14]

TABLE III
EFFECT OF α-Ketoisocaproate (KIC) on Acyl-CoA Levels in Isolated Rat Liver
Mitochondria[a]

Acyl-CoA	Acyl-CoA level (nmol/mg protein)			
	None	KIC	KIC + Succinate	KIC + Palmitoylcarnitine
CoASH	1.95	0.72	0.77	0.66
Acetyl	0.57	0.89	0.22	1.07
Isovaleryl	0	1.42	0.30	0.53
β-Methylcrotonyl	0	0.22	0.05	0.07
Succinyl	0.41	0.09	2.70	0.07

a Isolated rat liver mitochondria (1 mg/ml) were incubated at 30° for 8 min in
110 mM KCl, 5 mM KH$_2$CO$_3$, 20 mM HEPES, 2 mM MgCl$_2$, 5 mM KH$_2$PO$_4$,
0.5 mM EDTA, 0.3% dialyzed bovine serum albumin, pH 7.4.[10] When present,
substrates were added at the following concentrations: KIC, 1 mM; succinate,
10 mM, palmitoylcarnitine, 20 μM.

TABLE IV
ACYL-COA METABOLITES IN HEARTS PERFUSED WITH α-
Ketoisocaproate (α-KIC), α-Ketoisovalerate (α-KIV), or
Propionate

Acyl-CoA	Metabolites (nmol/g dry weight)			
	Control[a]	α-KIC[a]	Propionate[a]	α-KIV[b]
CoASH	399	217	103	165
Acetyl	26	37	13	
Succinyl	210	162	26	
Isovaleryl	0	134	0	18
β-Methylcrotonyl		32		
Propionyl			189	23
Methylmalonyl			448	23
Isobutyryl				115

a Hearts from 300 to 350 g rats were perfused at 37° by noncir-
culating Langendorff techniques.[10] After 20 min perfusion
with 5 mM glucose, 1 mM α-ketoisocaproate was added and
perfusion continued for a further 5 min. Propionate (1 mM)
was added similarly but perfusion was continued for 10 min.
b Values were expressed as nmol/g wet weight in the original
paper and converted to dry weight, assuming that 80% of the
heart was wet weight.[11] Hearts were perfused for 15 min with
0.5 mM α-ketoisovalerate with coinfusion of 0.25 mM pyruvate for 5 min prior to freeze-clamping. Data from three
separate hearts were averaged to obtain the value presented.

in Table IV. The dominant CoA moieties in heart are free CoASH and succinyl-CoA. Perfusion with α-ketoisovalerate or KIC causes redistribution of CoA esters into the BCKA pathway metabolites with accompanying decreases in free CoASH and succinyl-CoA. Propionate addition diminishes the same pools, with production of large amounts of propionyl- and methylmalonyl-CoA.

Relatively good agreement is obtained among workers with regard to the major CoA moieties; however, considerable scatter of the data is observed with regard to minor components. These differences may be due to the lability of CoA esters, and emphasize the need to ascertain accurately recovery values for each CoA ester under the precise conditions of extraction and assay.

[10] Determination of Short-Chain Coenzyme A Compounds by Reversed-Phase High-Performance Liquid Chromatography

By M. Todd King, Paul D. Reiss, and Neal W. Cornell

Coenzyme A (CoA) compounds play a central role in the metabolism of branched-chain amino acids. Catabolism of leucine, isoleucine, and valine is initiated by readily reversible transamination reactions which are followed by irreversible oxidative decarboxylations that result in the formation of acyl-CoA with one less carbon than the respective amino acid. Subsequent steps also involve CoA compounds and yield as end products CoA esters that can enter the tricarboxylic acid cycle. Among the CoA derivatives generated in the metabolism of these amino acids are acetyl-CoA, isobutyryl-CoA, isovaleryl-CoA, 3-hydroxy-3-methylglutaryl-CoA, 3-methycrotonyl-CoA, methylmalonyl-CoA, propionyl-CoA, and succinyl-CoA. Thus, a rapid, accurate means of measuring these CoA compounds in tissue or cell extracts would be useful in studies concerned with branched-chain amino acids as well as in other areas of metabolic research.[1]

Several reports describe the separation and measurement of CoASH, acyl-CoA esters, oxidized CoA, and mixed disulfides of CoA by high-per-

---

[1] See also B. E. Corkey, this volume [9]; K. Bartlett and A. G. Causey, this volume [11].

formance liquid chromatography.[2-5] However, these methods have several shortcomings including (1) the use of isocratic separation limiting the number of compounds identified per run, (2) large baseline shifts in gradient separations, (3) purification and use of corrosive ion-pair reagents, and (4) run times exceeding 2 hr, limiting the number of samples that can conveniently be analyzed. We have developed a reversed-phase HPLC procedure that avoids these problems and allows the separation and measurement of as many as 17 CoA compounds in tissue perchloric acid (PCA) extracts. All the compounds can be resolved in a single run requiring 45 min. CoA compounds separated include acetoacetyl-CoA, acetyl-CoA, butyryl-CoA, CoASH, crotonyl-CoA, dephospho-CoA, glutathione-CoA, 3-hydroxy-3-methylglutaryl-CoA (HMG-CoA), isobutyryl-CoA, isovaleryl-CoA, malonyl-CoA, 3-methylcrotonyl-CoA, methylmalonyl-CoA, oxidized CoA, propionyl-CoA, succinyl-CoA, and valeryl-CoA. Previously, reducing agents such as dithiothreitol or 2-mercaptoethanol have been used to prevent oxidation of CoASH, and tissue extracts were neutralized to near neutral (pH 6), a pH at which some CoA compounds are labile. We also describe here an evaluation of CoASH and CoA ester stability in partially neutralized biological extracts and specify conditions that do not require the addition of a reducing agent.

Standards and Reagents

Glutathione-CoA, methylmalonyl-CoA, and activated charcoal were purchased from Sigma Chemical Co., St. Louis, MO. All other CoA compounds were purchased from Pharmacia P-L Biochemicals, Piscataway, NJ. Solutions (5 mM) of each CoA compound were prepared by dissolving in 100 mM sodium phosphate buffer, pH 3.0. These solutions were stored at $-90°$ until time of use. Sodium dihydrogen phosphate was purchased from Mallinckrodt, Inc., St. Louis, MO. The deionized water used for preparing phosphate and other solutions was purified with a Milli-Q system (Millipore Corp., Bedford, MA) and had a resistivity of at least 15 MΩ. Anion (AG 1-X8) and cation (Chelex) exchange resins for the purification of the phosphate buffer were purchased from Bio-Rad Laboratories, Richmond, CA. Acetonitrile was obtained from American Burdick and Jackson Laboratories, Inc., Muskegon, MI.

[2] B. E. Corkey, M. Brandt, R. J. Williams, and J. R. Williamson, Anal. Biochem. 118, 30 (1981).
[3] M. S. DeBuysere and M. S. Olson, Anal. Biochem. 133, 373 (1983).
[4] F. C. Baker and D. A. Schooley, Anal. Biochem. 94, 417 (1979).
[5] Y. Hosokawa, Y. Shimomura, R. A. Harris, and T. Ozawa, Anal. Biochem. 153, 45 (1986).

HPLC Equipment

The analysis of standards and samples was performed using a modular liquid chromatograph from Waters, Milford, MA, including a WISP model 710A autosampler, modified to maintain sample temperature at approximately 8°, two model M-6000 pumps, a model 441 absorbance detector to monitor column effluent at 254 nm, and a model 720 system operation controller that also generates the gradients used for the chromatographic separation.

The separation for this study was effected using a 4.5 × 100 mm, 3 μm laureate series C_{18} column from IBM, Danbury, CT. The analytical column was preceded by a Guard-PAK C_{18}, precolumn from Waters. Analysis of extracts of freeze-clamped rat liver have also been performed using a 3 μm, C_{18}, 4.6 × 75 mm column from Altex Scientific, Inc., Berkeley, CA.[6] Data collection and postrun analysis were performed using an IBM 9000 computer with chromatography application program (CAP) software.

HPLC Analysis

Prior to use for CoA separation, the sodium phosphate stock solution is passed through ion-exchange resins and charcoal to remove 254-nm-absorbing impurities. It was previously found in this laboratory[7] that purification of phosphate was necessary to prevent unacceptable baseline drift in the HPLC analysis of nucleotides on an ion-exchange column. Likewise for the analysis of CoA compounds described here, purification of phosphate allows a more sensitive detector setting at 254 nm and appears, although it has not been systematically studied, to extend the column lifetime. The purification procedure is as follows. The stock (0.5 M) phosphate solution is passed through consecutive columns of anion (AG 1-X8) and cation (Chelex) exchange resins. Each column (10 × 15 cm) contains approximately 500 g of resin. The resins are used as supplied. Organic contaminants leaching from the ion-exchange resins are removed by a final column (5 × 25 cm) of 14-60 mesh activated charcoal. The stock phosphate is kept at 4° and constantly recirculated by means of a peristaltic pump. Approximately 100 liters of stock buffer is purified prior to regeneration of the ion-exchange resins, which is performed according to the manufacturer's specifications. The activated-charcoal column is replaced when the ion-exchange resins are regenerated. When the resins and charcoal are regenerated or replaced, approximately 2 liters of the stock phosphate buffer is passed through the system to waste. Additionally, the stock

[6] M. T. King and P. D. Reiss, *Anal. Biochem.* **146,** 173 (1985).
[7] P. D. Reiss, P. F. Zuurendonk, and R. L. Veech, *Anal. Biochem.* **140,** 162 (1984).

buffer is allowed to pass through the system two times before use. Separation of the CoA compounds is carried out using a two-pump gradient system. The A buffer (0.2 M sodium phosphate pH 5.0) is prepared from the prepurified stock phosphate solution. The B buffer is prepared by adding 200 ml acetonitrile to 800 ml of 0.25 M sodium phosphate, pH 5.0, diluted from the stock buffer. The acetonitrile must be added slowly with constant mixing to prevent precipitation of phosphate. The A buffer is filtered through a 0.45 μm HATF filter and the B buffer through a 0.22 μm GVWP filter from Millipore Corp., Bedford, MA. Dissolved air is removed by purging the solutions with helium. The separations of CoA compounds from isolated hepatocytes shown herein were performed at ambient temperature, using the gradient described below and a flow rate of 1 ml/min. Initial conditions (3% B) are maintained for 2.5 min. Over the next 5 min the buffer composition is changed to 18% B using a convex gradient (Waters curve #5). After that, the composition is changed linearly to reach 28% B at 10 min and is held constant until 12 min. From 12 to 15 min the B buffer is increased linearly to 37% and remains isocratic until 17 min. A linear gradient is then employed to change the composition to 90% B at 34 min. After washing at 90% B for 3 min, the buffer composition is returned to initial conditions at 37.5 min. The column is allowed to equilibrate until 45 min, after which the next injection is made. Data are collected for 40 min, and amounts of each compound are calculated using a peak height mode of integration.

Because the absorption characteristics of a compound can be affected by organic solvents, the millimolar absorption coefficient of each CoA compound was determined under the conditions prevailing when the compound passed through the detector. The millimolar absorption coefficient (ϵ_{LC}) under eluent conditions for each CoA compound is given in Table I along with the millimolar absorption coefficient at the wavelength of maximum absorbance (ϵ_m) in solution as specified by the supplier. The ϵ_{LC} for each compound is determined using the following expression:

$$\epsilon_{LC} = \epsilon_m (A_{lc}/A_m)$$

where A_{LC} is the absorbance of the standard at 254 nm and A_m is the absorbance of the compound at its maximum wavelength. To determine the composition of a standard solution the absorbance (254 nm) of the solution is first determined. Then the mixture is separated by HPLC as described above. The amount of any component (a) is calculated using the equation:

$$a \text{ (nmol)} = \frac{\%a}{100} \times \frac{(A_{254})_{\text{total}}}{\epsilon_{LC^a}} \times \text{injection volume } (\mu l)$$

TABLE I
COMPARISON OF MILLIMOLAR ABSORPTION
COEFFICIENTS[a]

Coenzyme A	ϵ_m	ϵ_{LC}
Malonyl	14.6	14.9
Glutathione	15.0 (257)	13.9
Free	14.6	14.1
Methylmalonyl	15.4 (259)	15.0
Succinyl	15.2	14.3
HMG	15.4	14.8
Dephospho	14.6	14.0
Acetyl	15.4 (259)	14.4
Acetoacetyl	15.4	15.0
Oxidized	14.6	14.8
Propionyl	15.4	14.5
Crotonyl	19.0	17.5
Isobutyryl	15.4	14.7
Butyryl	15.4	14.6
3-Methylcrotonyl	22.1	21.1
Isovaleryl	15.4	15.0
Valeryl	15.4	15.4

[a] Values given are the means of at least three independent determinations. ϵ_m is the millimolar absorption coefficient determined in solution as specified by the supplier at maximum wavelength (260 nm, unless otherwise noted in parentheses). ϵ_{LC} is the millimolar absorption coefficient determined under the conditions in which the compound is detected.

where %a is the percentage of compound a in the solution, $(A_{254})_{total}$ is the absorbance at 254 nm of the mixture determined in a 1-cm light path and ϵ_{LC^a} is the millimolar absorption coefficient of compound a at 254 nm.

Previously, the linearity and sensitivity of this method were investigated.[6] It was found that the method is linear ($r^2 \geq 0.994$) over the range of 12–700 pmol of CoA compound using 0.02 absorbance units full scale. As little as 12 pmol of CoA compound can be measured with a signal-to-noise ratio of 10.

Since some CoA compounds are labile at near neutral pH, the stability of the CoA compounds separated was studied at pH 3 and 6. The test was performed using a perchloric acid extract of liver powder from 48 hr starved rats. The extracts were adjusted to either pH 3 or 6 and spiked with a standard solution immediately before injection. By doing so, the lability

of each CoA compound could be determined versus pH in a biological matrix. Immediately after addition of the standard solution, the spiked extract was mixed and placed in the cooled autosampler ($\sim 8°$) where it remained for the duration of the experiment. Table II summarizes the results after 12 hr in the autosampler. As can be seen, adjustment of a liver extract to pH 6, as is typically done, causes a marked decrease in the levels of the CoA compounds listed. By contrast, at pH 3 dephospho-CoA is most labile, decreasing by 27% after 12 hr. Aside from glutathione-CoA the other compounds were not adversely affected by the pH to which the extract was adjusted. Because of the presence of glutathione in rat liver, formation of glutathione-CoA was not unexpected as the pH of the extract was increased. We observed a 276 ± 36% increase in glutathione-CoA after 12 hr at pH 6 while extracts adjusted to pH 3 increased by only 31 ± 6% over the same period. Our experience has shown that expedient extraction, adjustment of extract pH to 3, and prompt analysis of the sample results in lower contents of glutathione-CoA. Therefore, values for authentic gluta-thione-CoA in extracts may overestimate the content *in vivo* and a more accurate estimate of tissue CoASH level is represented as the sum of CoASH and glutathione-CoA. Coelution of other materials with any of the CoA compounds was tested by treating a control extract with base, heating, and readjusting the pH to 3.2 with 60% perchloric acid.[8] With the exception of a peak corresponding to glutathione-CoA (formed during base

TABLE II
STABILITY OF SELECTED CoA COMPOUNDS AT pH
3 OR 6[a]

Coenzyme A	Stability	
	pH 3	pH 6
Succinyl-CoA	79 ± 5	23 ± 2
Dephospho-CoA	73 ± 3	38 ± 5
Crotonyl-CoA	97 ± 2	61 ± 3
Acetoacetyl-CoA	98 ± 1	0

[a] Values given are percentages remaining after 12 hr at approximately 8°. CoA compounds not listed (except glutathione-CoA) were unchanged. Glutathione-CoA had increased as described in the text.

[8] J. R. Williamson and B. E. Corkey, this series, Vol. 13, p. 434.

hydrolysis) there was no evidence of non-base-labile compounds coeluting with the CoA compounds separated.

Figure 1 shows the separation of a standard solution of 17 CoA compounds. Figures 2 and 3 show the separation of CoA compounds in a partially neutralized acid extract (pH 3) of hepatocytes isolated from an *ad lib* fed rat. Hepatocytes were prepared and incubated according to the method of Berry and Friend[9] as modified by Cornell *et al.*[10] Figure 2 shows results from hepatocytes incubated for 20 min in the presence of 2 mM α-ketoisovaleric acid, whereas Fig. 3 represents the corresponding control. The chromatograms in Figs. 2 and 3 were obtained by injecting onto the column extracts of 1.9 mg wet weight of hepatocytes. As can be seen, the CoA compounds are easily observed using a 0.02 absorbance units full scale detector setting and baseline shift is minimal.

Concentrations of CoA compounds measured in the treated and control hepatocytes are shown in Table III. The values given are the averages

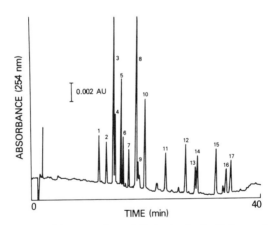

FIG. 1. Chromatographic separation of CoA standards. Column, 3 μm, C-18, 4.6 × 100 mm; temperature, ambient; detector sensitivity, 0.02 AUFS. Absorbance recorded at 254 nm. Flow rate, 1 ml/min; total run time 45 min. See the section HPLC Analysis for gradient. Compounds identified as (1) malonyl-CoA, (2) glutathione-CoA, (3) CoASH, (4) methylmalonyl-CoA, (5) succinyl-CoA, (6) 3-hydroxy-3-methylglutaryl-CoA, (7) dephospho-CoA, (8) acetyl-CoA, (9) acetoacetyl-CoA, (10) oxidized-CoA, (11) propionyl-CoA, (12) crotonyl-CoA, (13) isobutyryl-CoA, (14) butyryl-CoA, (15) 3-methylcrotonyl-CoA, (16) iso-valeryl-CoA, and (17) valeryl-CoA.

[9] M. N. Berry and D. S. Friend, *J. Cell Biol.* **43**, 506 (1969).
[10] N. W. Cornell, P. Lund, R. Hems, and H. A. Krebs, *Biochem. J.* **134**, 671 (1973).

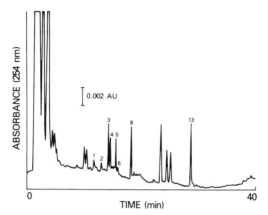

FIG. 2. Chromatographic separation of a partially neutralized (pH 3) perchloric acid extract of hepatocytes, from an *ad lib* fed rat, incubated for 20 min in the presence of 2 m*M* ketoisovaleric acid. The chromatogram represents an extract of 1.9 mg wet weight cells injected onto the column. Identification of peaks and amounts, in picomoles (brackets), is as follows: (1) malonyl-CoA [9.1]; (2) glutathione-CoA [5.0]; (3) CoASH [37.1]; (4) methylmalonyl-CoA [24.4]; (5) succinyl-CoA [21.1]; (6) 3-hydroxy-3-methylglutaryl-CoA [4.7]; (8) acetyl-CoA [41.8]; (13) isobutyryl-CoA [50.4]. Unknown peaks are not assigned. See HPLC Analysis for separation conditions.

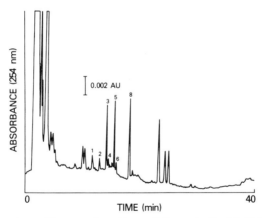

FIG. 3. Chromatographic separation of a partially neutralized (pH 3) perchloric acid extract of hepatocytes, from an *ad lib* fed rat, incubated for 20 min. The chromatogram represents an extract of 1.9 mg wet weight cells injected onto the column. Identification of peaks and amounts, in picomoles (brackets), is as follows: (1) malonyl-CoA [9.1]; (2) glutathione-CoA [7.7]; (3) CoASH [60.5]; (4) methylmalonyl-CoA [8.3]; (5) succinyl-CoA [51.7]; (6) 3-hydroxy-3-methylglutaryl-CoA [7.3]; (8) acetyl-CoA [66.2]. Unknown peaks are not assigned. See HPLC Analysis for separation conditions.

TABLE III
CoA Content of Control and
Ketoisovalerate-Treated Hepatocytes
from an *ad Lib* Fed Rat[a]

Coenzyme A	Content (nomol/g wet weight cells)	
	Control	Ketoisovalerate[c]
Malonyl	4.81	4.81
CoASH[b]	35.90	22.14
Methylmalonyl	4.39	12.85
Succinyl	27.20	11.10
HMG	3.85	2.46
Acetyl	34.83	21.98
Isobutyryl	NF	26.55

[a] Amounts reported are the averages of duplicate 20-min incubations of treated or control cells. NF, not found.
[b] The values given are the sum of the measured glutathione-CoA and CoASH.
[c] Cells were incubated in the presence of 2 mM α-ketoisovalerate.

of duplicate incubations of treated or control cells. As can be seen, after 20 min exposure to 2 mM α-ketoisovaleric acid, the amount of CoASH (sum of CoASH and glutathione-CoA), has decreased markedly, as has the succinyl-CoA concentration. The concentration of acetyl-CoA has also been lowered in the treated cells. α-Ketoisovalerate is the product of the initial transamination reaction with valine, and the final CoA ester in this metabolic sequence is succinyl-CoA. Therefore, it is somewhat surprising that succinyl-CoA is decreased. However, similar decreases in both succinyl-CoA and acetyl-CoA were seen by Corkey *et al.*[2] on incubating hepatocytes with α-ketoisovalerate. Conversely, the methylmalonyl-CoA concentration in the treated hepatocytes rose while, as expected, isobutyryl-CoA was increased to a level that was readily measurable.

A compound that elutes with a retention time close to that for propionyl-CoA is present at high levels in liver extracts. However, this compound appears to be something other than propionyl-CoA. Chromatograms of propionate-treated freeze-clamped rat liver extracts show two peaks in this region, one of which, propionyl-CoA, is absent in control extracts; the second peak is present in control and propionate-treated

animals, and its retention time is the same as the unknown peak seen in control and α-ketoisovalerate-treated hepatocytes (see Figs. 2 and 3). Therefore, determination of propionyl-CoA content in tissue extracts must be made with care.

Comments/Summary

Since several CoA compounds are involved in the catabolism of branched-chain amino acids, a method that allows the measurement of these intermediates would be extremely useful to research in branched-chain amino acid metabolism. The method presented here allows the user to measure as many as 17 CoA compounds in as little as 45 min using reversed-phase HPLC. Baseline shift is minimized using buffer salts purified by ion-exchange chromatography and acetonitrile as the organic modifier. In addition, the higher buffer concentration employed here retains the malonyl-CoA and glutathione-CoA on the column long enough to separate them from the 254-nm-absorbing compounds eluting at the front of the chromatographic run. It also should be noted that the pH of tissue extracts is an important factor in these measurements. Some CoA compounds that decompose rapidly at pH values near neutrality are much more stable at pH 3, the value to which extracts were adjusted for the analyses described here.

[11] Radiochemical High-Performance Liquid Chromatography Methods for the Study of Branched-Chain Amino Acid Metabolism

By KIM BARTLETT and ANTHONY G. CAUSEY

Most studies of the regulation of branched-chain amino acid metabolism have centered on the properties of the branched-chain 2-keto acid dehydrogenase and indeed, several chapters of the present volume are devoted to this topic. However, it has become apparent that there are other possible sites of regulation, for example, valine[1,2] and isoleucine[3,4] degrada-

[1] J. R. Williamson, E. Walajtys, and K. E. Coll, *J. Biol. Chem.* **254**, 11511 (1979).

[2] B. E. Corkey, A. Martin-Requero, E. Walajtys-Rode, R. J. Williams, and J. R. Williamson, *J. Biol. Chem.* **257**, 9668 (1982).

[3] A. G. Causey, L. Agius, and K. Bartlett, *Biochem. Soc. Trans.* **13**, 1220 (1985).

[4] A. G. Causey, B. Middleton, and K. Bartlett, *Biochem. J.* **235**, 343 (1986).

tion are also regulated at methylmalonyl-CoA mutase (L-methylmalonyl-CoA, CoA-carbonylmutase; EC 5.4.99.2).

These studies were made possible by the development of HPLC methods for the resolution of complex mixtures of acyl-CoA ester intermediates. Some investigators have utilized the strong light absorption of the adenine chromophore of CoA (A_{260} = 1.6 × 10⁴) and have detected CoA esters by UV absorption.[1,2,5] In our hands this did not provide sufficient sensitivity or specificity for the study of isoleucine metabolism by isolated rat liver mitochondrial fractions. Furthermore, we wished to extend the method to the analysis of intermediates generated by cultured cells derived from patients with inherited defects in isoleucine metabolism. The requirement for sensitivity and specificity is fulfilled by the use of radiolabeled substrates.

The detection of radioactivity in column eluants poses certain problems. One solution is to collect discrete fractions of the column eluant for manual scintillation counting. This creates unrealistic demands on counting time, as it would be necessary to collect 10- to 15-sec fractions to retain chromatographic resolution. The alternative approach is continuous on-line radiochemical monitoring of the column eluant. This can be achieved by the use of either solid scintillants (heterogeneous system) or by mixing the column eluant with a liquid scintillant (homogeneous system). Initial studies showed that heterogeneous counting resulted in irreversible binding of analytes to the solid scintillant and this approach was abandoned.

We describe here the radiochemical high-performance liquid chromatographic (RHPLC) methods we have developed for the measurement of acyl-CoA ester intermediates of isoleucine metabolism derived from the incubation of rat liver and muscle mitochondrial fractions and cultured human fibroblasts with 3-[U-¹⁴C]methyl-2-ketopentanoate (2-keto-3-methylvalerate). In some experimental situations, useful information can be gained from an analysis of organic acids, for example, to study the importance of acyl-CoA hydrolases or to confirm the identity of acyl-CoA esters after saponification. The β-oxidation of fatty acids has a profound effect on branched-chain amino acid degradation in which closely related intermediates are involved. We describe a method for the resolution of both short- and long-chain acyl-CoA esters in a single analysis.

Design of the Chromatographic System

The chromatographic system is shown in Fig. 1. HPLC was carried out with a Spectra-Physics SP 8700 solvent delivery system using a LiChrosorb

⁵ O. C. Ingebretsen and M. Farstad, *J. Chromatogr.* **202**, 439 (1980).

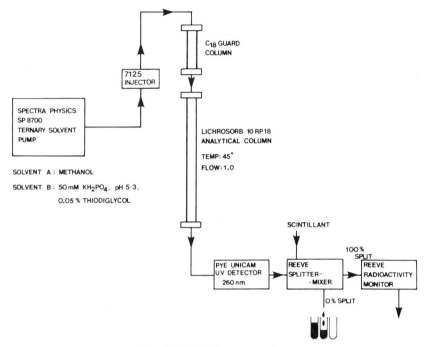

FIG. 1. RHPLC instrumentation.

10RP18 column (250 × 4.6 mm) at 45°. Samples were introduced with a manual Rheodyne 200 µl loop injector (model 7125) fitted with a guard column (75 × 2.1 mm packed with pellicular reversed-phase material; Chrompack, Middleburg, Holland). All solvents were continuously sparged with helium to degas. Resolved compounds were detected by a Pye Unicam variable-wavelength UV detector (model LC3; 8 µl flow cell; pathlength 10 mm) at 260 nm. The output was displayed using a Kipp and Zonen dual-channel strip-chart recorder (model BD9) and also a Spectra Physics SP4270 integrator. Radioactivity associated with eluted peaks was detected on-line with a Reeve Analytical radiochemical detector fitted with a 900 µl flow cell. The theory of operation of this detector has been fully described elsewhere.[4,6] A homogeneous counting system was used. The effluent from the UV detector (1 ml/min) was mixed with scintillator fluid (1.5 ml/min; 10 g- 2, 5-diphenyloxazole, 330 ml of Triton X-100, 670 ml of xylene, 150 ml of methanol) by a Reeve Analytical precision splitter/ mixer and then passed to the radiochemical detector. The radiochemical

---

[6] D. R. Reeve and A. Crozier, *J. Chromatogr.* **137**, 271 (1977).

detector output was displayed using a Hewlett-Packard 85B personal computer and the second channel of the strip chart recorder. The pens were offset to compensate for the dead volume between the UV and radiochemical detectors, to generate superimposable dual traces. This was achieved by chromatography of standard [1-^{14}C]acetyl-CoA and careful measurement of the retention times with respect to the radioactivity and UV absorption. The integrators were used to determine peak areas and accurate retention times. More recently we have used a Waters chromatography data station (model 840), which allows computer acquisition of the two detector signals and much more convenient manipulation of the time base.

Calculation of Results

The object of RHPLC is the determination of radioactivity per chromatographic peak; however, integration of the radioactive peak gives the total number of counts recorded per transit time (flow cell volume/rate of flow through cell). Counts per minute are calculated by dividing the integrated counts by the transit time. Counting efficiency can be calculated by RHPLC analysis of an aliquot of a radiolabeled compound of known disintegrations per minute (dpm). It is not always appreciated that, in order to determine counting efficiency, account must be taken of the transit time. This becomes obvious if one considers that the probability that a disintegration will be registered by the detector is dependent on the length of time a given molecule resides in the flow cell. In this particular instance, the total flow rate (column effluent plus scintillant) through the cell (900 μl volume) is 4.0 ml/min, the transit time is therefore 0.23 min, and thus the observed integrated counts must be multiplied by 4.35 to obtain counts per minute (cpm). Thus the accuracy of the determination of counts per minute is heavily dependent on the accuracy of measurement of the flow rate. Although the flow rate did not vary significantly either within or between runs on any given day, it was necessary to determine the flow rate at the start of each working day.

Factors Which Affect Precision and Accuracy of RHPLC

The single most important variable which influences the precision and accuracy of the estimation of disintegrations per minute/peak is the flow rate. It is for this reason that the delivery of scintillant must be precisely metered. Variations in viscosity due to changes in ambient temperature or column eluant composition may alter back-pressure and therefore residence time in the flow monitor. There is a requirement for careful monitoring of flow rate whenever the chromatographic system is changed.

Accuracy is also dependent on a precise value for the flow cell volume. The composition of the eluant/scintillant mixture may affect counting efficiency during the course of a gradient either because of phase changes or variations in chemical quenching. This can be checked by repeated isocratic elutions at different solvent concentrations. When ^{14}C-labeled analytes are chromatographed with methanol as the organic modifier this is not a problem and counting efficiencies of 70% were observed when the mobile phase was raised from 10 to 50% (v/v) methanol. The precision of any scintillation counting method depends on the total number of events registered and this applies to RHPLC as well as conventional liquid scintillation counting. In the latter case, low activity samples can be measured accurately simply by increasing the counting time. In the case of RHPLC counting time is fixed. Thus, although it is possible to detect low activity peaks (i.e., about 200 dpm/peak), precision is poor. We have obtained coefficients of variation of 4.1, 2.9, and 2.1% with 1028, 8447, and 25,330 dpm, respectively, in replicate analyses (n = 5) of [U-^{14}C]acetyl-CoA. In our hands contamination of the flow cell has also resulted in imprecision since this effectively introduces noise to the signal. Cleaning of glass flow cells is required about every 50 hr of use.

Gradient System for the Resolution of the Acyl-CoA Esters

Standard Acyl-CoA Esters

Acetyl-CoA and propionyl-CoA were prepared from their respective anhydrides by the method of Simon and Shemin.[7] Yields of 75–80% were obtained as judged by assay with 5,5'-dithiobis(2-nitrobenzoic acid).[8] The purity of the preparations was assessed by HPLC (see below). 2-Methylbutyryl-CoA was synthesized from the 1-acylimidazole[9] and purified by HPLC. 2-Methylacetoacetyl-CoA was synthesized and purified as previously described.[10] 2-Methyl-3-hydroxybutyryl-CoA was synthesized from tiglyl-CoA by the action of enoyl-CoA hydratase (EC 4.2.1.17). The incubation (1.0 ml) contained 20 μmol of KH$_2$PO$_4$, 0.5 μmol of tiglyl-CoA, and 0.4 units of crotonase. The pH was 5.3. The reaction (2 min, 37°) was stopped by freezing at −70°. The mixture was freeze-dried and 2-methyl-3-hydroxybutyryl-CoA purified by HPLC. Although the reaction goes to completion, the equilibrium greatly favors tiglyl-CoA so that the yield is about 10%. All other acyl-CoA esters were obtained from Sigma Chemical Co.

[7] E. S. Simon and D. Shemin, *J. Am. Chem. Soc.* **75**, 2520 (1953).
[8] K. Bartlett and D. Gompertz, *Biochem. Med.* **10**, (1974).
[9] A. Kawaguchi, T. Yoshimuara, and S. Okuda, *J. Biochem. (Tokyo)* **89**, (1981).
[10] B. Middleton and K. Bartlett, *Clin. Chim. Acta* **128**, 291 (1983).

Gradient System for the Intermediates of Isoleucine Degradation

Acyl-CoA esters were resolved by a gradient of methanol in water containing 50 mM KH$_2$PO$_4$ (pH 5.3) and 4 mM 2,2-thiodiethanol; isocratic 5% (v/v) (5 min), then isocratic 10% (v/v) (10 min), then linear to 17% (v/v) (18 min), then linear to 45% (v/v) (25 min), and finally isocratic 45% v/v (10 min). The flow rate was 1 ml/min. The order to maintain column performance, it is essential that the column is carefully reequilibrated at the end of each working day as follows: linear gradient to 100% phosphate buffer (2 min), linear gradient to 100% water (0.5 min), isocratic 100% water (3.5 min), linear gradient to 100% methanol (10 min), isocratic 100% methanol (10 min). The solvent system resolved all of the intermediates of KMV oxidation (Fig. 2), although methylmalonyl-CoA was poorly separated from free CoA.

Gradient System for the Intermediates of β-Oxidation

Several methodologies for the analysis of short-chain acyl-CoA esters are now established,[11-14] and recently a method for the analysis of long-

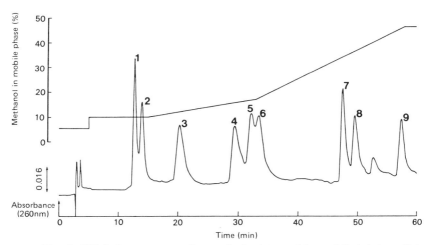

FIG. 2. HPLC chromatogram of a standard mixture of the acyl-CoA intermediates of isoleucine degradation (UV detection). Retention times in minutes are given in parentheses. The peaks are as follows: 1, methylmalonyl-CoA (12.6); 2, succinyl-CoA (14.0); 3, acetyl-CoA (20.7); 4, 2-methylacetoacetyl-CoA (29.4); 5, propionyl-CoA (32.6); 6, 2-methyl-3-hydroxy-butyryl-CoA (34.2); 7, 2-methylbut-2-enoyl-CoA (47.7); 8, 2-methylbutyryl-CoA (49.6); 9, hexanoyl-CoA (57.5).

[11] B. E. Corkey, M. Brandt, R. J. Williams, and J. R. Williamson, *Anal. Biochem.* **118,** 30 (1981).
[12] Y. Hosokawa, Y. Shimomura, R. A. Harris, and T. Ozawa, *Anal. Biochem.* **153,** 45 (1986).
[13] M. T. King and P. D. Reiss, *Anal. Biochem.* **146,** 173 (1985).

chain acyl-CoA esters in freeze-clamped liver has been reported.[15] To date, only one methodology has achieved good resolution of both long- and short-chain acy-CoA esters in one chromatogram.[16] This method is not popular as it uses a corrosive and expensive ion-pair agent. The most successful methods for the analysis of acyl-CoA esters in biological samples have used some form of phosphate buffer.

We have developed a method to resolve the homologous series of even-chain-length acyl-CoA esters from acetyl-CoA to *n*-hexadecanoyl-CoA within 50 min.[17] The method is a modification of the long-chain acyl-CoA methodology described by Woldegiorgis *et al.*[15] Resolution of short-chain from long-chain acyl-CoA esters was achieved by developing a gradient with a low organic solvent composition with higher buffer concentrations.

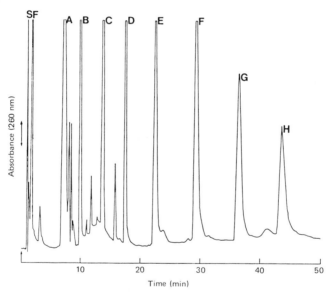

FIG. 3. Resolution of C_2- to C_{18}- acyl-CoA esters. Retention times in minutes are given in parentheses. Ten nanomoles of each standard was injected. SF, solvent front; A, acetyl-CoA (7.2); B, butyryl-CoA (10.2); C, hexanoyl-CoA (13.8); D, octanoyl-CoA (17.6); E, decanoyl-CoA (22.6); F, dodecanoyl-CoA (29.5); G, tetradecanoyl-CoA (36.5); H, hexadecanoyl-CoA (43.6).

[14] M. S. DeBuysere and M. S. Olson, *Anal. Biochem.* **133**, 373 (1983).
[15] G. Woldegiorgis, T. Spennetta, B. E. Corkey, J. R. Williamson, and E. Shrago, *Anal. Biochem.* **150**, 8 (1985).
[16] F. C. Baker and D. A. Schooley, *Anal. Biochem.* **94**, 417 (1979).
[17] A. G. Causey and K. Bartlett, *Biochem. Soc. Trans.* **14**, 1175 (1986).

Acyl-CoA esters were resolved by the following gradient of acetonitrile in water containing 50 mM KH$_2$PO$_4$ (pH 5.3): isocratic 5% (v/v) (5 min), then linear from 10% to 30% (v/v) (10 min), linear to 50% (v/v) (30 min). The column was reequilibrated by a linear gradient of 5% (v/v) (5 min). The flow rate was 2.0 ml/min. Figure 3 illustrates the baseline resolution of a mixture of even-chain-length acyl-CoA esters from acetyl-CoA to hexadecanoyl-CoA. For the analysis of biological samples it was found necessary to precede the gradient by a short isocratic elution at 5% (v/v) acetonitrile. This allowed the resolution of nucleotides and CoASH from the short-chain acyl-CoA esters.

Comment

The pH and buffer concentration of the mobile phase profoundly influence retention and resolution of acyl-CoA esters. Fig. 4A illustrates the effect of buffer concentration on the resolution of a mixture of short-chain acyl-CoAs. Increasing the buffer concentration increases the affinity of acyl-CoAs for the stationary phase and improves resolution between similar compounds. Fig. 4B shows the effects of pH on the selectivity of the column. Below pH 4 the resolved peaks were eluted as broad tailing peaks. Above pH 4 the resolved peaks were eluted earlier as sharp peaks with good symmetry. Gradient elution is used to achieve resolution of a wide range of compounds within the same chromatogram. Subtle manipulations of the gradient program give scope for the resolution of difficult analyte pairs (e.g., acetyl-CoA and acetoacetyl-CoA). We routinely use C$_{18}$-columns for studies of the intermediates of branched-chain amino acid metabolism; C$_8$-stationary phases do not have sufficient selectivity for the resolution of these closely related compounds. Acetonitrile has a higher extractive capacity than methanol and was used to resolve acyl-CoAs longer than C$_6$.

Sample Preparation Methods

Acyl-CoA Esters

There are two general approaches to the preparation of samples for HPLC analysis of acyl-CoA esters. Reactions or tissue extracts can be quenched with perchloric acid followed by concentration by, for example, lyophilization and reconstitution in a suitable solvent. Alternatively, CoA esters can be extracted directly into solvents provided that precautions are taken to ensure good recoveries. Both methods require internal standardization in order that correction can be made for sample to sample varia-

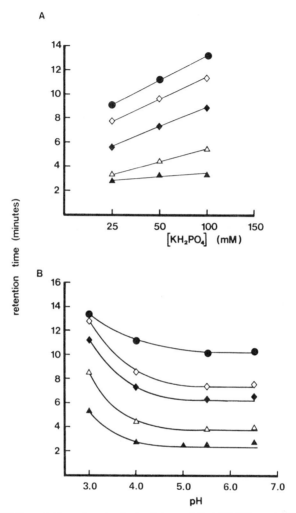

FIG. 4. HPLC resolution of a standard mixture of CoASH (▲), acetyl-CoA (△), pro-pionyl-CoA (◆), butyryl-CoA (◇), and crotonyl-CoA (●). (A) Effects of phosphate buffer concentration (linear gradient of 15–35% methanol (v/v) over 15 min). The aqueous buffer was 50 mM KH$_2$PO$_4$, pH 5.3. The flow rate was 1.5 ml/min. (B) Effect of pH on retention; gradient as in (A), phosphate concentration = 50 mM.

tions in extraction efficiency. In general, we have used the former method for our studies of the intermediates of isoleucine degradation and the latter for our studies of β-oxidation. The methods we have developed are de-scribed in detail below.

Protcol for Preparation of Isoleucine Metabolites

The isolation of mitochondrial fractions, the preparation of 3-[U-^{14}C] methyl-2-ketopentanoate, and the conditions of incubation were as previously described.[3] The reactions were stopped by the addition of 100 μl of 5 M perchloric acid. Internal standard (20 nmol of hexanoyl-CoA) was added and precipitated protein and KClO$_4$ removed by centrifugation. The supernatant was extracted three times with five volumes of diethylether to remove unused substrate, the pH adjusted to pH 7.0, and lyophilized. The residue was extracted with 400 μl of HPLC-grade water, centrifuged, and 200 μl analyzed by RHPLC. Provided that a guard column is used we found it unnecessary to filter samples prior to analysis. Recovery was determined by UV detection of the hexanoyl-CoA and the radiochemical results appropriately corrected. Recoveries varied from 85.4 to 99.2% and demonstrated the need for internal standardization.

Protocol for the Preparation of β-Oxidation Metabolites

Samples were quenched with glacial acetic acid and extracted with 3 × 5 volumes of diethyl ether. Saturated ammonium sulfate (100%; 100 μl) was then added to the aqueous phase and 6 ml of a mixture of chloroform : methanol (1 : 2 v/v) was slowly added. The mixture was left at room temperature for 20 min. The salt – protein complex was removed by centrifugation and the supernatant removed. The pellet was washed with 2 ml chloroform : methanol (1 : 2, v/v) and the supernatants combined. The organic solvents were removed at 50° under a gentle stream of nitrogen. The residue was resuspended in 400 μl KH$_2$ PO$_4$ (50 mM; pH 5.3) and a 200-μl aliquot analyzed by RHPLC. Because of the wide variation in chain length of the analytes, butyryl-CoA, octanoyl-CoA, and tetradecanoyl-CoA were used as internal standards. It was found that the addition of ammonium sulfate was essential and its omission greatly reduced extraction efficiency (Table I).

TABLE I
RECOVERY OF ACYL-CoA STANDARDS FROM MITOCHONDRIAL FRACTIONS.[a]

Extraction protocol[b]	Recovery (%)		
	Butyryl-CoA	Octanoyl-CoA	Tetradecanoyl-CoA
With ammonicen sulfate	99.9 ± 2.5	100.0 ± 34.0	68.1 ± 4.7
Minus ammonicen sulfate	73.6 ± 2.2	11.6 ± 4.0	6.7 ± 0.5

[a] Results shown are mean recoveries ± SD ; $n = 3$.
[b] See text.

Some Problems and Pitfalls in the Interpretation of RHPLC Results

The object of the experimental approach described in this chapter is the measurement of acyl-CoA intermediates in a variety of experimental situations. Typical radiochromatograms of isoleucine metabolites produced by incubations with rat liver and muscle mitochondrial fractions are shown in Fig. 5 and the results obtained from an incubation of rat liver mitochon-

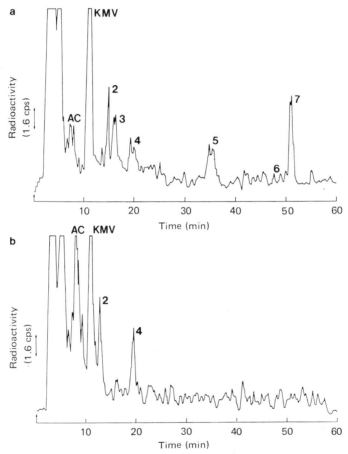

FIG. 5. HPLC chromatograms of extracts of (A) rat liver mitochondrial fractions and (B) rat muscle mitochondrial fractions incubated with 1 mM 3-[U-^{14}C]methyl-2-ketopentanoate (64.4 MBq/mmol). Incubations were made as described elsewhere.[4] For chromatographic conditions see text. KMV, unextracted substrate; AC, acylcarnitines; 2, methylmalonyl-CoA; 3, succinyl-CoA; 4, acetyl-CoA; 5, propionyl-CoA; 6, 2-methylbut-2-enoyl-CoA; 7, 2-methylbutyryl-CoA.

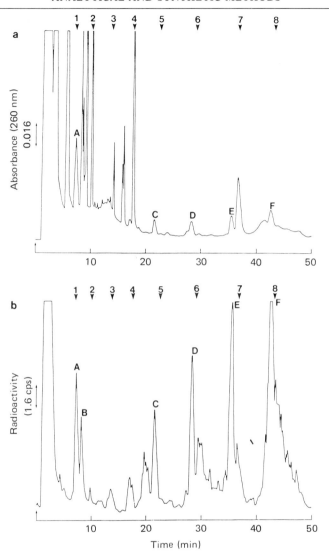

FIG. 6. (a) HPLC chromatogram and (b) HPLC chromatogram of extracts of rat liver mitochondrial fractions incubated with 120 μM [U-^{14}C]hexadecanoate (5 $\mu Ci/\mu mol$) in the presence of rotenone (10 μg). Incubations were as described previously[4] with the addition of 5 mM ATP, 1 mM 1-carnitine, and 0.1 mM CoASH. The hexadecanoate was complexed in a 5 : 1 ratio with bovine serum albumin. The numbers represent the elution position of standard saturated acyl-CoA esters: 1, acetyl-CoA; 2, butyryl-CoA; 3, hexanoyl-CoA; 4, octanoyl-CoA; 5, decanoyl-CoA; 6, dodecanoyl-CoA; 7, tetradecanoyl-CoA; 8, hexadecanoyl-CoA. The letters are A, acetyl-CoA; B, acetoacetyl-CoA; C, D, E, and F are the 2,3-enoyl-CoA derivatives of decanoyl-CoA, dodecanoyl-CoA, tetradecanoyl-CoA, and hexadecanoyl-CoA esters, respectively.

dria with [U-^{14}C]hexadecanoate in the presence of rotenone are shown in Fig. 6. An important point in the calculation of the concentration of intermediates is that the specific activities (dpm per mol) of the intermediates will not be the same when universally labeled substrates are used. This is not readily apparent from the isoleucine experiments but is obvious in the hexadecanoate incubation. Thus, from inspection of the radiochromatogram (Fig. 6a), it appears that the concentration of the 2,3-enoyl-CoA intermediates generated by the presence of rotenone decreases with decreasing chain length. However, from the UV trace (Fig. 6b), it can be seen that approximately equal amounts of these intermediates are present. The reason for this apparent discrepency is, of course, that the extinction coefficient of the CoA chromophore is independent of chain length of the acyl substituent, whereas the specific activity of [U-^{14}C]dec-2-enoyl-CoA is five-eighths of that of [U-^{14}C]hexadec-2-enoyl-CoA when both compounds are derived from [U-^{14}C]hexadecanoate. In order to arrive at an estimate of the concentration of an intermediate in the mitochondrial matrix, it is necessary to make a number of assumptions. First, an estimate of the matrix space must be assumed which is dependent on the nature of the impermeant marker used. Second, and more important from the point of view of radiochromatography, it must be assumed that there is no dilution of label by endogenous substrates. From our study of isoleucine metabolism we arrived at a value of 1 mM for the concentration of methylmalonyl-CoA under our standard incubation conditions.

Another more practical problem with radiochromatography is the precise registration of the UV and radioactivity traces to enable reliable identification of the radiolabeled analytes. This is complicated by the inevitable band broadening which occurs when the column effluent stream is mixed with the scintillant. If a dual-channel strip-chart recorder is used, the simple expedient of altering the pen offsets, after chromatography of a standard compound which is both radioactive and UV absorbing, overcomes this problem. Alternatively, with computer data acquisition, the same end is achieved by split-screen presentation of the two traces and computer manipulation of the time base.

Related Techniques

In the incubation of rat muscle mitochondria with [U-^{14}C]3-methyl-2-ketopentanoate (KMV) we noted a large radiolabeled peak which appeared early in the chromatogram and did not absorb light at 260 nm. We trapped this peak and were able to demonstrate that it was 2-methylbutyryl-1-carnitine by radio-HPTLC.[18] Thus it is possible to isolate radiolabeled ana-

---

[18] A. K. M. J. Bhuiyan, A. G. Causey, and K. Bartlett, *Biochem. Soc. Trans.* **14,** 1072 (1986).

lytes and then subject them to further analysis. In this instance RHPLC is being used preparatively.

Another application of RHPLC of relevance to branched-chain amino acid metabolism is the measurement of enzyme activities. This application makes use of the sensitivity inherent in radiochemical methods and is particularly useful in the assay of fibroblast and leukocyte enzymes in the diagnosis of inherited metabolic disorders. Thus we have devised an assay of methylmalonyl-CoA mutase using analysis of the organic acids liberated by alkaline hydrolysis of the CoA ester substrate and product.[19]

Conclusion

RHPLC appears to be a powerful analytical method which produces precise quantitation provided that careful attention is paid to experimental detail and can be applied to a variety of metabolic pathways. It has the advantages of specificity and sensitivity when compared to conventional HPLC detection methods and has been used to study the regulation of branched-chain amino acid metabolism.[4]

[19] A. G. Causey and K. Bartlett, *Clin. Chim. Acta.* **139,** 179 (1984).

[12] Synthesis of α-Isopropylmalate, β-Isopropylmalate, and Dimethylcitraconate

By JOHN V. SCHLOSS, RON MAGOLDA, and MARK EMPTAGE

(S)-α-Isopropylmalate   erythro-(R,S)-β-Isopropylmalate   Dimethylcitraconate

(S)-α-Isopropylmalate, erythro-(R,S)-β-isopropylmalate, and dimethylcitraconate are central metabolites of leucine biosynthesis.[1] All three are substrates for isopropylmalate isomerase. α-Isopropylmalate is the product of the first enzyme specific to leucine biosynthesis, α-isopropylmalate

[1] S. R. Gross, R. O. Burns, and H. E. Umbarger, *Biochemistry* **2,** 1046 (1963); J. M. Calvo, C. M. Stevens, M. G. Kalyanpur, and H. E. Umbarger, *Biochemistry* **3,** 2024 (1964); F. E. Cole, M. G. Kalyanpur, and C. M. Stevens, *Biochemistry* **12,** 3346 (1973).

synthase (EC 4.1.3.12), and β-isopropylmalate is a substrate for the penultimate enzyme, β-isopropylmalate dehydrogenase (EC 1.1.1.85). While α- and β-isopropylmalate can be prepared via fermentation by use of the appropriately constructed mutants of fungi[2] or bacteria,[3] dimethylcitraconate has only been prepared by organic synthesis.[4] The following protocols provide a ready means of preparing these three intermediates of leucine biosynthesis in quantity.

α-Isopropylmalate

The following protocol is adapted from the method reported by Yamashita.[5] To 100 g of ethyl isobutyrylacetate (0.63 mol, Aldrich Chemical Co.) and 50 g of sodium cyanide (1.0 mol) in a 1-liter round-bottom flask is added 114 ml of concentrated HCl. Acidification of the sodium cyanide is conducted at 4° in a well-ventilated fume hood.[6] Following addition of the HCl, the flask is sealed with a ground glass stopper and allowed to stir at room temperature for 3 hr. The flask is then chilled to 4° and equipped with a rubber stopper containing an inlet line connected to a polyethylene tube extending below the surface of the reaction mixture and an outlet tube connected to the air space above the reaction mixture. The outlet line is connected via latex tubing to a sparger containing 10% NaOH. After allowing the flask to warm to room temperature, N_2 is slowly purged through the reaction mixture for 2 days. Following elimination of the sodium cyanide, the sample is extracted three times with 300-ml portions of diethyl ether. Evaporation of the ether by rotary evaporation gives about 125 ml of an oil–water emulsion. The oil is transferred to a 2-liter round-bottom flask and the sample is allowed to stir at room temperature for 24 hr following the addition of 250 ml of concentrated HCl. One liter of water is added to the flask, and the contents are brought to reflux for 15 hr. The water is removed by rotary evaporation and the residue is suspended with about 1-liter of ethyl acetate. Ethyl acetate-insoluble material is removed by filtration and the solution taken to dryness by rotary evapora-

[2] J. M. Calvo and S. R. Gross, this series, Vol. 17, p. 791.
[3] P. N. Fultz, K. K. L. Choung, and J. Kemper, *J. Bacteriol.* **142**, 513 (1980).
[4] W. Ssemenow, *Zh. Ova. Veng. Khim.* **23**, 430 (1891); *Zh. Ova. Veng. Khim.* **30**, 1003 (1898); R. Fittig and H. Kraft, *Justus Liebigs Ann. Chem.* **304**, 195 (1899); W. R. Vaughan and K. S. Andersen, *J. Org. Chem.* **21**, 673 (1956); F. E. Cole, M. G. Kalyanpur, and C. M. Stevens, *Biochemistry* **12**, 3346 (1973).
[5] M. Yamashita, *J. Org. Chem.* **23**, 835 (1958).
[6] Exercise caution when working with hydrogen cyanide due to its extreme toxicity and volatility. A ready method for the detection of HCN is Cyantesmo test paper from Macherey-Nagel, Düren, Federal Republic of Germany.

tion. This residue is redissolved with about 500 ml of ethyl acetate, filtered through a fine sintered glass filter, and taken to dryness by rotary evaporation. After dissolving the residue in 450 ml of ethyl acetate, 450 ml of petroleum ether is added, and the mixture is incubated for 5 hr at room temperature, followed by 15 hr at 4°. The resultant crystals are collected by filtration, washed with cold petroleum ether, and air-dried to give 77.8 g of the free acid of α-isopropylmalate (0.49 mol, 77% yield).

Dimethylcitraconate

To 1 g of α-isopropylmalate (6.24 mmol, as prepared above) in 100 ml of $CHCl_3$ are added 1.8 ml of triethylamine (12.9 mmol) and two 0.57-ml portions of triflic (trifluoromethanesulfonic) anhydride (Aldrich Chemical Co.). After 1 hr at room temperature, an additional 1.8 ml of triethylamine and two 0.55-ml portions of triflic anhydride are added. Following 30 min at room temperature, the reaction mixture is chilled to 4° and extracted twice with an equal volume of cold 1 N H_2SO_4, saturated aqueous $NaHCO_3$, and water. Removal of the $CHCl_3$ by rotary evaporation gives 0.6 g of dimethylcitraconic anhydride (4.3 mmol). Dimethylcitraconic anhydride is applied to a 2.4 × 55 cm column of silica gel (Baker) and the column is eluted with 200 ml of hexane, 200 ml of 10% $CHCl_3$ in hexane, 200 ml of 50% $CHCl_3$ in hexane, 400 ml of $CHCl_3$, and 600 ml of CH_2Cl_2. The CH_2Cl_2 eluates are pooled, rotary evaporated to remove CH_2Cl_2, and neutralized with 1 N KOH after being suspended in about 100 ml of water. Lyophilization of this solution gives 0.76 g of dimethylcitraconate dipotassium salt (3.2 mmol, 51% yield).

β-Isopropylmalate

To 11.2 g of potassium *tert*-butoxide (0.1 mol, Aldrich Chemical Co.) in 100 ml of diethyl ether is added 13.6 ml of diethyloxalate (0.1 mol). After the addition of 15 ml of ethyl isovalerate (0.1 mol, Aldrich Chemical Co.), the dark yellow reaction mixture is allowed to stir at room temperature in a stoppered flask for 17 hr. The flask is cooled to 4° and, following 5 hr incubation in the cold, the precipitate is collected by filtration with a medium sintered glass funnel. After washing the precipitate with several 200-ml portions of cold diethyl ether, it is dried by suction, and 19.7 g of the potassium enolate (0.074 mol) is collected. To 11.5 g of the potassium enolate (0.043 mol) is added 50 ml of 1 N HCl. The mixture is then extracted with three 100-ml portions of diethyl ether. The ether extracts are pooled, dried with 30 g of anhydrous sodium sulfate, filtered, and dried overnight with 50 g of anhydrous sodium sulfate. Following filtration and

removal of the diethyl ether by rotary evaporation, 8.6 g of diethyl β-isopropyloxaloacetate (0.037 mol) is obtained. A portion of the diethyl β-isopropyloxaloacetate (4 g, 0.017 mol) is dissolved in 100 ml of absolute ethanol, and 0.25 g of platinum oxide is added. The reaction mixture is shaken on a Parr apparatus for 45 min under an atmosphere of 36 psi H_2. Approximately 20 mmol of H_2 is consumed in the first 30 min, after which consumption of H_2 is complete. Platinum is removed by filtration through Celite, and after rotary evaporation, 3.96 g of diethyl β-isopropylmalate (0.017 mol) is recovered. To a 250-ml round-bottom flask are added 1 g of diethyl β-isopropylmalate (4.31 mmol) and 100 ml of 1 N HCl. The mixture is refluxed for 18 hr, then rotary evaporated to a syrup. Following suspension of the syrup with a small portion of water (50 ml), neutralization with 1 N KOH, lyophilization, desiccation over P_2O_5, H_2SO_4, and KOH pellets *in vacuo,* 0.86 g of dipotassium β-isopropylmalate (3.41 mmol, 50% yield overall from ethyl isovalerate) is recovered.

Properties

α-Isopropylmalate, β-isopropylmalate, and dimethylcitraconate are analytically pure as judged by ^{1}H and ^{13}C NMR. Dimethylcitraconate has the following extinction constants (A mM^{-1} cm^{-1}, pH 8) at 210, 215, 220, 225, 230, 235, 240, 245, and 250 nm: 8.25, 7.44, 6.76, 6.03, 5.14, 4.10, 2.97, 1.95, 1.16, respectively. These values deviate by less than 10% from those previously reported.[1] End-point analyses with limiting dimethylcitraconate in a coupled assay consisting of isopropylmalate isomerase, isopropylmalate dehydrogenase, Mg^{2+}, and NAD, give >95% of the expected amount of NADH based on weight. The Michaelis constant for dimethylcitraconate with yeast isopropylmalate isomerase is 82 ± 4 μM in reasonable agreement with previously determined values of 180–220 μM.[7] α-Isopropylmalate is presumably a racemic mixture. The Michaelis constant for α-isopropylmalate with yeast isopropylmalate isomerase is 110 μM compared with previously determined values of 670–950 μM for the natural S-stereoisomer, obtained via fermentation.[7] β-Isopropylmalate is presumably a mixture of two diastereomeric pairs of enantiomers. The Michaelis constants for β-isopropylmalate with yeast isopropylmalate isomerase and *Salmonella typhimurium* isopropylmalate dehydrogenase are 96 and 14 μM, respectively, compared with previously determined values of 64–78

[7] S. R. Gross, R. O. Burns, and H. E. Umbarger, *Biochemistry* **2,** 1046 (1963); J. M. Calvo and S. R. Gross, this series, Vol. 17, p. 791; Y. S. Cho-Chung and H. E. Umbarger, this series, Vol. 17, p. 782; S. R. Gross, this series, Vol. 17, p. 786.

μM[7] and 19 μM,[8] obtained with the natural *erythro-R,S*-isomer. Inhibition of either enzyme by one of the unnatural stereoisomers would result in a lower K_m and a proportionately lower V_{max}.[9] Since the Michaelis constants obtained with synthetic α- and β-isopropylmalate are substantially lower than those obtained with the natural stereoisomers, the unnatural stereo-isomers appear to inhibit the enzymes, lowering both the apparent K_m and the maximal rate obtained. Therefore, caution should be exercised if the synthetic α- and β-isopropylmalate are used to determine enzymatic activity. While these substrates will be acceptable where relative measures of activity are required, the absolute rates obtained will be lower than those obtained with the natural stereoisomers. For the yeast isopropylmalate isomerase only about 8 and 36% of the expected absolute rate should be obtained with the synthetic α- and β-isopropylmalate, respectively. Only about 20% of the rate is expected with synthetic β-isopropylmalate and the dehydrogenase. Purified yeast isopropylmalate isomerase has a specific activity of 7.5 units/mg when assayed with the dimethylcitraconate prepared by the procedure described here, compared with the previously reported value of 6.2 units/mg.[7]

The *erythro* and *threo* diastereomers of β-isopropylmalate have previously been resolved by silica gel chromatography.[10] These diastereomers can also be resolved by ion-exchange chromatography on a Dionex HPIC-AS6 column. β-Isopropylmalate is resolved into two peaks, detectable by conductivity, which run at 15 min and 16.2 min (relative ratio 52:48) with an 8:1:1 mixture of water: 10 mM Na_2CO_3: 2 mM NaOH as a solvent system. Half of the later eluting isomer is consumed by extended incubation with isopropylmalate dehydrogenase and NAD. End-point analyses of limiting β-isopropylmalate with isopropylmalate dehydrogenase and excess NAD determined that *erythro-(R,S)-β*-isopropylmalate comprises 22% by weight of the mixture of stereoisomers, compared with 24% expected. α-Isopropylmalate and dimethylcitraconate run as single peaks in this chromatographic system with retention times of 17 and 19 min, respectively.

[8] S. J. Parsons and R. O. Burns, this series, Vol. 17, p. 799.
[9] C. E. Grimshaw and W. W. Cleland, *Biochemistry* **19**, 3153 (1980).
[10] J. M. Calvo, M. G. Kalyanpur, and C. M. Stevens, *Biochemistry* **1**, 1157 (1962).

[13] Utilization of Sulfometuron Methyl, an Acetolactate Synthase Inhibitor, in Molecular Biological and Metabolic Studies of Plants and Microbes

By ROBERT A. LaROSSA and TINA K. VAN DYK

Recently, attention has refocused upon acetolactate synthase (ALS) (EC 4.1.3.18). This enzyme normally catalyzes two reactions required for the biosynthesis of the branched-chain amino acids and pantothenate:

$$2 \text{ Pyruvate} \rightarrow \alpha\text{-acetolactate} + CO_2$$

$$\text{Pyruvate} + \alpha\text{-ketobutyrate} \rightarrow \alpha\text{-aceto-}\alpha\text{-hydroxybutyrate} + CO_2$$

Contributing to renewed interest in this enzyme and its substrates are: (1) identification of ALS as the target of sulfonylurea,[1] imidazolinone[2], and triazolo pyrimidine[2a] herbicides, (2) selection of resistant alleles of structural genes encoding acetolactate synthase,[1,3-5] (3) isolation of resistant and wild-type alleles in pure form by recombinant DNA technology providing important opportunities in plant genetic engineering,[4-13] and (4) toxicity of α-ketobutyrate (AKB) and its metabolite, propionyl-CoA, to bacteria coupled with the probability that AKB accumulation is also toxic to animals and plants.[14-17,34]

[1] R. A. LaRossa and J. V. Schloss, *J. Biol. Chem.* **259**, 8753 (1984).

[2] D. L. Shaner, P. C. Anderson, and M. A. Stidham, *Plant Physiol.* **76**, 545 (1984).

[2a] W. A. Kleschick, R. J. Ehr, B. C. Gerwick, W. T. Monte, N. R. Pearson, M. J. Costales, and R. W. Meikle, *Eur. Pat. Applic. 142152* (1984).

[3] R. S. Chaleff and T. B. Ray, *Science* **223**, 1148 (1984).

[4] S. C. Falco and K. S. Dumas, *Genetics* **109**, 21 (1985).

[5] N. S. Yadav, R. McDevitt, S. Bernard, and S. C. Falco, *Proc. Natl. Acad. Sci. U.S.A.,* **83**, 4418 (1986).

[6] T. Newman, P. Friden, A. Sutton, and M. Freundlich, *Mol. Gen. Genet.* **186**, 378 (1982).

[7] C. A. Hauser and G. W. Hatfield, *Nucleic Acids Res.* **11**, 127 (1983).

[8] R. A. Weinberg and R. O. Burns, *J. Bacteriol.* **160**, 833 (1984).

[9] R. P. Lawther, B. Nichols, G. Zurawski, and G. W. Hatfield, *Nucleic Acids Res.* **7**, 2289 (1979).

[10] D. L. Blazey, R. Kim, and R. O. Burns, *J. Bacteriol.* **147**, 452 (1981).

[11] C. H. Squires, M. De Felice, S. R. Wessler, and J. M. Calvo, *J. Bacteriol.* **147**, 797 (1981).

[12] C. H. Squires, M. De Felice, C. T. Lago, and J. M. Calvo, *J. Bacteriol.* **154**, 1054 (1983).

[13] B. J. Mazur, C. F. Chui, S. C. Falco, R. S. Chaleff, and C. J. Mauvais, *World Biotech. Rep.* **2**, 97 (1986).

[13a] B. J. Mazur, C. F. Chui, and J. K. Smith, *Pl. Physiol.* **85**, 1110 (1987).

[14] J. Daniel, L. Dondon, and A. Danchin, *Mol. Gen. Genet.* **190**, 452 (1983).

[15] T. K. Van Dyk and R. A. LaRossa, *J. Bacteriol.* **165**, 386 (1986).

[16] W. Yang and K. S. Roth, *Clin. Chim. Acta* **145**, 173 (1985).

[17] D. Wellner and A. Meister, *Annu. Rev. Biochem.* **50**, 911 (1981).

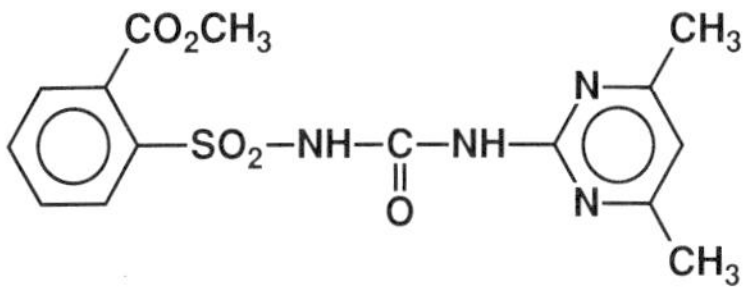

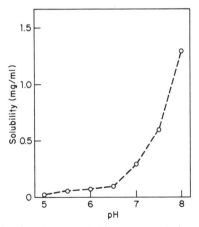

FIG. 1. Structure of sulfometuron methyl.

In this review, we describe some chemical and biological properties of the sulfonylurea herbicide sulfometuron methyl (SM), the active ingredient in the product Oust®.[18] The approaches that have been successful in isolation of structural genes encoding ALS from both prokaryotes and eukaryotes are enumerated. Methods for the isolation of sulfometuron methyl-resistant and hypersensitive mutants of the bacterium *Salmonella typhimurium* and resistant variants of plants and the yeast *Saccharomyces cerevisiae* are presented. The molecular bases underlying these phenotypes are summarized. Last, methods for the analysis of AKB and other isoleucine biosynthteic intermediates are outlined.

Sulfometuron Methyl

This complex molecule (Fig. 1) has a molecular weight of 364.4. It is acidic in aqueous solution having a pK_a of 5.2 and is unstable at acidic pH in aqueous solution. Both its solubility (Fig. 2) and stability increase with

FIG. 2. Solubility of sulfometuron methyl in aqueous solution as a function of pH.

[18] R. E. Sauers and G. Levitt, *in* "Pesticide Synthesis through Rational Approaches" (P. S. Magee, G. V. Kohn, and J. J. Mem, eds.), p. 21. Am. Chem. Soc., Washington, D.C., 1984.

pH. SM is also soluble in a variety of organic solvents including acetone (2.4 mg/ml), acetonitrile (1.5 mg/ml), methanol (0.14 mg/ml), xylene (0.037 mg/ml), and diethyl ether (0.032 mg/ml). Stock solutions of SM utilized in the preparation of microbiological media are prepared immediately prior to their use in acetone at a final concentration of 2 mg/ml (5.5 mM). When aqueous stock solutions are required for enzymological analyses, 10 mM SM is prepared by mixing 3.644 mg of SM with 1 ml of 0.01 N NaOH; this aqueous solution is stable at $-20°$ for at least 1 year.[19] Upon mixture of either dilute base or acetone with SM, a suspension is obtained that is converted to a solution by sonication in a ultrasonic bath.

Microbial Susceptibility to Sulfometuron Methyl

Two methods have been utilized to assess microbial susceptibility to SM.

Disk Diffusion Assay[1]

Microbes, grown overnight in rich media from a single colony, are collected by centrifugation, washed once with minimal medium, and resuspended in an equal volume of minimal medium. Then 0.1 ml (either 10^8 bacteria or 10^7 fungi) of the suspension is mixed by vortexing with 2.5 ml of molten (50°) 0.6% agar in minimal medium. The resulting suspension is evenly distributed over a 100-mm Petri dish containing 30 ml of minimal medium solidified with 1.5% agar. Twenty-microliter (40 μg) aliquots of a 2-mg/ml SM-acetone stock solution are applied to 6-mm Whatman paper disks placed in a glass Petri dish. Evaporation of the acetone is accomplished by placing the open Petri dish in a chemical hood for 20 min. The disks are aseptically transferred to lawns of bacteria or fungi. After overnight (16 hr) incubation at 30°, plates are inspected for zones of inhibition whose clarity is estimated and whose diameter is measured. Table I shows representative zones of inhibition obtained with a variety of organisms. Since multiple forms of ALS exist in enteric bacteria,[20] and since ALS isozyme I of both *S. typhimurium* and *Escherichia coli* is not inhibited substantially by SM *in vitro*[21,21a] (Table II), it is not surprising that many strains are insensitive to SM. Similarly, many bacterial strains are not

[19] J. V. Schloss, unpublished observations.
[20] M. De Felice, C. T. Lago, C. H. Squires, and J. M. Calvo, *Ann. Microbiol. (Paris)* **133A**, 251 (1982).
[21] R. A. LaRossa and D. R. Smulski, *J. Bacteriol.* **160**, 391 (1984).
[21a] J. V. Schloss, L. M. Ciskanik, and D. E. Van Dyk, *Nature (London)* **331**, 360 (1988).

TABLE I
ZONES OF INHIBITION GENERATED BY SM ON MINIMAL MEDIUM

Species	Genotype	Addition	Zone of inhibition (diameter in mm)
Acinetobacter sp.	+	−	13[a]
Citrobacter freundii	+	−	14[a]
Escherichia coli K12	+	−	<6[a]
E. coli K12	*ilvIH*	−	<6[b]
E. coli K12	*ilvB*	−	17[b]
Klebsiella pneumoniae	+	−	<6[a]
Pseudomonas fluorescens	+	−	<6[a]
Pseudomonas putida	+	−	<6[a]
Salmonella typimurium	+	−	<6[a]
S. typimurium	+	Valine	32[a]
S. typimurium	*ilvB*	−	40[b]

[a] From LaRossa and Schloss.[1]
[b] From LaRossa and Smulski.[21]

inhibited by L-valine since this amino acid does not inhibit ALS isozyme II.[22] End-product inhibition of ALS I activity by supplementing minimal media with L-valine can often increase the potency of SM.[1]

Minimal Inhibitory Concentrations[15]

These are estimated by spotting 5-μl aliquots of washed cultures containing 2×10^6 microbes on a series of minimal plates supplemented with

TABLE II
SENSITIVITY OF ALS ISOZYMES TO
SULFOMETURON METHYL

Isozyme	I_{50} (nM)
Bacterial I	>1,000,000[a]
Bacterial II	65[b]
Bacterial III	<1,000,000[a]
Yeast	120[c]
Pea	16[d]
Tobacco	8[e]

[a] From LaRossa and Smulski.[21]
[b] From LaRossa and Schloss.[1]
[c] From Falco and Dumas.[4]
[d] From T. Ray, *Plant Physiol.* **75,** 827 (1984).
[e] From Chaleff and Mauvais.[28]

SM at concentrations which typically range from 1 to 1000 μg/ml (2.7 to 2750 μM). The MIC is defined as the lowest SM concentration at which growth of the culture is significantly retarded after overnight incubation.[15]

Assay of Specific Bacterial Acetolactate Synthase Isozymes in Crude Extracts

The activity of individual ALS isozymes can be estimated in bacterial strains containing more than one ALS activity. Addition of 1.25 mM valine to cell extracts inhibits ALS I and III,[22] while addition of 1 mM SM will inhibit ALS II and III.[1,21] Synthesis of ALS III can be repressed by the inclusion of leucine in the growth medium.[22] Thus, extracts of *S. typhimurium* containing isozymes I and II are assayed in the presence of SM to measure isozyme I, or in the presence of valine to measure isozyme II. *E. coli* containing isozymes I and III are grown in the presence of leucine or extracts assayed in the presence of SM to quantitate isozyme I activity. The difference between the total ALS activity of an *E. coli* extract and that due to ALS I is attributed to ALS III.

Isolation of Acetolactate Synthase Structural Genes

Genes encoding ALS have been cloned from bacteria, yeast, and plants by a variety of methods.

Bacterial ALS Genes

The availability of *E. coli* strains deficient in all ALS activities allows the isolation of bacterial ALS structural genes by complementation. In addition, knowledge of the map locations of the genes and availability of F' factors, defective λ specialized transducing phages, and multicopy plasmids known to harbor neighboring genes facilitates molecular cloning. Table III summarizes methods that have been successfully used to isolate the structural genes encoding the ALS isozymes from *E. coli* and *S. typhimurium*.

Yeast ALS Gene

In the absence of available ALS-deficient strains, isolation of yeast genes encoding ALS can be accomplished by selection for segmental aneuploidy.[23] Overproduction of ALS from a high copy number plasmid

---

[22] M. De Felice, M. Levinthal, M. Iaccarino, and J. Guardiola, *Microbiol. Rev.* **43**, 42 (1979).

[23] J. Rine, W. Hansen, E. Hardeman, and R. W. Davis, *Proc. Natl. Acad. Sci. U.S.A.* **80**, 6750 (1983).

Table III

METHODS UTILIZED TO ISOLATE STRUCTURAL GENES ENCODING ALS ISOZYMES

Isozyme[a]	Genes	Bacterium	DNA source[b]	Selection
I	ilvBN	E. coli	F' ilvB4	Complementation[c]
I	ilvBN	E. coli	Multicopy plasmid carrying uhp†	Complementation[d]
I	ilvBN	S. typhimurium	F' ilvB	Complementation[e]
II*	ilvGM	E. coli	λdilv58	Overlap with a plasmid known to contain ilv genes[f]
II	ilvGM	S. typhimurium	Bacterial chromosome	Valine-resistant E. coli[g]
III	ilvIH	E. coli	λG4 leu†	Complementation[h]
III*	ilvIH	S. typhimurium	Multicopy plasmid carrying leu†	DNA homologous to E. coli ilvIH[i]

[a] Asterisk denotes normally cryptic isozymes.
[b] Dagger denotes genes neighboring ALS structural genes.
[c] From Newman et al.[6]
[d] From Hauser and Hatfield.[7]
[e] From Weinberg and Burns.[8]
[f] From Lawther et al.[9]
[g] From Blazey et al.[10]
[h] From Squires et al.[11]
[i] From Squires et al.[12]

confers resistance to low levels of SM.[4] Yeast are transformed to uracil prototrophy by yeast genomic libraries maintained in a high copy plasmid vector capable of autogenous replication. Pooled transformants at 1×10^6 cells per plate are spread on media lacking uracil and containing SM at 1 μg/ml (2.7 μM). Colonies resistant to SM appear at a frequency of 1–5×10^{-4} per transformant.[4] In one experiment,[4] four of six resistant colonies had plasmid-borne determinants of SM resistance. One transformant had a 4- to 5-fold increase in ALS activity over the untransformed parent, and mutations in this cloned gene gave rise to SM-resistant ALS activity, indicating that it contained the structural gene for ALS.[4] Unlike bacteria, there is no evidence for a multiple subunit structure of ALS in yeast.[4]

Plant ALS Genes

Isolation of plant ALS genes is possible using heterologous hybridization probes.[13] Phage plaques of plant genomic libraries are screened with the coding sequence of the yeast ALS gene under low stringency hybridization conditions. DNA from the hybridizing phage is prepared and analyzed by restriction enzyme digestion and Southern blot hybridization to show that specific sequences are obtained. DNA sequence analysis and compari-

son of the plant sequences to those of the bacterial and yeast ALS genes to show regions of homology indicate that plant ALS sequences are present in the clones. The cloning of ALS genes from *Anabaena, Arabidopsis thaliana,* and *Nicotiana tabacum* has been reported.[13,13a]

Localizing Cloned Genes by Tn5 Mutagenesis

The ability of Tn5 to act as a point mutation which destroys gene function is useful for physical mapping of cloned genes.[24] At high copy numbers, Tn5 confers resistance to higher concentrations of neomycin sulfate than does a single copy of Tn5,[25] allowing facile selection of transposition events from the bacterial chromosome to a multicopy plasmid.[26] *E. coli* strain DP2229,[26] containing a Tn5 insertion in the chromosome, is transformed with a multicopy plasmid containing a cloned gene. Transformants are purified on plates containing 30 μg/ml kanamycin and a selective agent to ensure maintenance (e.g., ampicillin). Many individual single colonies are streaked to plates containing 1.0 mg/ml neomycin sulfate[26] and the selective agent. One neomycin-resistant colony from each streak is chosen and repurified, ensuring the independent origin of each Tn5 transposition. Plasmid DNA from the neomycin-resistant isolates is isolated and digested with restriction endonucleases. Knowledge of the restriction map of the original plasmid, the restriction map of Tn5,[27] and the sizes of restriction fragments in the Tn5 containing plasmids allows mapping of the Tn5 insertion sites. The phenotype of the gene of interest conferred by the plasmids containing different Tn5 insertions is scored. Comparison of the mapped location of Tn5 in those plasmids in which Tn5 had inactivated gene function will give the location of the gene of interest within the cloned DNA. In one example, 21 of 40 Tn5 insertions into a plasmid containing an SM-resistant yeast ALS clone were within a 5.6-kb cloned fragment. Seven adjacent transpositions inactivated the ALS gene localizing it to a region of approximately 1.4 kb.[26]

Mutant Isolation

Herbicide-Resistant Mutants

Bacteria. Genes (Table III) for the three distinct ALS isozymes (Table II) exist in enteric bacteria. Isolation of herbicide resistant mutations in the

[24] N. J. Kleckner, J. Roth, and D. Botstein, *J. Mol. Biol.* **116**, 125 (1977).
[25] C. Sasakawa, J. B. Lowe, L. McDevitt, and D. E. Berg, *Proc. Natl. Acad. Sci. U.S.A.* **79**, 7450 (1982).
[26] T. Van Dyk, S. C. Falco, and R. A. LaRossa, *Appl. Environ. Microbiol.* **51**, 206 (1986).
[27] R. A. Jorgensen, S. J. Rothstein, and W. S. Reznikoff, *Mol. Gen. Genet.* **177**, 65 (1979).

structural gene for ALS II thus requires utilization of either (a) *S. typhimurium* or (b) variants of *E. coli* harboring a mutated *ilvGM* locus expressing the normally cryptic ALS II. In both instances, the other ALS isozymes (I and/or III) are inhibited by L-valine, allowing the isolation of variant forms of ALS II. Selection conditions are

1. 10^8 *S. typhimurium* LT2 cells are plated on minimal E medium supplemented with 0.2% glucose, 83 μg/ml L-valine (to inhibit ALS I activity), and 33 μg/ml SM (90 μM). Between 10 and 100 colonies arise on these plates after 2 days at 37°.[1]

2. 10^8 cells of the *E. coli* strain HB101/pB1 expressing ALS II from a weak promoter on a multicopy plasmid are plated on M9 medium supplemented with glucose, valine, 50 μg/ml ampicillin, and 20 μg/ml SM (55 μM). After 3 days at 37°, from 5 to 10 resistant colonies are obtained per plate.[5]

Both selections, which represent 10 independent spontaneous events, exclusively yielded alterations of *ilvGM,* the locus encoding ALS II. In nine cases, ALS II activity had been altered from the SM-sensitive parental forms to novel activities refractile to inhibition by SM.[1,5]

Yeast. 10^7 haploid cells of *S. cerevisiae* are plated on minimal media supplemented with 3 μg/ml (8.2 μM) of SM.[4] In one report, 66 independent mutants were isolated, 51 (77%) of which mapped to *ILV2,* the structural gene encoding ALS in this organism. These mutants, in contrast to wild type and the remaining 15 variants, harbored an ALS whose activity was insensitive to SM *in vitro.* Most of these structural gene mutants were resistant to more than 30 μg/ml (82 μM) of SM. The other 15 mutants, containing a herbicide-sensitive form of ALS, were resistant to low (3–10 μg/ml, 8.2–27.5 μM) but not high levels of SM. This second group of mutations map at two loci, *PDR1* and *SMR3.* These loci are thought to alter the permeability of yeast to a wide variety of compounds.[4]

Plants. Haploid callus cultures of *Nicotiana tabacum* cv. Xanthi (tobacco) maintained on solidified minimal medium are plated on the same medium supplemented with 2 ng/ml of either SM (5.5 nM) or its analog chlorsulfuron.[3] After regeneration of whole plants from callus culture,[3] linkage analysis indicates that two distinct genetic loci give rise to herbicide resistance.[3] Wild-type tobacco contains a herbicide-sensitive ALS activity. ALS activity of one homozygous tobacco mutant is at least 30-fold more resistant to SM than the parental activity.[28] Sulfonylurea-resistant mutants of another species, *Arabidopsis thaliana,* have been obtained at the whole

[28] R. S. Chaleff and C. J. Mauvais, *Science* **224,** 1443 (1984).

plant level by selection following chemical mutagenesis of seed. The ALS activity of one such mutant is resistant to SM *in vitro*.[28a]

Herbicide-Hypersensitive Mutants

The preceding section illustrates the ease with which herbicide-resistant mutants are obtained in a variety of organisms due to the powerful genetic selection imposed by SM. Since a positive selection scheme for herbicide-hypersensitive mutants is lacking, an efficient mutagenesis protocol is needed. The technology for such mutagenesis exists in systems based on transposable genetic elements in the enteric bacteria *S. typhimurium* and *E. coli*.[24] *S. typhimurium* LT2 is infected with a defective P22 transducing phage carrying the transposon Tn*10* (P22 *c2ts29, 12-amN11, 13-amH101, int-3*, Tn*10*)[29] at a multiplicity of infection of 0.8 phage particles per cell. Mutations in the phage prevent its lysogenization and replication; thus Tn*10* is inherited by transposition from the phage genome to the bacterial chromosome.[29] After phage absorption for 30 min at 37°, the transduction mixture is plated on rich medium containing 12.5 μg/ml of tetracycline and 10 m*M* ethylene glycol bis(β-aminoethyl ether)*N, N'*-tetraacetic acid (EGTA).[29] The tetracycline-resistant transductants, representing random insertions of Tn*10*, are screened for altered SM sensitivity by comparing their growth on solidified minimal media with growth on media containing 50 μg/ml of SM (135 μ*M*).[15] The parental *S. typhimurium* LT2 grows equally well on both media. Nineteen out of 5000 independent insertion mutants were inhibited by SM.[15] Many of these mutations have been mapped by conjugation and generalized transduction to specific genes on the bacterial chromosome including *poxA*,[30] *ack*,[31] *pta*,[31] *thiAC*,[32] and *aspC*.[15] Other previously described point mutations including *ilvB*,[32] *relA*,[32] *oxyR*,[33] and *ilvA*[34] also cause herbicide hypersensitive phenotypes. The gene-product correspondences[35,36] of these mutations are summarized in Table IV.

[28a] G. W. Haughn and C. Sommerville *Mol. Gen. Genet.* **204**, 430 (1986).

[29] R. W. Davis, D. Botstein, and J. R. Roth, "Advanced Bacterial Genetics." Cold Spring Harbor Lab., Cold Spring Harbor, New York, 1980.

[30] T. K. Van Dyk, D. R. Smulski, and Y. Y. Chang *J. Bacteriol.* **169**, 4540 (1987).

[31] T. K. Van Dyk and R. A. LaRossa, *Mol. Gen. Genet.* **207**, 435 (1987).

[32] T. K. Van Dyk, unpublished observations.

[33] R. A. LaRossa, unpublished observations.

[34] R. A. LaRossa, T. K. Van Dyk, and D. R. Smulski, *J. Bacteriol.* **169**, 1372 (1987).

[35] B. J. Bachmann, *Microbiol. Rev.* **47**, 180 (1983).

[36] K. E. Sanderson and J. R. Roth, *Microbiol. Rev.* **47**, 410 (1983).

TABLE IV
MUTATIONS CONFERRING SM HYPERSENSITIVITY ON SALMONELLA TYPHIMURIUM

Gene	Alteration
ack	Acetate kinase deficiency
aspC	Aspartate aminotransferase deficiency
ilvA	Threonine dehydratase insensitivity to isoleucine
ilvBN	ALS I deficiency
oxyR	Inability to respond to oxidative damage
poxA	Deficiency in *poxB* (pyruvate oxidase) positive regulator
pta	Phosphate acetyltransferase deficiency
thiAC	Thiamine pyrophosphate deficiency
relA	Stringent factor deficiency

Analysis of Isoleucine Biosynthetic Intermediates

Accumulation of Ketone Intermediates

L-Isoleucine is derived from L-threonine via four intermediates, three of which are ketones. Ketone accumulation is measured by administration of radioactive L-threonine to cells followed by treatment of the culture with acidic dinitrophenylhydrazine (DNP). Dinitrophenylhydrazones are formed by the condensation of ketones and DNP. After separation of the hydrophobic DNP and DNP derivatives from amino acids by extraction of the aqueous solution with toluene, the concentration of isoleucine intermediates is determined by scintillation counting.[14] The requisite materials for such experiments are listed in Table V.[15]

A culture of *S. typhimurium* is incubated aerobically at 37° in a gyratory water bath. After overnight growth, the culture is diluted 50-fold into fresh medium (solution G of Table V) and incubation is resumed under

TABLE V
SOLUTIONS REQUIRED FOR THE ANALYSIS OF ISOLEUCINE BIOSYNTHETIC INTERMEDIATES

A. 0.1% dinitrophenylhydrazine (Kodak) in 2 N HCl
B. 10 mM SM in 0.01 N NaOH
C. 1% L-isoleucine
D. 0.1% L-threonine
E. L-[U-^{14}C]threonine (New England Nuclear; 0.44 μM, 228 mCi/mmol, 100 μCi/ml)
F. 1% L-valine
G. Minimal medium [M9^a or E^a] supplemented with 0.2% dextrose
H. Toluene

^{a} From Davis *et al.*[29]

the same conditions until a culture density of about 3×10^8 cells per milliliter is achieved. Culture aliquots (2 ml) are placed in prewarmed test tubes containing L-threonine (24 μl of solution D and 20 μl of solution E yielding a final concentration of 100 μM and a final specific activity of 10 mCi/mmol; i.e., 1 μCi/ml). Incubation at 37° is continued for 2 hr with intermittent sampling.[15] Test tubes may also contain L-isoleucine (100 μg/ml; 20 μl of solution C), L-valine (100 μg/ml; 20 μl of solution F), and SM (100 μM; 20 μl of solution B); inhibitors of threonine dehydratase, ALS I, and ALS II, respectively.[15,34]

In 1.5-ml microfuge tubes, 0.2 ml samples are mixed with 0.5 ml of solution A. After at least 5 min at ambient temperature, during which the DNP derivatives form, the aqueous solution is mixed with 0.5 ml of toluene by vortexing for 30 sec. The aqueous and toluene phases are separated by a brief (20 sec) centrifugation. Radioactivity in 0.15 ml of the upper, toluene phase which contains DNP derivatives is determined by liquid scintillation counting.[15]

The identity of the resultant DNP derivatives can be determined by thin-layer chromatography.[34]

α-Ketobutyrate Degradation Rate Determination

The above assay has been modified to determine AKB turnover.[34] AKB accumulation is initiated by the inhibition of ALS activities. This is accomplished by addition of the culture to a test tube containing radiolabeled threonine, SM, and valine.[34] During an incubation period of 20 min about 25% of the available L-threonine is converted to AKB.[34] Further AKB synthesis is prevented by the addition of L-isoleucine to a final concentration of 100 μg/ml.[34] The kinetics of AKB conversion to material no longer extractable into the toluene phase after reaction with DNP is monitored for 120 min by liquid scintillation counting.[34]

Acknowledgments

We wish to thank our colleagues, especially S. C. Falco, J. V. Schloss, D. R. Smulski, and D. E. Van Dyk, who have contributed to the development of sulfometuron methyl methodology in our laboratory.

[14] Antibodies against Branched-Chain α-Keto Acid Dehydrogenase Proteins for Use in Defining Human Mutations and Gene Isolation

By Dean J. Danner, Louis J. Elsas, and Stuart Litwer

Mutations are known in the human genome which specifically decrease the activity of the mitochondrial multienzyme complex, branched-chain α-keto acid dehydrogenase (BCKD).[1] These mutations show inheritance patterns which follow Mendelian laws for an autosomal recessive trait, thus implying nuclear coding for these proteins. Specific mutations remain to be defined. As a means of defining the mutations, antibodies specific for the BCKD proteins are employed to study the proteins themselves through Western blot analysis. These same antibodies are used to isolate nucleotide probes for studying the genes directing synthesis of the BCKD proteins.

Principle

Specificity of antibodies allows these molecules to identify their antigenic determinants on proteins within a mixed population of proteins. Polyclonal antibodies made against purified proteins of the BCKD complex can therefore be used to recognize the specific antigens within mitochondria, despite the low abundance of the BCKD proteins in these organelles. If mitochondria from cells of patients with decreased BCKD activity are analyzed by this method, one can ask if the BCKD proteins are present and if they have mobility different from the wild-type proteins. In a similar manner, these same antibodies can be used to screen cDNA expression libraries for clones expressing the recognized antigen. Isolation of the cDNA from positive clones provides probes for studying the genes and their expression in normal and mutant cells.

Antibody Production[2]

Branched-chain α-keto acid dehydrogenase complex purified from bovine liver (100 μg) as previously described[3] is mixed with an equal volume

---

[1] D. J. Danner and L. J. Elsas, *in* "Metabolic Basis of Inherited Disorders" (C. R. Seriver, A. L. Beaudet, W. S. Sly, and D. Valle, eds.), 6th ed. McGraw-Hill, New York, 1988, in press.

[2] S. C. Heffelfinger, E. 1. Sewell, and D. J. Danner, *Biochem. J.* **213**, 339 (1983).

[3] D. J. Danner and S. C. Heffelfinger, this volume [38].

of Freund's complete adjuvant (Difco) and used for multiple subcutaneous injections into New Zealand white rabbits. Similarly, if the BCKD complex is first resolved by electrophoresis in a polyacrylamide gel containing sodium lauryl sulfate and 2-mercaptoethanol,[4] the individual protein bands can be excised and used as antigens in this system.[5] Booster injections with these antigens (100–300 μg) in incomplete adjuvant are given at 2-week intervals for 6 weeks and monthly thereafter as needed. After 3 months, blood from an ear vein puncture is collected and an immunoglobulin fraction prepared by 50% ammonium sulfate precipitation. Immunoglobulin is stored in 50 mM potassium phosphate, pH 7.0, at 25 mg protein/ml and −20°.

Mitochondria Preparation

Fibroblasts obtained from human skin biopsies are maintained in culture with Dulbecco and Vogt's medium supplemented with 15% fetal bovine serum.[6] Monolayers of cells from four 75 cm² culture flasks are removed by treatment with 0.25% trypsin at 37° for 30 min, washed into centrifugation tubes with phosphate-buffered saline, and pelleted by centrifugation at 1000 g for 10 min. The cell pellet is suspended in 0.25 M sucrose buffered with 20 mM potassium phosphate, pH 7.4, and treated with 6.25 ng protease VI (Sigma) per milligram wet weight of cells for 7 min at 4°. Cells are disrupted by grinding with a Teflon–glass homogenizer and nuclei removed by centrifugation at 800 g for 10 min. Mitochondria are pelleted by centrifugation of the supernatant at 10,000 g for 10 min and washed once with the buffered sucrose. All centefugation were done at 4°C. The final pellet of mitochondria is suspended in a minimal volume of 20 mM potassium phosphate, pH 6.5.

Western Blot Analysis

Mitochondrial proteins (approximately 20 μg per lane) are resolved by electrophoresis in a 10% polyacrylamide minigel in the presence of sodium lauryl sulfate and 2-mercaptoethanol[7] plus 2 mM ethylenediaminetetraacetic acid (EDTA) in all solutions. The marker dye Pyronin G (0.1% in 15% glycerol) is used so that transfer to nitrocellulose can be followed without protein staining. To electroblot the proteins, the gel is soaked 3 × 10 min in 25 mM Tris, 192 mM glycine, and 20% (v/v) methanol. A

---

[4] S. C. Heffelfinger and D. J. Danner, *Biochemistry* **23**, 2219 (1983).
[5] O. DeMarcucci and J. G. Lindsay, *Eur. J. Biochem.* **149**, 641 (1985).
[6] D. J. Danner and J. H. Priest, *Biochem. Genet.* **21**, 895 (1983).
[7] U. Laemmli, *Nature (London)* **227**, 680 (1970).

sandwich is made using two layers of Whatman 3MM, gel, nitrocellulose filter, two layers of Whatmann 3MM. This sandwich is then placed in the blotting chamber (Hoefer Mini Transphor) containing the Tris/glycine/methanol buffer. Transfer is complete in 2 hr at room temperature and 200 mA. After transfer, the nitrocellulose is soaked for 45 min in 50 mM Tris-HCl, pH 7.4, containing 0.15 M NaCl, 0.1% Tween 20, and 1% tin (225 Bloom, bovine skin), and then transferred to 50 mM Tris-HCl, pH 7.4, containing 0.15 M NaCl, 0.1% Tween 20, 5 mM EDTA, 0.25% gelatin, and 5% bovine serum for an additional 45 min. To analyze for the presence of antigens recognized by the BCKD antisera, 75 μl of the specific antisera is added to 100 ml of the second soak solution. This volume is sufficient for a 6.5 × 8 cm nitrocellulose filter containing eight lanes of resolved mitochondrial proteins. Binding is allowed to continue for 1 hr at room temperature and then the filter is washed 3 × 10 min with the 0.25% gelatin solution without bovine serum. Binding of goat antirabbit IgG-peroxidase (Bio-Rad) is done in the 0.25% gelatin solution plus 5% bovine serum (3 μl IgG/10ml) for 45 min at room temperature. The filter is washed 4 × 10 min in the 0.25% gelatin solution without serum. Detection of specifically bound peroxidase is carried out in 50 mM Tris-HCl, pH 7.5, containing 0.3 mg diaminobenzidine and 0.2 μl 50% H_2O_2 per milliliter. After color development is complete, usually in 5–45 min, the filter is washed in water and dried on Whatmann 3MM paper.[8,9]

Screening Expression Libraries

Several cDNA libraries in λgt11 are currently available from investigators or commercial sources. These libraries are made essentially by the method of Young and Davis[10] from various tissues and species. cDNA is inserted near the C-terminal end of the β-galactosidase gene such that a fusion protein is produced in response to stimulation of β-galactosidase synthesis. If the cDNA insert is in the proper reading frame, then the fusion protein may be recognized by specific antisera. Antisera for this type of selective screening must be freed of antibodies which recognize λgt11 and host *Escherichia coli* antigens. To prepare this antiserum, *E. coli* (BNN97) containing λgt11 is grown to stationary log phase in 1 liter of LB broth and harvested by centrifugation. Bacteria are washed in 100 ml of phosphate-buffered saline, pelleted by centrifugation, and suspended in 40 ml of

[8] D. J. Danner, N. Armstrong, S. C. Heffelfinger, E. T. Sewell, J. H. Priest, and L. J. Elsas, *J. Clin. Invest.* **75**, 858 (1985).
[9] S. Litwer and D. J. Danner, *Biochem. Biophys. Res. Commun.* **131**, 961 (1985).
[10] R. A. Young and R. W. Davis, *Science* **222**, 778 (1983).

0.1 M 3-(N-morpholino) propanesulfonic acid buffer (MOPS), pH 7.5, for lysis by sonication. Insoluble protein and cell debris is removed by low-speed centrifugation and soluble proteins in the supernatant are mixed with Bio-Rad Affi-Gel 10 and 15 according to the manufacturer's instructions. After overnight coupling of proteins and gel at 4°, 1 M glycine ethyl ester is added for 1 hr to block uncoupled sites on the Affi-Gel. The gel is then washed extensively with phosphate-buffered saline prior to overnight mixing with anti-BCKD serum at 4°. In the morning, the Affi-Gel is poured into a 0.7 × 10 cm column and unbound antibody collected. The procedure is repeated with fresh coupled Affi-Gel using the first unbound antibody solution. Unbound antibodies from the second column are used to screen the cDNA library.

$E.$ $coli$ strain Y1090 is infected with λgt11 containing the cDNA library. Initially, 1–3 × 10° recombinants are distributed at a density of 400,000 recombinants per 150 cm² plate. Plaque formation is induced by growing the recombinants lytically in top agarose on LB agar for 4 hr at 42°. At this time, the plates are overlayed with nitrocellulose filters impregnated with 10 mM isopropyl-β-D-thiogalactopyranoside (IPTG) and grown at 37° for 2 hr. Filters are keyed to the plates, removed, and treated as described above for Western blot analysis of proteins. Positive plaques are marked and picked using the large bore of a sterile Pasteur pipet.[11] Selected agarose plugs are placed in SM buffer[12] to elute the phage for subsequent rounds of selection. For each selecting round, plaque number per plate is reduced 10-fold until all plaques on a plate produce antigenically positive fusion protein. It is important that the fusion protein be isolated and tested for antigenicity. A single plate of confluent positive plaques is induced to produce fusion protein by overlaying the plate with a 2 mM IPTG solution and incubating the plate for 2 hr at 37°. Induced proteins are rinsed from the plate with Laemmli sample buffer and resolved on a polyacrylamide gel and Western blotted.

To select genes for BCKD proteins, we used polyclonal antisera containing antibodies for all four BCKD proteins. To identify which protein is detected, fusion protein produced by a single plaque-positive confluent plate is transferred to nitrocellulose and the antisera exposed to the filter as above. After 1 hr of binding, unbound antisera are removed and the filter washed in phosphate-buffered saline. Bound antibodies are eluted from the filter by washing with 1 ml of 5 mM glycine-HCl, pH 2.3, containing

---

[11] J. R. deWet, H. Fukushima, N. N. Dewji, E. Wilcox, J. S. O'Brien, and D. R. Helinski, DNA **3**, 437 (1984).
[12] T. Maniatis, E. F. Fritsch, and J. Sambrook, "Molecular Cloning: A Laboratory Manual," p. 443. Cold Spring Harbor Lab., Cold Spring Harbor, New York, 1982.

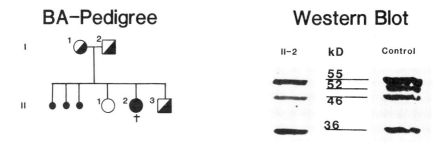

FIG. 1. Use of antibodies to demonstrate the absence of a protein from the BCKD complex. (A) The family pedigree where the half-filled symbols indicate the heterozygous status of an individual based on enzyme activity. The small filled circles indicate the loss of a pregnancy to spontaneous abortion. The filled circle, II-2, denotes the homozygous affected child who died at 13 days, denoted by the cross. Values contained in B are BCKD activity in cultured skin fibroblasts quantified by $^{14}CO_2$ released from [1-^{14}C]leucine as dpm/90 min/μg cell protein. (C) A Western blot analysis (see text) of mitochondrial proteins (20 μg per lane) derived from the homozygous affected child (II-2) and a control human skin fibroblast culture with full BCKD activity. The numbers indicate the mass of the stained bands based on known sized markers run on the same gel.

150 mM NaCl, 0.5% Triton X-100, and 100 μg bovine serum albumin. Filters are washed three times and each elution is immediately neutralized to pH 7.4 with 1 M Tris-HCl. The eluted antibody is used in a Western blot analysis of mitochondrial proteins to establish which BCKD protein is recognized. In this manner we have been able to select clones for the 52,000- and 47,500-Da subunits.

Results and Discussion

Western blot analysis of mitochondria from the cells of patients with decreased BCKD activity shows that generally the four proteins are present and have mobility identical to catalytically normal proteins. However, when major alterations in these proteins are present, this technique will identify the change. We have been able to identify one individual who antigenically lacks the 52,000-Da protein of the BCKD complex (Fig. 1).

As expected, BCKD activity for this individual is essentially zero and this absence of protein and activity apparently is lethal. As yet, with over 50 cell lines from different patients with impaired BCKD activity, no BCKD proteins with altered mobility have been found. A report from Japan has described three patients who antigenically lack the Elβ subunit (36,000 Da).[13]

To study the genetic mutations directly, gene analysis is required. Again the specificity of antibodies for the proteins of interest can be used to select clones of cDNA specific for the genes of interest. Expression libraries have been constructed which produce fusion proteins in quantities sufficient to be detected by antigen–antibody reactions if the recognized epitope is present. We have used these libraries to select clones for genes directing the synthesis of proteins for the BCKD complex. Figure 2 shows an example of

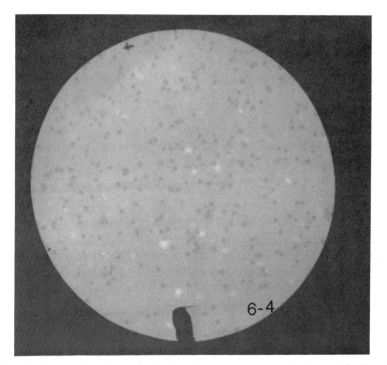

FIG. 2. Nitrocellulose filter from final (sixth) round of plaque selection purification. All plaques on the filter are stained positive with antibody to the proteins of the BCKD complex using a Western blot analysis.

[13] Y. Indo, F. Endo, I. Akaboshi, and I. Matsuda, *J. Clin. Invest.* **80,** 63 (1987).

clonal selection by plaque purification which required six rounds of selection. This clone is for the 52-kD protein of the complex. Confirmation of the identity of this cDNA clone was done by matching amino acid sequence of a 10-residue fragment from bovine kidney BCKD-E2 which contained the lipoate binding site with the amino acid sequence deduced from nucleic acid sequence determined for the cDNA. An exact match was found for this peptide between the bovine and human BCKD-E2.[14] This oligonucleotide can thus be used to study the genes and their transcripts in cells with wild-type proteins for comparison to these structures in cells with mutant genes.

[14] K. B. Hummel, S. Litwer, A. P. Bradford, A. Aitken, D. J. Danner, and S. J. Yeaman, *J. Biol. Chem.* **263** (1988), pp. 6165–6168.

[15] Inhibition of Branched-Chain α-Keto Acid Dehyrogenase Kinase by α-Chloroisocaproate

By ROBERT A. HARRIS, MARTHA J. KUNTZ, and RONALD SIMPSON

The branched-chain α-keto acid dehydrogenase (BCKDH) complex, the major regulatory enzyme of branched-chain amino acid metabolism, is subject to regulation by covalent modification.[1-3] BCKDH kinase phosphorylates two serine residues in the E1α subunit of BCKDH, resulting in inactivation of the complex; dephosphorylation of the same sites by BCKDH phosphatase induces activation. α-Chloroisocaproate[4] specifically inhibits BCKDH kinase activity.[5] Since inhibition of kinase activity perturbs the normal activity state of BCKDH (set by the opposing kinase and phosphatase activities), α-chloroisocaproate has proved to be a useful reagent for studies of the regulation of branched-chain amino acid metabolism.

[1] H. R. Fatania, K. S. Lau, and P. J. Randle, *FEBS Lett.* **132**, 285 (1981).
[2] R. Odessey, *Biochem. J.* **204**, 353 (1982).
[3] R. Paxton and R. A. Harris, *J. Biol. Chem.* **257**, 14433 (1982).
[4] 2-Chloro-4-methylpentanoic acid is the correct systematic name of α-chloroisocaproic acid. The name α-chloroisocaproate, referring to the sodium salt, is used here and in most publications.
[5] R. A. Harris, R. Paxton, and A. A. DePaoli-Roach, *J. Biol. Chem.* **257**, 13915 (1982).

The incentive to synthesize and test α-chloroisocaproate as an inhibitor of BCKDH kinase originated from known inhibition of pyruvate dehydrogenase (PDH) kinase by dichloroacetate and 2-chloropropionate.[6,7] Pyruvate, the primary substrate for the PDH complex, is a noncompetitive inhibitor of PDH kinase.[8] The two chloro analogs of pyruvate named above act in a similar manner to inhibit PDH kinase.[6-8] α-Ketoisocaproate, the oxidative deamination product of leucine and one of the primary substrates for the BCKDH complex, is a noncompetitive inhibitor of BCKDH kinase.[3,9] Since dichloroacetate did not prove very effective as an inhibitor of BCKDH kinase,[10] we synthesized and subsequently found α-chloroisocaproate to be a potent inhibitor of this kinase.[5]

Synthesis of α-Chloro Compounds

α-Chloroisocaproic Acid

Racemic α-chloroisocaproic acid may be prepared by the direct chlorination of isocaproic acid[11,12] or, particularly conveniently, by the diazotization of (R,S)-leucine.[13] The enantiomers may be separated by high-resolution gas chromatography[14,15] or, alternatively, (R)-[13] and (S)-[16,17] α-chloroisocaproic acids may be individually prepared by the generalized procedure shown below. The optically active acids so produced are somewhat labile and should be handled carefully. In biochemical experiments the compounds are most conveniently used as their sodium salts, which may be prepared by traditional methods as further exemplified.

General Procedure: (R)-α-Chloroisocaproic Acid

A solution of $NaNO_2$ (16.9 g, 0.245 M) in water (30 ml) was added dropwise over 3 hr to a well-stirred solution of (R)-leucine (20 g, 0.152 M)

---

[6] S. Whitehouse, R. H. Cooper, and P. J. Randle, *Biochem. J.* **141**, 761 (1973).

[7] D. W. Crabb, E. A. Yount, and R. A. Harris, *Metab. Clin. Exp.* **30**, 1024 (1981).

[8] M. L. Pratt and T. E. Roche, *J. Biol. Chem.* **254**, 7191 (1979).

[9] K. S. Lau, H. R. Fatania, and P. J. Randle, *FEBS Lett.* **144**, 57 (1982).

[10] R. Paxton and R. A. Harris, *Arch. Biochem. Biophys.* **231**, 58 (1984).

[11] Y. Ogata, T. Harada, K. Matsuyama, and T. Ikejiri, *J. Org. Chem.* **40**, 2960 (1975).

[12] Y. Ogata, T. Sugimoto, and M. Inaishi, *Org. Synth.* **59**, 20 (1980).

[13] R. Simpson and R. J. Strohschein, unpublished observations.

[14] E. Koch, G. J. Nicholson, and E. Bayer, *J. High Resol. Chromatogr. Chromatogr. Commun.* **7**, 398 (1984).

[15] K. Watabe and E. Gil-Av, *J. Chromatogr.* **318**, 235 (1985).

[16] M. Renard, *Bull. Soc. Chim. Biol.* **28**, 497 (1946).

[17] P. Karrer, H. Reschofsky, and W. Kaase, *Helv. Chim. Acta* **30**, 271 (1947).

in 6 M HCl (300 ml) at 0°. After stirring for a further 4 hr at 0°, the reaction mixture was extracted with *tert*-butyl methyl ether (3 × 150 ml); the combined organic extracts were then washed with saturated NaCl solution, dried (MgSO₄), and evaporated to an oil [22.9 g, GC (OV-1701 capillary) 88% purity] which was purified by flash chromatography[18] on silica gel, eluting with 50% CHCl₃/hexane and 100% CHCl₃. Fractions were analyzed by GC; combination of those greater than 94% purity afforded 13.74 g of *(R)*-α-chloroisocaproic acid $[\alpha]_D^{23}$ + 28.2° (C 2.5, methanol).

Sodium (R,S)-α-Chloroisocaproate

(R,S)-α-Chloroisocaproic acid (12.35 g, 82 mM) was added dropwise to a well-stirred 2 M NaOH solution (41 ml) at 0–5°. The resulting product mixture was filtered and evaporated to a solid which was dissolved in warm methanol (50 ml) and filtered into 2-propanol (100 ml). The resulting solution was concentrated almost to dryness and the semisolid residue was triturated with further 2-propanol (100 ml). Filtration, washing the solids with 2-propanol and then ether, and drying afforded sodium *(R,S)*-α-chloroisocaproate (8.05 g). (GC on the regenerated acid indicates 99% purity.)

Sodium Dichloroacetate

This material is commercially available[19] or may be prepared[13,20] from the readily available dichloroacetic acid.

Sodium 2-Chloropropionate

Racemic material may be prepared[13,20] from commercially available 2-chloropropionic acid. If required, the *(R)*- and *(S)*-enantiomers of 2-chloropropionic acid can be prepared from *(R)*- and *(S)*-alanine, respectively, using the diazotization process described above.[13]

Specificity of α-Chloro Compounds as Inhibitors of BCKDH Kinase

Dichloroacetate is a potent inhibitor of PDH kinase and a weak inhibitor of BCKDH kinase.[5] The converse is true for α-chloroisocaproate, a

---

[18] W. C. Still, M. Kahn, and A. Mitra, *J. Org. Chem.* **43**, 2923 (1978).

[19] Suppliers: CTC Organics (Atlanta, Georgia) and American Tokyo Kasei, Inc. (Portland, Oregon).

[20] Using the above sodium α-chloroisocaproate procedure, a 1 : 1 acid : base stoichiometry and crystallization from 2-propanol (minimal volume)/ether.

potent inhibitor of BCKDH kinase and a weak inhibitor of PDH kinase.[5] Both the *(R)*- and *(S)*-isomers of 2-chloropropionate are potent inhibitors of PDH kinase but show no activity toward BCKDH kinase.[21] *(R,S)*-α-Chloroisocaproate has been the compound used routinely as an inhibitor of BCKDH kinase activity. Limited studies indicate that both the *(R)*(+)- and the *(S)*(−)- isomers of the α-chloroisocaproate are BCKDH kinase inhibitors. The *(R)*(+)- isomer, however, is about twice as effective as the *(S)*(−)- isomer in studies with purified BCKDH kinase[21] and isolated hepatocytes (described below). *(S)*(−)-2-Chloro-3-methyl pentanoate, an isoleucine derivative, shows little acitivity as a BCKDH kinase inhibitor. Likewise, (R)(−)-2-chloroisovalerate, a valine derivative, has virtually no activity. This agrees well with the order found for the effectiveness of α-keto acids as inhibitors of BCKDH kinase activity: α-ketoisocaproate > α-keto-β-methylvalerate >> α-ketoisovalerate.[3,9]

Phosphorylation of glycogen synthase by six different protein kinases was shown to be insensitive to α-chloroisocaproate inhibition,[5] establishing this compound as a selective inhibitor of mitochondrial versus cytosolic protein kinases.[5]

Depending on experimental conditions used to assay BCKDH kinase activity, concentrations of α-chloroisocaproate as low as $5-10 \mu M$ will cause 50% inhibition of BCKDH phosphorylation.[5] Potassium phosphate at concentrations greater than 10 mM increases α-chloroisocaproate sensitivity.[22] Inhibition of PDH kinase by α-chloroisocaproate has been observed to require much higher concentrations (50% inhibition at 675 μM).[5]

BCKDH kinase activity can be assayed by ^{32}P incorporation into the BCKDH E1α subunit or ATP-mediated loss of dehydrogenase activity. Assuming the compound being tested as a potential kinase inhibitor does not inhibit dehydrogenase activity directly, the latter assay is the method of choice. This is because BCKDH is subject to phosphorylation at two sites, and the two sites are not of equal importance in regulating enzyme activity.[23,24] Furthermore, inhibitors of BCKDH kinase affect phosphorylation of the two sites disproportionately.[24] For this reason, a better estimation of the potential physiological importance or usefulness of a compound is obtained by assaying protection of dehydrogenase activity against ATP-mediated inactivation. Assays for BCKDH activity are given elsewhere in

[21] R. Paxton and R. A. Harris, unpublished observations.
[22] Y. Shimomura, M. J. Kuntz, and R. A. Harris, *Fed. Proc., Fed. Am. Soc. Exp. Biol.* **46,** Abstr. 1394 (1987).
[23] K. G. Cook, A. P. Bradford, S. J. Yeaman, A. Aitken, I. M. Fearnley, and J. E. Walker, *Eur. J. Biochem.* **145,** 587 (1984).
[24] R. Paxton, M. J. Kuntz, and R. A. Harris, *Arch. Biochem. Biophys.* **244,** 187 (1986).

this volume.[25] First-order rate constants for ATP inactivation over an appropriate range of inhibitor concentrations, calculated by least-squares linear regression analysis, should be compared to assess the effectiveness of a given compound as an inhibitor of BCKDH kinase.

α-Chloroisocaproate Activation of the BCKDH Complex in Intact Cell and Tissue Preparations

Activation of the BCKDH complex by α-chloroisocaproate has been demonstrated with isolated rat hepatocytes,[26,27] perfused rat heart,[5] perfused rat hindquarter,[28] and cultured fibroblasts.[29] Work from this laboratory has been primarily with hepatocytes prepared from rats fed a low-protein diet.[26,27] The BCKDH complex of liver of chow-fed rats, 48-hr starved rats, and streptozotocin- or alloxan-diabetic rats is nearly 100% active (completely dephosphorylated).[30,31] Hepatocytes isolated from such animals retain this activity state and show, therefore, no response of BCKDH activity state to incubation with α-chloroisocaproate. Low-protein fed rats have proved most useful since the enzyme is 15–30% active in the liver of these animals, the activity state is retained during hepatocyte isolation, and a very dramatic response to α-chloroisocaproate is observed.[26,27]

Hepatocytes are prepared by the procedure of Berry and Friend[32] with modifications.[33,34] Rats weighing 175–250 g and fed a low-protein diet for at least 4 days are used. Semisynthetic diets containing 8% casein as the source of protein can be purchased from various suppliers.[27,31] The diet currently used in this laboratory[31] is isocaloric with respect to the AIN-76A

[25] P. A. Patston, J. Espinal, M. Beggs, and P. J. Randle, this volume [22]; D. J. Danner and S. C. Heffelfinger, this volume [38]; K. G. Cook and S. J. Yeaman, this volume [39]; F. H. Pettit and L. J. Reed, this volume [40]; R. Paxton, this volume [41].

[26] R. A. Harris, R. Paxton, and P. A. Jenkins, *Fed. Proc., Fed. Am. Soc. Exp. Biol.* **44**, 2463 (1985).

[27] R. A. Harris, R. Paxton, G. W. Goodwin, and S. M. Powell, *Biochem. J.* **234**, 285 (1986).

[28] E. J. Davis and S.-H. C. Lee, *Biochem. J.* **229**, 19 (1985).

[29] K. Toshima, Y. Kurod, I. Yokota, E. Naito, M. Ito, T. Watanabe, E. Takeda, and M. Miyao, *Clin. Chim. Acta* **147**, 103 (1985).

[30] S. E. Gillim, R. Paxton, G. A. Cook, and R. A. Harris, *Biochem. Biophys. Res. Commun.* **111**, 74 (1983).

[31] R. A. Harris, S. M. Powell, R. Paxton, S. E. Gillim, and H. Nagae, *Arch. Biochem. Biophys.* **243**, 542 (1985).

[32] M. N. Berry and D. S. Friend, *J. Cell Biol.* **43**, 506 (1969).

[33] H. A. Krebs, N. W. Cornell, P. Lund, and R. Hems *in* "Regulation of Hepatic Metabolism." (F. Lundquist and N. Tygstrup, eds.), Alfred Benzon Symp. 6, p.726. Academic Press, New York, 1974.

[34] R. A. Harris, *Arch. Biochem. Biophys.* **169**, 168 (1975).

diet recommended by the American Institute of Nutrition.[35] Greater than 90% of the isolated hepatocytes must exclude Trypan blue, while ATP content should be greater than 2 μmol/g wet weight. We experienced considerable difficulty in initial attempts to obtain viable and stable hepatocytes from rats fed low-protein diets. We feel (but have not proved) that a major factor was low hepatic glutathione levels as a consequence of the low-protein diet.[26] Supplementing the 8% casein diet with 0.3% methionine correlated with a marked improvement in quality and stability of hepatocyte preparations, perhaps reflecting increased resistance to oxidative stress. The glutathione level of hepatocytes from rats fed 8% protein diet without supplemental methionine was approximately 1.5 μmol/g wet weight. Supplementation of the diet with 0.3% methionine increased the levels to those of normal liver (about 5 μmol/g wet weight).[26] Other factors we consider to be important in preparing good hepatocytes for metabolite studies include (1) addition of 20 mM glucose to the perfusion medium to preserve endogenous glycogen stores[34]; (2) monitoring the pH of the perfusion medium to ensure adequate rates of oxygenation and carbonation[26]; (3) use of very clean glass- and plasticware at all steps of the preparation. A rinse with methanol helps to ensure complete removal of detergent[26]; and (4) addition of bovine serum albumin to the wash and suspension media for the cells, as suggested by Krebs et al.[33]

The effect of α-chloroisocaproate on flux through the BCKDH complex of hepatocytes from low-protein rats is shown in Fig. 1. For this experiment, hepatocytes (30–40 mg wet weight) were incubated in 2 ml of Krebs–Henseleit buffer supplemented with 2.5% (w/v) bovine serum albumin (bovine albumin powder CRG-7; Armour Pharmaceutical Co.; dialyzed for 3 days against three changes of Krebs–Henseleit buffer) under an atmosphere of 95% O_2 – 5% CO_2 (v/v) in 25-ml Erlenmeyer flasks sealed with rubber serum caps fitted with hanging center wells (Kontes Glass Co.). Incubation was at 37° in a shaking water bath. Hepatocytes were preincubated for 15 min prior to the initiation of BCKDH flux determinations with the addition of 0.2 mM α-keto[1-^{14}C]isovalerate (100 cpm/nmol). The reaction was terminated 15 min later by the addition of 0.3 ml of β-phenylethylamine : methanol (1 : 1, v/v) to the center well and 1.0 ml of 5 M H_2SO_4 to the reaction mixture. Incubations were continued for 1 hr at room temperature to collect $^{14}CO_2$ in the hanging center wells. The wells were removed and counted for radioactivity in a xylene-based scintillation fluid supplemented with 10% (v/v) β-phenylethylamine to prevent loss of

[35] American Institute of Nutrition ad Hoc Committee Reports on Standards for Nutritional Studies, J. Nutr. 107, 1340 (1977); J. Nutr. 110, 1726 (1980).

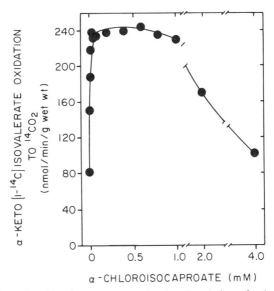

FIG. 1. Effect of α-chloroisocaproate on the decarboxylation of α-ketoisovalerate by hepatocytes isolated from a rat fed a low-protein diet. Adapted with permission from data published previously.[27]

$^{14}CO_2$ from the scintillation fluid. Half-maximal activation of flux through the BCKDH complex occurred at $10-20$ μM α-chloroisocaproate under the conditions of this experiment. As apparent in Fig. 1, very high concentrations of α-chloroisocaproate should not be used. α-Chloroisocaproate is a competitive inhibitor of BCKDH activity ($K_1 = 0.5$ mM)[5] and causes strong inhibition of BCKDH flux with isolated hepatocytes at concentrations greater than 1 mM.

That the increase in flux caused by α-chloroisocaproate corresponds to activation of the BCKDH complex can be demonstrated by direct assay of the complex in Triton X-100 extracts of hepatocytes. For this measurement hepatocytes are rapidly removed from the incubation medium by centrifugation (30 sec in an Eppendorf centrifuge). Pellets are quickly frozen in liquid nitrogen and stored at $-70°$ prior to assay. Extraction and assay of the complex is described elsewhere in this volume.[36] Preincubation with α-chloroisocaproate (100 μM) usually doubles the activity of the BCKDH complex. A lower basal activity state and correspondingly greater stimulatory effect of α-chloroisocaproate are achieved by preincubation of hepatocytes with DL-β-hydroxybutyrate (20 mM). The mechanism respon-

[36] See G. W. Goodwin, B. Zhang, R. Paxton, and R. A. Harris, this volume [23].

sible for inactivation of the complex by β-hydroxybutyrate has not been defined but may relate to the highly reduced NAD^+ redox state of the mitochondria caused by this substrate.

Site-Specific Inhibition of BCKDH Phosphorylation by α-Chloroisocaproate

Two serine residues separated by nine amino acids in the E1α subunit are subject to phosphorylation by BCKDH kinase.[23] The relative rates of phosphorylation of these two sites, designated site 1 and site 2, vary with the experimental conditions used to assay phosphorylation of the purified BCKDH complex. Similar rates of site 1 and site 2 labeling with $[\gamma\text{-}^{32}P]$ATP were reported by Paxton et al.,[24] whereas site 1 was labeled much faster than site 2 under the experimental conditions used by Cook et al.[23] Recent studies demonstrate that the ionic composition of the medium has dramatic effects on the relative rates of site 1 and site 2 phosphorylation.[37] Site 1 phosphorylation has been directly correlated to inactivation of the enzyme,[23,24] while the function of site 2 phosphorylation remains unclear. It has been shown in mitochondria from various rat tissues[38] and with adipocytes[39] that both sites become phosphorylated. Although site 1 appears to be most important for regulation of enzyme activity, α-chloroisocaproate is considerably more effective as an inhibitor of site 2 that site 1 phosphorylation,[24] and recent studies[37] with purified rat liver BCKDH have shown that the same is true for endogenous kinase inhibitors (e.g., α-ketoisocaproate). Characterization of the physiological significance of this preferential inhibition requires further study.

For the study of site-specific phosphorylation, purified rabbit liver BCKDH was phosphorylated and inactivated in the presence and absence of 1 mM α-chloroisocaproate at 37° for 240 min. Aliquots were precipitated with trichloroacetic acid, washed, digested with trypsin, and site-specific phosphate incorporation was determined by HPLC separation of ^{32}P-labeled peptides as described previously.[24] The rates of phosphorylation of both sites 1 and 2 were significantly decreased by the presence of α-chloroisocaproate; however, after 240 min the amount of phosphate incorporated into site 1 was almost the same as that seen in the absence of α-chloroisocaproate, and the enzyme was essentially inactive. In striking contrast, phosphorylation of site 2 was dramatically reduced, thereby pri-

[37] M. J. Kuntz, Y. Shimomura, and R. A. Harris, Fed. Proc., Fed. Am. Soc. Exp. Biol. **46,** Abstr. 1395 (1987).

[38] K. G. Cook, R. Lawson, and S. J. Yeaman, FEBS Lett. **164,** 85 (1983).

[39] S. M. A. Jones and S. J. Yeaman, Biochem. J. **236,** 209 (1986).

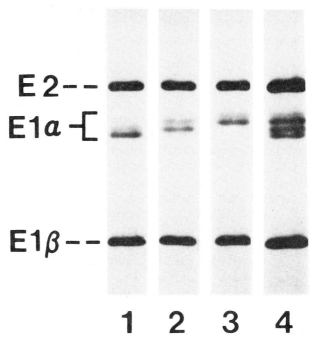

FIG. 2. Resolution of electrophoretically distinct phosphorylated forms of the $E1\alpha$ subunit of BCKDH. Lane 1, nonphosphorylated enzyme; lane 2, monophosphorylated enzyme generated by incubation with ATP (0.4 mM) and α-chloroisocaproate (1 mM) for 4 hr; lane 3, diphosphorylated enzyme generated by incubation with ATP (0.4 mM) only; lane 4, mixture of the forms of the enzyme shown in lanes 1–3. Proteins were visualized by silver stain.[41] Adapted with permission from data published previously.[40.]

marily generating a species (monophosphorylated enzyme) with phosphate incorporated only in site 1, along with a small amount of a diphosphorylated species. Recent studies in this laboratory[40] (Fig. 2) establish that the nonphosphorylated (lane 1), monophosphorylated (lane 2), and maximally (di-) phosphorylated (lane 3) species can be separated by 6–12% gradient polyacrylamide gel electrophoresis in the presence of sodium dodecyl sulfate.[42] This methodology, in combination with immunoblotting, provides a new approach for investigating the physiological importance of site 1 and site 2 phosphorylation of BCKDH in complex biological matrices (e.g., mitochondria, isolated cells, and *in vivo*).

[40] M. J. Kuntz, R. Paxton, Y. Shimomura, G. W. Goodwin, and R. A. Harris, *Biochem. Soc. Trans.* **14,** 1077 (1986).
[41] W. Wray, T. Boulikas, V. P. Wray, and R. Hancock, *Anal. Biochem.* **118,** 197 (1981).
[42] U. K. Laemmli, *Nature (London)* **227,** 680 (1970).

Acknowledgments

The work in this chapter was supported in part by U.S. Public Health Service Grant (DK 19259), the American Heart Association, Indiana Affiliate, and the Grace M. Showalter Residuary Trust.

Section II

Enzyme Assay

[16] Measurement of Relative Carbon Flux in α- and β-Keto Pathways of Leucine Metabolism

By J. MICHAEL POSTON

Leucine[1] may be metabolized by two routes. One, known as the α-keto pathway, involves transamination and subsequent decarboxylation of the resulting α-ketoisocaproic acid. The other, known as the β-keto pathway, may be primarily anabolic, but can also function in the catabolic degradation of leucine. This latter pathway involves a cobalamin-dependent migration of the amino group to the beta position, removal of the amino group, thioesterification with coenzyme A of the resulting β-ketoisocaproic acid, the thiolytic cleavage of the β-isocaproyl-S-coenzyme A. The β-keto pathway has been reviewed[2] and its first enzyme, leucine 2,3-aminomutase (EC 5.4.3.7), is discussed in this volume.[1]

Assay Method

Principle

The primary oxidation products of α-keto[1-^{14}C]isocaproic acid are [14]CO_2 and isovaleryl-S-coenzyme A. This step is unique to the α-keto pathway and is irreversible. By measuring the amount of [14]CO_2 and [1-^{14}C]acetate produced from L-[1-^{14}C]leucine, the relative flux through the two pathways may be estimated.[3]

Reagents

Potassium phosphate buffer, 0.02 M, pH 7.0
Triethanolamine-HCl buffer, 1.0 M, pH 8.5
Coenzyme A, 0.01 M
NAD^+, 0.01 M
Flavin adenine dinucleotide (FAD), 0.01 mM
Pyridoxal phosphate, 0.01 M
L-[1-^{14}C]leucine, 0.01 M, ~10[6] counts/μmol
KOH, 1 M
$HClO_4$, 20%
Scintillant

[1] See also D. J. Aberhart and J.-A. Cotting, this volume [3]; D. J. Aberhart, this volume, [4]; J. M. Poston, this volume [17].
[2] J. M. Poston, *Adv. Enzymol.* **58,** 173 (1986).
[3] J. M. Poston, *J. Biol. Chem.* **259,** 2059 (1984).

HCl, concentrated
HCl, 1 M
Dowex 50, hydrogen ion form
Dowex 2, hydroxyl ion form
NH_4OH, 2 M

Procedure

Crude slurries of rat tissues are prepared by disrupting the tissue in a volume of phosphate buffer equal to the weight of the tissue using a Potter-Elvehjem homogenizer equipped with a Teflon pestle.

The reactions, carried out in small, single-side-arm Warburg vessels, contain 0.5 ml tissue homogenate, 100 mM triethanolamine buffer, 0.5 mM coenzyme A, 0.5 mM NAD^+, 2.5 μM FAD, 0.5 mM pyridoxal phosphate, 1 mM L-[1-^{14}C]leucine; total volume is 1.0 ml. A fluted filter paper is placed in the center well of the vessel and moistened with 0.1 ml KOH and 0.25 ml $HClO_4$ is placed in the side arm. The vessel is stoppered with a tight-fitting serum cap and is incubated in a gently shaking water bath at 37°. The incubation is done in the dark in order to protect the light-sensitive cobalamin coenzyme of leucine 2,3-aminomutase. After 1 hr, the reaction is stopped by tipping the $HClO_4$ into the reaction mixture. The vessel is returned to the incubator for an additional 90 min in order to permit any CO_2 to be evolved from the acidified mixture and to be trapped on the basic filter paper.

Measurement of CO_2

The filter paper is removed from the center well and placed in a scintillation vial. The well is rinsed with 0.1 ml water and the rinse is added to the vial, which is then counted in a liquid scintillation system using Aquasol or similar scintillant.

Measurement of Anions and Cations in the Reaction Mixture

The acidified contents of the Warburg vessel are transferred to a graduated centrifuge tube. The vessel is rinsed carefully, and the rinse is added to the centrifuge tube. The pooled material is then made basic to phenolphthalein so as to hydrolyze any labeled CoA esters. (One hr is probably more than adequate to ensure this hydrolysis.) The material is then reacidified with HCl and centrifuged. An aliquot of the clear, protein-free supernatant solution is passed through a small (7 × 45 mm) column of Dowex- 50 in the hydrogen ion form to remove protonated leucine and other cations. The column is washed with water and the pooled wash and

effluent is made just basic to phenolphthalein (faintly pink) and passed through similar columns of Dowex- 2 in the hydroxyl ion form to absorb anions. The column is washed with water and the anions are eluted with 1 M HCl. The acid eluate is passed through a smaller (7 × 20 mm) column of Dowex- 50 to remove any unreacted leucine that may have escaped the first column. An aliquot of the anion-containing effluent is counted directly and a second aliquot is placed in a scintillation vial and dried under a stream of nitrogen. Approximately 0.1 ml water and 0.1 ml glacial acetic acid are added to the vial and again taken to dryness. The residue is taken up in 0.1 ml water, scintillant is added, and the vial is counted. The difference in the counts in the two aliquots represents the volatile non-CO_2 anionic fraction.

The labeled material which is absorbed on the Dowex- 50 columns represents, primarily, unreacted L-[1-^{14}C]leucine. It may be eluted with 2 M NH$_4$OH and counted. The labeled material which sticks to neither Dowex- 50 nor Dowex- 2 (the pass-through and wash from the Dowex- 2 column in the scheme above) represents neutral, uncharged material.

Separation and Estimation of the Anions in the Anionic Fraction

In general, measuring the label in the CO_2 and in the volatile non-CO_2 anionic fraction will give a reasonable estimate of the relative flux of carbon through the two pathways. However, since α-ketoisocaproic acid is volatile under the conditions described here, the flux through the β-keto pathway may be overestimated. Partition chromatography on a column of diatomaceous earth impregnated with 0.2 N H$_2$SO$_4$ and elution with hexane containing gradually increasing amounts of 1-butanol and chloro-

Table I

FLUX AND DISTRIBUTION OF RADIOACTIVITY
FROM THE METABOLISM OF L-[1-^{14}C]LEUCINE BY
RAT TISSUES[a]

Tissue	Flux through the β-keto pathway as percentage of total flux
Liver	0.40
Kidney	0.41
Heart	4.83
Brain	1.38
Testis	40.20

[a] From Poston.[3]

form[4,5] may be used to separate the various components of the anionic fraction.

Table I shows typical results of the measurement of flux in the catabolic direction in various tissues from the rat. The percentage of flux through the β-keto pathway is low in most tissues. The testis, however, seems to provide an exception to this pattern; the percentage of anions was isocaproic acid 1.6, α-ketoisocaproic acid 16.6, β-hydroxyisocaproic acid 72.5, and the acetic acid 9.2. This distribution in the anions reveals that some of these are from the α-keto pathway, α-ketoisocaproic acid especially. Isocaproic acid may result from the reduction of either the amino acid itself, or of any of its deaminated products. β-Hydroxyisocaproic acid is undoubtedly derived by reduction of the β-ketoisocaproic acid. Correcting for the distribution of anions, the flux through the β-keto pathway for the testis is 33.5%.

[4] J. E. Verner, this series, Vol. 30, p. 397.
[5] K. Kuratomi and E. R. Stadtman, J. Biol. Chem. 241, 4217 (1966).

[17] Assay of Leucine 2,3-Aminomutase

By J. MICHAEL POSTON

$$(CH_3)_2\ CH\text{-}CH_2\text{-}CH(NH_2)CO_2H \underset{}{\overset{AdoCbl}{\rightleftharpoons}} (CH_3)_2\ CH\ CH\ (NH_2)CH_2CO_2H$$

Assay Method

Principle

When incubated with cell-free extracts, β-leucine is converted to leucine,[1] which is then measured by amino acid analysis.[2]

Reagents

Potassium phosphate buffer, 0.02 M, pH 7.0
Triethanolamine-HCl buffer, 1 M, pH 8.5
Flavin adenine dinucleotide (FAD), 10 μM
NAD$^+$, 0.01 M
Pyridoxal phosphate, 0.01 M
β-Leucine, 0.1 M

[1] See also D. J. Aberhart and J.-A. Cotting, this volume [3]; D. J. Aberhart, this volume [4]; J. M. Poston, this volume [16].
[2] J. M. Poston, Adv. Enzymol. 58, 174 (1986).

Adenosylcobalamin (AdoCbl), 1 μM
Na_2WO_4, 10% (w/v)
HCl, 6 M

Procedure

Incubation mixtures contain in 1.0 ml: 100 mM triethanolamine buffer, 0.5 μM FAD, 0.5 mM coenzyme A, 0.5 mM NAD$^+$, 0.5 mM pyridoxal phosphate, 20 mM μ-leucine, 5 $\times$ 10^{-8} M AdoCbl, and 1 to 25 mg protein. Addition of AdoCbl is made in dim light and the incubation is carried out in the dark. Reactions are started by the addition of enzyme. After 60 min at 37°, the reactions are stopped by the addition of 0.05 ml Na_2WO_4 and 0.05 ml HCl and centrifuged. Aliquots of the protein-free supernatant solution are taken for analysis by gas chromatography or by other methods.

Synthesis of DL-β-Leucine

4-Methyl-2-pentenoic acid (20 ml) is placed in a 125-ml Kimler bottle to which is added 100 ml concentrated (28%) NH$_4$OH. The bottle is sealed with its screw cap and placed in an autoclave at 121°. After approximately 60 hr, the contents of the bottle are washed into a flask with methanol and the solvent and ammonia are removed on a rotary evaporator, yielding a thick, colorless oil. Addition of a few drops of acetone brings about copious precipitation. The precipitate can be recrystallized from methanol–ethyl ether solution. The yield is about 5°g DL-β-leucine, melting at 197°. Aberhart[1] cautions that explosions have occurred during heating with the ammonia and recommends conducting the heating in a Parr pressure reactor.

Synthesis of 4-Methyl-2-pentenoic Acid

Malonic acid (100 g, 1.055 mol) is dissolved in 300 ml of freshly opened pyridine containing 1.5 ml diethylamine. The mixture is cooled in an ice bath and 75 g (1.04 mol) of isobutyraldehyde is slowly added. After the addition is complete, the mixture is stirred magnetically at room temperature for approximately 20 hr. The mixture is then poured into an equal volume of cold water, the oily upper layer is removed, and the aqueous layer is shaken with approximately 100 ml of benzene. The benzene is added to the oily upper layer and the solution is washed first with 250 ml of 2 M HCl to remove pyridine and then with water. Solid MgSO$_4$ is added to dry the solution and the solvent is removed on a rotary evaporator. The yield of clear, viscous oil is about 35 g.[3]

[3] J. M. Poston, *J. Biol. Chem.* **251**, 1859 (1976).

Gas Chromatographic Analysis

Reagents

N_2 or other inert gas for drying sample
Bis(trimethylsilyl)trifluroracetamide (BSTFA)
Acetonitrile (CH_3CN), water free
Dichloromethane, (CH_2Cl_2)
Norleucine, 0.01 M
GLC column, 10% OV-11 on Supelcoport, 2 × 180 mm

Separation of trimethylsilyl derivatives of the branched-chain amino acids using the technique of Gehrke and Leimer[4] involves drying the sample, derivatization with BSTFA, and gas chromatographic separation. Samples (0.05-ml aliquots are convenient) are placed in heavy glass vials with conical interiors (Reactivials, Pierce Chemical Co., or similar) and norleucine is added as an internal standard (50 to 500 μmol depending on the anticipated concentration of leucine). The vial is placed on a warm plate (~ 50°) and exposed to a gentle jet of N_2. When the sample is dry, approximately 0.5 ml CH_2Cl_2 is added and the vial is taken to dryness again. The addition of CH_2Cl_2 ensures the removal of traces of water which may interfere with the derivatization. BSTFA, 0.10 ml, is added to the vial, followed by 0.10 ml CH_3CN and the vial is stoppered with a Teflon-faced silicon rubber septum. The sealed vial is placed in an oil bath at 150° for 15 min. The oil must cover only about half the vial since refluxing seems to be important in the derivatization. The vial is cooled and an aliquot of the derivatized amino acid solution in injected on the GLC column equipped with a flame ionization detector. Initial column temperature is 110° and, with a flow rate of 30 ml/min and a temperature increase of 5°/min, the branched-chain amino acids are eluted after about 7 to 13 min. The order of elution is valine, leucine, isoleucine, norleucine, β-leucine. There is no separation of DL-pairs nor do the amino acids yield more than one peak.

Amino Acid Analysis

Aliquots of the protein-free reaction mixture may be diluted with appropriate buffers and injected on an ion-exchange amino acid analyzer. However, if measurement of β-leucine is desired, it is important that the system be equipped with a fluorescence detector and utilize either *o*-phthalaldehyde or fluorescamine in the postcolumn derivatization. These reagents work very well with both the usual α-amino acids and β-leucine. Ninhydrin does not give color with β-leucine in liquid reactions, although

[4] C. W. Gehrke, and K. Leimer, *J. Chromatogr.* **57**, 219 (1971).

it does yield some on thin-layer or paper chromatograms. HPLC systems also may be used.[3,5] Both phenyl isothiocyanate and *o*-phthalaldehyde precolumn derivatization have been shown to work with all the amino acids concerned here, including norleucine and β-leucine.

Preparation of the Enzyme

The enzyme activity has been found in a wide variety of sources, but it has not been appreciably purified from any of them. Typically, extracts of tissues are made by homogenizing the tissue in a volume of buffer equal to its weight. Whereas the pH optimum for the enzyme is about 8.5, the extracts are usually made in 0.02 *M* potassium phosphate buffer, pH 7.0, because they seem to be more stable. Extracts of mammalian tissues are quite stable to freezing and thawing and may be stored at $-20°$ for several weeks with minimal loss of activity. This is true, as well, for plant extracts. On the other hand, extracts of clostridia and of yeast are less stable, perhaps owing to contaminating proteolytic activities.

Active enzymes have been prepared using simple potassium phosphate buffer, pH 7.0, or from buffers supplemented with 2-mercaptoethanol or dithiothreitol. No appreciable advantage seems to be gained by using such protective agents, nor does the addition of glycerol to the buffers offer increased stability to the activity. Assays conducted under inert atmospheres of nitrogen or argon are not different from those conducted in air.

Preparation from Hair Roots

Active enzyme may be extracted from hair roots by suspending about 20 roots containing both bulb and sheath in about 200 μl buffer and subjecting the suspension to several cycles of freezing at $-20°$ and thawing at room temperature. No further disruption of the tissue is required.[6] Activity has been extracted from freshly plucked hairs as well as from those allowed to dry for several days, taped to a card. Thus, it would be possible to sample a population at a distance by mail.

Cofactor Requirements

There is an absolute requirement for AdoCbl. This may be demonstrated by the addition of purified intrinsic factor to the enzyme. Intrinsic factor, a glycoprotein produced by mammalian gastric mucosa, binds cobalamin in food and it is only as the tightly bound complex that the

[5] D. J. Aberhart and J.-A. Cotting, this volume [3].
[6] J. M. Poston, *J. Biol. Chem.* **255**, 10067 (1980).

cobalamin is able to be absorbed by the ileum. When added to extracts, intrinsic factor combines with the cobalamin on the enzyme and renders it inactive. Addition of exogenous AdoCbl restores the activity of the enzyme. From these observations, it may be concluded that the binding of AdoCbl to the leucine 2,3-aminomutase (EC 5.4.3.7) is less tight than that of AboCbl to intrinsic factor.

One exception to this requirement appears to be the enzyme extracted from the plant *Andrographis planiculata*. No evidence for any cobalamin involvement has been observed for the enzyme from this source.[7]

In addition to the cobalamin, supplementation of the reaction mixtures with NAD^+, coenzyme A, and FAD seems to stimulate the reaction, sometimes quite profoundly. In no instance has it been possible to demonstrate an absolute dependence on these cofactors and it is probable that they act by providing appropriate conditions which favor the aminomutase reaction. Explanation of these cofactor effects awaits purification of the aminomutase.

Pyridoxal phosphate also seems to provide some slight stimulation and it has been routinely added to reaction mixtures, although evidence for its participation is minimal. It has been suggested that ω-aminomutases, however, might involve pyridoxal phosphate as a cofactor in addition to cobalamin.[8] These enzymes act on lysine and ornithine to cause the ω-amino group to shift to the penultimate carbons. *S*-Adenosylmethionine seems to be without effect on the leucine 2,3-aminomutase.

Occurrence and Distribution

Leucine 2,3-aminomutase has been detected in a wide variety of organisms: mammalian tissues; plants such as beans, potatoes, annual ryegrass, and spinach; and microorganisms such as the yeast, *Candida utilis,* and the bacteria *Clostridium lentoputrescens* and *Clostridium sporogenes.*

In humans, it is found in the white blood cells and the liver, but not in the erythrocytes. Rats have been shown to have the enzyme in all tissues examined (liver, heart, kidney, brain, and testis) and it is apparently cytosolic. Certainly, the mitochondria have only a minimal percentage of the total activity, but it is not certain what percentage might be associated with a small membrane fraction that could remain in suspension after 2 hr of centrifugation at ~50,000 g.[9]

[7] I. Freer, G. Pedrocchi-Fantoni, D. J. Picken, and K. H. Overton, *J. Chem. Soc. Chem. Commun.* **1981,** 80 (1980).

[8] J. J. Baker and T. C. Stadtman, *in* "Biochemistry and Medicine" (D. Dolphin, ed.), Vol. 2, p. 203. Wiley, New York, 1982.

[9] J. M. Poston, *Biochem. Biophys. Res. Commun.* **96,** 838 (1980).

Clinical Associations

Leucine 2,3-aminomutase has been observed to be diminished in activity in conditions where there is a deficiency of cobalamin in the body. In ailments caused by either nutritional lack of the vitamin or by inability to absorb cobalamin (due, for instance, to pernicious anemia and the lack of intrinsic factor or to gastric resection and its resulting deficiency of intrinsic factor), there are elevated circulating levels of β-leucine in the serum which are restored to normal upon repletion of the body's cobalamin stores. This change in β-leucine is mirrored by lowered levels of circulating leucine in deficient states and normal levels upon supplementation with cobalamin. No other condition has yet been identified which can be associated with an abnormal activity of the enzyme.[6]

[18] Enzyme Assays with Mutant Cell Lines of Maple Syrup Urine Disease

By DAVID T. CHUANG and RODY P. COX

Maple syrup urine disease (MSUD) is an autosomal recessive inborn error of metabolism in which the primary defect is the oxidative decarboxylation of branched-chain α-keto acids (BCKA).[1] The latter consist of α-ketoisovalerate (KIV), α-ketoisocaproate (KIC), and α-keto-β-methylvalerate (KMV) that are derived by transamination from the branched-chain amino acids (BCAA), valine, leucine, and isoleucine, respectively. MSUD has been classified into classical, intermittent, intermediate, and thiamin-responsive types as well as the type associated with dihydrolipoyl dehydrogenase (EC 1.6.4.3) deficiency.[1,2] The classification is based on rapidity of onset, severity of the disease, tolerance for dietary protein, response to thiamin supplements, and enzyme analysis data.

The enzyme deficient in MSUD, the BCKA dehydrogenase complex, consists of three catalytic components, i.e., BCKA decarboxylase (EC 1.2.4.4) or E1, dihydrolipoyl acyltransferase or E2, and dihydrolipoyl dehydrogenase or E3.[3,4] The E3 component is common among mammalian

[1] K. Tanaka and L. E. Rosenberg, in "The Metabolic Basis of Inherited Disease" (J. B. Stanbury, J. B. Wyngaarden, D. S. Fredrickson, J. L. Goldstein, and M. S. Brown, eds.), p. 440. MCGraw-Hill, New York, 1983.
[2] J. Dancis, J. Hutzler, S. E. Snyderman, and R. P. Cox, J. Pediatr. 81, 312 (1972).
[3] S. C. Heffelfinger, E. T. Sewell, and D. J. Danner, Biochemistry 22, 5519 (1983).
[4] F. H. Pettit, S. J. Yeaman, and L. J. Reed, Proc. Natl. Acad. Sci. U.S.A. 75, 4881 (1978).

pyruvate, α-ketoglutarate, and BCKA dehydrogenase complexes. The BCKA complex also contains two regulatory enzymes, i.e., a specific kinase and a specific phosphatase that modulate activity of the complex by a phosphorylation–dephosphorylation cycle.[5]

The BCKA dehydrogenase complex is expressed in peripheral leukocyte preparation[6] and skin fibroblasts.[7] Thus cultured diploid fibroblasts derived from patients with various forms of MSUD have provided a useful system for studying enzyme deficiency in this disease.[2,8-11] In this chapter, enzyme assays for activities of the BCKA dehydrogenase complex and its components in normal and MSUD cell culture are described. These methods may be useful for elucidating biochemical mechanisms of MSUD at the enzymatic level.

Assays with Intact Cells

Principle

The method measures the flux of BCKA and BCAA through the BCKA dehydrogenase complex in intact fibroblasts. Harvested cells are incubated with $[1-^{14}C]KIV$ in a balanced salt solution. The $^{14}CO_2$ evolved is trapped into 2 N NaOH and the latter counted for radioactivity. In MSUD cells, the rate of decarboxylation of $1-^{14}C$-labeled α-keto acids is greatly reduced compared with that in normal cells, as a result of the metabolic block in the mutants at the BCKA dehydrogenase complex step. This property in mutant cells provides a basis for diagnosis of MSUD using the intact-cell assay.

Procedure

Cell Culture. All steps are carried out under sterile conditions. Fibroblasts are grown from explants of skin biopsies[12] derived from normal individuals and MSUD patients. Explants are immobilized with sterile silicon grease inside a T-25 plastic culture flask (Corning), and overlaid

[5] P. J. Randle, H. R. Fatania, and K. S. Lau, *Mol. Aspects Cell. Regul.* **3**, 1 (1984).

[6] J. Dancis, J. Hutzler, and M. Levitz, *Biochim. Biophys. Acta* **77**, 523 (1963).

[7] J. Dancis, J. Hutzler, and R. P. Cox, *Biochem. Med.* **2**, 407 (1969).

[8] L. B. Lyons, R. P. Cox, and J. Dancis *Nature (London)* **243**, 533 (1973).

[9] J. Frézal, O. Amédéé-Manesme, G. Mitchell, S. Heuertz, F. Rey, and J. M. Saudubray, *Hum. Genet.* **71**, 89 (1985).

[10] I. Yoshida, L. Sweetman, and W. L. Nyhan, *Pediatr. Res.* **20**, 169 (1986).

[11] D. J. Danner, N. Armstrong, S. C. Heffelfinger, E. T. Sewell, J. H. Priest, and L. J. Elsas, *J. Clin. Invest.* **75**, 858 (1985).

[12] W. W. Sly and J. Grubb, this series, Vol. 58, p. 444.

with coverslips. To the flask is added 2 ml of Waymouth medium[13] (GIBCO) containing 20% fetal calf serum (GIBCO) and antibiotics (50 units/ml of penicillin, 50 μg/ml of streptomycin, and 30 μ/ml of kanamycin). The flask is gassed with CO_2, tightly capped, and incubated at 37°. The medium is changed twice a week. Within 7–11 days the cells that have migrated out of the explant are dispersed or passaged by trypsinization. The growth medium is removed with a pipet from the T-25 flask, and the attached cells rinsed once with 4 ml of Puck's saline A without glucose (w/v)[14] (0.8% NaCl, 0.04% KCl, 0.075% NaHCO₃, and phenol red). The rinsed monolayer culture is incubated at 37° for 3 to 5 min in 2 ml of the Versene–trypsin buffer [Puck's saline A containing (w/v) 0.02% ethylenediaminetetraacetic acid (EDTA) and 0.04% trypsin]. At the completion of incubation, cells are detached by shaking the culture flask gently. Cell clumps are broken up by pipeting the suspension up and down several times. The dispersed cell suspension in 1.5 ml of the Versene-trypsin buffer is transferred to a new T-25 flask, to which 4 ml of Waymouth medium with 15% fetal calf serum is added. Cells are grown to confluence with medium changed every week. The passages are repeated until adequate numbers of cells are available from the original explant.

Fibroblasts for the intact-cell assay are transferred to a T-75 flask containing 10 ml of Waymouth medium with 10% fetal calf serum as described above. The medium is changed every 4 to 7 days depending on the rate of fibroblast multiplication. Confluent cultures are passaged as described above using 4 ml of the Versene–trypsin buffer to remove cells from the plastic surface. The dispersed cell suspension (4 ml) is divided by a 1 to 2 split. Each 2-ml portion is transferred to a T-75 culture flask to which 10 ml of Waymouth medium with 10% fetal calf serum is added. The date and number of passage are recorded. Both normal and MSUD cells are passaged at 1 to 2 split and reach confluence in approximately 1 week.

To preserve cells for storage, the medium is removed from confluent cells in a T-75 flask, and cultures are rinsed with 5 ml of Puck's saline A without glucose. Fibroblasts are harvested by trypsinization in 4.5 ml of Versene–trypsin buffer as described. Trypsinization is terminated by adding 0.5 ml of fetal calf serum to the cell suspension. Cells are sedimented by centrifuging at 800–1000 rpm at 25° for 5 min in a Sorvall GLC-2 centrifuge. The cell pellet is suspended in 1 ml of Waymouth medium containing 10% dimethyl sulfoxide and 20% fetal calf serum. The cell suspension is transferred to a 1-ml cryotube (Vangard) and frozen in a

[13] C. Waymouth, *J. Natl. Cancer Inst.* **22**, 1003 (1959).
[14] P. I. Marcus, S. J. Cieciura, and T. T. Puck, *J. Exp. Med.* **104**, 615 (1956).

controlled-rate freezer (Union Carbide, Type BF-5) in liquid nitrogen vapor phase. The frozen ampules are stored in racks in a liquid nitrogen freezer. To recover cells from storage, the frozen ampule is rapidly warmed to 37° in a water bath, immediately upon removal from the liquid nitrogen freezer. The thawed cell suspension is placed in a T-75 flask, to which is added 10 ml of Waymouth medium with 15% fetal calf serum. The remaining steps for cell culture are as described above.

Intact-cell Assay. The method is a modification of that described by Dancis *et al.*[15] To measure rates of decarboxylation with [1-^{14}C]BCKA, confluent normal and MSUD cells are harvested by trypsinization. Cell suspensions from eight T-75 flasks are combined. An aliquot of the cell suspension is counted in a hemocytometer.[16] Approximately 3 to 5 × 10^6 cells are harvested per flask. Cells are collected by low-speed centrifugation. Cell pellets are combined and washed once with 10 ml of Krebs–Ringer phosphate buffer (w/v)[17] (0.018 M sodium phosphate, pH 7.38; 0.68% NaCl, 0.045% KCl, 0.03% MgSO$_4$, and 0.02% CaCl$_2$) containing 10% fetal calf serum, and once with Krebs buffer alone. The washed cells are suspended in Krebs buffer at a density of 1 × 10^6 cells/0.05 ml for assays.

The intact-cell assay is carried out in a Kontes pear-shaped flask. The reaction mixture contains 0.25 ml of Krebs buffer, 0.005 ml of thiamin hydrochloride (100 mg/ml), 1 × 10^6 cells (equivalent to 0.3 mg of protein) in 0.05 ml Krebs buffer and H$_2$O in a final volume of 0.35 ml. The reaction flask is capped with a rubber serum stopper that holds a plastic center well (Kontes) containing 0.1 ml of 2 N NaOH. The reaction is initiated by an injection of 0.02 ml of 37 mM [1-^{14}C]KIV with a specific radioactivity of 172 cpm/nmol. The reaction mixture is incubated at 37° in a shaking water bath for 80 min. The ^{14}CO$_2$ evolved is trapped in the NaOH solution. The reaction is terminated by injection of 0.05 ml of 15% trichloroacetic acid into the mixture. The incubation is continued for an additional 60 min, and the NaOH solution then transferred into 10 ml of a scintillation cocktail to count the radioactivity. The same incubation medium containing 0.3 mg of bovine serum albumin in place of fibroblasts serves as a blank. The blank values range from 120 to 150 cpm with the substrate [1-^{14}C]KIV used. The detailed procedure and the preparation of 1-^{14}C-labeled α-keto acid are as described for the E1 assay of BCKA dehydrogenase complex (see chapter [19], this volume). To serve as a control for cell viability, 0.5 mM [1-^{14}C]pyruvate (specific radioactivity 1,275 cpm/nmol) is incubated with 1 × 10^6 MSUD or normal cells for 10 min at 37°, and ^{14}CO$_2$ evolved determined as described above.

[15] J. Dancis, J. Hutzler, and R. P. Cox, *Am. J. Hum. Genet.* **29**, 272 (1977).
[16] M. K. Patterson, this series, Vol. 58, p. 142.
[17] H. A. Krebs, *Z. Physiol. Chem.* **217**, 191 (1933).

Definition of the Rate of Decarboxylation. The 1-^{14}C-labeled atom of KIV decarboxylated by intact fibroblasts is expressed as nanomoles of CO_2 evolved/minute/milligram of protein. To calculate the amount of CO_2 released, specific radioactivity of the prepared substrate [1-^{14}C]KIV is used without correction for the pool size of the α-keto acid in cells. Protein is determined by the method of Lowry *et al.*[18] using bovine serum albumin as standard.

Remarks. The intact whole-cell assay permits an estimate of the rate of decarboxylation of BCKA in cultured fibroblasts under defined assay conditions. The assay also can be readily applied to measure the oxidation of BCKA or BCAA in leukocytes[6] or cultured amnion cells.[19] Cell suspensions offer a reasonable control over the number of cells or amount of protein used for each assay, thereby yielding relatively reproducible activity for each cell strain. The close correlations between the patient's phenotype and tolerance for dietary protein with the degree of enzyme deficiency determined by the intact-cell assay[2,20] suggest that this method approximates the situation *in vivo.* Thus for diagnostic purpose, the use of intact cells is preferred over the disrupted-cell system[1] (see the following section).

The choice of [1-^{14}C]BCKA over [1-^{14}C]BCAA as substrate in the intact-cell assay avoids the interposed step of transamination. This allows a more direct analysis of enzyme deficiency in the BCKA dehydrogenase complex with MSUD cells. KIV is the preferred substrate because it is decarboxylated at the highest rate among the BCKAs tested. For routine assays, a relatively high concentration (2 mM) of [1-^{14}C]KIV is used. This is based on the observation that the concentration for half-maximal decarboxylation ($K_{0.5}$ value) is 0.3 mM KIV when assayed with intact cells.[21] At 2 mM concentration, normal fibroblasts decarboxylate [1-^{14}C]KIV at a rate of 0.2 to 0.3 nmol/min/mg of protein.[21] Classical MSUD fibroblasts exhibit no decarboxylating activity.[21] The complete failure to decarboxylate KIV may, in part, be due to the isotopic dilution of the substrate by a high concentration of the α-keto acid accumulated in classical MSUD cells. Residual activities (0.5 to 36% of the normal) are observed with intact cells from patients with variant forms of MSUD.[22,23] The sensitivity of the intact-cell assay can be enhanced by increasing specific radioactivity of the substrate [1-^{14}C]KIV; however, the substrate concentration should be con-

---

[18] O. H. Lowry, N. J. Rosebrough, A. L. Farr, and R. J. Randall, *J. Biol. Chem.* **193**, 265 (1951).

[19] R. P. Cox, J. Hutzler, and J. Dancis, *Lancet* **2**, 212 (1978).

[20] U. Wendel, H. Wentrup, and H. W. Rüdiger, *Pediatr. Res.* **9**, 709 (1975).

[21] D. T. Chuang, W.-L. Niu, and R. P. Cox, *Biochem. J.* **200**, 59 (1981).

[22] D. T. Chuang, L. S. Ku, and R. P. Cox, *Proc. Natl. Acad. Sci. U.S.A.* **79**, 3300 (1982).

[23] D. T. Chuang, unpublished observations.

comitantly reduced to maintain low back grounds for the assay. Thus the intact-cell assay can be carried out by incubating 2×10^5 cells with 0.2 mM [1-^{14}C[KIV with specific radioactivity of 1,720 cpm/nmol.[23]

Other Methods

A similar assay in which fibroblast suspensions are incubated with [1-^{14}C]BCAA was developed by Elsas et al.[24] Microassays using monolayer cultures grown in microtiter plates instead of cell suspensions were described by Wendel et al.[25] and Fensom et al.[26] In the latter two methods, the cell material per assay is greatly reduced to 5,000 to 10,000 cells by using microtiter plates with [1-^{14}C]BCKA or [1-^{14}C]BCAA as substrate.

Assay for the BCKA Dehydrogenase Complex

$$\text{RCO-}^{14}\text{COOH} + \text{CoASH} + \text{NAD}^+ \xrightarrow{\text{TPP, Mg}^{2+}} \text{RCO-SCoA} + \text{NADH} + \text{H}^+ + {}^{14}\text{CO}_2{\uparrow} \quad (1)$$

Principle

The radiochemical assay is based on the determination of $^{14}\text{CO}_2$ evolved from [1-^{14}C]BCKA through the overall reaction [Eq. (1)] catalyzed by the BCKA dehydrogenase complex. To carry out the assay, harvested fibroblasts or amnion cells are disrupted by freezing and thawing. The activity of the enzyme complex in the disrupted-cell suspension is assayed with [1-^{14}C] KIV in the presence of other substrates and cofactors. The $^{14}\text{CO}_2$ evolved is trapped in 2 N NaOH. The radiochemical assay can be employed to measure the actual as well as the total activity of BCKA dehydrogenase complex in normal and MSUD cells.

Procedure

Actual Activity. Cultured fibroblasts or amnion cells are harvested and washed as described in the intact-cell assay. Washed cells are suspended in Krebs–Ringer buffer[17] at a density of 1×10^6 cells/0.05 ml. The cell suspension is rapidly frozen in a Revco freezer ($-75°$) and kept for 15 hr or longer and then thawed for assays. Cells thus treated are fully disrupted as observed under a phase-contrast microscope. The activity of BCKA dehy-

[24] L. J. Elsas, A. B. Pask, F. B. Wheeler, D. P. Perl, and S. Trusler, *Metabolism* **21,** 929 (1972).
[25] U. Wendel, W. Wohler, H. W. Goedde, U. Langenbeck, E. Passarge, and H. W. Rudiger, *Clin. Chim. Acta* **45,** 433 (1973).
[26] A. H. Fensom, P. F. Benson, and J. E. Baker, *Clin. Chim. Acta* **87,** 169 (1978).

drogenase complex in frozen cells is stable for at least 3 years at $-75°$.[23] Repeated freezing and thawing results in reduced enzyme activity.

The assay is carried out in Kontes pear-shaped reaction flasks. The reaction mixture is a modification of that described by Danner and Elsas.[27] It contains in a final volume of 0.37 ml: 50 mM Tris-HCl (pH 7.5), 0.2 mM EDTA, 0.35 mM MgCl$_2$, 0.2 mM CoA, 0.2 mM NAD$^+$, 0.2 mM-thiamin pyrophosphate (TPP), 1.4% fetal calf serum, 1 mM [1-^{14}C]KIV and the freeze/thaw disrupted cell suspension (1 × 10^6 cells equivalent to 0.3 mg of protein) in 0.05 ml of Krebs buffer. The reaction flask is capped with a rubber serum stopper. The reaction is started by an injection of 0.01 ml of 37 mM [1-^{14}C]KIV in 15 mM HCl (specific radioactivity, 172 cpm/nmol) to the assay mixture without the keto acid. The reaction mixture is incubated at 37° in a gyrotory shaking water bath (New Brunswick) for 80 min. The ^{14}CO$_2$ evolved is trapped in 0.1 ml of 2 N NaOH contained in a center well. The termination of reaction and counting of radioactivity are as described in the E1 assay of the BCKA dehydrogenase complex (see chapter [19], this volume). The blank value with Krebs buffer in place of cells is ~80–90 cpm. In a parallel experiment, 0.5 mM of α-keto[1-^{14}C]glutarate (specific radioactivity, 1100 cpm/nmol) substitutes for [1-^{14}C]KIV in the reaction mixture in order to measure the activity of α-ketoglutarate dehydrogenase complex. The incubation is at 37° for 10 min. The results serve as a positive control for viability of MSUD mutant cells.

Total Activity. As described above, the BCKA dehydrogenase complex is regulated by a phosphorylation–dephosphorylation cycle,[5] with phosphorylation resulting in inhibition and dephosphorylation in stimulation. In cultured cells, the BCKA dehydrogenase complex may not be in the entirely active state or the fully dephosphorylated form. Thus to measure the total activity, it is essential to convert the enzyme complex into the fully active state by dephosphorylation. This can be accomplished by incubating the fibroblast suspension with a kinase inhibitor, α-chloroisocaproate,[28] prior to enzyme assays. The preincubation results in dephosphorylation of the enzyme complex by the intrinsic phosphatase while the kinase reaction is blocked. A method using α-chloroisocaproate to activate the BCKA dehydrogenase complex in fibroblasts has recently been documented by Toshima et al.[29] The procedure described here differs from the previous method primarily in that harvested fibroblasts are incubated with α-chloroisocaproate (1 mM) in Waymouth medium rather than Krebs

[27] D. J. Danner and L. J. Elsas II, *Biochem. Med.* **13**, 7 (1975).
[28] R. A. Harris, R. Paxton, and A. A. DePaoli-Roach, *J. Biol. Chem.* **257**, 13915 (1982).
[29] K. Toshima, Y. Kuroda, I. Yokota, E. Naito, M. Ito, T. Watanabe, E. Takeda, and M. Miyao, *Clin. Chim. Acta* **147**, 103 (1985).

buffer. In culture medium, the activation of the enzyme complex reaches a plateau in 15 min without a rapid leveling off of the increased enzyme activity.

The steps for enzyme activation are as follows. Cultured cells are harvested by trypsinization and washed as described in the intact-cell assay. Washed cells are suspended in a test tube in Waymouth medium with 10% fetal calf serum at a density of 1×10^6 cells/0.05 ml. To the cell suspension is added 0.148 M α-chloroisocaproate to a final concentration of 1 mM. The compound α-chloroisocaproate was kindly provided by Dr. R. Simpson of Sandoz, Inc. The incubation is for 15 min at 37°. At the completion of incubation, cells are sedimented by low-speed centrifugation. The cell pellet is resuspended in Krebs–Ringer buffer at the density of 1×10^6 cells/0.05 ml. The cell suspension is rapidly frozen at $-75°$ in a Revco Ultralow freezer. The activity of the α-chloroisocaproate-activated BCKA dehydrogenase complex is assayed with 1 mM [1-^{14}C]KIV (specific radioactivity, 172 cpm/nmol) using freeze/thaw disrupted fibroblast suspensions as described above. The reaction mixture containing Krebs buffer instead of cells serves as a blank.

Definition of Unit and Specific Activity. One milliunit is the amount of BCKA dehydrogenase complex in cells that catalyzes the evolution of 1 nmol of $^{14}CO_2$ per minute under the defined assay conditions. Specific activity is expressed as milliunit per milligram of protein. Cellular protein is determined by the Lowry method[18] using bovine serum albumin as standard.

Remarks. The assay methods of actual and total BCKA dehydrogenase complex activities in cultured cells allow an investigation of enzyme deficiency in MSUD at the level of the enzyme complex. MSUD diagnosed by the intact-cell assay must be confirmed by the enzyme assay, since other factors such as defects in transport of BCKA or BCAA, or transamination of BCAA, or isovaleric acidemia[1] may also contribute to reduced decarboxylation of BCKA or BCAA by intact cells.

The activity of BCKA dehydrogenase complex is highly sensitive to protease because the E2 component is susceptible to proteolysis resulting in disassembly of the enzyme complex,[3,30] Thus it is important to inhibit any residual trypsin activity that may be present in cultured cells used for enzyme assays. This is effected by an inclusion of fetal calf serum in the wash medium for trypsinized cells and in the assay mixture. The radiochemical assay can be readily used to measure activity of the purified BCKA dehydrogenase complex in the presence of 2 units per assay of the E3 component from pig heart (Sigma). With freeze/thaw disrupted cells,

---

[30] D. T. Chuang, C.-W. C. Hu, L. S. Ku, P. J. Markovitz, and R. P. Cox, *J. Biol. Chem.* **260**, 13779 (1985).

no added E3 is required for the assay since excess E3 is present in cultured cells (see the following section).

The apparent K_m for KIV for the BCKA dehydrogenase complex of normal human fibroblasts is estimated to be 0.05 to 0.1 mM by the disrupted-cell assay.[21] In contrast, the mutant enzyme complex of MSUD cells shows much reduced affinity for KIV with K_m values in the range of 2–7 mM.[21,22] Significant residual activity (40–60% of the normal) of the BCKA dehydrogenase complex is observed with MSUD cells at 5 mM KIV.[21,22] Thus, to differentiate MSUD from normal cells, low KIV concentrations (1 mM or lower) must be used for the assay. At 1 mM KIV, untreated normal fibroblasts show the measured activity of BCKA dehydrogenase in the range of 0.06 to 0.09 m/mg protein. Under the same assay condition, no detectable activity of the enzyme complex is observed with classical MSUD, cells,[21,22] while 0.01 to 0.03 m/mg protein of residual activity is measured in cells from variant MSUD patients.[22]

The preincubation of cell suspensions with 1 mM α-chloroisocaproate results in a 2- to 3-fold increase in the activity of BCKA dehydrogenase complex in normal fibroblasts to the range of 0.12 to 0.25 m/mg protein. In contrast, the activity of the enzyme complex in cultured fibroblasts from classical MSUD cells,[21,22] while 0.01 to 0.03 m/mg protein of residual caproate treatment. The data are in good agreement with those reported by Toshima et al.[29]

It should be mentioned that the assay with α-chloroisocaproate-treated cells only provides an estimate of the total or absolute activity of BCKA dehydrogenase complex. It remains to be established that the complex is 100% dephosphorylated after the incubation of cells with the kinase inhibitor. A method for obtaining the fully dephosphorylated form of the complex with a broad-specificity protein kinase has been described.[31] However, this approach is not feasible with cultured fibroblasts or 3T3-L1 cells, since solubilization of these cells with Triton X-100 results in a complete loss of the BCKA dehydrogenase complex activity[23]

Assays for Components of the BCKA Dehydrogenase Complex

$$RCO—{}^{14}COOH + 2,6\text{-dichlorophenolindophenol (oxidized)} \xrightarrow{\text{TPP, Mg}^{2+}}$$
$$RCOOH + {}^{14}CO_2\uparrow + 2,6\text{-dichlorophenolindophenol (reduced)} \quad (2)$$

$$R\text{-}{}^{14}CO\text{-}S\text{-}CoA + Lip(SH)_2 \rightleftharpoons R\text{-}{}^{14}CO\text{-}S\text{-}LipSH + CoA\text{-}SH \quad (3)$$

$$LipS_2 + NADH + H^+ \rightleftharpoons Lip(SH)_2 + NAD^+ \quad (4)$$

[31] S. E. Gillim, R. Paxton, G. A. Cook, and R. A. Harris, *Biochem. Biophys. Res. Commun.* **111**, 74 (1982).

Principles

The assays described here are used to measure activities of the components of BCKA dehydrogenase complex separately in cultured cells. Radiochemical assays for E1 and E2 are based on the decarboxylation [Eq. (2)] and transacylation [Eq. (3)][32] reactions catalyzed by E1 and E2 components, respectively using freeze/thaw disrupted fibroblasts. The E3 assay is based on the NADH-linked reduction of DL-lipoamide [Eq. (4)] catalyzed by this component. The oxidation of NADH is monitored spectrophotometrically at 340 nm.

Procedures

E1 Assay. Cultured fibroblasts are harvested and washed, and cells incubated with 1 mM α-chloroisocaproate for 15 min as described above. α-Chloroisocaproate-treated cells are suspended in 0.05 M potassium phosphate buffer, pH 7.0, at the density of 1×10^6 cells/0.05 ml, and disrupted by freezing and thawing as also described above. The E1 assay is carried out in a Kontes reaction flask. The assay mixture contains in 0.37 ml: 100 mM potassium phosphate, pH 6.5, 0.54 mM EDTA, 1 mM MgCl$_2$, 0.2 mM TPP, 2 mM NaAsO$_2$, 0.005 ml fetal calf serum, 0.1 mM 2,6-dichlorophenolindophenol, 0.2 mM [1-^{14}C]KIV, and 0.05 ml of freeze/thaw disrupted cell suspension (1×10^6 cell or 0.3 mg of protein). The reaction flask containing the reaction mixture minus [1-^{14}C]KIV is capped with a rubber serum stopper with the center well charged with 0.1 ml 2 N NaOH. The reaction is initiated with the injection of 0.02 ml of 3.7 mM [1-^{14}C]KIV (specific radioactivity, 1,700 cpm/nmol). The incubation is at 37° for 40 min. The remaining steps, including the counting of ^{14}CO$_2$ released are described in the E1 assay for purified BCKA dehydrogenase complex (Chapter [19], this volume). The reaction mixture containing 0.3 mg of bovine serum albumin in place of cells in 0.05 ml of 0.05 M potassium phosphate, pH 7.0, serves as a blank. The blank count is ~80 cpm.

E2 and E3 Assays. The E2 assay is essentially that described for measuring transacylation activity of the E2 component and purified BCKA dehydrogenase complex.[12] The assay mixture contains in 0.37 ml: 25 mM morpholinepropane sulfonate (MOPS) buffer, pH 7.4, 2 mM DL-dihydrolipoamide, 12.5 μM leupeptin, 1.4% fetal calf serum, 0.4 mM [1-^{14}C]isobutyryl-CoA, and 0.05 ml of freeze/thaw disrupted fibroblast suspension (1×10^6 cell or 0.3 mg protein in Krebs buffer). The reaction is initiated with the addition of 0.06 ml of 2.5 mM [1-^{14}C]isobutyryl-CoA (specific

[32] D. T. Chuang, C.-W. C. Hu, L. S. Ku, W.-L. Niu, D. E. Myers, and R. P. Cox, *J. Biol. Chem.* **259**, 9277 (1984).

radioactivity, 400 cpm/nmol) to the assay mixture without substrate. The incubation is for 20 min at 37°. The assay mixture without cells serves as blank. The remaining details of the E2 assay are described in Chapter [19] in this volume.

The E3 assay is essentially that described by Stumpf and Park.[33] The reaction mixture contains in 1 ml: 50 mM potassium phosphate buffer, pH 6.5, 1.5 mM EDTA, 0.2 mM NADH, 0.15% Triton X-100, 1 mM DL-lipoamide (Sigma), and 1 to 5 × 10⁵ freeze/thaw disrupted cells in Krebs buffer (0.03 to 0.16 mg of protein). The reaction is started with the addition of 0.06 ml of 0.017 M lipoamide (dissolved in 95% ethanol) to the mixture without substrate that has been preincubated for 4 min at 37° in a 1-ml cuvette. The oxidation of NADH at 37° is monitored spectrophotometrically at 340 nm. The extinction coefficient of 6.22 OD μmol is used. The complete assay mixture without lipoamide serves as a blank.

Definition of Unit and Specific Activity. Same as those described in the assay method for activity of the BCKA dehydrogenase complex.

Remarks. The assay methods for catalytic components of the BCKA dehydrogenase complex allow direct and independent measurements of the activities of these components in cultured cells from normal individuals and MSUD patients. This approach is useful in locating the catalytic component of the complex that is deficient in patients with various forms of MSUD.[21] Incubation of normal cells with α-chloroisocaproate results in the activation of E1 activity, similar to that observed in the overall reaction catalyzed by the complex. Thus α-chloroisocaproate-treated cells are better enzyme sources for measurement of E1 activity compared to untreated cells.

The E1 activity in α-chloroisocaproate-treated normal fibroblasts is in the range of 0.011 to 0.018 mU/mg protein when assayed with 0.2 mM [1-¹⁴C]KIV. In 9 classical and 3 variant MSUD cell lines that have been studied in this laboratory, the E1 activity is deficient, ranging from the trace level to 0.004 mU/mg protein. The E1 activity in these MSUD mutant cells is not stimulated by the α-chloroisocaproate treatment. The latter property provides a useful criterion to differentiate normal from MSUD cells. Moreover, the E1 activity data indicate that, in the above MSUD cell lines, the enzymatic lesion involves the E1 component of BCKA dehydrogenase complex. A reduced affinity of the mutant E1 for KIV has been shown to cause E1 deficiency in several classical MSUD patients.[21,34,35]

[33] D. A. Stumpf and J. K. Parks, *Ann. Neurol.* **4**, 366 (1978).
[34] H. W. Rüdiger, U. Langenbeck, M. Schulze-Schencking, and H. W. Goedde, *Humangenetik* **14**, 257 (1972).
[35] D. T. Chuang, L. S. Ku, D. S. Kerr, and R. P. Cox, *Am. J. Hum. Genet.* **34**, 416 (1982).

The E2 activity as measured by the transacylation reaction is in the range of 0.16 to 0.31 mU in normal fibroblasts. In MSUD cells with E1 deficiency the activity of E2 is unaffected.[21] A recent report by Danner et al.[11] has shown the absence of both E2 polypeptide and activity in an MSUD patient. The E3 activity in normal fibroblasts ranges from 36.1 to 56.8 mU/mg protein.[35] E3 activity is normal in homozygous and heterozygous MSUD cell strains that we have studied.[35] A defect in the E3 component has been shown to be the cause of a combined enzyme deficiency in pyruvate, α-ketoglutarate, and BCKA dehydrogenase complexes.[36]

Acknowledgment

This work is supported by Grant 1-796 from March of Dimes Birth Defects Foundation and Grants DK 26758 and DK 37373 from National Institutes of Health.

[36] B. H. Robinson, J. Taylor, S. G. Kahler, and H. N. Krikman, *Eur. J. Pediatr.* **136,** 35 (1981).

[19] Assays for E1 and E2 Components of the Branched-Chain Keto Acid Dehydrogenase Complex

By DAVID T. CHUANG

The branched-chain keto acid (BCKA) dehydrogenase complex catalyzes the oxidative decarboxylation of BCKAs as shown in Reaction (1). The BCKAs consist of α-ketoisovalerate (KIV), α-ketoisocaproate (KIC), and α-keto-β-methylvalerate (KMV) that are derived from the branched-chain amino acids, valine, leucine, and isoleucine, respectively. The highly purified BCKA dehydrogenase complex consists of multiple copies of three catalytic components, i.e., BCKA decarboxylase or E1, dihydrolipoyl acyltransferase or E2, and dihydrolipoyl dehydrogenase or E3.[1-8] The BCKA

[1] S. C. Heffelfinger, E. T. Sewell, and D. J. Danner, *Biochemistry* **22,** 5519 (1983).
[2] F. H. Pettit, S. J. Yeaman, and L. J. Reed, *Proc. Natl. Acad. Sci. U.S.A.* **75,** 4881 (1978).
[3] R. Paxton and R. A. Harris, *J. Biol. Chem.* **257,** 14433 (1982).
[4] K. S. Lau, H. R. Fatania, and P. J. Randle, *FEBS Lett.* **144,** 57 (1982).
[5] R. Odessey, *Biochem. J.* **204,** 353 (1982).
[6] K. C. Cook, A. P. Braford, and S. J. Yeaman, *Biochem. J.* **225,** 731 (1985).
[7] D. T. Chuang, C.-W. C. Hu, L. S. Ku, W.-L. Niu, D. E. Myers, and R. P. Cox, *J. Biol. Chem.* **259,** 9277 (1984).
[8] J. R. Sokatch, V. McCully, and C. M. Roberts, *J. Bacteriol.* **148,** 647 (1981).

dehydrogenase complex is structurally and mechanistically similar to pyruvate and α-ketoglutarate dehydrogenase complexes.[9] Based largely on the mechanisms elucidated with the pyruvate dehydrogenase complex, the following reaction sequence, catalyzed by the above enzyme components and leading to the oxidative decarboxylation of BCKA [Reaction (1)], can be proposed [Reactions (2–6)].

$$\text{RCO-COOH} + \text{CoASH} + \text{NAD}^+ \xrightarrow{\text{TPP, Mg}^{2+}} \text{RCO-SCoA} + \text{NADH} + \text{H}^+ + \text{CO}_2 \uparrow \quad (1)$$

$$\text{RCO-COOH} + \text{E1-TPP} \xrightarrow{\text{Mg}^{2+}} \text{E1-TPP-RCHOH} + \text{CO}_2 \uparrow \quad (2)$$

$$\text{E1-TPP-RCHOH} + \text{LipS}_2\text{-E2} \rightleftharpoons \text{RCO-SLipSH-E2} + \text{E1-TPP} \quad (3)$$

$$\text{RCO-SLipSH-E2} + \text{CoASH} \rightleftharpoons \text{RCO-SCoA} + \text{Lip(SH)}_2\text{-E2} \quad (4)$$

$$\text{Lip(SH)}_2\text{-E2} + \text{FAD-E3} \rightleftharpoons \text{LipS}_2\text{-E2} + \text{FADH}_2\text{-E3} \quad (5)$$

$$\text{FADH}_2\text{-E3} + \text{NAD}^+ \rightleftharpoons \text{FAD-E3} + \text{NADH} + \text{H}^+ \quad (6)$$

Based on the above reaction sequence for each component, radiochemical assays for the E1 and E2 components of the BCKA dehydrogenase complex as described below have been developed. These assays are highly sensitive and specific, which allows independent measurements of E1 and E2 activities in the purified complex[7,10] and isolated components[11] as well as in cultured cells.[12,13]

Branched-Chain α-Keto Acid Decarboxylase

$$\text{RCO-}^{14}\text{COOH} + \text{electron acceptor (oxidized)} \xrightarrow{\text{TPP, Mg}^{2+}}$$
$$\text{RCOOH} + {}^{14}\text{CO}_2 \uparrow + \text{electron acceptor (reduced)} \quad (7)$$

Assay Method

Principle. The radiochemical assay is based on the sum of Reactions (2) and (3) except that an electron acceptor substitutes for the lipoyl-bearing E2 component of BCKA dehydrogenase complex. The model reaction for the E1 assay is shown in Eq. (7) with either ferricyanide[7,14] or 2,6-dichloro-

[9] L. J. Reed, F. H. Pettit, S. J. Yeaman, W. M. Teague, and D. M. Bleile, *in* "Enzyme Regulation and Mechanism of Action" (P. Mildner and B. Ries, eds.), p. 47. Pergamon, Oxford, England, 1980.

[10] C.-W. C. Hu, T. A. Griffin, K. S. Lau, R. P. Cox, and D. T. Chuang, *J. Biol. Chem.* **261**, 343 (1986).

[11] D. T. Chuang, C.-W. C. Hu, L. S. Ku, P. J. Markovitz, and R. P. Cox, *J. Biol. Chem.* **260**, 13779 (1985).

[12] D. T. Chuang, W.-L. Niu, and R. P. Cox, *Biochem. J.* **200**, 59 (1981).

[13] D. T. Chuang, C.-W. C. Hu, and M. S. Patel, *Biochem. J.* **214**, 177 (1983).

[14] S. G. Sullivan, J. Dancis, and R. P. Cox, *Arch. Biochem. Biophys.* **176**, 225 (1976).

phenolindophenol[15,16] as the electron acceptor. The radioactive CO_2 evolved is collected in 2 N NaOH and the latter counted for radioactivity. Moreover, 2 mM sodium arsenite is included in the assay mixture to inhibit the E2 [Eq. (4)] and, consequently, the overall reaction [Eq. (1)] of the enzyme complex. The inhibition nullifies contribution to CO_2 release by the overall reaction without affecting E1 activity. The latter step is essential when E1 activity in crude mitochondrial or cellular extracts is measured by this assay. The assay is independent of E2 for activity.[11]

Reagents

Potassium phosphate buffer, 0.25 M, pH 6.5
EDTA (ethylenediaminetetraacetic acid), 0.1 M, disodium salt
MgCl$_2$, 0.1 M, stored at $-20°$
K$_3$Fe(CN)$_6$, 63 mM, freshly prepared
2,6-dichlorophenolindophenol, 3.7 mM, freshly prepared
TPP (thiamin pyrophosphate), 3.7 mM
Fetal calf serum (GIBCO)
NaAsO$_2$, 37 mM
Enzyme: Purified BCKA dehydrogenase complex[7,10] or E1 component[11] is diluted with 0.05 M phosphate buffer, pH 7.5, such that the aliquot catalyzes the evolution of 0.2–3.6 nmol of CO_2 during incubation
α-Keto[1-^{14}C]isovalerate or α-keto[1-^{14}C]isocaproate, 3.7 mM in 15 mM HCl (specific activity, 1,700 cpm/nmol). The preparation of the [1-^{14}C]-α-keto acids is described in a later subsection

Procedure. The method is a modification of that used to measure the rate of leucine decarboxylation by intact fibroblasts suspension.[17] To a 10-ml pear-shaped reaction flask (Kontes) are added 0.148 ml of the phosphate buffer, pH 6.5, 0.002 ml of EDTA, 0.004 ml of MgCl$_2$, 0.02 ml of TPP, 0.02 ml of NaAsO$_2$, 0.01 ml of 2,6-dichlorophenolindophenol or 0.02 ml of K$_3$Fe(CN)$_6$, 0.005 ml of fetal calf serum, and 0.01 to 0.05 ml of the BCKA dehydrogenase complex or isolated E1. The same incubation medium with 0.01 to 0.05 ml of 0.05 M phosphate buffer, pH 7.5, in place of the enzyme solution serves as a blank. After bringing the mixture to a volume of 0.35 ml with H$_2$O (Millipore), the reaction flask is capped with a rubber serum stopper. On the latter is mounted a plastic center well with stem (Kontes) holding 0.1 ml of 2 N NaOH. Optionally, a small filter paper strip is placed in the well to soak up the NaOH solution. The

[15] P. J. Sykes, J. Menard, V. McCully, and J. R. Sokatch, *J. Bacteriol.* **162**, 203 (1985).
[16] H. Holzer and H. W. Goedde, *Biochem. J.* **329**, 192 (1957).
[17] J. Dancis, J. Hutzler, and R. P. Cox, *Am. J. Hum. Genet.* **29**, 272 (1977).

reaction is initiated by the addition of 0.02 ml of α-keto[1-^{14}C]isovalerate or α-keto[1-^{14}C]isocaproate to the mixture. This is accomplished by injection through the serum stopper with a Hamilton syringe (0.05 ml capacity). During injection of the substrate, the thumb of the flask-holding hand presses against the exposed stem of the center well to prevent spillage of NaOH into the reaction mixture. The reaction mixture is incubated in a 37° water bath (New Brunswick) with gyratory shaking for 10 to 40 min and $^{14}CO_2$ evolved is collected in the NaOH solution. The reaction is terminated by injection of 0.05 ml of 15% trichloroacetic acid into the reaction mixture. Care is again taken to avoid spillage of NaOH from the center well. The incubation is allowed to continue for 1 hr to recover residual $^{14}CO_2$ from the reaction mixture. At the completion of incubation, the NaOH solution or the impregnated filter strip is transferred to 10 ml of a scintillation cocktail to determine radioactivity. The scintillation cocktail contains in 1 liter: 640 ml of methanol, 320 ml of toluene, 20 ml of phenethylamine, 20 ml of glycerol, 6 g of methylbenzethonium chloride, 6 g of 2,5-diphenyloxazole, and 0.12 g of 1,4-bis(4-methyl-5-phenyl-2-oxazolyl)benzene. The blanks for a 40-min incubation are ~ 100 and 200 cpm with 2,6-dichlorophenolindophenol and ferricyanide, respectively, as electron acceptor. Higher blanks result if dithioerythritol is included in the reaction mixture.

Definition of Unit and Specific Activity. One unit is the amount of E1 that catalyzes the evolution of 1 μmol of $^{14}CO_2$ per minute under the defined assay conditions. Specific activity is expressed as units per milligram of protein. Protein is determined by the Lowry method[18] using bovine serum albumin as standard.

Preparation of the 1-^{14}C-Labeled α-Keto Acids. The labeled keto acid is synthesized enzymatically from the corresponding 1-^{14}C-labeled L-amino acid using L-amino-acid oxidase[19] [Eq. (8)].

$$RCH(NH_2)\text{-}^{14}COOH + O_2 \longrightarrow RCO\text{-}^{14}COOH + H_2O_2 + NH_3 \uparrow \qquad (8)$$

To drive the reaction to completion, H_2O_2 produced is degraded by catalase and NH_3 removed by absorption with Nessler's ammonia reagent (Sigma). L-[1-^{14}C]Valine or L-[1-^{14}C]leucine (New England Nuclear, 40–60 mCi/mmol) in 100 μCi is dried *in vacuo* in a pear-shaped reaction flask (Kontes). The dried amino acid is dissolved in 0.5 ml of 1 M Tris-HCl, pH 7.7. To the reaction mixture is added, in sequence, 65,000 units (in 0.05 ml of H_2O) of catalase (Boehringer-Mannheim) and 4.7 units (in 0.2 ml of H_2O) of L-amino-acid oxidase (Sigma, Type 1). The reaction flask is gassed

[18] O. H. Lowry, N. J. Rosebrough, A. L. Farr, and R. J. Randall, *J. Biol. Chem.* **193**, 265 (1951).

[19] A. Meister, *Biochem. Prep.* **3**, 66 (1953).

with O_2 and capped with a serum stopper. The latter holds a stemmed center well containing 0.1 ml of Nessler's reagent. The reaction mixture is incubated at 37° with shaking. At 1 hr-intervals, the reaction flask is regassed with O_2 and recapped with serum stopper holding a fresh well of Nessler's reagent. At the end of 4.5 hr incubation, the reaction is terminated by applying the mixture to an AG 50W-X8 column (1 × 16 cm) equilibrated in H_2O. The synthesized 1-^{14}C-labeled α-keto acid is eluted with H_2O and collected in the void volume. The radioactive purity of 1-^{14}C-labeled α-keto acids is >99% as judged by ascending thin-layer chromatography.[12] The labeled α-keto acid is stored at −20° in 30 mM HCl. Before assay, the thawed 1-^{14}C-labeled α-keto acid is diluted with sodium salt of the corresponding acid (Sigma) to a desired specific activity. The prepared 1-^{14}C-labeled substrate is incubated with shaking for 1 hr at 37° in a reaction flask capped with a serum stopper. The [14]CO_2 evolved is absorbed in 0.1 ml of 2 N NaOH contained in a mounted center well. After incubation, the labeled keto acid is ready for use in the E1 assay.

Properties

Substrate Specificity. The E1 component of BCKA dehydrogenase complex catalyzes decarboxylation of 1-^{14}C-labeled α-keto acids in the order of KIV > KIC > pyruvate. KMV and α-ketobutyrate have not been tested. The rate of decarboxylation of [1-^{14}C]KIV with 2,6-dichlorophenol indophenol as electron acceptor is twice that obtained with potassium ferricyanide. In the latter study, purified BCKA dehydrogenase complex was used for the E1 assay.

K_m *Values.* At pH 6.5, the K_m values of the E1 reaction for KIV, KIC, and pyruvate are 10.6 ± 3.2, 6.2 ± 1.6, and 173 μM, respectively.[20] There are no significant differences in these K_m values whether an isolated E1 or the purified BCKA dehydrogenase complex is used for the E1 assay. The K_m values for E1 for the above α-keto acids are significantly lower than those determined for the overall reaction of the complex.[2,21]

Inhibitors. The E1 reaction is strongly inhibited by clofibrate[20] (I_{50} = 40 μM) and arylidene pyruvates[22] such as furfurylidene pyruvate (K_i = 0.5 μM), 4-(3-thienyl)-2-ketobutenoate (K_i = 150 μM), and cinnamal pyruvate (K_i = 500 μM). Similar inhibitions by these compounds are observed in the overall reaction catalyzed by the BCKA dehydrogenase complex.[20,22,23]

[20] D. T. Chuang, K. S. Lau, and C.-W. C. Hu, unpublished observations.
[21] D. J. Danner, S. K. Lemmon, and L. J. Elsas II, *Biochem. Med.* **19**, 27 (1978).
[22] K. S. Lau, A. J. L. Cooper, and D. T. Chuang, submitted for publication.
[23] D. J. Danner, E. T. Sewell, and L. J. Elsas, *J. Biol. Chem.* **257**, 659 (1982).

Protein Subunits. To determine which protein subunits catalyze the oxidative decarboxylation of BCKA, the BCKA dehydrogenase complex was dissociated in 1 M NaCl at pH 8.3, and its enzyme components were resolved by Sepharose 4B column chromatography.[11] Figure 1 shows that E1 and E2 (see the section Dihydrolipoyl Acyltransferase, below) activities are completely separated from each other. SDS-polyacrylamide gel electrophoresis indicates that the peak fraction (No. 64) for E1 activity contains E1-α ($M_r = 47,000$), and E1-β ($M_r = 37,000$) subunits (Fig. 1 inset). The data establish that the oxidative decarboxylation of α-keto acids [Eq. (7)] is catalyzed by E1-α and/or E1-β polypeptides, and that the presence of E2 is not required for E1 activity.

Other Assay Methods

A similar radiochemical assay for E1 activity using 2,6-dichlorophenolindophenol coupled with phenazine methosulfate as electron acceptor has been developed by Sykes *et al.*[15] A spectrophotometric method has been described in which the activity of isolated E1 is assayed in the presence of excess E2 and E3 in a reconstituted system.[6]

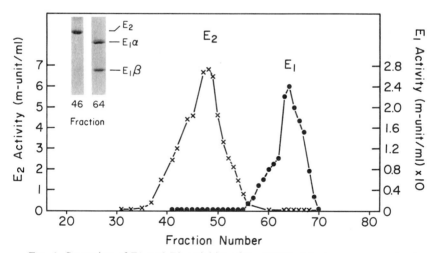

FIG. 1. Separation of E1 and E2 activities of the BCKA dehydrogenase complex by Sepharose 4B column chromatography. The purified complex from bovine liver was dissociated into E1 and E2 components by incubating the complex with 1 M NaCl and 10 mM dithioerythritol at pH 8.3 and 0° for 3 hr. The enzyme components were separated by Sepharose 4B column chromatography. Fractions were assayed for E1 [Eq. (7)] and E2 [Eq. (9)] activities. Inset depicts SDS-polyacrylamide gels of protein subunits present in the peak fractions for E1 and E2 activities (Nos. 64 and 46, respectively). Reproduced from Chuang *et al.*[11] with permission.

Dihydrolipoyl Acyltransferase

$$R\text{-}^{14}CO\text{-}S\text{-}CoA + Lip(SH)_2 \rightleftharpoons R\text{-}^{14}CO\text{-}S\text{-}LipSH + CoA\text{-}SH \qquad (9)$$

Assay Method

Principle. The assay is based on a model reaction [Eq. (9)] and analogous to that catalyzed by the acetyltransferase component of the pyruvate dehydrogenase complex.[24] In the radiochemical assay which is measured in the reversed physiological direction [Eq. (4)], exogenous dihydrolipoamide substitutes for the protein (E2)-bound lipoyl moiety, and enzymatically synthesized [1-¹⁴C]isobutyryl-CoA or [1-¹⁴C]isovaleryl-CoA is the acyl-CoA substrate. The labeled *S*-acyldihydrolipoamide formed is extracted with benzene, and the extract counted for radioactivity. The E2 assay is independent of E1 for activity.[11]

Reagents

4-Morpholinopropane sulfonate (MOPS) buffer, 0.1 *M*, pH 7.4
DL-Dihydrolipoamide, 10 m*M* in 25% ethanol. DL-Dihydrolipoamide is prepared from DL-lipoamide (DL-6,8-thioctic acid amide, Sigma) according to the previously described method[25]
Leupeptin (Sigma), 1 m*M*
Fetal calf serum (GIBCO)
[1-¹⁴C]Isobutyryl-CoA or [1-¹⁴C]isovaleryl-CoA, 2.5 m*M* (specific activity, 400 cpm/nmol). The preparation of radiolabeled branched-chain acyl-CoA is described in a later subsection
Enzyme, the purified BCKA dehydrogenase complex or isolated E2 is diluted with 0.05 *M* phosphate buffer, pH 7.5, such that the diluted aliquot will catalyze the formation of 2–20 nmol of *S*-acyldihydro-lipoamide during the incubation

Procedure. To a 1.5 × 10 cm Pyrex test tube are added 0.186 ml of the MOPS buffer, pH 7.4, 0.074 ml of DL-dihydrolipoamide, 0.005 ml of leupeptin, 0.005 ml of fetal calf serum, 0.01 to 0.05 ml of purified BCKA dehydrogenase complex or isolated E2, and water to a final volume of 0.31 ml. The reaction is initiated by adding 0.06 ml of [1-¹⁴C]isobutyryl-CoA or [1-¹⁴C]isovaleryl-CoA to the mixture. The incubation is at 37° for 20 min. To stop the reaction, 0.75 ml of benzene is added and the mixture is mixed for 10 sec on a Vortex mixer to extract the labeled *S*-acyldihydro-lipoamide. The benzene/aqueous solution mixture is allowed to stand

[24] P. J. Butterworth, C. S. Tsai, M. H. Eley, T. E. Roche, and L. J. Reed, *J. Biol. Chem.* **250,** 1921 (1975).
[25] L. J. Reed, M. Koike, M. E. Levitch, and F. R. Leach, *J. Biol. Chem.* **232,** 143 (1958).

briefly in an ice bath. The separated top benzene layer is then transferred to a 15-ml conical glass centrifuge tube. A 0.2-ml aliquot of the benzene extract is withdrawn from the centrifuge tube and counted for radioactivity as described for the E1 assay. The recovery of labeled thioesters in the benzene extract is approximately 95%.[7] Enzyme is omitted from control tubes to serve as a blank ($\sim$ 50 to 80 cpm). Higher blanks are obtained if dithioerythritol or higher concentrations of dihydrolipoamide are included in the reaction mixture.

Definition of Unit and Specific Activity. One unit is the amount of E2 that catalyzes the formation of 1 μmol of *S*-acyldihydrolipoamide per minute under the defined assay conditions. Specific activity is expressed as units per milligram of protein. Protein is determined by the Lowry method.[18]

Enzymatic Synthesis of [1-^{14}C]Acyl-CoA. The medium-chain (C_{4-11}) fatty acid:CoA ligase (AMP-forming) (EC 6.2.1.2, butyrate–CoA ligase) is isolated from beef liver as described previously.[26] Partially purified enzyme is used to synthesize radiolabeled acyl-CoA according to Eq. (10).

$$R\text{-}^{14}COOH + CoA\text{-}SH + ATP \rightleftharpoons R\text{-}^{14}CO\text{-}S\text{-}CoA + AMP + PP_i \qquad (10)$$

The reaction mixture contains in 3.5 ml: 200 μmol of K/HEPES, pH 8.0, 18 μmol of ATP, pH 7.0, 24 μmol of MgCl$_2$, 1 μmol of CoA, 1 μmol of [1-^{14}C]isobutyric or [1-^{14}C]isovaleric acid (specific activity 50 mCi/mmol and 4.8 mCi/mmol, respectively; Research Products International), 500 μg of the ligase, and 10 units of inorganic pyrophosphatase to drive the reaction in the direction of acyl-CoA synthesis. Incubation is at 37° for 3 to 4 hr, and the decrease in CoA is followed spectrophotometrically by the 5,5'-dithiobis(2-nitrobenzoic acid) method.[27] Additional 1-^{14}C-labeled acid is added to the reaction mixture until CoA is exhausted. The remaining ATP is converted to ADP by addition of hexokinase (4 units) and 20 μmol of glucose. The incubation mixture is eluted from a Sephadex G-25 column (1 $\times$ 50 cm) with water. The radioactivity within an A_{260} peak containing the synthesized [1-^{14}C]acyl-CoA is pooled and stored at $-75°$. The radioactive purity of the labeled acyl-CoA is evaluated by thin-layer chromatography.[26] The labeled acyl-CoA can be either purified further by DEAE-cellulose column chromatography[26] to remove contaminants from the synthesis, or used directly to prepare the radioactive substrate. The desired specific radioactivity is obtained by diluting the [1-^{14}C]acyl-CoA of known counts with the unlabeled lithium salt (Sigma).

[26] D. E. Myers and M. F. Utter, *Anal. Biochem.* **112**, 23 (1981).
[27] P. A. Srere, this series, Vol. 13, p. 2.

Properties

Substrate Specificity and Inhibitors.[7] Rates of the E2-catalyzed trans-acylation reaction with various [1-^{14}C]acyl-CoAs are in the order of [1-^{14}C]isobutyryl-CoA > [1-^{14}C]isovaleryl-CoA > [1-^{14}C]acetyl-CoA. The activity with acetyl-CoA is 15% of that with isobutyryl-CoA. The E2 activity is strongly inhibited by 2 mM NaAsO$_2$ and CoA.

K_m *Values.*[7] The K_m values determined by common intersects of the double reciprocal plots are 0.05, 0.10, and 2 mM for isovaleryl-CoA, iso-butyryl-CoA, and dihydrolipoamide, respectively. By the same method, a K_m value of 0.11 mM is determined for acetyl-CoA.

Protein Subunit.[11] As described earlier, transacylation (E2) activity was completely separated from E1 by Sepharose 4B column chromatography of the dissociated BCKA dehydrogenase complex (Fig. 1). SDS-gel electro-phoresis indicates that the E2 (M_r = 52,000) subunit is present in the peak fraction (No. 46) containing E2 activity (Fig. 1 inset). The results provide direct evidence that the transacylation activity [Eq. (9)] resides in the E2 subunit. Moreover, the model reaction [Eq. (9)] for E2 assay is indepen-dent of E1 for activity.

Other Assay Method

A spectrophotometric assay for E2 based on Eq. (9) has been de-scribed.[15] In this method, CoA formed is determined by its conversion into acetoacetyl-CoA with diketene, followed by an assay of acetoacetyl-CoA with hydroxyacyl-CoA dehydrogenase.

Acknowledgment

This work is supported by Grant 1-796 from March of Dimes Birth Defects Foundation and Grants DK 37373 and DK 26758 from National Institutes of Health.

[20] Mutant Isovaleryl-CoA Dehydrogenase in Isovaleric Acidemia Cells: Assay of Activity and Molecular Characterization

By YASUYUKI IKEDA and KAY TANAKA

Isovaleric acidemia is an inherited disorder of leucine metabolism.[1] It is clinically characterized by episodic vomiting, lethargy, coma, and ketoacidosis.[2,3] Isovaleric acid and its derivatives accumulate in the serum and urine of affected patients. In patients with this disease, no other short-chain fatty acids accumulate in the body fluids. Results from our initial and recent biochemical studies on isovaleric acidemia unequivocally indicate that this disease is caused by a deficiency in the activity of isovaleryl-CoA dehydrogenase (EC 1.3.99.10).[4-7]

Isovaleryl-CoA dehydrogenase (IVDHase) is a mitochondrial flavoenzyme. It is a homotetramer of a 43-kDa subunit[6] and is located in the mitochondrial matrix. The IVDHase is cytoplasmically synthesized as a precursor polypeptide that is 2 kDa larger than its mature subunit. The precursor of IVDHase is then translocated into mitochondria by an energy-dependent process and processed with its extended sequence clipped.[8] The synthesis of acyl-CoA dehydrogenases can be studied using [35S]methionine-labeling, immunoprecipitation with a monospecific antibody, and gel electrophoretic analysis of the immunoprecipitates. Both the precursor and mature forms of acyl-CoA dehydrogenase can be demonstrated in normal hepatocytes by labeling the cells with [35S]methionine in the presence and absence, respectively, of an inhibitor of mitochondrial energy metabolism such as dinitrophenol (DNP) or rhodamine.[8]

When isovaleric acidemia was first described in 1966, the dye reduction assay, which utilizes dichloroindophenol or tetrazolium as an electron acceptor, was the only available method for assaying acyl-CoA dehydro-

---

[1] See also Y. Ikeda and K. Tanaka, this volume [47].

[2] K. Tanaka, M. A. Budd, M. L. Efron, and K. J. Isselbacher, *Proc. Natl. Acad. Sci. U.S.A.* **56**, 236 (1966).

[3] K. Tanaka and K. J. Isselbacher, *J. Biol. Chem.* **242**, 2960 (1967).

[4] K. Tanaka and L. E. Rosenberg, *in* "The Metabolic Basis of Inherited Disease" (J. B. Stanbury, J. B. Wyngaarden, D. S. Fredrickson, J. L. Goldstein, and M. S. Brown, eds.), 5th Ed., pp. 440–473. McGraw-Hill, New York, 1982.

[5] W. R. Rhead and K. Tanaka, *Proc. Natl. Acad. Sci. U.S.A.* **77**, 580 (1980).

[6] Y. Ikeda and K. Tanaka, *J. Biol. Chem.* **258**, 1077 (1983).

[7] D. B. Hyman and K. Tanaka, *Pediatr. Res.* **20**, 59 (1986).

[8] Y. Ikeda, S. M. Keese, W. A. Fenton, and K. Tanaka, *Arch. Biochem. Biophys.*, **252**, 662 (1987).

genases, but it could not be used for determining IVDHase activity in cells from patients due to a strong nonspecific interference. In order to characterize IVDHase deficiency in isovaleric acidemia cells, we devised the tritium release assay.[1,5,9] This assay was further modified to enhance its sensitivity. We were then able to accurately measure the residual activity in numerous cell lines from isovaleric acidemia patients with varying clinical severity.[7] Furthermore, we studied the biosynthesis of variant IVDHase in fibroblasts from patients with isovaleric acidemia, using [35S]methionine-labeling, immunoprecipitation, and sodium dodecyl sulfate-polyacrylamide gel electrophoresis (SDS-PAGE).[10] The biosynthesis study of variant IVDHase demonstrated extensive molecular heterogeneity. Similar techniques have been used successfully in the study of the molecular basis of other genetic enzyme defects, such as medium-chain acyl-CoA dehydrogenase deficiency[11] and glutaric aciduria type II.[12]

General Methodology

Cell Culture

Fibroblasts from skin biopsy specimens of 15 isovaleric acidemia (IVA) patients, 6 parents of these patients, and 4 normal individuals are cultured in Eagle's minimum essential medium supplemented with nonessential amino acids, 10% fetal calf serum, and kanamycin. They are routinely maintained at 37° in a 5% CO_2/95% air atmosphere.

Enzyme Assay

Principle. IVDHase activity is assayed by modifying the tritium release assay.[1,5,9] The main point of this modification is that the assay is done in a pair. Tube A contains a complete assay system. Tube B contains a complete assay system plus 300 μM (methylenecyclopropyl)acetyl-CoA (MCPA-CoA), a strong inhibitor of IVDHase.[6,13] The activity detected in tube B is subtracted as nonspecific background from the activity detected in tube A, so that the true IVDHase-dependent activity can be accurately determined.

Reagents. MCPA-CoA: MCPA is prepared from hypoglycin via oxidative deamination/decarboxylation using snake venom L-amino-acid oxi-

[9] W. J. Rhead, C. L. Hall, and K. Tanaka, *J. Biol. Chem.* **256**, 1616 (1981).
[10] Y. Ikeda, S. M. Keese, and K. Tanaka, *Proc. Natl. Acad. Sci. U.S.A.* **82**, 7081 (1985).
[11] Y. Ikeda, D. E. Hale, S. M. Keese, P. M. Coates, and K. Tanaka, *Pediatr. Res.* **20**, 843 (1986).
[12] Y. Ikeda, S. M. Keese, and K. Tanaka, *J. Clin. Invest.* **78**, 997 (1986).
[13] K. Tanaka, E. M. Miller, and K. J. Isselbacher, *Proc. Natl. Acad. Sci. U.S.A.* **68**, 20 (1971).

dase.[14] Coenzyme A ester of MCPA is synthesized by a modified mixed-anhydride synthesis.[9,15] The synthesized MCPA-CoA is purified by ascending paper chromatography using ethanol:0.1 M potassium acetate (1:1), pH 4.5, as a developing agent.

The other reagents are the same as those described in the chapter on IVDHase.[1]

Assay Procedure. The assay is carried out in two tubes as follows: Tube A contains a complete reaction mixture including 80 μM [2,3-^{3}H]IVD-Hase (10 mCi/mmol), 1 mM phenazine methosulfate, 0.1 mM flavin adenine dinucleotide (FAD), and 30 mM potassium phosphate, pH 7.5. Tube B contains 300 μM MCPA-CoA, in addition to the complete reaction mixture in tube A. The reaction is started by the addition of cell homogenates. The final volume is 0.1 ml. The termination of reaction and isolation of tritiated water are done according to the method described in another chapter.[1] The activity detected in tube B represents nonenzymatically released ^{3}H$_2$O, and is subtracted from the activity detected in tube A.

Preparation of Rabbit Anti-Rat IVDHase IgG

In order to raise monospecific antibody against IVDHase, 1 mg of pure IVDHase from rat liver mitochondria, emulified with Freund's complete adjuvant, was injected into a New Zealand white rabbit weighing 3.0 kg. After 3 weeks, a booster of 1 mg of the pure IVDHase was injected again. Blood was obtained from the auricular vein 10 days after the final booster injection. Antiserum was partially purified by twice fractionating with 50% saturated ammonium sulfate. The final volume of the antiserum preparation was adjusted to the volume of the original serum. Serum from a rabbit that had not been immunized was processed in the same way and used in control experiments.

Study of IVDHase Synthesis Using the Whole Cells

Principle. All cellular proteins are labeled by incubating the cells with [^{35}S]methionine. The labeled IVDHase is isolated via immune–complex formation using the monospecific anti-rat IVDHase antibody, precipitation of the immune complex by the addition of *Staphylococcus aureus* cells, and subsequent centrifugation. The complex of an antigen with the Fe region of IgG is bound to protein A in *S. aureus* cells. The precipitated immune–complex is then analyzed by SDS-PAGE. When the precursor IVDHase is to be demonstrated, [^{35}S]methionine labeling is done in the presence of 2.1 μM rhodamine 6G. Rhodamine 6G inhibits mitochondrial

[14] K. Tanaka, *J. Biol. Chem.* **247,** 7465 (1972).
[15] T. Wieland and L. Rueff, *Angew. Chem.* **65,** 186 (1953).

uptake of the precursor, so that the precursor IVDHase is not processed further in the presence of rhodamine 6G.[16]

Reagents

L-[35S]Methionine (> 600 Ci/mmol)
Rhodamine 6G
Labeling medium: contains 60% Puck's saline F (v/v), 15% fetal calf serum (v/v), 10% 0.5 M glucose (v/v), and 15% H_2O (v/v).
NETS solution: contains 150 mM NaCl, 10 mM ethylenediaminetetraacetic acid (EDTA), 0.5% Triton X-100, and 0.25% SDS and pH adjusted to 7.2.

Procedure. Labeling of the cells: Fibroblasts are grown to confluence in 6-cm Petri dishes. The growth medium is removed and each dish is washed twice with 2 ml of phosphate-buffered saline (PBS), pH 7.4. The cells are incubated for 1 hr with 5 ml of labeling medium. The medium is aspirated and replaced with 2.5 ml of fresh labeling medium containing L-[35S]methionine (50 – 100 μCi). Again, the dishes are incubated at 37° for 60 min. The medium is removed, the cell layers are rinsed twice with PBS, and then solubilized by the addition of 1.0 ml of NETS solution containing 2% unlabeled methionine (w/v). The resulting disrupted-cell homogenates are centrifuged at 105,000 g for 30 min at 4°. The supernatant solution is then subjected to immunoprecipitation.

When rhodamine 6G is used, it is added at the final concentration of 2.1 μM, 30 min before the medium is changed to the labeling medium containing the same concentration of rhodamine 6G. Cells are then labeled and extracted as described above.

Immunoprecipitation of radiolabeled IVDHase: The cell extract from one dish, 1.0 ml, is mixed with 10 μl of rabbit anti-rat IVDHase antibody and is incubated at 4° for 12 hr. The antigen–antibody complexes are recovered by adding 10 volumes of *S. aureus* cell suspension (10% w/v) per volume of antibody, followed by centrifugation at 1700 g for 10 min. The pellets were first washed four times with 2 ml of H_2O to remove labeled actin bound to *S. aureus* cells. The washed immunoprecipitates were solubilized in 20 μl of 0.1 M Tris-HCl buffer, pH 6.8, containing 3% SDS, 37% glycerol, 15% 2-mercaptoethanol, and 0.02% bromphenol blue (w/v), and boiled for 5 min. The sample mixtures were centrifuged at 4000 g for 10 min to precipitate *S. aureus* cells, and the supernatant is then subjected to slab SDS-PAGE.

Gel electrophoresis and fluorography: Slab SDS-PAGE is performed using 9% gels (0.8 mm thick) according to the method of Laemmli.[17] The

[16] W. A. Fenton, A. M. Hack, D. Helfgott, and L. E. Rosenberg, *J. Biol. Chem.* **259**, 6616 (1984).
[17] U. K. Laemmli, *Nature (London)* **227**, 680 (1970).

TABLE I

RESIDUAL ACTIVITIES OF ISOVALERYL-CoA DEHYDROGENASE IN
ISOVALERIC ACIDEMIA CELLS[a]

| | Tritium release from [³H]isovaleryl-CoA | | | |
| | pmol ³H₂O/min/mg | Percentage | Variant | Clinical |
Cell line	protein	of control	type	course
262	0	0	I	Mild
502	0	0	I	Mild
778	0	0	IV	Mild
834	−0.01 ± 0.25	0	I × II	Severe
766	0.42 ± 0.56	2.2 ± 2.9	III	Severe
763	0.44 ± 0.50	2.3 ± 2.6	I	Severe
501	0.52 ± 0.20	2.3 ± 1.0	I	Severe
765	0.58 ± 0.33	3.0 ± 1.7	III	Severe
747	0.68 ± 0.08	3.5 ± 0.4	II	Severe
Controls (7)[b]	19.4 ± 8.0	100		

[a] Adopted from Hyman and Tanaka[7] and Ikeda et al.[10]
[b] The number of control cell lines used.

gels are stained with Coomassie Brilliant Blue and destained with 10% acetic acid and 10% methanol. The destained gel is treated for fluorography with Autofluor (National Diagnostics, NJ), dried, and fluorographed according to the supplier's directions. X-Ray film (Kodak, XAR-5) is used to detect the labeled protein bands from the fluorographed gels. The following ¹⁴C-labeled proteins are used as molecular size markers: phosphorylase *b* (94 kDa), bovine serum albumin (68 kDa), ovalbumin (46 kDa), and α-chymotrypsinogen (26 kDa).

Isovaleryl-CoA Dehydrogenase Activity in Normal and Isovaleric Acidemia Fibroblasts

With the modification of the tritium release assay, the sensitivity of the method is greatly enhanced. Using this method, IVDHase activity in low ranges can be accurately measured in whole-cell homogenates without resorting to the isolation of mitochondria. As shown in Table I, the residual activity determined in nine isovaleric acidemia cell lines is uniformly low, ranging from 0 to 3.5% of the mean of normal controls. These values agree with the previous data that the ability of isovaleric acidemia cells to oxidize [2-¹⁴C]leucine was only 0–2% of the mean of controls.[18]

[18] K. Tanaka, R. Mandell, and V. E. Shih, *J. Clin. Invest.* **58**, 164 (1976).

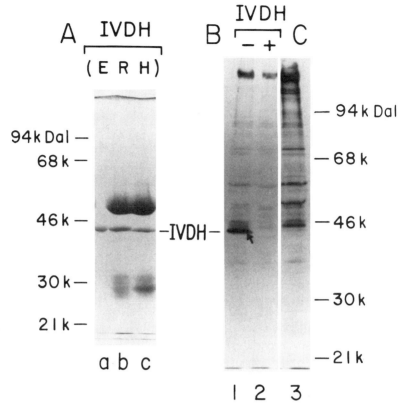

FIG. 1. Gel electrophoresis analysis of IVDHase immunoprecipitated from rat and human liver homogenates (A), and from extracts of cultured human fibroblast labeled with [^{35}S]methionine (B) using anti-rat IVDHase antiserum. (A) Freshly prepared 50% liver homogenates are solubilized in a final volume of 1.0 ml of NETS solution. The solubilized preparation is incubated with 10 μl of the anti-rat IVDHase antiserum and the resulting antibody–antigen complexes are analyzed by SDS-PAGE. Lane a, pure rat IVDHase; lane b, IVDHase from rat liver homogenate; lane c, IVDHase from human liver homogenate. (B) The confluent fibroblast monolayers are labeled with [^{35}S]methionine and then solubilized by NETS solution. The extracts (1 ml each) are subjected to the immunoprecipitation with anti-rat IVDHase antiserum and the immunoprecipitates are analyzed by SDS-PAGE. Lane 1, IVDHase immunoprecipitated; lane 2, same as in lane 1, except that 7 μg of unlabeled rat liver IVDHase was added before the addition of antiserum; lane 3, control immunoprecipitates with 10 μl of nonimmune rabbit serum. The arrow indicates the position of IVDHase polypeptides.

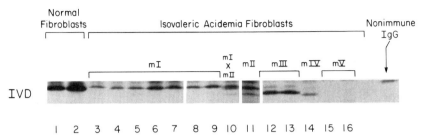

FIG. 2. Electrophoretic demonstration of five distinct types of mutant IvDHase from isovaleric acidemia fibroblasts. All experiments were carried out under the conditions described in Fig. 2. Lanes 1 and 2, normal fibroblasts; lanes 3–16, isovaleric acidemia fibroblasts from different patients. The cell lines used in lanes 1–16 are numbered 1074, 1159, 501, 502, 743, 763, 262, 1302, 1310, 834, 747, 765, 766, 778, 1339, and 1311, respectively, in order of lane number. The last lane immunoprecipitates with nonimmune serum.

Analysis of Normal and Variant Human IVDHase in Liver and Cultured Fibroblasts

Immunocross-Reactivity of Anti-Rat IVDHase Antibody to Human IVDHase

Figure 1A shows the SDS-PAGE analysis of the immunoprecipitates from crude homogenates of human and rat liver that are solubilized in NETS solution[19] using the antiserum raised against rat IVDHase. A sharp single band with a molecular size of 43K was detected from human liver homogenate (lane c). Its molecular size is identical to that of pure rat IVDHase (lane a). These data indicated that the antiserum against rat liver IVDHase was monospecific, and specifically cross-reacted with human liver IVDHase. The antiserum was capable of precipitating approximately 15 μg of the pure IVDHase per 100 μl of antiserum.

When human fibroblast extracts were labeled with [35S]methionine and immunoprecipitated with the anti-rat IVDHase antiserum, a labeled peptide of 43K, behaving identically to IVDHase, was detected (Fig. 1B, lane 1). This band disappeared when excess unlabeled rat IVDHase was added to the immunoprecipitation reaction mixture (lane 2) or when labeled cell extract was treated with nonimmunized rabbit serum (lane 3). These results confirm the identity of the immunoprecipitated IVDHase, and indicate that cultured fibroblasts are suitable for molecular studies of IVD-Hase.

[19] Y. Ikeda and K. Tanaka, *Biochem. Med.*, **37**, 329 (1987).

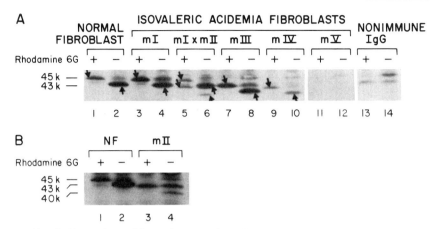

FIG. 3. Comparison of the molecular weight of precursor isovaleryl-CoA dehydrogenase in normal and isovaleric acidemia fibroblasts to their mature counterparts. Extracts of fibroblasts labeled with [^{35}S]methionine in the presence and absence of rhodamine 6G were prepared and analyzed as described in Materials and Methods. Equal volumes of the extracts were treated with anti-IVD antibody or nonimmune serum. (A) The cell line number analyzed in lanes 1 and 2 was 1074; lanes 3 and 4, 502; lanes 5 and 6, 834; lanes 7 and 8, 765; lanes 9 and 10, 778; lanes 11 and 12, 1311. Lanes 13 and 14 were immunoprecipitated with nonimmune serum. (B) Cell lines used: lanes 1 and 2, 1074; and lanes 3 and 4, 747. Precursor and mature IVDs are indicated by arrows.

Heterogeneity of the Molecular Weight of Variant IVDHase

The biosynthesis of IVDHase in fibroblasts derived from 2 normal persons and 15 individuals with isovaleric acidemia is studied using [^{35}S]methionine labeling, immunoprecipitation, and SDS-PAGE as shown in Fig. 2.

Normal IVDHase (w) (M_r 43K) from two normal cell lines are shown in lanes 1 and 2. The molecular weights of IVDHase immunoprecipitated from seven IVA cell lines, shown in lanes 3 through 9, are also 43K and indistinguishable from normal IVDHase, although these cell lines are as deficient in IVDHase activity as any of the other IVA cell lines (Table I). We designate this variant I. In lane 11, one major band with a molecular weight of 42K and a faint smaller band (M_r 40K) are detected. We designate this variant II. In lanes 12 and 13, there is only one major IVDHase band (M_r 41K) which is 2K smaller than normal IVDHase. We named this variant III. These two cell lines are from two Senegalese sibs.[20] In lane 14, there is a major band (M_r 40K) which is 3K smaller than normal IVDHase

[20] J. M. Saudubray, M. Sorin, E. Depondt, C. Herauin, C. Charpentier, and J. L. Possett, *Arch. Fr. Pediatr.* **33,** 795 (1976).

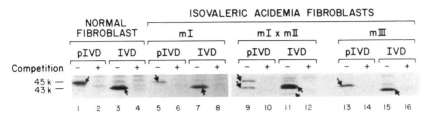

FIG. 4. Competitive inhibition of the immunoprecipitation of the labeled precursor (pIVD) and mature (IVD) forms of normal and mutant human IVDHases by pure unlabeled rat IVDHase. Cell lines used were lanes 1–4, 1074; lanes 5–8, 743; lanes 9–12, 834; and lanes 13–16, 765. Seven micrograms of pure rat IVDHase was added in those indicated with +. Other experimental conditions are described in Materials and Methods. Precursor and mature IVDHases are indicated by arrows.

and it is designated as variant IV. In lanes 15 and 16, no cross-reactive material is detected. We named this mutant variant V. In the cells shown in lane 8, both the normal-size IVDHase and the 42K one with a 40K minor band were detected, indicating that this patient is a compound heterozygote for variant I and variant II alleles.

Precursor and Mature Forms of Five Distinct Variant IVDHases

The precursor forms of the variant IVDHases are studied using rhodamine 6G, as shown in Figs. 3–5. The identities of the variant IVDHase precursors were confirmed by the competition experiment. When excess pure IVDHase was added to the media after incubation with [^{35}S]methionine, but before the addition of anti-IVDHase antiserum, the IVDHase precursors were not detectable (Fig. 4A, lanes 1, 3, 5, 7, and 9 and Fig. 4B, lanes 1 and 3). The molecular features of these variant IVDHase are summarized in Table II.

Variant I. The precursor of variant I IVDHase (45 kDa) is 2K larger than mature variant I (Fig. 3A, lane 3), as in the case of the normal

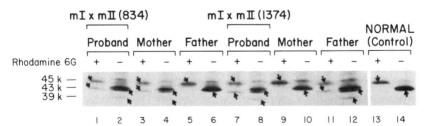

FIG. 5. Analysis of the precursor and mature forms of [^{35}S]methionine-labeled IVDHase in cells from two mI × mII compound heterozygotes (Nos. 834 and 1374) and those from their parents. Precursor and mature IVDHases are indicated by arrows.

TABLE II
CLASSIFICATION AND MOLECULAR FEATURES OF NORMAL AND VARIANT HUMAN IVDHASE[a]

| | Bands synthesized (kDa) | | |
Class	No rhodamine 6G (mature enzyme)	Rhodamine 6G (precursor)	Cell line(s)
Normal IVDHase	43	45	1074, 1159
Variant IVDHase			
Homozygotes			
Variant I	43	45	262, 501, 502, 743, 763, 1302, 1310
Variant II	42 (major) 40 (minor)	42	747
Variant III	41	43	765, 766
Variant IV	40	42	778
Variant V	b	b	1311, 1339
Compound heterozygote			
Variant I – Variant II	43, 42, and 40	45 and 42	834, 1374
Heterozygotes			
Normal – Variant I	43	45	1389 (father of 834), 1376 (mother of 1374)
Normal – Variant II	43, 42, and 40	45 and 42	1375 (father of 1374), 1390 (mother of 834)

[a] From Ikeda et al.[10]
[b] No cross-reactive material was precipitated with the antiserum.

counterpart. Variant I is probably due to a point mutation which occurred in the region encoding the mature IVDHase sequence. It is important to note that there is still a possibility that further heterogeneity exists among cell lines of variant I type; the location of the point mutation in the nucleotide sequence in some of the variant I cells may differ from that in other variant I cells.

Variant II. When variant II homozygote cells are labeled in the presence of rhodamine 6G (Fig. 3B, lanes 3 and 4), only a single band of 42K peptide is detected. This band is identical in size to the major variant II band that is synthesized in the absence of rhodamine 6G. The minor band disappeared when labeling was done in the presence of rhodamine 6G. These data suggest that the precursor of variant II IVDHase is poorly processed, with only a small portion processed in the absence of rhodamine 6G. Thus, it appears that variant II is due to a mutation which causes truncation of the main sequence. It is synthesized with a normal extended leader sequence. However, the conformation of this mutant IVDHase is altered in such a way that either this mutant enzyme is hardly transported

into mitochondria or, alternatively, it resists the action of the mitochondrial protease which cleaves the extended sequence. In either case, it is likely that variant II is due to a deletion of a small nucleotide sequence somewhat downstream of the point that corresponds to the amino terminus of the mature IVDHase, rather than premature termination of translation. Intracellular location of the unprocessed variant II, whether in the cytosol or in the mitochondria, is not known at present.

Variants III and IV. The precursors of variant III (M_r 41K) (Fig. 3A, lanes 7 and 8) and variant IV (M_r 40K) (Fig. 3A, lanes 9 and 10) are 2K larger than their respective mature forms. These precursors are normally processed to the corresponding mature enzyme, although the variant precursors are 4–5K smaller than that of the normal precursor IVDHase. Both variants III and IV appear to be due to a mutation which causes a premature termination of translation. The sequence at the amino terminus, including the extended leader sequence, is not altered in the precursors of these mutant enzymes.

Variant V. When variant V homozygote cells are labeled either in the presence or absence of rhodamine 6G, no cross-reactive material is synthesized (Fig. 3A, lanes 11 and 12). Thus, a deletion or a nonsense mutation close to the N-terminus, or an extremely labile mRNA are the possible mechanisms for variant V.

Compound Heterozygotes and Their Parents. Among the 15 isovaleric acidemia cell lines which we studied, two cell lines (Nos. 834 and 1374) are found to be variant I/variant II compound heterozygotes (Figs. 4 and 5). When these cells were labeled in the presence or absence of rhodamine 6G, both variant I and II proteins were detected as the precursor and mature forms, respectively (Figs. 3–5). When the cells from the parents of these patients were studied, cells from the mother of No. 834 and the father of No. 1374 exhibited the same pattern as those of the respective proband cells, indicating that they are wild-type/variant II heterozygotes. The cells from the father of No. 834 and those of the mother of No. 1374 are indistinguishable from normal controls, indicating that they are wild-type/variant I heterozygotes. No other obvious compound heterozygotes were detected. However, it is possible that some of the cells that were considered homozygotes of variant I, II, III, or IV, may be compound heterozygotes of one of these alleles and variant V.

Relationship between Genotype, Phenotype, and Residual Activity

As shown in Table I, in all the nine cell lines tested, residual IVDHase activity was extremely low, ranging from 0 to 3.5%, regardless of the genotype or phenotype. Among these nine cell lines, variants I, II, III, and

IV are included, but no variant V cell lines were assayed for the IVDHase activity, due to loss of the cells. Undoubtedly, however, the residual IVD-Hase activity in the two variant V cell lines are expected to be extremely low in view of the complete absence of cross-reactive material. It must be mentioned that there have been reports of isovaleric acidemia patients with extremely mild or no clinical manifestations (Dr. Beatte Steinmann, Zurich, Switzerland, personal communication). Since cells from these patients with extremely mild presentation are not included in our studies, there remains the possibility of additional variant types.

[21] Assay of Branched-Chain α-Keto Acid Dehydrogenase Kinase in Mitochondrial Extracts and Purified Branched-Chain α-Keto Acid Dehydrogenase Complexes

By JOSEPH ESPINAL, MARK BEGGS, and PHILIP J. RANDLE

$$[\text{BCDH complex}]E1\alpha + 2ATP \rightarrow [\text{BCDH complex}]E1\alpha(\text{seryl-P})_2 + 2ADP$$

$$(\text{BCDH} = \text{branched-chain } \alpha\text{-keto acid dehydrogenase}) \qquad (1)$$

In eukaryotic cells studied the mitochondrial branched-chain α-keto acid dehydrogenase complex (abbreviated branched-chain complex) is phosphorylated with MgATP and inactivated.[1,2] The reaction [Eq. (1)] results in phosphorylation of two seryl residues in an α-chain of E1 component (branched-chain α-keto acid dehydrogenase); the amino acid sequences of the tryptic phosphopeptides (ox kidney) are known.[3] Inactivation is due mainly and possibly wholly to phosphorylation of one seryl residue (site 1).[3,4] Phosphorylation is assumed to be catalyzed by a branched-chain α-ketoacid dehydrogenase kinase (branched-chain kinase) which copurifies with the complex under favorable conditions and which may copurify with the E2 component when the complex is disrupted and E1 and E2 separated.[5] Phosphorylation of the α-chain of the E1 component with concurrent inactivation of branched-chain complex has been

[1] H. R. Fatania, K. S. Lau, and P. J. Randle, *FEBS Lett.* **132**, 285 (1981).

[2] R. Odessey, *Biochem. J.* **204**, 353 (1982).

[3] K. G. Cook, A. P. Bradford, S. J. Yeaman, A. Aitken, I. M. Fearnley, and J. E. Walker, *Eur. J. Biochem.* **145**, 587 (1984).

[4] K. S. Lau, C. Phillips, and P. J. Randle, *FEBS Lett.* **160**, 149 (1983).

[5] S. J. Yeaman, K. G. Cook, R. W. Boyd, and R. Lawson, *FEBS Lett.* **172**, 38 (1984).

demonstrated in rat heart, liver, kidney, and adipocyte mitochondria,[2,6-8] in rat heart and liver mitochondrial extracts,[9] and in purified complexes from ox kidney,[1] rat liver,[9] rat kidney,[2] and rabbit liver.[10]

Branched-chain kinase activity has been assayed by the rate of MgATP-dependent inactivation of branched-chain complex and by the rate of incorporation of ^{32}P from [γ-^{32}P]ATP into the complex (or, more specifically, the E1α component). Both methods are described here. The rate of MgATP-dependent inactivation is the most convenient because the assay for branched-chain complex based on NADH formation is specific; and because MgATP-dependent inactivation is usually a pseudo-first-order reaction and branched-chain kinase activity may be expressed as the apparent first-order rate constant. Nevertheless, it is important to confirm that changes in branched-chain kinase activity, measured by the rate of ATP-dependent inactivation, are associated with comparable changes in the rate of ^{32}P phosphorylation of E1α. Before describing the methods, potential sources of error in assays of branched-chain kinase are defined in detail.

Potential Sources of Error in Assays for Branched-Chain α-Keto Acid Dehydrogenase Kinase

General

The K_m for MgATP in the ox kidney branched-chain kinase reaction is $\sim 13 \ \mu M$[11] and the reaction is usually carried out with a saturating concentration of MgATP (0.3–0.5 mM). Nevertheless if ATPases are present they may interfere because ADP is inhibitory (K_i, 270 μM; competitive with ATP).[11] ATPase activities should be measured (a method is given); in mitochondrial extracts oligomycin may be effective as an inhibitor of ATPases. Other inhibitors of branched-chain kinase include branched-chain α-keto acids and thiamin pyrophosphate,[11] and α-ketovalerate, α-ketoadipate, n-octanoate, acetoacetyl-CoA, and methylmalonyl-CoA.[12] Low M_r effectors of branched-chain kinase may be removed by gel filtration and a method is described. The kinase reaction is also inhibited by

[6] P. J. Randle, K. S. Lau, and P. J. Parker, in "Metabolism and Clinical Implications of Branched Chain Amino and Ketoacids" (M. Walser and J. R. Williamson, eds.), p. 13. Elsevier/North-Holland, New York, 1981.
[7] K. S. Lau, H. R. Fatania, and P. J. Randle, FEBS Lett. 126, 66 (1981).
[8] W. A. Hughes and A. P. Halestrap, Biochem. Soc. Trans. 8, 374 (1980).
[9] J. Espinal, M. Beggs, H. Patel, and P. J. Randle, Biochem. J. 237, 285 (1986).
[10] R. Paxton and R. A. Harris, J. Biol. Chem. 257, 144533 (1982).
[11] K. S. Lau, H. R. Fatania, and P. J. Randle, FEBS Lett. 144, 57 (1982).
[12] R. Paxton and R. A. Harris, Arch. Biochem. Biophys. 231, 48 and 58 (1984).

2-(N-morpholino)propane sulfonate buffers, which are best avoided. We have not investigated different buffers systematically but phosphate has proved satisfactory. Phosphatases which may dephosphorylate phosphorylated branched-chain complex will interfere, and means of excluding this source of error are given. The general precautions applicable to the stability of branched-chain complex are to be observed (i.e., dithiothreitol or mercaptoethanol in buffers; use of protease inhibitors). Thiamin pyrophosphate has sometimes been used to improve the stability of branched-chain complex but its use is to be avoided as it inhibits branched-chain kinase. Precipitation of branched-chain complex at acid pH and especially at pH < 6.5 may result in specific loss of branched-chain kinase activity[1] and is best avoided.

ATP-Dependent Inactivation

The assay is based on ATP-dependent inactivation and it is essential to include controls incubated under otherwise identical conditions but minus ATP. With the precautions described in Methods, branched-chain complex activity is generally maintained in the absence of ATP. For the assay to be valid for branched-chain kinase activity, the E1 reaction must be rate limiting for the holocomplex reaction [the latter is shown in Eq. (2)].

$$\text{RCO-COOH} + \text{NAD}^+ + \text{CoA} \xrightarrow{\text{Mg}^{2+}, \text{TPP}} \text{R-CO-CoA} + \text{NADH} + \text{H}^+ + \text{CO}_2 \quad (2)$$

If E1 is not rate limiting, inactivation will lag behind phosphorylation in the early stages of the branched-chain kinase reaction. The possibility may be suspected: (1) if there is a lag in the onset of inactivation following addition of ATP or (2) if the y intercept deviates significantly from ln 100% (4.61) in least-squares linear regression analysis of ln[active complex as percent of minus ATP control] = (y) against time (x) following addition of ATP (the first-order plot). Confirmation may be sought: (1) by addition of E2 (dihydrolipoamide acyltransferase) which will increase the assayable activity of branched chain complex if E2 is rate limiting for the holocomplex reaction (it is assumed that E3, dihydrolipoamide reductase, has been added to saturating activity), and (2) by correlation of ^{32}P phosphorylation of branched-chain complex and loss of activity which should be concurrent in the initial stages.

First-order plots are generally linear up to $\sim 80\%$ inactivation and timing of samples should take account of this. When branched-chain kinase activity is high (apparent first-order rate constant $> 4 \text{ min}^{-1}$), the reaction may be slowed by lowering the temperature of incubation with ATP (assay of branched-chain complex activity is at 30°). The Q_{10} of

branched-chain kinase in purified rat liver branched-chain complex was 2.0 ± 0.08 (mean $\pm$ SEM, 8 observations at $10°$, $20°$, and $30°$).[19] For guidance the time for 50% inactivation in minutes is $0.69/K$, where K is the apparent first-order rate constant (min^{-1}). This stricture applies also to measurement of ^{32}P phosphorylation (see below).

[32]P Phosphorylation

There are two sites of phosphorylation in E1α of branched-chain complex; inactivation is correlated with phosphorylation of site 1 and the relative rates of phosphorylation are sites $1 > 2$. The individual phosphorylation reactions are presumably pseudo-first-order, but in order to calculate the apparent first-order rate constants, it is necessary to analyze occupancy of individual sites and to determine full occupancy. This is inconvenient, although suitable analytical methods have been described.[3,14] A more convenient technique is to measure overall ^{32}P phosphorylation and to compute the initial rate. This method is described as it is a convenient check on results obtained for branched-chain kinase by assay of ATP-dependent inactivation. The simplest method is to measure protein-bound ^{32}P following incubation with [γ-^{32}P]ATP. This method is only valid if branched-chain complex E1α is the only protein incorporating [32]P. In mitochondrial extracts the pyruvate dehydrogenase (PDH) complex is also phosphorylated in its E1α component and is present in approximately 5- to 15-fold excess, depending on tissue. There is the possibility of other ATP-dependent phosphorylations. PDH complex may be present in purified branched-chain complex. Methods of ascertaining the specificity of ^{32}P phosphorylation and of overcoming these difficulties are described.

Methods and Procedures

Spectrophotometric Assay of Branched-Chain Complex

The method, reagents, and procedure are given in this volume by Patston *et al.*[15] Triton X-100 and KCN are unnecessary with purified branched-chain complex.

---

[13] J. D. Corbin and R. M. Reimann, this series, Vol. 38, p. 387.
[14] K. G. Cook, R. Lawson, and S. J. Yeaman, *FEBS Lett.* **164**, 85 (1983).
[15] P. A. Patston, J. Espinal, M. Beggs, and P. J. Randle, this volume [22].

Assay of Branched-Chain Kinase by ATP-Dependent Inactivation

The method is based on the rate of ATP-dependent inactivation of branched-chain complex, employing spectrophotometric assay of branched-chain complex activity.[15]

Reagents

Potassium phosphate buffer 1 M, pH 7.5
Dithiothreitol
EGTA, 0.2 M, pH 7.5
$MgSO_4$, 1 M
ATP (disodium), 10 mM
Oligomycin (high oligomycin B content), 5 mg/ml in ethanol

Procedure. Branched-chain kinase assay buffer is 30 mM potassium phosphate/5 mM EGTA/2 mM dithiothreitol/1 mM $MgSO_4$, pH 7.5, and is warmed to assay temperature. In performing assays with mitochondrial extracts or purified branched-chain complex from rat tissues we have added these to 25 mU of branched-chain complex/milliliter in a total incubation volume of 2 ml. Higher concentrations of complex are advantageous if sufficient material is available and ATPase activity is sufficiently low. After warming to assay temperature (3 min), reaction is initiated by adding ATP to 0.5 mM with mixing, the stopwatch started, and samples equivalent to 10 mU total complex taken for assay of branched-chain complex at 30° at four time intervals (usually 15, 30, 45, 60 sec). Incubations minus ATP are handled in the same way. Oligomycin was found to be unnecessary with purified complex or extracts prepared from mitochondria treated with digitonin during preparation. If required, oligomycin is 25 μg/ml.

It is important to exclude interference by branched-chain phosphatase, especially in mitochondrial extracts. We have checked this by examining the effect of 100 mM NaF (which completely inhibits branched-chain phosphatase). No effect of NaF has been seen.

Units. Units of branched-chain kinase activity are expressed as the apparent first-order rate constant (min^{-1}) computed from least-squares linear regression analysis of $\ln[100(A/B)] = (y)$ against time (min) $= (x)$ where A and B are branched-chain complex activity in the presence and absence of ATP, respectively. The plot should be examined and the y intercept [$100(A/B)$ at zero time] computed and considered in relation to the sources of error discussed previously. If $100(A/B)$ is <20% at the longest period of incubation, this point may be omitted and recalculation performed to ascertain whether a better fit is obtained. Alternatively (and preferably) the assay may be repeated with shorter time intervals or at a lower temperature.

Assay of Adenosinetriphosphatase (ATPase) Activity

The method is based on the selective adsorption of $[\gamma\text{-}^{32}P]ATP$ by acid charcoal, $[^{32}P]P_i$ being assayed in the supernatant (based on the method of Cooper *et al.*[16]).

Reagents. As for assay of branched chain kinase (see Assay of Branched-Chain Kinase by ATP-Dependent Inactivation) plus $[\gamma\text{-}^{32}P]ATP$, 1–4 mCi/μmol (Amersham International) and charcoal (Norit GSX, 10 mg/ml in 1 M HCl).

Procedure. The incubations are carried out under the precise conditions used for assay of branched-chain kinase except that $[\gamma\text{-}^{32}P]ATP$ (specific radioactivity ~50 uCi/μmol or 100 dpm/pmol) is used. The exact specific radioactivity of the ATP is determined by measurement of absorbance at 257 nm in 1 M HCl and samples taken from the cuvette for assay of radioactivity by liquid scintillation spectrometry. Samples of incubation medium (20 μl) are removed and mixed with 0.5 ml of acid charcoal (from a continuously stirred suspension) in Eppendorf centrifuge tubes and kept on ice. It is essential to avoid contamination of the incubate with acid charcoal when taking serial samples. The tubes are then centrifuged in an Eppendorf centrifuge, and samples of supernatant (50 μl) taken for liquid scintillation spectrometry of $[^{32}P]P_i$. At the end of the incubation, samples of incubate are taken for assay of total ^{32}P. A blank incubation (minus mitochondrial extract or branched-chain complex) gives $[^{32}P]P_i$ in the ATP.

Units. The rate of ATP hydrolysis is conveniently computed as the percentage of the ATP present in branched chain kinase assay/minute. The extent to which ATP hydrolysis may interfere with the branched-chain kinase assay may be calculated from the experimentally determined K_m for ATP and K_i for ADP.

Assay of Branched-Chain Kinase by ^{32}P Phosphorylation of Branched-Chain Complex E1α with $[\gamma\text{-}^{32}P]ATP$

The method is based upon the incorporation of ^{32}P into protein employing the paper squares method of Corbin and Reimann.[13] It is only suitable for use when phosphorylation is confined to branched-chain complex or when corrections may be applied for incorporation into other proteins (usually the PDH complex). The method is not suitable for use with mitochondrial extracts. Specific precipitation of ^{32}P-phosphorylated branched-chain complex E1α by antibodies is likely to be the method of choice when available.

---

[16] R. H. Cooper, P. J. Randle, and R. M. Denton, *Biochem. J.* **143**, 625 (1974).

Reagents. The reagents are as for assay of branched-chain kinase (see Assay of Branched-Chain Kinase) plus [γ-^{32}P]ATP (see Assay of ATPase Activity). The specific radioactivity should be such as to give accurate liquid scintillation spectrometry on the basis of an incorporation of ^{32}P into ^{32}P-phosphorylated branched-chain complex of ~ 0.5 nmol ^{32}P/unit of complex inactivated. Paper squares (2×1 cm) are cut from Whatman 3 MM paper which has been dipped in 5 mM ATP/1 mM MgSO$_4$ and dried in air at room temperature (heating should be avoided as this may form charcoal in the paper). This reduces blank disintegrations per minute.

Branched-Chain Complex. It is essential to determine whether branched-chain complex contains PDH complex. On SDS-polyacrylamide gel electrophoresis by the method of Laemmli[17] the approximate M_r values of subunits are branched-chain complex, 52K (E2), 46K (E1α), 35K (E1β); PDH complex, 74K (E2), 55K (E3), 50K (band 5), 42K (E1α), 36K (E1β). The E1α subunits of the two complexes are generally resolved and ^{32}P phosphorylation(s) may be detected by autoradiography.[6,7] Contamination with PDH complex may also be assessed by spectrophotometric assay (see Spectrophotometric Assay of Branched-Chain Complex). Ketoleucine is not a substrate for PDH complex, whereas pyruvate is a substrate for both PDH and branched-chain complexes. By use of relative V_{max} and K_m (e.g., for rat liver complexes, see Fatania *et al.*[18]) likely contamination with PDH complex can be assessed.

Procedure. The incubations are carried out as described for branched-chain kinase assay by ATP-dependent inactivation except that a higher concentration of branched-chain complex is used (0.4–2 units/ml). The reaction is initiated with 0.5 mM [γ-^{32}P]ATP and samples (25 μl maximum) are removed at specified time intervals to paper squares which are immediately immersed in aqueous 10% (w/v) trichloroacetic acid at 0° and processed as described by Corbin and Reimann.[13] [γ-^{32}P]ATP blanks are obtained by separate additions to paper squares (equivalent to above) of branched-chain complex in assay buffer (10 μl) followed by 10 μl of [γ-^{32}P]ATP [in 10% (w/v) trichloroacetic acid at 0°]. These paper squares are then processed as above.

There is no readily available method of correcting for ^{32}P phosphorylation of PDH complex when this is present. This is likely to be best accomplished in future by use of antibodies which specifically precipitate E1α from branched-chain complex under conditions of arrested phosphorylation. The resolution of E1α of branched-chain and PDH complexes on

[17] U. K. Laemmli, *Nature (London)* **227,** 680 (1970).
[18] H. R. Fatania, T. C. Vary, and P. J. Randle, *Biochem. J.* **234,** 233 (1986).

SDS gels is usually insufficient to allow scanning of autoradiographs to be used.

Units. Incorporations are computed as picomoles or nanomoles of protein-bound P from specific radioactivity of $[\gamma\text{-}^{32}P]ATP$. For progress curves, units are conveniently nmol P/unit of branched-chain complex incubated and initial rates may be obtained graphically. It is informative to calculate nmol P incorporated/unit of branched-chain complex inactivated through simultaneous or parallel assays of branched-chain complex activities.

Preparation of Mitochondrial Extracts for Assay of Branched-Chain Kinase Activity

The procedure to be described yields extracts of rat liver and heart mitochondria in which branched-chain complex and kinase activities are stable and ATPase activity low (<4% hydrolysis of ATP/min at 30°). All operations are at 0° unless indicated otherwise.

Procedure. Tissues are disrupted in a Teflon–glass Potter–Elvehjem homogenizer or a Polytron homogenizer as appropriate in sucrose medium (0.25 M sucrose/5 mM Tris/2 mM EGTA, pH 7.5) (50–100 mg tissue/ml). Homogenates are centrifuged (600 g, 10 min); the supernatant collected, and the pellets reextracted and centrifuged. The combined supernatants are centrifuged (10,000 g, 10 min), the mitochondria taken up in the same medium (4 mg protein/ml) and incubated (2 min, 0°) with digitonin (0.4 mg/ml). Digitonin is precipitated by centrifugation (600 g, 5 min) and mitochondria separated from the supernatant by centrifugation 10,000 g, 10 min). Mitochondria are taken up and incubated (4 mg protein/ml) at 30° in KCl medium containing 10 μM carbonyl cyanide m-chlorophenyl-hydrazone/5 mM ketoleucine (KCl medium is 0.12 M KCl/20 mM Tris-HCl/5 mM potassium phosphate/2 mM EGTA, pH 7.4). The incubation is to effect complete conversion of inactive phosphorylated complex into active dephosphorylated complex. The time required has to be determined and has ranged from 30 min (normal rats, normal diet) to 120 min (normal rats, 0% casein diet). Mitochondria are then separated (10,000 g, 20 min) and washed (three times) by resuspension in KCl medium (no ketoleucine) and centrifugation, and the pellets frozen in liquid nitrogen. Extracts (25–50 mg protein/ml) are prepared by freezing and thawing (three times) in 50 mM potassium phosphate/10 mM EGTA/2 mM dithiothreitol/ 1 mM benzamidine/1 mM phenylmethanesulfonyl fluoride/0.3 mM-to-syllysyl chloromethyl ketone/2% (v/v) ox serum, pH 7. The extracts are clarified by centrifugation (36,000 g, 10 min).

Gel Filtration of Mitochondrial Extracts for Assay of Branched-Chain Kinase Activity

The procedure described yields a clear filtrate containing concentrated branched-chain complex and kinase activities and effects separation from dissociable effectors of lower M_r. All operations are conducted at 0° unless otherwise indicated.

Procedure. A column (bed volume ~ 180 ml) of Sephacryl S300 (Pharmacia) is equilibrated overnight with 30 mM potassium phosphate/5 mM EGTA/2 mM dithiothreitol (DTT)/pH 7.5. The clarified mitochondrial extracts (36,000 g supernatant; see above, Preparation of Mitochondrial Extracts) are concentrated to approximately 1 unit of branched-chain complex/ml by the use of Centriflo cones (Amicon Inc.; 3000 g for 15–20 min will achieve a twofold concentration). Approximately 1.5–2 units of branched-chain complex is loaded and gel filtered at a flow rate of 0.3 ml/min using the equilibration buffer. Fractions of about 2–2.5 ml are collected. Branched-chain complex is eluted in the void volume (typically 30–40 ml) and recovered in a total volume of approximately 15 ml. Fractions containing branched-chain complex are concentrated fivefold in Centriflo cones and complex and kinase activities are assayed as described above (see Assay of Branched-Chain Kinase by ATP-Dependent Inactivation).

Purification of Branched-Chain Complex for Kinase Assays

The method described is essentially that of Fatania et al.[1] for ox kidney but omitting the final precipitation at pH 6.4, which appears to result in specific loss of kinase activity. All procedures are carried out at 4° unless otherwise stated.

Procedures. Livers from eight rats are disrupted in a Teflon–glass Potter–Elvehjem homogenizer in sucrose medium (0.25 M/5 mM Tris/2 mM EGTA, pH 7.5) (50 mg tissue/ml) and mitochondria prepared, incubated, and extracted as described in Preparation of Mitochondrial Extracts. Following the final centrifugation at 36,000 g for 10 min for clarification of extracts, MgCl$_2$ is added to a final concentration of 10 mM and the pH adjusted to 7.5. Polyethylene glycol 6000 (PEG 6000) (0.04 volume of 50% solution) is slowly added to precipitate the branched-chain complex (PDH is not precipitated at this PEG concentration). It is important to maintain the pH at 7.5 throughout this step and to make up the solution to 10 mM MgCl$_2$ prior to PEG precipitation. The solution is kept stirring on ice while adding PEG and subsequently for a further 15 min, and then for a further 45 min without stirring before centrifuging at 49,600 g for 15 min. The pellet, containing branched-chain complex, is

taken up in extraction buffer (see above) and clarified by a further centrifugation at 36,000 g for 10 min. The supernatant is centrifuged at 100,000 g for 7 min to remove brown pigmented materials. Branched-chain complex is precipitated by centrifugation at 157,000 g for 1.5 hr, and taken up in extraction buffer (no ox serum) at ~2–3 units branched-chain complex/ml.

Concluding Remarks

Employing the methods described, the activity of branched-chain kinase in extracts of mitochondria (normal rats, normal diet) was 1.3 min^{-1} (liver) and 3.4 min^{-1} (heart),[9] (apparent first-order rate constants for ATP-dependent inactivation at 30°). These correspond to $t_{0.5}$ (time for 50% inactivation) of 32 sec and 12 sec, respectively. Gel filtration led to a slight increase[9] (not usually statistically significant), attributed to removal of low M_r inhibitors of the kinase such as thiamin pyrophosphate and ADP. The kinase activity of purified rat liver branched-chain complex was 1.3 min^{-1}. Ox kidney branched-chain complex, purified by the method of Fatania et al.,[1] has given kinase activities ($t_{0.5}$) of ~0.7 min (range 0.5–6.9 min). Published $t_{0.5}$ for ox kidney[2] and rabbit liver[10] complexes were 1.5 and 25 min, respectively. With different methods of purification of branched-chain chain complex, variable loss of branched-chain kinase during purification may be an important factor in variability of branched-chain kinase activity. Branched-chain kinase activity can also be affected by diet; a recent study[9] showed an approximately 4-fold increase in liver and an approximately 1.7-fold increase in heart in rats fed 0% casein diet for 6–10 days.

[22] Assay of Total Complex and Activity State of Branched-Chain α-Keto Acid Dehydrogenase Complex and of Activator Protein in Mitochondria, Cells, and Tissues

By PHILIP A. PATSTON, JOSEPH ESPINAL, MARK BEGGS, and PHILIP J. RANDLE

$$\text{R-CO-COOH} + \text{CoA} + \text{NAD}^+ \xrightarrow[\text{TPP, Mg}^{2+}]{} \text{R-CO-CoA} + \text{CO}_2 + \text{NADH} + \text{H}^+ \quad (1)$$

In all eukaryotic cells studied, the branched-chain α-keto acid dehydrogenase complex (abbreviated to branched-chain complex) is located in

mitochondria.[1,2] It comprises three enzymes: branched-chain α-keto acid dehydrogenase (E1), dihydrolipoamide acyltransferase (E2), and dihydrolipoamide reductase (NAD^+) (E3). The complex exhibits broad specificity toward α-keto acid substrates ($R-CO-COOH$), including ketoleucine, D- and L-ketoisoleucine, ketovaline, and ketomethionine. [It is convenient to name these α-keto acids by reference to the amino acids from which they are formed by transamination; their chemical names are, respectively, 4-methyl-2-ketopentanoate, 3-methyl-2-ketopentanoate, 3-methyl-2-keto-butyrate, 4-(methylthio)-2-ketobutyrate.] Other substrates are 2-ketobutyr-ate and pyruvate. The holocomplex reaction [Eq. (1)] forms the basis of assays which measure formation of NADH or of $^{14}CO_2$ from 1-^{14}C-labeled α-keto acids. Our description is confined to the method based on NADH formation and to an immunoassay. There are important potential sources of error in assays of the branched-chain complex in extracts of cells, tissues, and mitochondria and these are defined in some detail.

Potential Sources of Error in Assays of Branched-Chain Complex in Extracts of Cells, Tissues, or Mitochondria

Substrate Kinetics

Apparent K_m and V_{max} values for rat liver and rat heart branched-chain complexes are shown in Table I. Substrate inhibition by α-keto acids has been observed at concentrations above 0.2–0.5 mM with partially purified complexes,[3-5] though not with extracts of rat liver mitochondria. As a precaution we have routinely used 0.2 mM-ketoleucine in assaying branched-chain complex as this gives rates close to V_{max}, and substrate inhibition has not been observed at this concentration. Higher rates (approximately 1.3×) could be achieved with ketovaline as the V_{max} is greater (Table I). In studies with branched-chain complexes where the α-keto acid substrate kinetics are not known, it is necessary to define them in detail to devise suitable assay conditions.

End-Product Inhibition

The branched-chain complex is inhibited by NADH (competitive with NAD^+) and by branched-chain acyl-CoA (competitive with CoA); K_i

---

[1] See also K. P. Block, R. P. Aftring, M. G. Buse, and A. E. Harper, this volume [24].
[2] See also G. W. Goodwin, B. Zhang, R. Paxton, and R. A. Harris, this volume [23].
[3] P. J. Parker and P. J. Randle, *FEBS Lett.* **90**, 183 (1978).
[4] P. J. Parker and P. J. Randle, *FEBS Lett.* **95**, 153 (1978).
[5] H. R. Fatania, T. C. Vary, and P. J. Randle, *Biochem. J.* **234**, 233 (1986).

TABLE I
SUBSTRATE KINETICS OF RAT LIVER AND HEART
BRANCHED-CHAIN COMPLEXES[a]

Substrate	Apparent K_m (μM) Liver	Apparent K_m (μM) Heart	Relative V_{max} Liver
Ketoleucine	14.6	14.5	1.0
Ketoisoleucine	10.5	—	0.64
Ketovaline	13.0	—	1.29
Pyruvate	734	—	0.2
CoA	10.1	10.6	—
Thiamin pyrophosphate	1.2	—	—
NAD⁺	42.3	40	—

[a] From Refs. 3–5.

values for rat liver complex with ketoleucine as substrate were 14.6 μM (NADH), and 13.6 μM (isovaleryl-CoA).[3] In spectrophotometric assays based on NADH formation it is necessary to inhibit NADH oxidases (we have used CN⁻) and end-product inhibition can develop in the course of the reaction. It is essential to record the progress curve, which is usually linear for at least 1–2 min, and to compute the initial rate. In assays based on $^{14}CO_2$ formation from ^{14}C-labeled keto acids by extracts of cells, tissues, or mitochondria, end-product inhibition by NADH or branched-chain acyl-CoA is unlikely to be a problem because NADH is removed by NADH oxidases (inhibitors of NADH oxidases are unnecessary in such assays). Branched-chain acyl-CoA are likely to be removed by the action of acyl-CoA hydrolases in extracts of cells, tissues, or mitochondria. Removal of branched-chain acyl-CoA may also be effected, if necessary, by addition of carnitine and carnitine acetyl transferase[6] (e.g., in determining K_m for CoA).

Interference by Other Enzymes

In mouse and rat spermatozoa, mitochondria and cytosol contain lactate dehydrogenase isozyme X which catalyzes the reduction of branched-chain α-keto acids[7] [Eq. (2)].

$$\text{R-CO-COOH} + \text{NADH} + \text{H}^+ \rightleftharpoons \text{R-CHOHCOOH} + \text{NAD}^+ \qquad (2)$$

[6] P. J. Parker and P. J. Randle, *Biochem. J.* **171,** 751 (1978).
[7] A. Blanco, C. Burgos, N. M. G. de Burgos, and E. E. Montamat, *Biochem. J.* **153,** 165 (1976).

In assays based on NADH formation it is necessary therefore to establish that extracts of cells, tissues, or mitochondria do not catalyze branched-chain α-keto acid-dependent oxidation of NADH. This enzyme activity has not been detected in extracts of rat liver, heart, or kidney mitochondria. Supernatant fractions (28,000 g or 70,000 g) of rat liver or kidney contain an oxidase which catalyzes the O_2-dependent decarboxylation of ketoleucine[8] [Eq. (3)].

$$\text{Ketoleucine} + O_2 \longrightarrow \text{3-hydroxyisovalerate} + CO_2 \qquad (3)$$

The oxidase does not require NAD^+ and CoA and does not utilize ketoisoleucine or ketovaline as substrates. If present, the oxidase is a source of error in assays of branched-chain complex based upon the production of $^{14}CO_2$ from keto[1-^{14}C]leucine. This stricture may not apply to assays utilizing keto[1-^{14}C]isoleucine or keto[1-^{14}C]valine as substrates. It should not interfere with assays based on NADH formation.

Transaminases are potentially a source of error in assays of branched-chain complex based on $^{14}CO_2$ production because they decrease the specific radioactivity of the ^{14}C-labeled α-keto acid by exchange reactions, if tissue extracts contain significant amounts of branched-chain amino acids. If tissue extracts contain glutamate, then the combined effects of transaminases and of the 2-oxoglutarate dehydrogenase complex are a potential source of error in assays based on NADH formation. Calculations based on published estimates of tissue concentrations and enzyme activities and kinetic constants (K_m and V_{max}) will indicate whether this may be a problem.

Holocomplex Reaction; Stability of Complex

Assays which measure production of NADH or of $^{14}CO_2$ (from 1-^{14}C-labeled α-keto acid) utilize the holocomplex reaction [Eq. (1)] which involves all three component enzymes. Dihydrolipoamide dehydrogenase (NAD^+) (E3; EC 1.6.4.3) is common to all three α-keto acid dehydrogenase complexes. It is not tightly bound by the branched-chain complex and it is necessary, therefore, to ensure that a saturating concentration of dihydrolipoamide dehydrogenase is present. The branched-chain complex is sensitive to agents which oxidize or bind SH groups and to proteolysis. It is necessary to include dithiothreitol or mercaptoethanol in extraction and assay buffers. These may generate NADH slowly by reduction of lipoamide in branched-chain, 2-oxoglutarate, and pyruvate dehydrogenase complexes, but this can be corrected by a blank rate determination as described in the methods. Inactivation of branched-chain complex by proteolysis can occur during isolation of intact mitochondria and, especially following

[8] P. J. Sabourin and L. L. Bieber, this volume [37].

disruption of mitochondria and protease inhibitors, should be included in solutions used for subcellular fractionation or disruption of mitochondria. We have used routinely benzamidine, phenylmethylsulfonyl fluoride, tosyllysyl chloromethyl ketone, and ox serum. It is essential to define the stability of branched-chain complex in extracts of cells, tissues, or mitochondria by repeat assays over $1-2$ hr at $0°$ (and over several days at $-10°$ to $-80°$ if storage in a deep freeze is contemplated) and to plan accordingly.

Choice of Assay Method; Sensitivity and Reproducibility

In spectrophotometric assays based on NADH formation (340 nm, 1 cm light path, 1.4 ml assay volume) the lower limit of detection is approximately 0.5 mU; the optimum range for assay in mitochondrial extracts is approximately $2.5-10$ mU; and the agreement between duplicates is approximately $\pm 5\%$. Greater sensitivity could be achieved by fluorimetry (continuous recording) or by luminometry with NADH luciferase (intermittent samples into ethanolic KOH), but we have had no occasion to use these methods for branched-chain complex. Employing published methods,[9] luminometry is potentially capable of assaying $0.03-0.06$ mU in a convenient sample volume (20 μl). The sensitivity of assays based on $^{14}CO_2$ formation from 1-^{14}C-labeled α-keto acids depends on the specific radioactivity of the α-keto acid, the period of linear $^{14}CO_2$ formation, and the blank (^{14}C released to the CO_2 trapping reagent from acidified 1-^{14}C-labeled α-keto acid in the absence of branched-chain complex). With keto[1-^{14}C]leucine the blank rate was approximately 1% of the total radioactivity. Thus, for example, with 0.2 mM ketoleucine (5 uCi/μmol) the blank rate was approximately 4000 cpm/ml and $^{14}CO_2$ formation was approximately 4000 cpm/min of incubation/mU of branched-chain complex (counting efficiency approximately 70%). Agreement between replicates was approximately $\pm 5\%$ (unpublished observations).

For assay of branched-chain complex in tissues which are generally available in sufficient quantity, we have used only the spectrophotometric assay (NADH formation) because a progress curve is obtained in each single assay; the blank rate is determined within the assay except when ketoleucine is present in extracts; the assays are rapid, less laborious, and capable of automation.

Assay in Tissues; Preparation of Extracts

The concentration of branched-chain complex in mitochondria in rat tissues studied is approximately $3-10$ mU/mg mitochondrial protein, de-

---

[9] A. L. Kerbey, P. M. Radcliffe, and P. J. Randle, *Biochem. J.* **164**, 509 (1977).

pending on the tissue. The concentration of complex in tissues depends mainly on the concentration of mitochondria, which ranges from approximately 80–100 in heart and liver to approximately 18 in skeletal muscle (mg mitochondrial protein/g fresh wt tissue). For spectrophotometric assay it is necessary to concentrate the complex from tissue extracts before assay. Although differential centrifugation has been used,[10,11] losses may occur and there is no accurate method of measuring recovery. The complex may conveniently be concentrated by isolating mitochondria. Recovery may then be monitored accurately by assay of citrate synthase, which is confined to mitochondria, easily assayed, and present at high activity (approximately 150–1000 mU/mg mitochondrial protein, depending on the tissue). Additional advantages are that lysosomes (proteases) can be largely removed during subcellular fractionation; conversion of phosphorylated complex to dephosphorylated complex can be effected in mitochondria by branched-chain phosphatase (see below) and extracts can be rapidly prepared from frozen mitochondria immediately before assay.

Potential Sources of Error in Assays of Activity State of Branched-Chain Complex in Cells, Tissues, or Mitochondria

In all animal tissues studied, branched-chain complex is inactivated by phosphorylation of $E1\alpha$ with MgATP (branched-chain α-keto acid dehydrogenase kinase) and reactivated by dephosphorylation (corresponding phosphatase). Assay of activity state requires separate assays of active (dephosphorylated) complex and of total complex [sum of active (dephosphorylated) form and inactive (phosphorylated) form]. The active form may be assayed spectrophotometrically in extracts prepared from mitochondria isolated from tissues in the presence of ketoleucine and NaF (to inhibit kinase and phosphatase, respectively). Total complex may be assayed in one of two ways. Inactive complex may be converted into active complex before spectrophotometric assay by incubation of mitochondria with an uncoupler (to inhibit respiratory ATP synthesis) and ketoleucine (to inhibit branched-chain kinase). Alternatively, immunoassay may be used, with an antibody which fails to discriminate between phosphorylated and dephosphorylated complex (we have used an antibody to E2).

　　Rat liver and kidney mitochondria from normal rats fed a normal diet contain a protein (activator protein) which reactivates phosphorylated

[10] P. J. Parker and P. J. Randle, *FEBS Lett.* **112**, 186 (1980).
[11] S. E. Gillim, R. Paxton, G. A. Cook, and R. A. Harris, *Biochem. Biophys. Res. Commun.* **111**, 74 (1983).

branched-chain complex without dephosphorylation. It has been purified to apparent homogeneity and is either free E1 or an isoenzyme of E1.[12,13] Activator protein (free E1) has not been detected in heart or skeletal muscle mitochondria. A method for assay of activator protein (free E1) is described. There is evidence that liver mitochondria may contain an inactive form of activator protein; this is presumably phosphorylated E1, as free E1 may be phosphorylated by ATP in the presence of phosphorylated branched-chain complex. An assay for phosphorylated activator protein (free E1) has yet to be developed. For reasons which are not known, rat liver activator protein (free E1) does not reactivate inactive complex in extracts of rat liver mitochondria, although it reactivates phosphorylated purified branched-chain complex, inactive complex in extracts of rat heart mitochondria, and complex inactivated in rat liver mitochondrial membranes by incubation with MgATP. Reactivation of phosphorylated complex by activator protein (free E1) is inhibited by NaF. There is no evidence that activator protein (free E1) interferes with assays of the activity state of branched-chain complex with the methods to be described.

Methods and Procedures

Spectrophotometric Assay of Branched-Chain Complex

The method is based on the measurement at 30° of NADH production by the increase in absorbance at 340 nm with a recording spectrophotometer fitted with a temperature-controlled cuvette chamber.

Reagents

Potassium phosphate buffer, 1 M, pH 7.5; MgSO$_4$, 1 M; dithiothreitol; Triton X-100; KCN, 154 mM

28 mM thiamin pyrophosphate/28 mM CoA/70 mM NAD$^+$, in a single stock solution

Dihydrolipoamide dehydrogenase (Boehringer, pig heart diaphorase Type 11), 19 mg/ml (1520 units/ml)

Ketoleucine (Sigma, sodium α-ketoisocaproate), 28 mM

Procedure. Assay buffer contains 30 mM potassium phosphate/2 mM MgSO$_4$/2 mM dithiothreitol/0.1% (v/v) Triton X-100/final pH 7.5 and is warmed to 30°. The reaction mixture contains 1.4 ml of assay buffer, 30 units of dihydrolipoamide dehydrogenase (20 μl), and, in micromoles,

[12] J. Espinal, P. A. Patston, H. R. Fatania, K. S. Lau, and P. J. Randle, *Biochem. J.* **225**, 509 (1985).

[13] S. J. Yeaman, K. G. Cook, R. W. Boyd, and R. Lawson, *FEBS Lett.* **172**, 38 (1984).

thiamin pyrophosphate, 0.56; CoA, 0.56; NAD$^+$, 1.4; KCN, 1.54. Mito-chondrial extract is then added (equivalent to 0.5–2.5 mg mitochondrial protein) and the blank rate (absence of ketoleucine) recorded until linearity is achieved (1–2 min). Ketoleucine, 0.28 μmol (10 μl) is then added and the progress curve recorded while linearity is maintained (1–2 min). When extracts from mitochondria incubated with ketoleucine are assayed for total complex, ketoleucine is added to the cuvette and reaction initiated with mitochondrial extract. The blank rate is then determined separately employing extracts of mitochondria incubated with CCCP alone (see later section). The blank rate is dependent on the presence of thiamin pyrophosphate + CoA + NAD$^+$ but its origin is not fully defined. An increase in absorbance of 0.002–0.04 per minute during the initial phase of the holocomplex reaction is a linear function of extract concentration. The concentration of dihydrolipoamide dehydrogenase gives maximum rates for the holocomplex reaction, but it is advisable to check this with each preparation. Other approximate concentrations as a function of K_m are 14× (ketoleucine), 300× (thiamin pyrophosphate), 160× (CoA), and 20× (NAD$^+$).

Units. Units are expressed as micromoles of NADH produced per minute at 30° computed from the increase in rate over the blank. Specific activities are units/g of mitochondrial protein (determined by the Biuret method) or units of branched-chain complex/g fresh weight of tissue which may be calculated from [units branched-chain complex/unit of citrate synthase in mitochondrial extract] × [units of citrate synthase/g fresh weight of tissue]. Citrate synthase is assayed in the same mitochondrial extract and in extracts of tissues by a modification[14] of the method of Srere *et al.*[15]

Isolation of Mitochondria; Preparation of Mitochondrial Extracts

Mitochondria. Mitochondria are prepared at 2° by conventional methods of differential centrifugation following disruption of tissue with a Teflon–glass Potter–Elvehjem homogenizer or Polytron homogenizer as appropriate (final concentration 60–160 mg tissue/ml).[16] Nagarse (subtilisin) or trypsin should not be used as they may inactivate branched-chain complex in mitochondria. For assay of the activity state of branched-chain complex tissue, samples are divided. One part is disrupted immediately in medium supplemented with 5 mM ketoleucine/100 mM NaF; the other in medium lacking these additions. Mitochondria are then isolated and washed (two times) by resuspension and centrifugation in the appropriate

[14] H. G. Coore, R. M. Denton, B. R. Martin, and P. J. Randle, *Biochem. J.* **125,** 115 (1971).
[15] P. A. Srere, H. Barzil, and L. Conen, *Acta Chem. Scand.* **17,** S129 (1963).
[16] P. A. Patston, J. Espinal, and P. J. Randle, *Biochem. J.* **222,** 711 (1984).

medium. The fluffy layer (containing lysosomes) is carefully removed. The pellets are then resuspended in the appropriate medium to approximately 50 mg mitochondrial protein/ml and the protein assayed by the Biuret method with serum albumin standards.[17] Samples of mitochondria isolated with ketoleucine/NaF (1–5 mg protein) are added to 0.5 ml of KCl medium (120 mM KCl/20 mM Tris-HCl/5 mM potassium phosphate/ 2 mM EGTA/pH 7.4) containing 100 mM NaF in Eppendorf centrifuge tubes, sedimented (Eppendorf centrifuge, 20–60 sec as appropriate for firm pellets), the supernatant aspirated, and pellets stored in liquid nitrogen. Samples of mitochondria prepared in the absence of ketoleucine/NaF (1–5 mg protein) are mixed with 0.5 ml of KCl medium containing 10 μM carbonylcyanide m-chlorophenylhydrazone (CCCP)/5 mM ketoleucine and incubated at 30° to effect conversion of inactive branched-chain complex to active complex. The time of incubation has to be determined for each tissue and condition of the animal. With rat liver mitochondria it has varied from 30 min (normal rats, normal diet) to 2 hr (normal rats, 0% casein diet). Mitochondria are separated by centrifugation and the pellets stored in liquid nitrogen.

Mitochondrial Extracts. The extraction buffer is 50 mM potassium phosphate/10 mM EGTA/2 mM dithiothreitol/1 mM benzamidine/ 1 mM phenylmethylsulfonyl fluoride/0.3 mM tosyllysyl chloromethyl ketone/2% (v/v) ox serum/1% (v/v) ethanol/pH 7.4. Mitochondrial pellets are thawed by mixing (Hamilton syringe) with extraction buffer (25– 50 mg protein/ml) and frozen (liquid nitrogen) and thawed twice with whirlimixing. For mitochondria prepared in ketoleucine/NaF the extraction buffer contains 100 mM NaF. The final thaw is achieved immediately before assay of branched-chain complex and citrate synthase, which should be assayed simultaneously if possible. An alternative procedure is to use extraction buffer containing 50% (v/v) glycerol and stored at −10° and to disrupt the pellets by ultrasonic disintegration of the frozen pellet (we have not used this procedure for branched-chain complex but it has been used with advantage for pyruvate dehydrogenase complex[9]). Citrate synthase should always be assayed unless the protein concentration of mitochondrial extracts used for assays is known accurately, as the recovery of mitochondria following centrifugation and aspiration of supernatant is not always complete.

Assay of Activator Protein (Free E1)

The assay is based on the reactivation of phosphorylated ox kidney branched-chain complex employing spectrophotometric assay of branched-chain complex by NADH formation.

---

[17] H. G. Gornall, C. J. Bardawill, and M. M. David, *J. Biol. Chem.* **177**, 751 (1949).

Reagents

Phosphorylated branched-chain complex: Branched-chain complex is purified from ox kidney mitochondria[18] and inactivated by phosphorylation by incubation (1 unit/ml) at 30° for 5 min (or longer if necessary) with 0.5 mM ATP in 30 mM potassium phosphate/ 2 mM dithiothreitol/5 mM EGTA/10 mM MgCl$_2$/pH 7.5. A control incubation (minus ATP) is included. Inactivation by ATP should exceed 95% and inactivation in the absence of ATP should be <5%. Phosphorylated complex should be prepared freshly as required.

Mitochondrial extract, high-speed supernatant fraction: Extracts of mitochondria (prepared from mitochondria isolated in the absence of ketoleucine/NaF as in Isolation of Mitochondria; Preparation of Mitochondrial Extracts, above) (50 mg mitochondrial protein/ml) are centrifuged for 90 min at 150,000 g and the supernatant aspirated.

Assay of branched-chain complex: The reagents are as in Spectrophotometric Assay of Branched-Chain Complex, above, except that Triton X-100 is not required.

Procedure. The reaction mixture (1.4 ml) is as described in Spectrophotometric Assay of Branched-Chain Complex, but omitting Triton X-100. Each assay requires two cuvettes. To one (A) is added 10 μl of phosphorylated branched-chain complex and to the other (B) 10–100 μl (depending on activity) of the high-speed supernatant fraction of mitochondrial extract. The increase in absorbance is then measured until a linear progress curve is achieved. This gives the blank rates in the absence of phosphorylated complex (B) or mitochondrial supernatant (A). Phosphorylated complex is then added to (B) and mitochondrial supernatant to (A) and the increase in absorbance measured over a further 2 min. The effect of activator protein (free E1) reaches a maximum in approximately 10 sec and the progress curve is linear for 1–2 min. The initial rate after the maximum effect of activator protein is achieved is recorded.

Units. The activating effect of activator protein (free E1) is analogous kinetically to that of a substrate, i.e., it conforms to Michaelis–Menten kinetics.[18] One unit ($K_{0.5}$) has been defined as milligrams of mitochondrial protein (not high-speed supernatant fraction protein) giving 50% reactivation of phosphorylated branched-chain complex[16] (i.e., approximately 5 mU in the cuvette). Ideally, a range of concentrations of supernatant fractions should be assayed and $K_{0.5}$ and V_{max} computed; V_{max} should not

[18] H. R. Fatania, K. S. Lau, and P. J. Randle, *FEBS Lett.* **147**, 35 (1982).

differ significantly from the concentration of phosphorylated branched-chain complex in the cuvette (this may be greater than that generated by phosphorylation with MgATP if phosphorylated branched-chain complex is present in the purified ox kidney complex). In practice, it may suffice to obtain replicate estimates with a concentration of supernatant fraction which gives approximately 50% reactivation, and to compute $K_{0.5}$ by assuming V_{max}.

Alternative Procedure. It is possible to assay activator protein (free E1) by measuring the activity of branched-chain complex reconstituted in the presence of excess E2 and E3. This method is less convenient than the one described because of the need to separate E2 from purified complex. The relevant basic information is given in articles by Espinal *et al.*[12] and by Cook *et al.*[19]

Immunoassay of Total Complex in Mitochondrial Extracts

The method described utilizes an antibody to the E2 component. The assay is performed against standards of purified ox kidney branched-chain complex by rocket immunoelectrophoresis. Our experience of the assay is confined to extracts of rat liver and rat heart mitochondria and satisfactory rockets were not achieved with extracts of rat kidney mitochondria.

Reagents

Antibodies to E2: Antibodies were produced in half Sandylop eared rabbits by six subcutaneous injections of ox kidney branched-chain complex[18] (200 μg/rabbit) in Freund's complete adjuvant. Booster injections of 200 μg of complex in PBS (140 mM NaCl/2.7 mM KCl/1.5 mM KH$_2$PO$_4$/8.1 mM Na$_2$HPO$_4$/pH 7.4) were given by ear vein after 2 and 4 weeks. Blood was collected 3 days after the final injection. For further collections, booster injections in PBS were given 3 days before. With this strain of rabbit only antibodies to E2 were produced with complex; no antibodies to E1 were produced either with complex or with purified E1. A more certain method of generating antibodies specific for E2 could be to use E2 separated from the complex by published procedures.[20] The antibody was characterized by Ouchterlony diffusion gels; Western blots of SDS-polyacrylamide gels of complex; E2, E1, and mitochondrial extracts; and immunoprecipitation. Immunoprecipitation removed only E2; E1 (identified by the assay of activator

[19] K. Cook, A. P. Bradford, and S. J. Yeaman, *Biochem. J.* **225,** 731 (1985).
[20] P. A. Patston, J. Espinal, J. M. Shaw, and P. J. Randle, *Biochem. J.* **225,** 429 (1986).

protein, see above) or E1 [^{32}P]phosphate (from ^{32}P-phosphorylated complex) remained in the supernatant after centrifugation.[20]

Extracts and standards: Extracts of mitochondria are prepared (50 mg mitochondrial protein/ml) as described above, in 30 mM potassium phosphate/5 mM EGTA/2 mM dithiothreitol/5% (v/v) Triton X-100/pH 7.4. Standards are purified ox kidney branched-chain complex.

Procedure. Rocket immunoelectrophoresis is carried out in agarose gels containing 2% (v/v) Triton X-100 using an LKB Multiphor flat-bed system by the method of Laurell.[21] The concentration of antibody incorporated into the gel is determined for each batch of antiserum.[20] Each plate incorporates four standards (2.5, 5, 12, and 24 mU of ox kidney complex) and two amounts of mitochondrial extract (usually equivalent to 225 and 450 μg of mitochondrial protein). Electrophoresis is for 16 hr at a constant 10 V/cm. The plates are pressed between filter paper sheets (Whatman 3MM), washed with 0.1 M NaCl, dried, and stained with Coomassie Blue. The amount of complex in mitochondrial extracts is calculated from linear standard curves of ln[complex] against rocket height. It is important to incorporate more than one amount of mitochondrial extract to confirm that standards and extracts exhibit parallel slopes. The same technique is utilized to establish equivalence of phosphorylated and dephosphorylated branched-chain complex.

Units. The specific activity of branched-chain complex in units/g of mitochondrial protein is computed directly from the standard curve.

Concluding Remarks

The results of assays of total complex and of activity state in tissues of normal rats fed normal diet by the methods described here are compared with results employing other methods in Table II.[22] The methods which we have described here (methods 1 and 2 in Table II) give higher values for total branched-chain complex activity than assays based on extraction of whole tissue, precipitation of complex by centrifugation, and conversion of inactive complex into active complex by a broad specificity rat liver phosphoprotein phosphatase (methods 3 and 4, Table II). The reason for this has not been systematically investigated, but it seems to us that losses are likely with the lengthy centrifugation involved in methods 3 and 4 (Table

[21] C. B. Laurell, *Scand. J. Clin. Lab. Invest.* **29** (Suppl. 124), 21 (1972).

[22] A. J. M. Wagenmakers, J. T. G. Schepens, J. A. M. Veldhuizen, and J. H. Veerkamp, *Biochem. J.* **200,** 273 (1984).

TABLE II
BRANCHED-CHAIN α-KETO ACID DEHYDROGENASE COMPLEX IN RAT TISSUES:
TOTAL ACTIVITY AND ACTIVITY STATE

Parameter (units)	Method[a]	Tissue			
		Liver	Heart	Kidney	Skeletal muscle
Total activity	1	0.93	0.58	—	—
(units/g fresh wt)	2	0.92	0.57	0.77	0.04[b]
	3	0.65	0.27	0.34	—
	4	0.39	0.22	0.20	0.03[c]
Activity state	2	55	5	71	<20[b]
(% of total complex)	3	94	48	77	—
	4	93	12	49	6[c]

[a] Method 1: Immunoassay of mitochondrial extracts.[20] Method 2: Spectrophotometric assay (NADH formation) of mitochondrial extracts (combined results from Refs. 16 and 20 and unpublished work of the authors). Method 3: Spectrophotometric assay (NADH formation); extracts of frozen tissue, complex concentrated by precipitation at 180,000 g; inactive complex converted into active complex with a phosphoprotein phosphatase from liver.[11] Method 4: Method as in Ref. 5 but from Wagenmakers et al.[22]
[b] Hind limb muscle mitochondria; activity of active form below lower limit of assay.
[c] Assayed in homogenates of quadriceps muscle, by $^{14}CO_2$ production from keto[1-^{14}C]leucine.[22]

II) in which lysosomes may have been disrupted. There was excellent agreement between immunoassay of total complex, which does not involve conversion of inactive complex into active complex (method 1, Table II), and spectrophotometric assay, which does involve conversion within mitochondria (method 2, Table II). With method 2, the percentage of active complex in liver and heart was lower than with methods 3 and 4 (Table II). This may be due in part to the higher recovery of total complex with our method and in part to the presence throughout of ketoleucine/NaF to prevent interconversion in our method.[16] We have not referred here to methods in which the activity of branched-chain complex activity is measured in intact mitochondria contained within tissue homogenates. In our view, it is doubtful whether activity of branched-chain complex (active form) is being studied under V_{max} conditions with such methods. Also, it has been shown that branched-chain complex in mitochondria can be inactivated in tissue homogenates by Nagarse,[16] and it seems possible that endogenous proteases could have a similar effect in the absence of protease inhibitors.

Addendum

Since this paper was written Zhang *et al.*[23] have criticized our method for assay of activity state of branched chain complex in rat liver on the following grounds. They find that inclusion of 100 mM-NAF plus 5 mM-ketoleucine during isolation of rat liver mitochondria leads to loss of citrate synthase (by approximately 60%). No such loss of citrate synthase has been seen in our own studies. We have recently improved our method by use of 1 mM-dichloroacetate in place of 5 mM-ketoleucine as an inhibitor of branched chain kinase. This simplifies determination of blank rates in assays of branched chain complex. In studies with this improved method activities of citrate synthase (munits/mg protein, mean ± sem for 50 rat liver mitochondrial preparations) were: 164 ± 6.3 (mitochondria isolated in sucrose medium) and 159 ± 6.1 (mitochondria isolated in sucrose medium containing 100 mM-NaF plus 1 mM-dichloroacetate).

Zhang *et al.*[23] also report that 100 mM-NaF plus 5 mM-ketoleucine induced a progressive loss of branched chain complex activity in rat liver mitochondria incubated at 0° C in sucrose/Hepes medium. The loss was approximately 45% after 3 h. In strictly comparable studies in this laboratory no such change in activity was detected after 30, 90, or 180 min in the sucrose/tris/EGTA medium which we use routinely.[14,18] For example, after 180 min of incubation, branched chain complex activities (munits/mg protein: mean ± sem for 4 mitochondrial preparations and 16 assays) were: 9.5 ± 0.9 (sucrose medium), 9.2 ± 1.1 (sucrose/100 mM-NaF/ 5 mM-ketoleucine), and 8.8 ± 1.1 (sucrose/100 mM-NaF/1 mM-dichloroacetate). The zero time value was 9.2 ± 0.6 with 83% of complex in the active form. Citrate synthase activity was also unchanged (overall mean ± sem, munits/mg protein: 216 ± 2.1). The reason for the different results of Zhang *et al.* is not apparent, but in using our method it would seem advisable to isolate mitochondria in the sucrose/tris/EGTA medium which we have used.[14]

Zhang *et al.*[23] have also criticized our use of citrate synthase as a mitochondrial marker enzyme because it has sometimes given higher estimates of mg mitochondrial protein/g fresh liver (up to 100) than those generally accepted (approximately 60–70). This criticism is based on a misunderstanding of the use of citrate synthase. In our method citrate synthase is used as an index of mitochondrial recovery and not as a means of computing the liver content of mitochondrial proteins. The two are not the same. Different preparations of mitochondria can vary in purity and in recovery following centrifugation of individual samples.

[23] B. Zhang, R. Paxton, G. W. Goodwin, Y. Shimomura, and R. A. Harris, *Biochem. J.* **246,** 625 (1987).

The agreement between the results obtained with the method described here and those based on radioactive assays with extracts of frozen powdered rat tissues[23,24] are generally excellent when due allowance is made for the protein contents of the different chow diets fed.[14,18,23,24] It is at least as good as the agreement between results of different studies employing radioactive assays on extracts of frozen powdered tissue.[23-25] The published exception is starvation.[14,23-25] More recent studies in this laboratory with the method described here (as yet unpublished), have confirmed that the percentage of active complex in rat liver is not decreased by starvation.[23-25]

[24] R. A. Harris, S. M. Powell, R. Paxton, S. E. Gillim, and H. Nagae, *Arch. Biochem. Biophys.* **243**, 542 (1985).
[25] M. Solomon, K. G. Cook, and S. J. Yeaman. *Biochim. Biophys. Acta* **931**, 335 (1987).

[23] Determination of Activity and Activity State of Branched-Chain α-Keto Acid Dehydrogenase in Rat Tissues

By Gary W. Goodwin, Bei Zhang, Ralph Paxton, and Robert A. Harris

Branched-chain α-keto acid dehydrogenase[1,2] (BCKDH, EC 1.2.4.4 + no EC number for the acyltransferase + E.C. 1.8.1.4) is an intramitochondrial assembly of three enzymes which together catalyze the following irreversible reaction:

$$RCOCOO^- + CoA + NAD^+ \xrightarrow{Mg^{2+}, \ TPP} RCO\text{-}CoA + CO_2 + NADH$$

Thiamin pyrophosphate (TPP) is a dissociable cofactor for E1, the dehydrogenase component of the complex. Like pyruvate dehydrogenase, BCKDH is reversibly inactivated by phosphorylation of E1 by a specific kinase associated with the complex. BCKDH, pyruvate dehydrogenase, and α-ketoglutarate dehydrogenase constitute a family of enzymes which differ in the nature of their respective substrates. For BCKDH, the branched-chain α-keto acids (α-ketoisocaproate, α-ketoisovalerate, and α-keto-β-methylvalerate) derived from transamination of the corresponding amino acids (leucine, valine, and isoleucine), are the best known substrates, with half-maximal activity in the low micromolar range. Maxi-

[1] See also K. P. Block, R. P. Aftring, M. G. Buse, and A. E. Harper, this volume [24].
[2] See also P. A. Patston, J. Espinal, M. Beggs, and P. J. Randle, this volume [22].

mal activity is highest with α-ketoisovalerate, making this the substrate of choice for activity determination because of enhanced sensitivity. The present article describes radiochemical and spectrophotometric assays for the enzyme in extracts of freeze-clamped rat liver, kidney, and heart. The radiochemical assay employs $^{14}CO_2$ collection from 1-^{14}C-labeled α-keto acid substrates using whole-tissue homogenates, while the spectrophotometric assay measures NADH formation following partial purification of BCKDH by polyethylene glycol precipitation. The two assays are of comparable economy; the most expensive ingredients are radioactive substrate for the radiochemical assay, and coenzyme A and E3 for the spectrophotometric assay. The radiochemical assay has the advantage of enhanced sensitivity ($10 \times$ over the spectrophotometric assay). The spectrophotometric assay is preferred where sensitivity is not a problem because it is quicker and easier to execute, and can be applied to heart. A linear end-point assay that yields a sufficient number of counts per minute over the blank is not obtained radiochemically with heart extracts (see below).

The active form of BCKDH is that activity determined when the *in vivo* phosphorylation state is preserved. Tissues are quickly freeze-clamped at liquid nitrogen temperature and homogenized at ice temperature in the presence of Triton X-100 to solubilize the enzyme, KCl to promote extraction and stabilization of the enzyme,[3] protease inhibitors to protect against the unusual sensitivity of E2 to proteolysis, plus ethylenediaminetetraacetic acid (EDTA) and α-chloroisocaproate[4,5] to protect against MgATP inactivation. Dichloroacetate[1] and α-ketoisocaproate[6] have also been used to inhibit BCKDH kinase during homogenization. However, α-ketoisocaproate can spuriously contribute to the blank in a spectrophotometric assay, and irreversibly inactivate BCKDH in the absence of the required cofactors for the enzyme.[7] EDTA is employed to inhibit phosphatases. The necessity of freeze-clamping the tissue, as compared to prior isolation of mitochondria, and the adequacy by which the activity state and total activity are preserved are discussed below.

The total activity of BCKDH is that activity obtained after complete dephosphorylation, or activation. Activity state is the percent of the total activity which is in the active form when the phosphorylation state has been preserved. To achieve complete activation in a reasonable period of

[3] Y. Shimomura and R. A. Harris, unpublished observations (1986).
[4] R. A. Harris, R. Paxton, and A. A. DePaoli-Roach, *J. Biol. Chem.* **257**, 13915 (1982).
[5] R. A. Harris, M. J. Kuntz, and R. Simpson, this volume [15].
[6] S. E. Gillim, R. Paxton, G. A. Cook, and R. A. Harris, *Biochem. Biophys. Res. Commun.* **111**, 74 (1983).
[7] R. Paxton and R. A. Harris, *J. Biol. Chem.* **257**, 14433 (1982).

time (<1 hr), it has been necessary to add exogenous phosphatase to the homogenate (slow activation occurs by preincubation with Mg^{2+} only). We have employed a partially purified broad-specificity phosphoprotein phosphatase preparation from rabbit liver for this purpose; it is an analytical tool and probably not the physiologic phosphatase. The dephosphorylation reaction is stimulated by Mg^{2+}. The amount and incubation time with the phosphatase for complete activation should be determined for each phosphatase preparation. We utilize tissue homogenates from rats maintained on a low-protein diet for this purpose, a condition in which the enzyme is more inactive, or phosphorylated relative to other nutritional states. In general, more phosphatase or a longer incubation time is required to achieve complete activation in tissues where BCKDH is more inactive.

The E3 component of BCKDH, or dihydrolipoamide reductase is loosely associated with the complex and may become rate limiting.[8] It is commercially available. BCKDH activity is Mg^{2+} dependent, with maximal stimulation around 1 mM free Mg^{2+}. The standard assay, therefore, incorporates excess E3, as well as excess Mg^{2+} to offset chelation by EDTA present in the tissue extract.

Sources of Materials

Reagents were obtained as previously given,[4,7] or from Sigma Chemical Company, Saint Louis, MO. α-Keto[1-^{14}C]isovalerate was prepared from L-[1-^{14}C]valine obtained from RPI Corp. (Mount Prospect, IL) with L-amino-acid oxidase,[9] and stored at $-70°$ in 0.2-ml (5 μCi, 45 mCi/mmol) aliquots. Rat liver BCKDH (7.4 U/mg) was purified by a modification[10] of the basic procedure.[7,11] α-Chloroisocaproate was a kind gift from Dr. Robert J. Strohscheim and Dr. Ronald Simpson of Sandoz Inc., East Hanover, NJ. The broad-specificity phosphoprotein phosphatase was prepared as previously described.[7,12] The phosphatase, purified through the DEAE-Sephacel chromatography step, was concentrated with 75% $(NH_4)_2SO_4$, dialyzed against 50 mM imidazole-Cl, 0.5 mM EDTA, 0.5 mM dithiothreitol, and 40% (v/v) glycerol, pH 7.5, and stored (unfrozen at $-20°$) with a protein concentration of 15 mg/ml. The phosphatase is stable for several months under these conditions. Male Wistar rats with initial body weights of 170–185 g were purchased from Harlan Industries, Indianapolis, IN. Rats were given free access to water and starved for 48 hr

[8] R. Odessey, *Biochem. J.* **192**, 155 (1980).
[9] H. W. Rüdiger, U. Langenbeck, and H. W. Goedde, *Biochem. J.* **126**, 445 (1972).
[10] Y. Shimomura, R. Paxton, T. Ozawa, and R. A. Harris, *Anal. Biochem.*, **163**, 74 (1987).
[11] R. Paxton, this volume [41].
[12] R. Paxton and R. A. Harris, *Arch. Biochem. Biophys.* **231**, 48 (1984).

or fed *ad libitum* a standard lab chow diet or a low-protein (8%) diet described previously[13] for 7 days. Following cervical dislocation, livers (< 10 sec), hearts (< 20 sec), and kidneys (< 30 sec) were removed, freeze-clamped in Wollenberger clamps[14] precooled in liquid nitrogen, and stored at −70°. Mitochondria were prepared by the method of Johnson and Lardy[15] either in a sucrose medium (pH 7.4 with Tris) or the same medium supplemented with 2 mM EGTA, 0.1 M NaF, and 5 mM α-ketoisocaproate, which corresponds to that described by Patston *et al.*[2,16] Mitochondrial pellets were frozen in liquid nitrogen, extracted in homogenization buffer by vortexing, and BCKDH activity assayed spectrophotometrically essentially as described below. Citrate synthase was assayed as described by Shepherd and Garland.[17] Mitochondrial protein was determined by the biuret method.[18]

Preparation of Tissue Homogenates

The homogenization buffer is prepared fresh, and consists of 50 mM K-N-2-hydroxyethyl piperazine-N'-2-ethanesulfonic acid (K-HEPES), 0.2 M KCl, 3 mM EDTA, 5 mM dithiothreitol, 0.1 mM α-chloroisocaproate, 0.1 mM N-α-tosyl-L-lysine chloromethyl ketone (TLCK), 0.1 mg/ml trypsin inhibitor from egg white, 0.5 μM leupeptin, 0.5 μM pepstatin A, 1 μg/ml aprotinin, 2% (v/v) rat serum, and 0.5% (v/v) Triton X-100, pH 7.5, at 20°. Frozen tissue is pulverized to a fine powder under liquid nitrogen with a mortar and pestle. About 0.2 g of powdered tissue is transferred to a tared, precooled tube and suspended in ice-cold homogenization buffer by vigorous mixing with a Teflon stirring rod. A 10% (w/v) homogenate (i.e., 9 volumes of homogenization buffer) is routinely prepared for the spectrophotometric assay. Three to 17% (w/v) homogenates are prepared for the radiochemical assay, depending on the activity of the tissue. The tissue suspension is frozen in liquid nitrogen and returned to room temperature until the icy slurry can be mixed again with a Teflon stirring rod. When the suspension is nearly completely thawed, it is placed on an ice bath and homogenized for 30 sec at a low setting of a Polytron PT-10 homogenizer. More vigorous homogenization does not alter activ-

[13] R. A. Harris, S. M. Powell, R. Paxton, S. E. Gillim, and H. Nagae, *Arch. Biochem. Biophys.* **243**, 542 (1985).
[14] A. Wollenberger, O. Ristau, and G. Schoffa, *Pfluegers Arch. Gesamte Physiol. Menschen Tiere* **270**, 399 (1960).
[15] D. Johnson and H. Lardy, this series, Vol. 10, p. 94.
[16] P. A. Patston, J. Espinal, and P. J. Randle, *Biochem. J.* **222**, 711 (1984).
[17] D. Shepherd and P. B. Garland, this series, Vol. 13, p. 11.
[18] G. L. Peterson, *Anal. Biochem.* **83**, 346 (1977).

ity. However, for the radiochemical assay in which the crude homogenate is assayed directly, a low setting is desirable to minimize formation of foam from the Triton X-100.

Spectrophotometric determination of BCKDH involves clarification of insoluble material from the crude homogenate, followed by polyethylene glycol precipitation of BCKDH away from NADH oxidase activity, as well as much of the material that gives rise to a backround rate in the clarified extract. Following clarification of the crude homogenate at 12,000 g for 5 min, a known portion of the supernatant is transferred to a preweighed tube, and made 9% (w/v) polyethylene glycol by addition of 0.5 volume of ice-cold 27% (w/v) polyethylene glycol 8000. Following incubation on ice for 20 min, the extract is centrifuged at 12,000 g for 10 min, the supernatant discarded, and the tube plus pellet weighed to determine the volume of the pellet by difference. The pellet is then homogenized in ice-cold homogenization buffer with a Potter-Elvehjem type homogenizer to a final concentration of from 50 to 100 mg initial wet weight of the tissue per milliliter, depending on the activity. The homogenate is transferred back and forth between the homogenizer and the centrifuge tube, scraping down the inside of the tube with a spatula, for quantitative recovery of the pellet.

The homogenates (crude for the radiochemical assay, partially purified for the spectrophotometric assay), maintained on ice, are either assayed directly for the active form of BCKDH, or preincubated with the phosphatase for determination of total activity. In the latter case, an aliquot is made 10 mM in added Mg^{2+} with 0.1 M MgCl$_2$ and incubated with the phosphatase at 30° before proceeding to the basic assay.

Determination of BCKDH Activity

Assays are conducted in (final concentration) 30 mM KP$_i$, 0.4 mM thiamine pyrophosphate, 0.4 mM coenzyme A, 2 mM dithiothreitol, 3 mM NAD$^+$, 0.1% (v/v) Triton X-100, 7.5 U/ml porcine heart dihydrolipoamide reductase (E3, EC 1.8.1.4,), and 2 mM MgCl$_2$ for the spectrophotometric assay, or 5 mM for the radiochemical assay (excess necessary to offset chelation by EDTA present in the homogenization buffer), pH 7.5, at 20°. The assay cocktail is prepared in twice-concentrated form, lacking MgCl$_2$ and E3, and may be stored at −20° for at least 1 week. Storage of the cocktail with Mg^{2+} results in the slow formation of a precipitate, requiring that a complete assay cocktail containing Mg^{2+} and E3 be prepared just prior to assay. BCKDH complex is largely freed of endogenous E3 during polyethylene glycol precipitation. Maximal activity is achieved at 5 U/ml added E3, though this should be checked with each batch. E3 is

partially freed of $(NH_4)_2SO_4$ by sedimenting in an Eppendorf centrifuge for 1 min, and resuspended in assay cocktail before addition to the assay.

Radiochemical assays are conducted in 1.5-ml Eppendorf microcentrifuge tubes with the caps cut off, or another suitable reaction tube, containing 0.2 ml complete, twice-concentrated assay cocktail, and homogenization buffer such that the final volume after addition of tissue extract and substrate is 0.4 ml. Sensitivity is inversely proportional to the assay volume with a given amount of radioactivity; 0.4 ml is just large enough for comfortable addition volumes. Sensitivity is inversely proportional to substrate concentration with a given amount of radioactivity so long as the enzyme is nearly saturated with substrate. Sensitivity is not increased simply by adding more radioactivity beyond about 50,000 cpm at a given substrate concentration because of a commensurate increase in the blank. Whole-tissue extract is added to the reaction tube, which has been prewarmed in a 30° water bath, mixed by pipetting up and down, and returned to the water bath. After a 5- to 10-min preincubation for thermal equilibration, reactions are initiated with 50 μl of 4 mM α-keto[1-^{14}C]isovalerate (200 μCi/mmol, final concentration of 0.5 mM) which has been prewarmed to 30°, and mixed by pipetting up and down. The reaction tube is gently placed in a miniscintillation vial (Stockwell Scientific, Walnut, CA) or other suitable vial, containing 1 ml of 1.2 M KOH prewarmed to 30°, fitted with a serum cap, and the reaction is allowed to proceed at 30°. Reactions are terminated by injection of 0.5 ml of 2 M acetic acid containing 2% (w/v) sodium dodecyl sulfate into the reaction medium, and CO_2 collection is continued at 40° on a shaking water bath. Fifty percent of the $^{14}CO_2$ is collected in the KOH in 20 min, 95% in 3 hr. CO_2 collection is less efficient with smaller bore reaction tubes, and should be checked by the release of $^{14}CO_2$ from a known quantity of $NaH^{14}CO_3$. Excessive CO_2 collection time increases the blank. The Eppendorf tube is removed and any adherent KOH is washed into the vial with 0.5 ml of β-phenylethylamine : methanol (1 : 2 v/v). The contents of the vial are mixed with 5 ml of Safety-Solve (RPI Corp., Mount Prospect, IL) or another suitable aqueous scintillant, allowed to sit in the dark until fluorescence has subsided (about 1 hr as indicated by a low, stable blank), and counted. It is convenient to perform the preincubation in a water bath, while the scintillation vials containing KOH and a tube containing substrate are located in a nearby heating block prewarmed to 30°. With practice, reactions can be initiated every 30 sec.

The blanks consist of vials in which homogenization buffer replaces tissue extract. Routinely, the first and last vial in every series of determinations are blanks, and are generally 0.2 to 0.5% of the total number of

counts added. The average is subtracted from all other vials. A fresh, disposable needle should be used for each series of determinations, as an acid-contaminated needle will accumulate corrosion products which promote decarboxylation of the radioactive substrate. We routinely place the tube containing substrate over KOH solution in a larger sealed container for a few hours prior to assay to trap a small amount of volatile radioactivity which would otherwise contribute to the blank. A small amount of radioactive substrate is counted like all other vials for determination of specific activity. One unit of BCKDH activity results in the formation of 1 μmol of CO_2 or NADH per minute at 30°. With a 10-min preincubation, CO_2 production is linear with incubation time for at least 15 min, and is linear with amount of tissue extract used in the assay. In general, a linear assay can be expected with decarboxylation of up to 10% of the substrate, though 5% is considered optimal. The intraassay coefficient of variation was determined with a liver from a starved rat and kidney from a chow-fed rat (1.2 and 0.52 U/g wet weight tissue, respectively) with from 0.6 to 4 mg tissue, and incubation times from 3 to 19 min, and found to be 3.2% for liver ($n = 9$) and 4.4% for kidney ($n = 11$). As mentioned, the radiochemical assay is not linear in heart extracts for unknown reasons (before or after polyethylene glycol precipitation, or if 1 mM L-carnitine is added to the assay).

Total assay volume for the spectrophotometric assay is 1 ml using 2 cuvettes (one as reaction cell, and the other as reference cell to subtract out the blank), each containing 0.5 ml twice-concentrated complete assay cocktail, sample (usually 30 μl), and water to 0.95 ml, prewarmed to 30°. The reference cell is initiated with 50 μl of water, while the reaction cell is initiated with 50 μl of 10 mM α-ketoisovalerate (final concentration of 0.5 mM) prewarmed to 30°. Assays are routinely performed with a split-beam spectrophotometer, though the two cells may also be run side by side with an automatic sample changer, and the blank rate subtracted manually. In more active tissues (liver from chow-fed and starved animals), the blank is frequently negligible. It may be possible to determine the blank rate, if any, and the activity in a single cuvette by initiating the reaction with a small volume so that the blank does not change appreciably upon volume expansion. Assays are linear for several minutes or more. The apparent K_m for α-ketoisovalerate in liver extracts was 43 μM (10 concentrations between 0.01 and 0.5 mM). The intraassay coefficient of variation was determined in liver, kidney, and heart from a chow-fed animal (1.51, 0.90, and 0.15 U/g wet weight, respectively) using five tissue concentrations over a 3-fold range in concentration, and found to be 2.7, 4.2, and 5.4%, respectively.

Potential Interfering Substances

Cytosolic α-ketoisocaproate oxidase[19] will interfere with the radiochemical assay if α-keto[1-^{14}C]isocaproate is used as substrate in whole-tissue homogenates of liver and kidney.

If significant amounts of BCKDH substrates (branched-chain α-keto acids, α-ketobutyrate, γ-methylthio-α-ketobutyrate, pyruvate[20]) are present in the cuvette before initiating the spectrophotometric assay, they will spuriously contribute to the blank.

NADH oxidase activity, as well as lactate dehydrogenase isoenzyme X (which also utilizes branched-chain α-keto acids[21]) if present, will lead to an underestimation of BCKDH activity obtained spectrophotometrically. These activities were not detected in partially purified extracts of heart, liver, or kidney; the extracts do not catalyze significant NADH oxidation when 50 μM NADH is added to the standard assay system, nor is there NADH oxidation with or without α-ketoisovalerate when 50 μM NADH replaces NAD$^+$ and E3 in the standard assay system. Commercial E3 contains NADH oxidase activity which is completely suppressed at a high NAD$^+$/NADH ratio (i.e., 3 mM NAD$^+$, 50 μM NADH). Other potential interferences include transamination followed by NADH production from α-ketoglutarate dehydrogenase, NADH production by oxidative steps subsequent to BCKDH, and reduction of specific activity of the radioactive substrate upon introduction of branched-chain keto acids and amino acids (via transamination) present in the tissue homogenate. There is no detectable NADH production in partially purified homogenates of heart, liver, or kidney from isobutyryl-CoA (i.e., the oxidative decarboxylation product of α-ketoisovalerate). Furthermore, NADH production can be completely inhibited by antisera which specifically inhibit BCKDH. Finally, the two assays yield nearly identical activities; the percent difference between the two assays in the same tissue samples was 2.6 $\pm$ 4.3% ($n = 10$, 5 livers and 5 kidneys).

The possibility of interconversion of the activity state of BCKDH during homogenization and assay was checked, first, by investigating the effect of fluoride (50 mM KF in the homogenization buffer, 25 mM in the assay buffer) on the spectrophotometric determination of the active form (Table I). This potent phosphatase inhibitor was of little effect (though 40 mM or more in the assay buffer inhibited the dehydrogenase, presumably from complexation with Mg$^+$), arguing that the enzyme is not subject

[19] P. J. Sabourin and L. L. Bieber, *J. Biol. Chem.* **257**, 7460 (1982).
[20] R. Paxton, P. W. D. Scislowski, E. J. Davis, and R. A. Harris, *Biochem. J.* **234**, 295 (1986).
[21] A. Blanco, C. Burgos, N. M. G. deBurgos, and E. E. Montamat, *Biochem. J.* **153**, 165 (1976).

TABLE I
ACTIVE FORM OF BCKDH IN THE PRESENCE AND
ABSENCE OF FLUORIDE[a]

Tissue	Nutritional state	BCKDH activity (U/g wet wt)	
		Without fluoride	With fluoride
Liver	Low-protein	0.062 ± 0.001	0.066 ± 0.001
Liver	Starved	1.97 ± 0.05	1.88 ± 0.03
Kidney	Low-protein	0.44 ± 0.01	0.44 ± 0.01
Kidney	Starved	0.59 ± 0.01	0.59 ± 0.01
Heart	Low-protein	0.063 ± 0.001	0.065 ± 0.01

[a] Values are the mean ± SEM in units per gram of wet weight tissue determined spectrophotometrically in freeze-clamped tissues in triplicate. Fluoride refers to the addition of 50 mM KF to the homogenization buffer, and 25 mM to the assay buffer.

to dephosphorylation during the procedure. Similar results have been obtained using the radiochemical assay, with and without phosphate in addition to fluoride to inhibit phosphatases[13] (data not shown). Second, powdered liver from chow-fed, starved, and low-protein fed rats was spiked with purified ^{32}P-labeled rat liver BCKDH, homogenized, and precipitated with perchloric acid. Practically no acid-soluble radioactivity was detected, indicating that the enzyme is not subject to activation by dephosphorylation during homogenization, and also that phosphate groups do not undergo significant turnover under these conditions. Finally, powdered liver from starved rats (1.31 ± 0.14 U/g wet wt with no additions, $n = 3$) was spiked with 1 U/g tissue of purified BCKDH$_a$, or the same amount of enzyme that had been inactivated with ATP to give BCKDH$_b$, and homogenates were prepared for the spectrophotometric assay. Nearly 100% recovery of enzyme activity was observed either before or after dephosphorylation (BCKDH$_a$ and BCKDH$_b$ spiked samples yielded, respectively, 2.31 ± 0.17 and 1.43 ± 0.08 U/g before phosphatase, and 2.34 ± 0.12 and 2.32 ± 0.12 U/g after preincubation with the phosphatase, $n = 3$), indicating not only that the enzyme is not subject to interconversion, but also that the procedure yields complete recovery of total activity (i.e., BCKDH is adequately protected from proteolysis, and the dephosphorylation reaction goes to completion). In lieu of the above evidence that BCKDH is not subject to activation during preparation of homogenates and assay of freeze-clamped tissue, another reason must be sought for the discrepancy between the high activity states for hepatic BCKDH in chow-fed and

TABLE II
TOTAL ACTIVITY AND ACTIVITY STATE OF HEPATIC BCKDH BY VARIOUS
METHODS[a]

Nutritional state	BCKDH activity following mitochondrial isolation[b]	BCKDH activity in freeze-clamped liver	
		Assayed radiochemically	Assayed spectrophotometrically
Chow-fed	1.09 ± 0.04	1.04 ± 0.05	1.32 ± 0.06
	(103 ± 3)	(100 ± 2)	(96 ± 3)
Starved	1.34 ± 0.05	1.43 ± 0.09	1.69 ± 0.13
	(93 ± 3)	(100 ± 3)	(102 ± 2)
Low-protein	0.31 ± 0.10	0.59 ± 0.07	0.79 ± 0.04
	(26 ± 5)	(28 ± 6)	(13 ± 3)

[a] Values are the mean $\pm$ SEM in U/g wet wt tissue for 5 to 12 animals. Activity states (percentage of total activity in the active form) are in parentheses.
[b] Activities in U/mg mitochondrial protein were converted to U/g wet wt by multiplying by the ratio of units citrate synthase/g wet wt/units citrate synthase/mg mitochondrial protein.

starved animals obtained with the methods described in this article (see Refs. 13 and 22 and Table II) and low values obtained by isolation of mitochondria in the presence of α-ketoisocaproate and fluoride.[2,16,23] In addition, there is no evidence for the occurrence of activator protein (free E1 which can activate phosphorylated BCKDH[2,16,24,25]) in homogenates of freeze-clamped liver, as indicated by experiments in which powdered liver from chow-fed and low-protein fed animals were mixed and the expected and observed activities were compared.[13] The former should be rich in activator protein, while the latter should contain none.[16] Consequently, one would expect activation of phosphorylated BCKDH in liver from low-protein fed animals by activator protein present in liver from chow-fed animals, though this was not found to be the case.

Table II compares the total activity and activity state of hepatic BCKDH obtained in freeze-clamped liver by the two methods described in this article, and also following prior isolation of mitochondria. Essentially, all of the hepatic enzyme is in the active form in chow-fed and starved

[22] G. W. Goodwin, R. Paxton, S. E. Gillim, and R. A. Harris, *Biochem. J.* **236**, 111 (1986).
[23] J. Espinal, P. A. Patston, H. R. Fantania, K. S. Lau, and P. J. Randle, *Biochem. J.* **225**, 509 (1985).
[24] J. Patston, J. Espinal, J. M. Shaw, and P. J. Randle, *Biochem. J.* **235**, 429 (1986).
[25] H. R. Fantania, K. S. Lau, and P. J. Randle, *FEBS Lett.* **147**, 35 (1982).

animals in freeze-clamped liver (regardless of the assay method) as well as in mitochondria isolated by a conventional method.[15] Lower activity states are obtained when mitochondria are isolated in the presence of α-ketoisocaproate and fluoride as described by Patston *et al.*[2,16] ($73 \pm 3\%$ in the fed state, $n = 6$; $47 \pm 3\%$ in the fasted state, $n = 12$). Similar results are obtained in the presence of fluoride, ADP, and EDTA (data not shown),

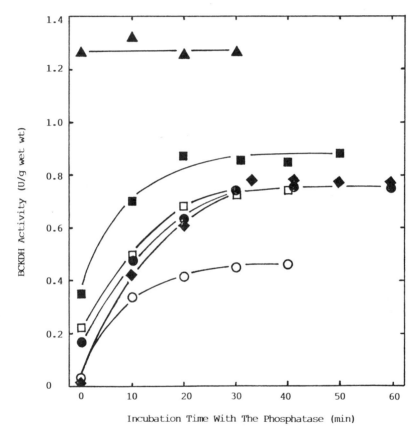

Incubation Time With The Phosphatase (min)

FIG. 1. Activation of BCKDH in liver, kidney, and heart homogenates by the phosphatase. Crude homogenates (open symbols) and homogenates partially purified by polyethylene glycol precipitation (closed symbols) of freeze-clamped tissue from low-protein fed animals were made 10 mM in added $MgCl_2$, incubated with the phosphatase at 30°, and, at the indicated times, assayed for BCKDH activity either radiochemically or spectrophotometrically, respectively. Final concentrations of the phosphatase were (mg phosphatase protein/ml): liver (○) 2.9, (●) 2.3; kidney (□) 3.2, (■) 1.5; heart (◆) 2.3. A liver homogenate from a chow-fed animal (▲) was treated the same as the low-protein fed animal.

TABLE III
BCKDH ACTIVE FORM, TOTAL ACTIVITY, AND ACTIVITY STATE IN
VARIOUS TISSUES AND NUTRITIONAL STATES[a]

Tissue	Nutritional state	Active form	Total activity	Activity state
Liver	Starved	1.72 ± 0.14	1.69 ± 0.13	102 ± 2
	Low-protein	0.11 ± 0.02^b	0.79 ± 0.04^b	13 ± 3^b
Kidney	Starved	0.71 ± 0.05	1.00 ± 0.03	71 ± 5
	Low-protein	0.41 ± 0.04^b	0.99 ± 0.08	42 ± 7^b
Heart	Starved	0.40 ± 0.005	0.57 ± 0.07	7.2 ± 1.0
	Low-protein	0.064 ± 0.017	0.78 ± 0.02	8.3 ± 2.2

[a] Values are the mean $\pm$ SEM determined spectrophotometrically
in freeze-clamped tissues from five animals, in U/g wet wt tissue.
[b] $p < 0.01$ versus starved.

corresponding to the medium of Wagenmakers et al.[26] Considerably lower
activity states have been reported when mitochondria are isolated in the
presence of α-ketoisocaproate and fluoride, especially in the starved
state.[2,16,23,24] One interpretation is that inclusion of α-ketoisocaproate and
fluoride results in inactivation of the complex, apparently because inhibition of phosphatases by fluoride is more effective than BCKDH kinase
inhibition by α-ketoisocaproate or ADP in intact mitochondria.[27]

Activities and activity states in mitochondria isolated in the absence of
BCKDH kinase and phosphatase effectors agree fairly well with values
obtained in freeze-clamped tissue (Table II). Thus, the activity state of
BCKDH appears to be fairly stable in the absence of additions which may
differentially affect the kinase and phosphatase activities. Since it seems
that the activity state can be subject to alterations in intact mitochondria,
the freeze-clamping technique remains the method of choice for estimation
of the active form and activity state of the BCKDH complex.

BCKDH Activities and Activity States in Various Tissues and Nutritional States

Figure 1 shows typical time courses of activation during preincubation
with the phosphatase. Generally, no more than 5% of the total activity of
BCKDH is lost during this preincubation (presumably from proteolysis) as
determined when fluoride is included, or when BCKDH is already com-

[26] A. J. M. Wagenmakers, J. T. G. Schepens, J. A. M. Valdhuizen, and J. H. Veerkamp,
Biochem. J. 220, 273 (1984).
[27] B. Zhang, R. Paxton, G. W. Goodwin, and R. A. Harris, Biochem. J., 246, 625 (1987).

pletely active, or dephosphorylated (i.e., liver from a chow-fed animal, Fig. 1). Table III shows BCKDH activities and activity states determined spectrophotometrically in freeze-clamped tissues under various nutritional states. Low-protein feeding is associated with a marked reduction in the active form and activity state of the hepatic enzyme. Activities and activity states determined spectrophotometrically in freeze-clamped kidney and heart under the same nutritional states are also shown.

Acknowledgments

The work in this chapter was supported in part by U.S. Public Health Service grant (DK19259), the American Heart Association, Indiana Affiliate, and the Grace M. Showalter Residuary Trust.

[24] Estimation of Branched-Chain α-Keto Acid Dehydrogenase Activation in Mammalian Tissues

By Kevin P. Block, R. Paul Aftring, Maria G. Buse, and Alfred E. Harper

Introduction

The first committed step in branched-chain amino acid (BCAA) catabolism is catalyzed by branched-chain α-keto acid dehydrogenase (BCKAD), a multienzyme complex located on the inner surface of the inner mitochondrial membrane of most mammalian cells.[1-4] BCKAD catalyzes the irreversible oxidative decarboxylation of branched-chain α-keto acids (BCKA) to acyl-CoA derivatives with one less carbon:

$$BCKA + NAD^+ + CoASH \rightarrow acyl\text{-}CoA + NADH + H^+ + CO_2$$

Like the pyruvate dehydrogenase complex, BCKAD ($M_r \sim 2 \times 10^6$) is a complex composed of three subunits: (1) E1b (BCKA decarboxylase) arranged in an $\alpha_2\beta_2$ substructure with thiamin pyrophosphate as the coenzyme, (2) E2b (dihydrolipoamide acetyltransferase) with lipoate as the

---

[1] See also P. A. Patston, J. Espinal, M. Beggs, and P. J. Randle, this volume [22].
[2] See also G. W. Goodwin, B. Zhang, R. Paxton, and R. A. Harris, this volume [23].
[3] P. J. Randle, H. R. Fatania, and K. S. Lau, *Mol. Aspects Cell. Regul.* **3**, 1 (1984).
[4] A. E. Harper, R. H. Miller, and K. P. Block, *Annu. Rev. Nutr.* **4**, 409 (1984).

coenzyme, and (3) E3b (dihydrolipoamide dehydrogenase) with flavin adenine dinucleotide as the coenzyme. NAD^+ and CoA are also required as cofactors. BCKAD is unusual among amino acid-degrading enzymes in that it exists in active (dephosphorylated) and inactive (phosphorylated) forms. A protein phosphatase and kinase are presumed to be associated with the complex *in vivo*. Phosphorylation occurs on two serine residues of the E1b α subunit.[5,6] Inactivation of the complex is complete when site 1 is phosphorylated; phosphorylation of site 2 may make activation of the complex more difficult. The physiological importance of BCKAD modulation by phosphorylation remains to be established, but it may play a role in the disposal of excesses of BCAA (i.e., high-protein feeding) or conservation of essential BCAA carbon (i.e., low-protein intake).

Numerous reports of the degree of activation of BCKAD have appeared in the literature (Table I). Whole homogenates, mitochondria, and extracted partially purified enzyme from these fractions have been used in studies of BCKAD activation. Detergent or calcium ions may be added to the tissue preparation to disrupt mitochondrial membranes and eliminate any transport barriers which may exist. Also, inhibitors of the kinase (dichloroacetate, α-ketoisocaproate, or α-chloroisocaproate) and phosphatase (F-, sodium or potassium salt) have been used to maintain the complex in its *in vivo* (also called: basal, initial, actual, or expressed) phosphorylation state. Estimation of total (dephosphorylated) BCKAD activity has been accomplished by one of three methods: (1) measuring activity after preincubation of enzyme preparation, (2) measuring activity after treatment of the preparation with a broad-specificity phosphatase, and (3) immunoassay. While the keto acid derivatives of all three BCAA are substrates for the enzyme, α-ketoisocaproate (ketoleucine) or α-ketoisovalerate (ketovaline) are routinely used to estimate BCKAD activities. Rates of the BCKAD reaction may be monitored either by measuring NADH appearance at 340 nm or $^{14}CO_2$ release from [1-^{14}C]BCKA. Considering the differences in assay methods, it is not surprising that values reported for the proportions of active and inactive enzyme in tissues differ among laboratories (Table I). In this chapter we describe assays for the estimation of the degree of activation of rat liver[7] and muscle[8] BCKAD which we have used for dietary and hormonal studies of BCAA metabolism.

[5] K. G. Cook, A. P. Bradford, S. J. Yeaman, A. Aitken, I. M. Fearnley, and J. E. Walker, *Eur. J. Biochem.* **145,** 587 (1984).
[6] R. Paxton, M. Kuntz, and R. A. Harris, *Arch. Biochem. Biophys.* **244,** 187 (1986).
[7] K. P. Block, B. W. Heywood, M. G. Buse, and A. E. Harper, *Biochem. J.* **232,** 593 (1985).
[8] R. P. Aftring, K. P. Block, and M. G. Buse, *Am. J. Physiol.* **250,** E599 (1986).

TABLE I

RAT LIVER AND MUSCLE BCKAD ACTIVITIES REPORTED BY DIFFERENT INVESTIGATORS[a]

Tissue and treatment	BCKAD Activity (mU/g tissue)		Percentage active	Comments	Ref.
	Basal	Total			
Liver					
1. Normal	612	649	94	Whole tissue extract; exogenous	17
Low protein	65[b]	189[b]	33[b]	phosphatase for total	
2. Normal	167	171	98	Homogenate with intact mitochondria;	33
Starved	279[b]	283[b]	99	preincubation for total	
3. Normal	100	230	43	Intact mitochondria; preincubation for	25
Diabetic	170[b]	210	81[b]	total; no kinase/phosphatase	
Starved	150[b]	223	67[b]	inhibitors	
4. Normal	450	820	55	Mitochondrial extract; preincubation	16
Low protein	40[b]	320[b]	13[b]	for total	
Starved	60[b]	610[b]	10[b]		
5. Normal	1509	2177	69	Homogenate with disrupted mitochon-	7
Low protein	285[b]	1669[b]	17[b]	dria; preincubation for total	
6. Normal	787	816	96	Hepatocyte extract; exogenous	19
Low protein	152[b]	302[b]	50[b]	phosphatase for total	
7. Normal	310	930	33	Mitochondrial extract; immunoassay	15
Low protein	30[b]	910	3[b]	for total	
Muscle					
1. Normal	<7	34	<20	Assay like liver, treatment 4; basal below limit of detection; diet effects unknown	16
2. Normal	1.8	22	8	Assay like liver, treatment 2; increase in percentage but no change in basal activity	33
Exercised	2.3	18	13[b]		
3. Normal	4.4	9	49	Assay like liver, treatment 3	25
Diabetic	7.6[b]	9	85[b]		
Starved	7.3[b]	9	82[b]		
4. Normal	2.4	43	5.6	Homogenate with disrupted mitochon-	34
Exercised	27.1[b]	41	66[b]	dria; preincubation for total; phosphatase but no kinase inhibitors	
5. Normal	0.84	28	3	Whole tissue extract (PEG ppt.);	8
Leucine infused	6.16[b]	28	22[b]	preincubation for total	

[a] Liver BCKAD assayed with α-ketoisocaproate[15,16,25,32] or α-ketoisovalerate.[7,17,19] Muscle BCKAD assayed with α-ketoisocaproate.
[b] Significantly different from normal controls.

Assay Method

Principle

Both assays employ [1-^{14}C]BCKA as the substrate, and enzyme rates are determined by quantification of $^{14}CO_2$ released. Tissues are isolated in the presence of phosphatase and kinase inhibitors. Furthermore, in all assays detergent is employed to eliminate transport barriers. In both assays, total BCKAD activities are estimated by preincubation of the enzyme preparation with Mg^{2+}.

Reagents

α-Keto[1-^{14}C]isovalerate and α-keto[1-^{14}C]isocaproate were synthesized from L-[1-^{14}C]valine and L-[1-^{14}C]leucine, respectively, according to the method of Rüdiger *et al.*[9] Radiolabeled amino acids were purchased from Research Products International Corporation. Polyoxyethylene ether (W1) detergent, CoA (sodium salt), thiamin pyrophosphate, dithiothreitol, α-ketoisovalerate, α-ketoisocaproate, methylbenzethonium (Hyamine) hydroxide, and polyoxyethylene 20 cetyl ether (Brij 58) detergent were purchased from Sigma Chemical Company. N-2-hydroxyethylpiperazine-N'-2-ethanesulfonic acid (HEPES), N-2-p-tosyl-L-lysine chloromethyl ketone (TLCK), and diets used in the protein studies were purchased from United States Biochemical Corporation. NAD$^+$ and leupeptin were from Boehringer Mannheim. Potassium fluoride, KH_2PO_4, ethylenediaminetetraacetic acid (EDTA), ethylene glycol bis(β-aminoethyl ether)N,N'-tetraacetic acid (EGTA), sodium arsenite, $MgSO_4$, Na_2CO_3, trichloroacetate, D-mannitol, sucrose, dichloroacetic acid, and polyethylene glycol (PEG) 6000 were of reagent quality or better and were purchased from Sigma or Fisher. α-Chloroisocaproate was a gift from Dr. Ronald Simpson, Sandoz, East Hanover, NJ. Stock solutions of the following reagents were prepared: potassium fluoride, 2.5 M; $MgSO_4$, 1 M; trichloroacetic acid, 25% (w/v); dichloroacetate, 0.5 M (pH 7.4 with 1 N KOH); PEG 6000, 27% (w/v); sodium arsenite, 100 mM; leupeptin, 20 mg/ml (in ethanol); and α-chloroisocaproate, 50 mM. The latter two solutions were kept frozen in 0.2-ml portions. Assay solutions were brought to pH 7.4 with KOH.

Liver BCKAD

Assay Solutions

Basal homogenization buffer (pH 7.4 at 4°): D-mannitol, 200 mM; sucrose, 70 mM; HEPES, 4 mM; EGTA, 0.2 mM; potassium fluoride 100 mM; α-chloroisocaproate, 0.5 mM (added fresh daily)

Total homogenization buffer: same as above with potassium fluoride and α-chloroisocaproate omitted

Basal assay buffer (pH 7.4 at 37°): HEPES, 13.3 mM; W1 detergent, 1.3%; Na$_2$CO$_3$, 13.3 mM; EGTA, 5.3 mM

Total assay buffer: same as above with potassium fluoride, 60 mM, and α-chloroisocaproate, 0.26 mM, added fresh daily

Cofactor solution (pH 7.4 at 37°): NAD$^+$, 6 mM; CoA, 4 mM; thiamin pyrophosphate, 1.6 mM; dithiothreitol, 20 mM; TLCK, 4 mM; leupeptin, 80 μg/ml. Made up in basal assay buffer immediately prior to assay

α-Keto[1-^{14}C]isovalerate solution, 16 mM (1,600,000 cpm/ml) made up in basal assay buffer fresh each day

Procedure

Livers are excised rapidly and washed in ice-cold saline. Two separate 10% homogenates are immediately prepared in a Potter-Elvehjem type homogenizer using a Teflon pestle (three strokes). One homogenate is prepared in basal homogenization buffer containing potassium fluoride and α-chloroisocaproate. The enzyme activity measured in this homogenate is used to estimate the *in vivo* activity of the complex. The second homogenate is prepared in total homogenization buffer and is used to estimate total (dephosphorylated) activity. The basal homogenate is made first to minimize interconversion of the complex. Both homogenates are kept on ice and stored in a cold room until time of analysis (usually within 2 hrs).

The following are added to each assay tube: 0.15 ml assay buffer; 0.10 ml cofactor solution, and 0.05 ml α-ketoisovalerate solution. α-Ketoisovalerate, rather than α-ketoisocaproate, is used as the substrate to avoid overestimates of BCKAD due to the cytosolic ketoleucine-preferring oxidase (~ 15% of liver ketoleucine oxidizing capacity).[10,11] After a 3-min preincubation period at 37° to attain thermal equilibrium, reactions are initiated by adding 0.1 ml of liver homogenate that has been mixed thoroughly using a Vortex agitator. The final reaction mixture (pH 7.4 at 37°) contains, in micromoles: α-keto[1-^{14}C]isovalerate, 0.8 (~ 100 cpm/nmol); NAD$^+$, 0.6; CoA, 0.4; thiamin pyrophosphate, 0.16; dithiothreitol, 2; HEPES, 20; EGTA, 0.8; Na$_2$CO$_3$, 2; leupeptin, 16; TLCK, 0.4; and 2 mg of W1 detergent and 0.1 ml of liver homogenate in a total volume of 0.4 ml. The assay medium for basal BCKAD estimation contains, in addition to

[9] H. W. Rüdiger, U. Langenbeck, and H. W. Goedde, *Biochem. J.* **126**, 445 (1972).
[10] P. J. Sabourin and L. L. Bieber, *J. Biol. Chem.* **257**, 7460 (1982).
[11] M. E. May, J. J. Mancusi, R. P. Aftring, and M. G. Buse, *Am. J. Physiol.* **239**, E215 (1980).

these components, 9 μmol potassium fluoride and 0.04 μmol α-chloroiso-caproate from the homogenization step. These compounds are included in the total assay buffer so the final assay media for both basal and total enzyme measurements are identical. Reactions are terminated after 2 min with 0.2 ml 25% trichloroacetic acid. A 0.6-ml portion of homogenate is heated in a boiling water bath for 5 min and is used as the blank (~ 100 cpm). $^{14}CO_2$ is collected in Hyamine hydroxide and quantified by liquid scintillation spectrometry. Enzyme rates are calculated as described previously.[11] Each homogenate is assayed in triplicate and routinely releases greater than 1000 cpm above background. Assays are linear with respect to time (up to 4 min) and enzyme amount (20–100 μl). Routinely, $<5\%$ of the substrate is oxidized during the reaction.

Estimation of Total Activity

For estimation of total enzyme activity, samples of homogenates made in buffer without potassium fluoride and α-chloroisocaproate are incubated at 37° for 60 min before assay. Addition of 100 mM KCl to the total homogenization buffer, to mimic the osmolarity of the basal homogenization buffer, has no effect on total BCKAD activity. MgSO$_4$ (2 mM) is routinely added to the preincubation medium to stimulate activation of the complex.[12-14] Randle and co-workers[15] have shown, in contrast to other reports, that the total amount of BCKAD, as measured by immunoassay, is not decreased by low-protein feeding (see Table I). Hence, total BCKAD activities in livers of rats fed low-protein diets may be greater than previously reported.[7,12,16-19] Total BCKAD activities, as measured by the present assay, are modestly (10–20%) but significantly depressed when expressed per gram of liver in rats fed low-protein diets[7] (Table II). Efforts to abolish this difference by adding α-chloroisocaproate (0.4, 1, 2, or 5 mM), MgSO$_4$ (2, 15, or 20 mM), and/or dinitrophenol (0.1 mM) to the preincubation mixture have been unsuccessful.[19a] Total liver BCKAD activities of adequate and low-protein fed rats are not significantly different

[12] K. P. Block, S. Soemitro, B. W. Heywood, and A. E. Harper, *J. Nutr.* **115**, 1550 (1985).
[13] R. Odessey, *Biochem. J.* **192**, 155 (1980).
[14] H. R. Fatania, P. A. Patston, and P. J. Randle, *FEBS Lett.* **158**, 234 (1983).
[15] P. A. Patston, J. Espinal, J. M. Shaw, and P. J. Randle, *Biochem. J.* **235**, 429 (1986).
[16] P. A. Patston, J. Espinal, and P. J. Randle, *Biochem. J.* **222**, 711 (1984).
[17] S. E. Gillim, R. Paxton, G. A. Cook, and R. Harris, *Biochem. Biophys. Res. Commun.* **111**, 74 (1983).
[18] R. A. Harris, S. M. Powell, R. Paxton, S. E. Gillim, and H. Nagae, *Arch. Biochem. Biophys.* **243**, 542 (1985).
[19] R. A. Harris, R. Paxton, G. W. Goodwin, and S. M. Powell, *Biochem. J.* **234**, 285 (1986).
[19a] K. P. Block, R. P. Aftring, and M. G. Buse, unpublished observations.

TABLE II
EFFECTS OF DIETARY PROTEIN AND HORMONES ON LIVER BCKAD

	BCKAD activity (mU/g tissue)		Percentage active
Treatment	Basal	Total	
Experiment 1[a]			
0% casein	239 ± 68	1360 ± 104	17 ± 4
9% casein	395 ± 77	1443 ± 48	28 ± 6
25% casein	1531 ± 103	1645 ± 122	93 ± 2
50% casein	1737 ± 84	1967 ± 16	89 ± 5
Experiment 2[b]			
9% casein	285 ± 33	1669 ± 87	17 ± 2
9% casein + glucagon	1210 ± 36	1912 ± 111	64 ± 4
9% casein + insulin	842 ± 94	1583 ± 65	53 ± 5

[a] Rats were trained to consume a 25% casein diet for 6 hrs/day for 2 weeks. After this period, they were meal-fed one of the four diets listed for 10 days. Rats were killed 3 hrs postprandially and BCKAD activities measured using 2 mM α-keto[1-^{14}C]isovalerate. Numbers are mean $\pm$ SEM for 5 rats/diet.[19a]

[b] Rats were fed a 9% casein low-protein diet *ad libitum* and administered saline, glucagon (200 μg/100 g body weight), or insulin (1 U/100 g body weight). BCKAD was assayed with 2 mM α-keto[1-^{14}C]isovalerate. Numbers are mean $\pm$ SEM for 4–5 rats/group.[7]

when expressed per gram of tissue protein.[19a] Whether the decrease in total BCKAD activity in low-protein fed rats we and others have observed represents a decrease in BCKAD protein remains to be established.

Kinetics of Liver BCKAD

K_m and V_{max} values of the fully activated complex from liver isolated from normally fed rats were 58 ± 4 μM and 1528 ± 21 mU/g tissue, respectively (mean $\pm$ SEM from four preparations using nine concentrations of α-ketoisovalerate from 5 μM to 4 mM).[19a]

Stability

Once homogenized, basal and total BCKAD activities from adequate or low-protein fed rats remain stable for at least 5 hr. Storage of basal or total homogenates for 24 hr at $-20°$ or $4°$ results in decreases in basal and total activities. The decrease in total activity after 1 day of storage is particularly marked in homogenates isolated from low-protein fed rats.

Preincubation of homogenates for greater than 2 hr at 37° results in decreases in total activity.

Comments

The yield of measurable BCKAD using the present assay is greater than that previously reported (Table I). Our value of 1.5–2 mU/g tissue for total liver BCKAD is quite similar to the total activities reported for the pyruvate dehydrogenase complex of rat liver (1.6–2.4 mU/g).[20] The high yield of BCKAD is probably due to the use of whole homogenates rather than tissue or mitochondrial extracts. Zhang *et al.*[20a] have reported that inclusion of sodium fluoride and α-ketoisocaproate in the isolation medium of liver mitochondria results in partial inactivation of BCKAD. The use of homogenates and the rapid processing of samples may account for the lack of an inhibitory effect of fluoride and α-chloroisocaproate on BCKAD in the present assay.[7] The use of homogenates, rather than extracts, also allows one to analyze a large number of samples in a relatively short time. The ease of sample preparation makes this assay attractive, particularly for large-scale dietary studies.

The contribution of the mitochondrial "activator protein" described by Randle and co-workers[21] to the total BCKAD activity cannot be assessed using the present assay. Responses of rat liver BCKAD to dietary protein and hormonal treatments using the present assay are given in Table II.

This assay system can also be used to measure the proportions of active and inactive kidney BCKAD. In normally fed rats, basal and total kidney BCKAD activities are 350 ± 22 and 1160 ± 67 mU/g tissue, respectively (mean $\pm$ SEM of 12 rats; $31 \pm 3\%$ active). The basal BCKAD activity measured in kidney using this assay is comparable to those reported by others.[16,17,22] However, total BCKAD activity measured in kidney using the present assay is 2–5 times greater than previously reported. Indeed, Hutson[23] has suggested that estimates of tissue BCKAD activities may be low. Reappraisal of tissue BCKAD capacity is needed to improve understanding of the contributions of different tissues to whole-body BCAA metabolism. Quantification of solubilized muscle BCKAD activity requires a different assay system (see below).

[20] J. G. McCormack, *Biochem. J.* **231**, 597 (1985).
[20a] B. Zhang, R. Paxton, G. W. Goodwin, Y. Shimomura, and R. A. Harris, *Biochem. J.* **246**, 625 (1987).
[21] J. Espinal, P. A. Patston, H. R. Fatania, K. S. Lau, and P. J. Randle, *Biochem. J.* **225**, 509 (1985).
[22] A. J. M. Wagenmakers, J. T. G. Schepens, J. A. M. Veldhuizen, and J. H. Veerkamp, *Biochem. J.* **220**, 273 (1984).
[23] S. M. Hutson, *J. Biol. Chem.* **261**, 4420 (1986).

Muscle BCKAD

Assay Solutions

Extraction buffer (pH 7.4 at 4°): KH_2PO_4, 100 mM; W1 detergent, 5% (w/v); EDTA, 5 mM; dithiothreitol, 5 mM; thiamin pyrophosphate, 0.4 mM; TLCK, 1 mM; leupeptin, 20 μg/ml; potassium fluoride, 50 mM; dichloroacetate, 5 mM (last six added fresh daily)

Resuspension buffer (pH 7.4 at 37°): HEPES, 25 mM; Brij 58, 0.2%; EDTA, 0.2 mM; dithiothreitol, 1 mM; thiamin pyrophosphate, 0.4 mM (last 2 added fresh daily)

Cofactor solution (pH 7.4 at 37°): NAD$^+$, 1.75 mM; CoA, 1.75 mM; leupeptin, 34.6 μg/ml; MgSO$_4$, 3.5 mM. Made up in resuspension buffer immediately prior to assay

[1-^{14}C]BCKA solution, 0.7 mM (2 × 10^6 cpm/ml) made up in resuspension buffer fresh each day.

Procedure

Hindlimb muscles (primarily gastrocnemius) are frozen *in situ* with Wollenberger tongs precooled in liquid nitrogen. Frozen muscles are ground in a porcelain mortar and pestle which have been cooled in liquid nitrogen. Cold temperatures are used to prevent interconversion of the complex prior to addition of phosphatase and kinase inhibitors. The powdered sample is then homogenized in ice-cold extraction buffer (5 ml/g muscle) containing potassium fluoride and dichloroacetate. Routinely, a Potter-Elvehjem homogenizer fitted with a Teflon pestle (10 strokes) is used; however, a Polytron may be used if desired (30 sec at a setting of 5). The homogenate is centrifuged at 27,000 g_{av} for 5 min to remove cell debris. The supernatant volume is determined and 0.5 volumes of 27% PEG 6000 is then added to precipitate large-molecular-weight proteins (final PEG concentration is 9%). After standing on ice for 20 min, the mixture is centrifuged at 12,000 g_{av} for 10 min. The supernatant is discarded and the pellet resuspended with resuspension buffer (1 ml/g muscle). Leupeptin is added to the resuspended pellet for a final concentration of 20 μg/ml.

The following are added to assay tubes used to estimate basal activities: 0.05 ml resuspension buffer; 0.1 ml cofactor solution; 0.15 ml enzyme extract. After a 3-min preincubation period at 37° to reach thermal equilibrium, reactions are begun by adding 0.05 ml of the 0.7 mM [1-^{14}C]BCKA solution. The basal final reaction mixture (pH 7.4 at 37°) contains, in micromoles: [1-^{14}C]BCKA, 0.035 (~2900 cpm/nmol); NAD$^+$, 0.175; CoA, 0.175; thiamin pyrophosphate, 0.14; dithiothreitol, 0.35;

HEPES, 8.75; EDTA, 0.07; MgSO$_4$, 0.35; leupeptin, 14; and 0.7 mg Brij 58 in a total volume of 0.35 ml. Increasing the concentration of BCKA beyond 0.1 mM requires increasing the specific radioactivity of the substrate. In addition to the prohibitive cost, use of more radioactivity decreases the sensitivity of the assay by increasing background counts. Blanks are prepared by adding 25 μl of a 100 mM sodium arsenite solution to the tubes prior to addition of enzyme extract. Samples heated in a boiling water bath may also be used; however, these tend to gel and are therefore difficult to pipet. Reactions are terminated after 5 min with 0.2 ml of 25% trichloroacetic acid, and enzyme rates are calculated as described previously. Each sample is assayed in triplicate, greater than 1000 cpm above background are released, and less than 10% of the BCKA added is used during the reaction.

Estimation of Total Activity

Maximal total activities are obtained after 40 min of preincubation of the enzyme preparation from rats fed 25% (normal) or 50% (high) casein diets. However, a lag in BCKAD activation occurs in muscle preparations from rats fed low-protein diets.[24] Maximal total activities in the latter group were achieved after 60 min of preincubation. Whether this is due to a greater kinase and/or a lower phosphatase activity remains to be established.

Total BCKAD activities are routinely measured after preincubating the enzyme extract in resuspension buffer (1 part extract: 1 part buffer) with 15 mM MgSO$_4$ at 37° for 60 min before assay. The following are added to assay tubes used to estimate total activities: 0.175 ml resuspension buffer; 0.1 ml cofactor solution; and 0.05 ml [1-^{14}C]BCKA solution. Reactions are started by adding 0.025 ml of preincubated extract for a total final volume of 0.35 ml. After 1 min, reactions are terminated with 0.2 ml of 25% trichloroacetate.

Kinetics of Muscle BCKAD

Due to the low basal activities of muscle BCKAD, kinetic analysis is only possible for the fully activated complex. For α-ketoisocaproate, a K_m of 90 ± 10 μM and V_{max} of 79 ± 2 mU/g tissue were derived (mean ± SEM from five preparations using eight concentrations of α-ketoisocaproate from 10 to 600 μM).[8] For α-ketoisovalerate, a K_m of 154 ± 8 μM and V_{max} of 160 ± 12 mU/g tissue were derived.[24a]

[24] K. P. Block, R. P. Aftring, W. B. Mehard, and M. G. Buse, *J. Clin. Invest.* **79**, 1349 (1987).
[24a] K. P. Block, W. B. Richmond, W. B. Mehard, and M. G. Buse, *Am. J. Physiol.* **252**, E396 (1987).
[25] H. S. Paul and S. A. Adibi, *J. Biol. Chem.* **257**, 4875 (1982).

PEG Precipitation

In agreement with most other investigators,[4,16,25-27] we were unable to detect BCKAD activity in detergent-solubilized muscle homogenates or with mitochondria prepared in Chappel Perry buffer and treated with detergent or calcium.[8,28] Addition of E3, which is easily lost upon purification,[29] had no effect. However, after partial "purification" with PEG, activity of the complex was measurable. The effect of detergent or calcium on muscle BCKAD may be due to release of an inhibitor.[26,30] Precipitation of the complex would be expected to result in dilution of this putative inhibitor. Alternatively, PEG precipitation may result in measurable basal and total BCKAD activities as the result of (1) concentration of the complex ($\sim$ 5-fold) and (2) coprecipitation of a phosphatase. The latter hypothesis is strengthened by the observation that BCKAD phosphatase is easily lost from the complex during purification.[31]

Precipitation with PEG may result in some loss of BCKAD from muscle. However, our calculated V_{max} values of 79 and 160 mU/g tissue for α-ketoisocaproate and α-ketoisovalerate, respectively, are similar to those reported by Hutson using highly coupled skeletal muscle mitochondria.[23]

Stability

Basal activity is stable for at least 90 min when stored on ice and for at least 75 min when incubated at 37°. The muscle complex showed signs of activation only when extra Mg^{2+} was included in the preincubation buffer. In the present assay, activation was not affected by $MnCl_2$, protamine sulfate, or insulin.[8] However, chelation of Mg^{2+} by addition of 14 mM EDTA blocks activation.[24a] Also, addition of 50 mM KF completely blocks activation of muscle BCKAD.[24a] The effect of F^- on activation is probably due to precipitation of free Mg^{2+}. Purified BCKAD phosphatase from bovine kidney, unlike pyruvate dehydrogenase phosphatase, does not appear to require Mg^{2+} or Ca^{2+} for activity.[31] However, we and others have found Mg^{2+} to be essential for activation of muscle BCKAD.[8,12,13,14,26] An inhibitor protein of the phosphatase has been isolated from mitochondria.[32] Mg^{2+} has been shown to inhibit this protein. Also, Mg^{2+} reverses

[26] R. Odessey and A. L. Goldberg, *Biochem. J.* **178**, 475 (1979).
[27] V. W. M. van Hinsbergh, J. H. Veerkamp, and J. F. C. Glatz, *Biochem. J.* **182**, 353 (1979).
[28] P. H. Crowell and A. E. Harper, unpublished observations (1985).
[29] F. H. Pettit, S. J. Yeaman, and L. J. Reed, *Proc. Natl. Acad. Sci. U.S.A.* **75**, 4881 (1978).
[30] H. S. Paul and S. A. Adibi, *J. Biol. Chem.* **257**, 12581 (1982).
[31] Z. Damuni, M. L. Merryfield, J. S. Humphreys, and L. J. Reed, *Proc. Natl. Acad. Sci. U.S.A.* **81**, 4335 (1984).
[32] Z. Damuni, J. S. Humphreys, and L. J. Reed, *Proc. Natl. Acad. Sci. U.S.A.* **83**, 285 (1986).

nucleotide-mediated inhibition of BCKAD phosphatase.[31] Hence, Mg^{2+} may work indirectly by allowing the phosphatase to function.

BCKAD activity of powdered muscle remains stable for at least 24 hr when stored at liquid nitrogen temperatures.

Comments

The use of solubilized muscle preparations appears necessary to detect changes in the activity state of BCKAD. Using intact muscle mitochondria, we have been unable to detect alterations in muscle BCKAD after: high-protein feeding, glucocorticoid treatment, high-leucine feeding, diabetes, and portacaval anastomosis.[28] However, each of these treatments results in activation of muscle BCKAD when a solubilized preparation is

TABLE III
EFFECTS OF DIETARY PROTEIN AND
GLUCOCORTICOIDS ON MUSCLE BCKAD

Treatment	BCKAD activity (mU/g tissue)		Percentage active
	Basal	Total	
Experiment 1[a]			
0% casein	2.0 ± 0.4	69 ± 6	3.0 ± 0.8
9% casein	1.2 ± 0.1	70 ± 8	1.9 ± 0.2
25% casein	10.0 ± 2.1	76 ± 6	12.7 ± 2.2
50% casein	31.6 ± 6.6	79 ± 6	43.2 ± 9.6
Experiment 2[b]			
Saline	1.1 ± 0.3	65 ± 7	2.1 ± 0.2
Glucocorticoid	6.3 ± 2.2	62 ± 8	9.9 ± 0.9

[a] Rats were treated as described in Table II for Experiment 1. Values shown are 3 hr postprandial BCKAD activities. The percentage active BCKAD before consumption of the 0, 9, 25, and 50% casein diets were 4.4 ± 0.9, 2.0 ± 0.8, 1.8 ± 0.2, and $7.3 \pm 1.9\%$, respectively (no change in total activity). BCKAD was assayed with 0.1 mM α-keto[1-^{14}C]isovalerate. Numbers are mean $\pm$ SEM for 5 (postabsorptive) or 8–10 (postprandial) rats/diet.[24]

[b] Rats were fed a normal protein diet *ad libitum*. After an overnight fast, rats were administered saline or the glucocorticoid 6α-methylprednisolone (125 mg/kg) and killed 4 hrs later. BCKAD was assayed with 0.1 mM α-keto[1-^{14}C]isovalerate. Numbers are mean $\pm$ SEM for 11 rats/group.[24a]

used as the enzyme source.[19a,24,24a,32a] Wagenmakers *et al.*,[33] using intact mitochondria, were unable to demonstrate an effect of exercise on muscle BCKAD. However, Kasperek *et al.*,[34] using detergent-treated muscle tissue, measured a marked activation of muscle BCKAD after rats were exercised (see Table I). A similar discrepancy between the effects of dietary protein on BCKAD from intact and disrupted liver mitochondria has been described.[4] A ΔpH-responsive mitochondrial transport system for BCKA has been characterized in heart and skeletal muscle.[23] In addition to influences of substrate transport and intramitochondrial BCAA aminotransferase activity on flux through BCKAD,[35] the actual precursor specific radioactivity of BCKA in intact mitochondria is unknown and may differ between treatment groups. Also, isolation of intact mitochondria may lead to partial activation of the enzyme. In the present assay, only approximately 2% of muscle BCKAD in normal rats is in the active form. This is probably due to the retardation of activation by freeze clamping the muscle *in situ* and the use of fluoride during extraction of the enzyme. As activation during isolation is prevented, effects of different stimuli may be more readily detected. Responses of rat muscle BCKAD to dietary protein and glucocorticoid treatment using the present assay are given in Table III.

Acknowledgments

The work described in this chapter was supported by grants from the U.S. Public Health Service, National Institutes of Health (AM 02001, AM 10748 and P30 AM 26659). K. P. B. was a recipient of postdoctoral research fellowships from the American Liver Foundation and the Medical University of South Carolina. R. P. A. was a recipient of a Medical Scientist Traineeship from the Medical University of South Carolina. The authors thank Barbara Whitlock for expert secretarial assistance.

[32a] R. P. Aftring, W. J. Miller, and M. G. Buse, *Am. J. Physiol.* **254**, E292 (1988).

[33] A. J. M. Wagenmakers, J. T. G. Schepens, and J. H. Veerkamp, *Biochem. J.* **223**, 815 (1984).

[34] G. J. Kasperek, G. L. Dohm, and R. D. Snider, *Am. J. Physiol.* **248**, R166 (1985).

[35] S. M. Hutson, D. Fenstermacher, and C. Maher, *J. Biol. Chem.* **263**, 3618 (1988).

[25] Assay of 3-Methylglutaconyl-CoA Hydratase

By KENNETH M. GIBSON

Introduction

3-Methylglutaconyl-CoA hydratase (3-MG-CoA hydratase, 3-hydroxy-3-methylglutaryl-CoA hydro-lyase, E.C. 4.2.1.18) catalyzes the reversible hydration of 3-methylglutaconyl-CoA to 3-hydroxy-3-methylglutaryl-CoA, the penultimate reaction in leucine catabolism. The reaction is shown in Eq. (1).

$$HOOC - CH_2 - \underset{\underset{CH_3}{|}}{C} = CH - \underset{\underset{}{||}}{\overset{O}{C}} - SCoA \quad + \quad H_2O \; \rightleftharpoons \; HOOC - CH_2 - \underset{\underset{CH_3}{|}}{\overset{OH}{\underset{}{C}}} - CH_2 - \overset{O}{\overset{||}{C}} - SCoA$$

$$\textbf{3-MG-CoA} \qquad\qquad\qquad\qquad\qquad\qquad \textbf{HMG-CoA} \quad (1)$$

Using stereospecifically tritiated substrates, Messner and co-workers[1] have shown that the action of 3-MG-CoA hydratase from sheep liver is stereospecific with the addition–elimination of water being *syn*. Further evidence from the same study indicated that the 3-MG-CoA substrate for this enzyme is geometrically the *E* form. The product of the hydration reaction is the diastereomer with the *(S)* configuration at the newly generated chiral carbon.

Assay Method

Principle

A new method for the assay of 3-MG-CoA hydratase has recently been developed.[2] In this procedure, the substrate for the reaction, 3-methyl[5-^{14}C]glutaconyl-CoA ([5-^{14}C]-labeled 3-MG-CoA) is synthesized using 3-methylcrotonyl-CoA carboxylase purified from bovine kidney. In this way the specific *trans* isomer (*E* form) of 3-MG-CoA, which is the natural substrate, is produced, and difficulties associated with mixtures of isomers *(E, Z)* are avoided.[3] The products of the reaction following alkaline hydrol-

[1] B. Messner, H. Eggerer, J. W. Cornforth, and R. Mallaby, *Eur. J. Biochem.* **53**, 255 (1975).
[2] K. Narisawa, K. M. Gibson, L. Sweetman, W. L. Nyhan, M. Duran, and S. K. Wadman, *J. Clin. Invest.* **77**, 1148 (1986).
[3] L. K. Massey, R. S. Conrad, and J. R. Sokatch, *J. Bacteriol.* **118**, 112 (1974).

ysis of coenzyme-A esters, 3-hydroxy-3-methyl[1-^{14}C]glutaric acid and 3-hydroxy[1-^{14}C]butyric acid, are quantified by reversed-phase high-performance liquid chromatography.

Substrate Preparation and Characterization

For the enzymatic synthesis of 5-^{14}C-labeled 3-MG-CoA, 3-methyl-crotonyl-CoA carboxylase (MCC) is purified from frozen bovine kidney by modification of the method of Lau et al.[4] MCC in the crude extract is precipitated with polyethylene glycol 6000 and chromatographed on DEAE-cellulose. After hydroxylapatite chromatography, gel filtration is performed on a column of Sepharose 6B (1 cm i.d. × 47.5 cm) equilibrated and eluted with 20 mM potassium phosphate, pH 7.0, containing 0.1 mM ethylenediaminetetraacetic acid (EDTA), 0.1 mM dithiothreitol, and 10% glycerol. The specific activity of the 600-fold purified MCC was 1.6 μmol/min mg protein. MCC activity was determined using a modification of the procedure of Weyler et al.,[5] by fixation of NaH^{14}CO$_3$ into acid nonvolatile products using 3-methylcrotonyl-CoA and ATP as substrates.

The product of MCC is the *trans* isomer (*E* form) of 3-MG-CoA.[6] That this is the isomer which is the substrate for 3-MG-CoA hydratase is indicated by the fact that *E*-3-MG-CoA is the product of the reverse reaction, the dehydration of HMG-CoA catalyzed by 3-MG-CoA hydratase.[1] Purified MCC (approximately 0.025 U) is added to 1 ml of solution containing the following: 100 mM Tris(Cl$-$), pH 8.0; 50 mM KCl; 8 mM MgCl$_2$; 0.25 mM EDTA; 3 mM ATP; 1.8 mM 3-methylcrotonyl-CoA; 10mM NaH^{14}CO$_3$ (New England Nuclear, Boston, MA; 9.2 mCi/mmol). The mixture is incubated for 2 hr at 30°. The pH is adjusted to 2–3 with 5 M HCl, and unreacted H^{14}CO$_3^-$ removed as ^{14}CO$_2$ by the addition of solid CO$_2$. The product 5-^{14}C-labeled 3-MG-CoA is purified by chromatography on a DEAE-Sephadex A-25 column (0.6 cm i.d. × 30 cm) using a linear gradient of LiCl (140 ml 0.065 M to 140 ml 0.2 M) in 5 mM HCl, modified from the method of Bartlett and Gompertz.[7] Chromatography is performed at room temperature and a flow rate of 0.7 ml/min. The fractions containing 5-^{14}C-labeled 3-MG-CoA are pooled, lyophilized, and LiCl is removed by chromatography on a Sephadex G-10 column (0.6 cm i.d. × 50 cm), equilibrated and eluted with distilled water.

[4] E. P. Lau, B. C. Cochran, and R. R. Fall, *Arch. Biochem. Biophys.* **205**, 352 (1980).
[5] W. Weyler, L. Sweetman, D. C. Maggio, and W. L. Nyhan, *Clin. Chim. Acta* **76**, 321 (1977).
[6] F. Lynen, J. Knappe, E. Lorch, G. Jutting, E. Ringelmann, and J. P. Lachance, *Biochem. Z.* **335**, 123 (1961).
[7] K. Bartlett and D. Gompertz, *Biochem. Med.* **10**, 15 (1974).

The specific activity of the purified 5-[14C]-labeled 3-MG-CoA is determined by enzymatic hydration to 3-hydroxy-3-methyl[5-[14C]glutaryl-CoA (5-[14C]HMG-CoA) with an excess of crotonase (Sigma, St. Louis, MO) and comparison of this with authentic [3-[14C]-HMG-CoA of known specific activity (New England Nuclear). Both compounds are analyzed by reversed-phase high-performance liquid chromatography.[8] The quantities are determined from chromatographic peak areas following UV detection at 254 nm and radioactivity in collected fractions measured by liquid scintillation counting. Results are corrected for recoveries of authentic [3-[14C]HMG-CoA. The mean value of duplicate determinations of the specific activity of [5-[14C]HMG-CoA derived from 5-[14C]-labeled 3-MG-CoA was 5.8 mCi/mmol, approximately 63% of the specific activity of the precursor $NaH^{14}CO_3$. The data indicate a considerable dilution by atmospheric CO_2 in the pH 8 Tris buffer used for the enzymatic synthesis of 5-[14C]-labeled 3-MG-CoA. The mean radiochemical purity on reversed-phase high-performance liquid chromatography of the [5-[14C]HMG-CoA derived from 5-[14C]-labeled 3-MG-CoA was 69%. Essentially all of the rest of the radioactivity is present as 3-methyl[5-[14C]glutaconic acid, which presumably arises from nonspecific hydrolysis of 5-[14C]-labeled 3-MG-CoA during enzymatic preparation with MCC. The concentration of substrate 5-[14C]-labeled 3-MG-CoA is determined by measuring its radioactivity and correcting for chemical and radiochemical purity. The solution containing 5-[14C]- labeled 3-MG-CoA is concentrated to about 0.25 mM and stored at $-20°$.

Procedure

Cultured human fibroblasts and lymphocytes derived from whole blood are obtained and washed as described in the next chapter, with the exception that pelleted cells are resuspended in 50 mM Tris(Cl$-$), pH 7.5, supplemented with 0.75 mM Na$_2$EDTA prior to sonic disruption. The standard assay contains (final volume, 0.1 ml): cell extract (20–30 μg protein), 100 mM potassium phosphate, pH 7.0, 0.055 mM 5-[14C]-labeled 3-MG-CoA (approximately 0.032 μCi), 3-hydroxybutyrate dehydrogenase (2.5 U/ml), and 2 mM NADH. Incubation is carried out for 30 min at 30°, and the reaction terminated by addition of 10 μl ice-cold 4.2 M HClO$_4$. Precipitated protein is removed by centrifugation, and the resulting supernatant neutralized by addition of 1.5 M potassium carbonate containing 0.25 M triethanolamine. Insoluble potassium perchlorate is removed by centrifugation, and the subsequent supernatant adjusted to pH 11–12 by

[8] M. S. DeBuysere and M. S. Olson, *Anal. Biochem.* **133**, 373 (1983).

addition of 1 M KOH. After cooling on ice and centrifuging, the supernatant is heated in an Eppendorf tube (1.5 ml capacity) for 30 min at 60° to hydrolyze coenzyme-A esters. Following hydrolysis, the pH is adjusted to 2–3 by addition of 1 M phosphoric acid and 0.1 ml is analyzed by reversed-phase high-performance liquid chromatography.

The chromatographic system employed is identical to that described in the next chapter, with the following differences: from 0 to 35 min of analysis, the column is eluted with buffer 1, from 35 min to 60 min 55% buffer 1 and 45% methanol are passed through the column, followed by elution with buffer 1 from 60 min to 90 min to return to the starting conditions. The flow rate is 0.44 ml/min, and the isotope content of collected 1-min fractions is quantified by liquid scintillation counting with 5 ml of scintillant. Using these conditions, 3-hydroxy[1-^{14}C]butyric and 3-hydroxy-3-methyl[1-^{14}C]glutaric acids elute at 20 and 26 min, respectively, whereas 3-methyl[5-^{14}C]glutaconic acid is eluted at 46 min. Studies on the separation of the substrate and products as acyl-CoA esters by reversed-phase high-performance liquid chromatography indicate a small amount of isomerization of substrate 5-^{14}C-labeled E-3-MG-CoA to 5-^{14}C-labeled Z-3-MG-CoA. The Z isomer coelutes with [5-^{14}C]HMG-CoA, the product of 3-MG-CoA hydratase. As a result, acyl-CoA intermediates are hydrolyzed and the chromatographic conditions modified to separate the free acids. The action of endogenous HMG-CoA lyase leaves a small amount of product [5-^{14}C]HMG-CoA to [1-^{14}C]acetoacetic acid, and for this reason 3-hydroxybutyrate dehydrogenase and NADH are included in the assay to generate the more stable 3-hydroxy[1-^{14}C]butyric acid. When calculated as the sum of isotope in 3-hydroxybutyric acid and HMG, 3-MG-CoA hydratase activity in fibroblast extracts is proportional to the time of incubation up to at least 90 min and to protein concentration up to 150 μg of protein per assay. In lymphocyte sonicates, enzyme activity is proportional to the time of incubation up to 60 min and to protein concentration up to 90 μg protein per assay.

Properties of the Human Enzyme

The Michaelis constant (K_m) in extracts of cultured human fibroblasts is 6.9 μM and the V_{max} value approximately 0.6 nmol/min mg protein.[2] The same values in extracts of human lymphocytes are 9.4 μM and approximately 1.1–1.4 nmol/min mg protein, respectively. The total activity of 3-MG-CoA hydratase in fibroblast extracts shows little variation as a function of pH from 6.0 to 8.0, although there is an optimum between pH 7.0 and 7.5.

Alternative Methods

3-MG-CoA hydratase activity has been estimated in biopsied muscle in forward and reverse (dehydration) directions, although specific methodology has not been presented.[9] Similarly, there are preliminary reports in the literature of assay of 3-MG-CoA hydratase in extracts of cultured human fibroblasts[10,11] without description of the methods employed. 3-MG-CoA hydratase has been purified approximately 100-fold with 30% yield[12] and 50-fold with 22% yield[13] from sheep liver. In these studies, 3-MG-CoA hydratase activity was estimated by ultraviolet absorption of the 2,3-unsaturation in 3-methylglutaconyl-CoA at 275 nm.

Concluding Remarks

The assay described is highly sensitive, and the use of radiolabeled substrate allows reliable determinations of enzyme activity in crude extracts. Given the high activity of HMG-CoA lyase in comparison to 3-MG-CoA hydratase in both fibroblast and lymphocyte extracts, it would appear difficult to carry out the 3-MG-CoA hydratase reaction in reverse (the dehydration of HMG-CoA) as a method of quantification of hydratase activity. The method for preparation of 5-^{14}C-labeled 3-MG-CoA is lengthy and time-comsuming, employing purification of enzyme and substrate. However, the use of a mammalian source of MCC appears at the present time to be the only way of obtaining the specific *(E)* isomer needed for the assay of 3-MG-CoA hydratase in mammalian tissues.

[9] B. H. Robinson, W. G. Sherwood, M. Lampty, and J. A. Lowden, *Pediatr. Res.* 10, 371 (1976).
[10] S. J. Wysocki, R. Hahnel, R. J. W. Truscott, B. Halpern, and B. Wilcken, *Lancet* 2, 371 (1979).
[11] R. J. W. Truscott, B. Halpern, S. J. Wysocki, R. Hahnel, and B. Wilcken, *Clin. Chim. Acta* 95, 11 (1979).
[12] V. R. Villanueva and F. Lynen, *C. R. Hebd. Seances Acad. Sci., Ser. D* 270, 3318 (1970).
[13] H. Hilz, J. Knappe, E. Ringelmann, and F. Lynen, *Biochem. Z.* 329, 476 (1958).

[26] Assay of 3-Hydroxy-3-Methylglutaryl-CoA Lyase

By KENNETH M. GIBSON

Introduction

3-Hydroxy-3-methylglutaryl-CoA (HMG-CoA) lyase (EC 4.1.3.4) catalyzes the cleavage of HMG-CoA to acetoacetic acid and acetyl-CoA by a Claisen-type retrocondensation reaction.[1] The reaction is shown in Eq. (1).

$$\underset{\textbf{HMG-CoA}}{\overset{\displaystyle \overset{OH}{\underset{\underset{CH_3}{|}}{\overset{|}{C}}}}{HOOC - CH_2 - \overset{O}{\overset{\|}{C}} - CH_2 - \overset{O}{\overset{\|}{C}} - SCoA}} \longrightarrow \underset{\textbf{Acetoacetic Acid}}{HOOC - CH_2 - \overset{O}{\overset{\|}{C}} - CH_3} \quad + \quad \underset{\textbf{Acetyl-CoA}}{CH_3 - \overset{O}{\overset{\|}{C}} - SCoA} \quad (1)$$

The enzyme acts only on the diastereomer with the (S) configuration at the chiral carbon.[2] HMG-CoA is an important intermediate in cholesterogenesis, ketogenesis, and leucine catabolism,[3] and the lyase reaction has a direct role in the latter two processes.

Assay Methods

Principle

Two radiometric procedures are described. The two differ essentially in the method of measurement of the product; both employ 3-hydroxy-3-methyl[3-^{14}C]glutaryl-CoA ([3-^{14}C]HMG-CoA) as the precursor. In the method of Clinkenbeard et al.[4] [3-^{14}C]acetoacetic acid production is monitored by acidification and volatilization of the labeled product. In essence, the product is estimated by the decrease of total radioactivity following volatilization. In the assay described by Gibson and co-workers,[5] the pro-

---

[1] P. R. Kramer and H. M. Miziorko, Biochemistry 22, 2353 (1983).
[2] L. D. Stegink and M. J. Coon, this series, Vol. 17, p. 823.
[3] R. Deana, R. Meneghello, L. Manzi, and C. Gregolin, Biochem. J. 138, 481 (1974).
[4] K. D. Clinkenbeard, W. D. Reed, R. A. Mooney, and M. D. Lane, J. Biol. Chem. 250, 3108 (1975).
[5] K. M. Gibson, L. Sweetman, W. L. Nyhan, T. M. Page, C. Greene, and H. M. Cann, Clin. Chim. Acta 126, 171 (1982).

duction of [3-^{14}C]acetoacetic acid is monitored directly. Various methods are employed, including liquid partition chromatography following enzymatic transformation of the labeled acetoacetic acid into 3-hydroxybutyric acid, and both anion-exchange and reversed-phase high-performance liquid chromatography.

Procedures

The method of Clinkenbeard *et al.*,[4] originally applied to the assay of chicken and rat liver preparations, is widely employed and has been adapted to the assay of extracts of cultured human fibroblasts,[6] human lymphocytes,[7-9] and biopsied human liver.[10] Advantage is taken of the fact that, in contrast to [3-^{14}C]acetoacetic acid, [3-^{14}C]HMG-CoA is not volatile when taken to dryness in 6 M HCl at 95°.

The reaction is initiated by the addition of 0.5 to 3.5 mU (nanomoles of substrate converted/minute) of HMG-CoA lyase to a reaction mixture containing 40 nmol of (R,S)-[3-^{14}C] HMG-CoA (specific activity, 5 to 20 × 10^5 dpm/μmol) in a total volume of 0.2 ml 100 mM Tris(Cl$^-$), pH 8.2. After 2, 4, 6, and 8 min of incubation at 30°, 0.04-ml aliquots are transferred to glass vials containing 0.1 ml of 6 M HCl and taken to dryness at 95°. Water and scintillant are added and nonvolatile ^{14}C content determined using a scintillation spectrophotometer. Formation of product is linear with time up to 8 min in the presence of as much as 14 μg of mitochondrial matrix protein. The addition of 5 mM dithiothreitol to the assay mixture increases HMG-CoA lyase activity by approximately 1.5-fold.[4]

The assay of HMG-CoA lyase described by Gibson *et al.*[5] has been applied to extracts of cultured human fibroblasts and isolated human lymphocytes.[11] Following trypsinization, harvested fibroblasts are washed with phosphate-buffered saline. Lymphocytes from whole human blood are washed following isolation by density centrifugation.[12] Pelleted cells are resuspended to a concentration of 5 × 10^6/ml in 100 mM Tris(Cl$^-$), pH 8.0, and disrupted by sonication at 24 W with two 10-sec bursts and a

[6] T. E. Stacey, C. de Sousa, B. M. Tracey, A. Whitelaw, J. Mistry, P. Timbrell, and R. A. Chalmers, *Eur. J. Pediatr.* **144,** 177 (1985).

[7] D. Leupold, M. Bojasch, and C. Jakobs, *Eur. J. Pediatr.* **138,** 73 (1982).

[8] M. Duran, R. B. H. Schutgens, A. Ketel, H. Heymans, M. W. J. Berntssen, D. Ketting, and S. K. Wadman, *J. Pediatr.* **95,** 1004 (1979).

[9] S. J. Wysocki and R. Hahnel, *Clin. Chim. Acta* **73,** 373 (1976).

[10] R. B. H. Schutgens, H. Heymans, A. Ketel, H. A. Veder, M. Duran, D. Ketting, and S. K. Wadman, *J. Pediatr.* **94,** 89 (1979).

[11] W. G. Wilson, M. B. Cass, O. Sovik, K. M. Gibson, and L. Sweetman, *Eur. J. Pediatr.* **142,** 289 (1984).

[12] A. Boyum, *Scand. J. Clin. Lab. Invest.* **77** (Suppl. 97), 77 (1968).

10-sec pause between. The cell extract is centrifuged briefly to remove debris and unbroken cells. The standard reaction mixture contains the following: 100 mM Tris(Cl$^-$), pH 8.0; 5 mM MgCl$_2$; 5 mM NADH; 0.09 mM (S isomer only) [3-^{14}C]HMG-CoA (diluted to a specific activity of 5mCi/mmol), and 1 unit of 3-hydroxybutyrate dehydrogenase. Cell extract (20 μl) is added to the reaction mixture and incubated for 10 min at 37°. The final assay volume is 0.1 ml. NADH and 3-hydroxybutyrate dehydrogenase are omitted when anion-exchange high-performance liquid chromatography is used for the measurement of product.[5] The reaction is terminated by the addition of 10 μl of ice-cold 4.2 M HClO$_4$ and, following centrifugation to remove precipitated protein, the supernatant neutralized with 6 μl of 6 M KOH. Insoluble potassium perchlorate is removed by centrifugation. For analysis of the product by liquid partition chromatography,[13] the reaction is terminated with 10 μl 2.5 M H$_2$SO$_4$ without neutralization. HMG-CoA lyase activity, expressed as pmol [3-^{14}C]acetoacetic acid produced/min mg crude fibroblast protein, is linear with time up to 20 min in the presence of as much as 200 μg of protein.

The instrument employed for quantification and identification of [3-^{14}C]acetoacetic acid by anion-exchange high-performance liquid chromatography has been described.[14] Substrate and product are separated on a stainless steel column (0.16 cm i.d. × 70 cm) containing Aminex A-25 anion-exchange resin, Cl$^-$ form (Bio-Rad Laboratories, Richmond, CA). Column effluent is first analyzed for UV-absorbing components at 254 nm, after which radioactive components are quantified by continuous liquid scintillation counting. A stepwise elution system is employed, using as starting buffer 0.2 M NH$_4$Cl with 10% (v/v) CH$_3$CN, pH 9.0 (buffer A) followed by 0.5 M NH$_4$Cl with 10% (v/v) CH$_3$CN, pH 9.0 (buffer B). Chromatographic analysis is initiated by injection of 40 μl of neutralized supernatant onto the column previously equilibrated with buffer A. The column is washed with buffer A for 30 min and [3-^{14}C]acetoacetate is eluted at 10 min and [3-^{14}C]HMG at 22 min. The flow rate is 0.54 ml/min. This [3-^{14}C]HMG is produced nonenzymatically from substrate [3-^{14}C]HMG-CoA by passage of the reaction mixture over the Aminex A-25 column bed. At 30 min, buffer B is passed through the column to elute unreacted [3-^{14}C]HMG-CoA after approximately 60 min.

In the liquid-partition chromatographic procedure for the quantification of 3-hydroxy [3-^{14}C]butyric acid on silicic acid columns,[13,15] the acidified reaction mixture is adsorbed onto dried silicic acid and placed

[13] L. Sweetman, in "Heritable Disorders of Amino Acid Metabolism" (W. L. Nyhan, ed.), p. 730. Wiley, New York, 1976.

[14] B. Bakay, E. Nissinen, and L. Sweetman, *Anal. Biochem.* **86**, 65 (1978).

[15] K. M. Gibson, L. Sweetman, W. L. Nyhan, C. Jakobs, D. Rating, H. Siemes, and F. Hanefeld, *Clin. Chim. Acta* **133**, 42 (1983).

onto a 0.6 cm i.d. × 15 cm column of hydrated silicic acid (50 ml of 0.05 M H_2SO_4 per 92 g dried 100 mesh silicic acid).[13] The column is eluted isocratically with 5% (v/v) 2-methyl-2-butanol in $CHCl_3$. The flow rate is 2.5 ml/min, and 3-hydroxybutyric acid is eluted after 20 min. Fractions containing 3-hydroxy[3-^{14}C]butyric acid are pooled, solvent evaporated under nitrogen stream, and radioactivity determined by liquid scintillation counting. Radiolabeled HMG and unreacted [3-^{14}C]HMG-CoA remain on the silicic acid column, which is discarded. More recently, reversed-phase high-performance liquid chromatography has been applied to the separation and quantification of substrate and product to assay HMG-CoA lyase.[5,11] A column (0.41 cm i.d. × 2 cm) packed with SC C18 reversed-phase pellicular particles (30–40 μm) is connected to a 0.46 cm i.d. × 25 cm Spherisorb ODS II, 5 μm column (Custom LC, Inc., Houston, TX). The guard column is connected to a Rheodyne model 7125 syringe loading sample injector equipped with a 200-μl sample loop. The mobile-phase solvents are 0.05 M potassium phosphate buffer, pH 2.1 (buffer 1) from 0 to 10 min of elution, 55% buffer 1 and 45% methanol from 10 to 25 min of elution, followed by 15 min of buffer 1 to return to the starting conditions. The flow rate is 1.5 ml/min. Using this elution system, 3-hydroxy[3-^{14}C]butyric acid elutes at 5 min and 3-[^{14}C]HMG at 7 min. Unreacted [3-^{14}C]HMG-CoA is eluted at 18 min. One-minute fractions are collected in 8-ml scintillation vials, 5 ml of scintillant is added, and the isotope content determined by liquid scintillation counting. Up to 100 μl of neutralized assay supernatant can be analyzed. This procedure is probably the method of choice, in that product quantification as 3-hydroxy[3-^{14}C]butyric acid is achieved within 10 min. It is also possible to account quantitatively for all of the radioactive material initially used in the assay.

Levels of Activity

Both procedures have been employed in the documentation of deficient HMG-CoA lyase activity in lysates of cultured human fibroblasts and isolated lymphocytes. The assay of Clinkenbeard et al.[4] has yielded activities in normal fibroblast extracts ranging from approximately 4.0 up to 9.0 nmol/min mg protein.[6,16] Higher values, ranging from 6.0 up to 28.0 nmol/min mg protein, have been demonstrated in isolated leukocytes using the same procedure.[7–9] After a correction for buffer interference in

[16] S. J. Wysocki and R. Hahnel, *Clin. Chim. Acta* **71**, 349 (1976).
[17] K. M. Gibson, L. Sweetman, W. L. Nyhan, T. M. Page, C. Greene, and H. M. Cann, *Clin. Chim. Acta* **130**, 397 (1983).
[18] R. J. A. Wanders, R. B. H. Schutgens, and P. H. M. Zoeters, *Clin. Chim. Acta* 171, 95 (1988)

the estimation of protein,[17] the assay of Gibson and co-workers has yielded activities ranging from 1.0 up to 5.0 nmol/min mg protein[11] in extracts of both tissues. The higher levels of activity afforded by the assay of HMG-CoA lyase described by Clinkenbeard et al.[4] may be due in part to the presence of reduced sulfhydryls, although, in the assay described by Gibson and co-workers,[5] the presence of either reduced dithiothreitol or glutathione did not yield an appreciable stimulation of HMG-CoA lyase activity in fibroblast or lymphocyte extracts. Wanders and coworkers[18] have demonstrated higher activities (16–32 nmol/min-mg protein) for HMG-CoA lyase in extracts of cultured human fibroblasts with a final assay pH of 9.25. Such high activities enabled the use of spectrophotometric methods for the estimation of enzyme activity in crude cell extracts, although there was no mention of the stability of precursor HMG-CoA at the pH employed.

Alternative Methods

Methods in Which HMG-CoA Lyase Is not Rate Limiting

HMG-CoA lyase activity has been estimated by oxidation of [1-^{14}C]isovaleric acid[19] and [U-^{14}C]leucine[20,21] in intact cultured human fibroblasts. The reactions catalyzed by acetoacetyl-CoA thiolase and HMG-CoA synthase have been coupled to HMG-CoA lyase using unlabeled, [1-^{14}C] or [2-^{14}C]acetyl-CoA as precursors.[22-26] The content of isotope in the resulting acetoacetate is determined following decarboxylation to acetone,[22,24,25] while unlabeled acetoacetate is quantified spectrophotometrically as described below. Estimation of HMG-CoA lyase activity has also been accomplished by conversion of [3-^{14}C]HMG to [3-^{14}C]acetoacetic acid in cell-free rat liver mitochondria preparations.[3,27]

[19] O. Sovik, L. Sweetman, K. M. Gibson, and W. L. Nyhan, *Am. J. Hum. Genet.* **36**, 791 (1984).
[20] I. Yoshida, O. Sovik, L. Sweetman, and W. L. Nyhan, *J. Neurogenet.* **2**, 413 (1985).
[21] P. Divry, M. O. Rolland, J. Teyssier, J. Cotte, M. C. Formosinho Fernandes, I. Tavares de Almeida, and C. da Silveira, *J. Inher. Metab. Dis.* **4**, 173 (1981).
[22] I. Mulder, E. A. de Vries-Akkerman, and S. G. van den Bergh, *Int. J. Biochem.* **8**, 237 (1977).
[23] J. B. Allred, *Biochim. Biophys. Acta* **297**, 22 (1973).
[24] C. A. Stanley, E. Gonzales, and L. Baker, *Pediatr. Res.* **17**, 224 (1983).
[25] I. Mulder and S. G. van den Bergh, *Int. J. Biochem.* **8**, 227 (1977).
[26] I. Mulder and S. G. van den Bergh, *Int. J. Biochem.* **13**, 411 (1981).
[27] R. Deana, M. Fabbro, and F. Rigoni, *Biochem. J.* **172**, 371 (1978).

Methods in Which HMG-CoA Lyase Is Rate Limiting

Assay methods for the quantification of acetyl-CoA produced in the HMG-CoA lyase reaction are generally enzymatic techniques. Acetyl-CoA has been measured by monitoring the synthesis of citrate from oxalo-acetate in the presence of citrate synthase (EC 4.1.3.7), as described by Decker[28] and Stegink and Coon.[2,29] A regenerating system for oxaloacetate production is established by the inclusion of L-malate, NAD$^+$, and malate dehydrogenase (EC 1.1.1.37), and the reduction of pyridine nucleotide followed spectrophotometrically. Fluorometric methods have been described for the measurement of citrate produced in the preceding assay system.[30] The conversion of citrate to 2-oxoglutaric acid is accomplished by the consecutive actions of aconitate hydratase (aconitase, EC 4.2.1.3) and isocitrate dehydrogenase (EC 1.1.1.42). The stoichiometric reduction of NADP$^+$ is monitored spectrophotometrically. Conversely, citrate levels can be determined[30] by the coupled actions of citrate lyase (EC 4.1.3.6) and malate dehydrogenase. In this method, the oxidation of NADH is measured spectrophotometrically. These techniques have been utilized to document inherited deficiencies of HMG-CoA lyase.[31,32]

The other product of HMG- CoA cleavage, acetoacetate, may be measured enzymatically or colorimetrically. The enzymatic determination of acetoacetate is based on the reaction with 3-hydroxybutyrate dehydrogenase (EC 1.1.1.30) and the accompanying oxidation of NADH is monitored spectrophotometrically.[33-35] Walker[36] described a colorimetric procedure for acetoacetate based on the reaction of acetoacetate with diazotized *p*-nitroaniline. This procedure has been widely accepted over the years, although it has been shown that thiols[37] interfere with this method.

[28] K. Decker, *in* "Methods of Enzymatic Analysis" (H. U. Bergmeyer, ed.), Vol. 4, p. 1988. Academic Press, New York, 1974.

[29] L. D. Stegink and M. J. Coon, *J. Biol. Chem.* **243**, 5272 (1968).

[30] J. R. Williamson and B. E. Corkey, this series, Vol. 13, p. 446.

[31] B. H. Robinson, J. Oei, W. G. Sherwood, A. H. Slyper, J. Heininger, and O. A. Mamer, *Neurology* **30**, 714 (1980).

[32] C. L. Greene, H. M. Cann, B. H. Robinson, K. M. Gibson, L. Sweetman, J. Holm, and W. L. Nyhan, *J. Neurogenet.* **1**, 165 (1984).

[33] J. Mellanby and D. H. Williamson, *in* "Methods of Enzymatic Analysis"(H. U. Bergmeyer, ed.), p. 454. Academic Press, New York, 1965.

[34] D. H. Williamson, J. Mellanby, and H. A. Krebs, *Biochem. J.* **82**, 90 (1962).

[35] J. R. Williamson and B. E. Corkey, this series, Vol. 13, p. 478.

[36] P. G. Walker, *Biochem. J.* **58**, 699 (1954).

[37] J. B. Allred, *Anal. Biochem.* **11**, 138 (1965).

Concluding Remarks

Because of the important role of HMG-CoA in intermediary metabolism, a variety of procedures have been employed for the estimation of HMG-CoA lyase activity in tissues, and have included spectrophotometric, colorimetric, and radiometric techniques, both direct and indirect. For the investigator whose goal is the monitoring of a purification procedure for HMG-CoA lyase, the use of a sensitive radiometric assay is both costly and time consuming, although such procedures are advantageous when unpurified extracts are studied. The assay of Gibson and co-workers,[5] which employs direct radiolabeled product quantification and has a sensitivity of product measurement of less than 10 pmol, is particularly useful for determining residual levels of activity in patients with inherited defects in the enzyme. On the other hand, a reliable radiometric assay which requires little manipulation and no chromatography, such as that of Clinkenbeard et al.,[4] has been widely accepted.

Acknowledgment

This work was supported in part by U.S. Public Health Services Grant No. HD-04608 from the National Institute of Child Health and Human Development. Support during preparation of the manuscript was provided by a Fellowship from the Bank of America-Giannini Foundation.

[27] Permeabilized Cell and Radiochemical Assays for β-Isopropylmalate Dehydrogenase

By LILLIE L. SEARLES and JOSEPH M. CALVO

Introduction

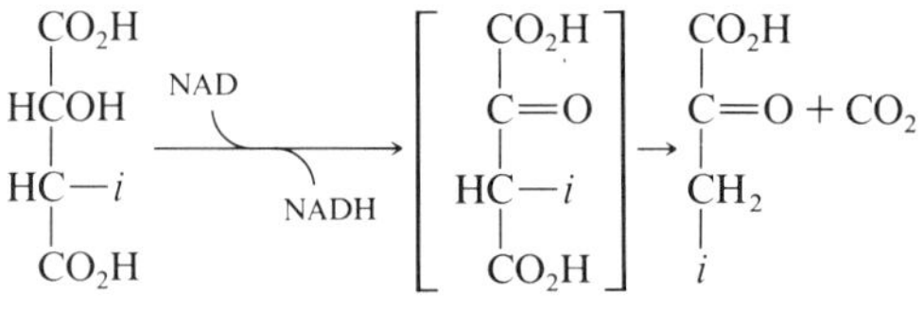

$$\begin{array}{c} CO_2H \\ | \\ HCOH \\ | \\ HC\!-\!i \\ | \\ CO_2H \end{array} \xrightarrow[\;\;NADH\;\;]{NAD} \left[\begin{array}{c} CO_2H \\ | \\ C\!=\!O \\ | \\ HC\!-\!i \\ | \\ CO_2H \end{array} \right] \rightarrow \begin{array}{c} CO_2H \\ | \\ C\!=\!O + CO_2 \\ | \\ CH_2 \\ | \\ i \end{array}$$

where i stands for an isopropyl group.

β-Isopropylmalate dehydrogenase (2-hydroxy-4-methyl-3-carboxyvalerate: NAD⁺ oxidoreductase, EC 1.1.1.85), one of the enzymes involved in leucine biosynthesis in bacteria and fungi, is usually assayed by incubating

an extract with β-isopropylmalate and NAD and measuring the α-ketoisocaproate formed after converting the latter to a 2,4-dinitrophenylhydrazone derivative. A rapid and simple permeabilized cell assay based on the dinitrophenylhydrazone procedure is described here. This assay is ideal for physiological studies in which a large number of small samples of enteric bacteria need to be analyzed. In addition, a radiochemical assay is described that measures $^{14}CO_2$ derived from radiolabeled β-isopropylmalate. This assay can be used with crude extracts prepared from any bacterium or fungi. Furthermore, it is sufficiently sensitive so that it can be used to measure β-isopropylmalate dehydrogenase formed *de novo* in a coupled transcription–translation system.

Permeabilized Cell Assay

Principle

The procedure here is a modification of the spectrophotometric assay of Friedmann and Haugan.[1] In the indirect procedure that they describe, the 2,4-dinitrophenylhydrazone of α-ketoisocaproate is extracted first into ether and then into basic aqueous solution. Here, the two extraction steps are omitted and the volume of the assay was reduced to increase the sensitivity of the assay.

Growth and Permeabilization of Cells

Cells of *Salmonella typhimurium* or *Escherichia coli* are grown at 37° in SSA medium[2] containing 0.2% glucose to an A_{550} of 0.6–0.8. Samples of cells (1–5 ml) are centrifuged and resuspended in the same volume of either SSA or 50 mM potassium phosphate, pH 7.2. After resuspension, cells can be stored in the refrigerator for at least 1 week without loss of activity. Tris-HCl (50 mM, pH 8) can be used to resuspend cells if assays are done immediately, but cells cannot be stored in this buffer without loss of activity. To permeabilize cells, 15 μl of $CHCl_3$ and 50 μl of a 1% (w/v) solution of sodium deoxycholate are added per milliliter of cells and the sample is incubated on ice for 5 min. These conditions are suitable for permeabilizing samples containing as many as 10^{10} cells per milliliter.

Reagents

Tris-HCl buffer, 1.57 M, pH 8.0
KCl, 1.57 M

[1] T. E. Friedmann and G. E. Haugan, *J. Biol. Chem.* **147**, 415 (1943).
[2] Per liter of distilled water: K_2HPO_4, 10.5 g; KH_2PO_4, 4.5 g; $(NH_4)_2SO_4$, 1.0 g; sodium citrate dihydrate, 0.97 g; $MgSO_4$, 0.05 g.

NAD, $10^{-2}\,M$
β-Isopropylmalate,[3] $2 \times 10^{-2}\,M$
$MnCl_2$, 0.0157 M
2,4-Dinitrophenylhydrazine, $3 \times 10^{-3}\,M$ in 1.5 M HCl
KOH, 40% (w/v)

Procedure

To a 1.4-ml Eppendorf centrifuge tube add 0.09 ml of Tris-HCl buffer, 0.030 ml of KCl, 0.03 ml of $MnCl_2$, 0.06 ml of β-isopropylmalate, 0.06 ml of NAD, 0–0.2 ml of permeabilized cells, and water to bring the total volume to 0.47 ml. Assays are performed in duplicate together with a control lacking β-isopropylmalate. After incubation for 15 min at 37°, 0.25 ml of 2,4 dinitrophenylhydrazine is added. After 15 min at room temperature, or 8 min at 45°, 0.15 ml of 40% KOH is added, samples are centrifuged for 15 min at 4°, and the A_{540} determined. The color is stable for at least 2 hr at room temperature. If absorbancies are above 1, samples can be diluted with 7% KOH. Standard curves are prepared using α-keto-isocaproate. NAD and 2,4-dinitrophenylhydrazine contribute about equally to the background of this assay, which is about 0.2 against water.

With the permeabilized cell assay, the formation of α-ketoisocaproate is linear with time up to at least 30 min and an A_{540} of 1.2. The activity is linear with the number of cells per assay up to at least 5×10^8 cells. Furthermore, the permeabilized cell assay and a standard assay employing extracts prepared by sonication give comparable results when applied to cells in different stages of a growth curve and to strains having different levels of β-isopropylmalate dehydrogenase.

Radiochemical Assay

Principle

This assay is a modification of the procedures described by Parsons and Burns[4] and Kung and Weisbach.[5] β-Isopropylmalate labeled with ^{14}C specifically in carbon 4 (carboxyl group that is furthest from the hydroxyl group) is used as a substrate. This labeled carboxyl group is released as CO_2 during the reaction catalyzed by β-isopropylmalate dehydrogenase.

---

[3] J. M. Calvo and S. R. Gross, this series, Vol. 17, p. 791.
[4] S. J. Parsons and R. O. Burns, *J. Biol. Chem.* **244**, 996 (1969).
[5] H. Kung and H. Weisbach, *J. Biol. Chem.* **253**, 2078 (1978).

Preparation of ^{14}C-Labeled β-Isopropylmalate

Incubation of a *Neurospora crassa* leu-1 strain[3] in medium containing [1-^{14}C]valine under appropriate conditions leads to the accumulation of α- and β-isopropylmalate labeled at carbon 4. A spore suspension of A_{550} 0.5–1.0 is prepared from a 7-day-old slant by adding 5 ml of sterile H_2O, vortexing, and filtering through sterile glass wool. Each 100 ml of Vogel's minimal medium[6] containing 1% sucrose and 15 μg/ml of L-leucine is inoculated with the equivalent of 0.25 ml of spore suspension having an A_{550} of 1. Two-liter Erlenmeyer flasks, each containing 500 ml of medium, are shaken at 30° for 20–24 hr. The mycelium is concentrated by filtering the culture through a sterile Whatman #1 filter. After washing with several milliliters of sterile Vogel's salt solution, the filter is added to one-tenth the original volume of Vogel's minimal medium containing 1% sucrose, but lacking leucine. The flask is shaken for 10 min at 30°. After removal of the filter paper, L[1-^{14}C]valine (40–60 mCi/mmol) is added to a final concentration of 0.5 μCi/ml and the flask is shaken at 30° for 4 hr. The uptake of valine and its conversion to α- and β-isopropylmalate can be monitored by analyzing filtered samples on 0.5-ml DEAE-cellulose columns. Immediately after addition of the label (time zero), all of the counts are eluted with a low salt wash (2 ml of 0.02 M sodium acetate, pH 5.5). As the valine is converted to isopropylmalates, counts are eluted with 2 ml of 0.4 M sodium acetate. Under optimal conditions, after 3–4 hr virtually all of the input radioactivity is converted to anions. The culture is filtered through a Whatman #1 filter, and after adjusting the pH of the filtrate to 1.6 with concentrated HCl, the filtrate is extracted four times with an equal volume of anhydrous ether. Following removal of the ether by evaporation, the flask containing the residue is heated to 67° under a stream of nitrogen gas for 15 min. For unknown reasons, this step significantly reduces the background of the assay. The residue is resuspended in a small volume of 0.1 M potassium phosphate, pH 8.6 (0.25–0.5 ml per microcurie of radioactivity added). The resulting solution is extracted four times with ether, and after traces of ether are removed by heating to 45° under a gentle stream of nitrogen, the solution is filtered through a 0.45 μm Millipore filter. Aliquots are transferred to smalll tubes and stored at −20°.

This crude preparation contains a mixture of α- and β-isopropylmalate, only 5–10% of which is the β compound. The presence of radioactive α-isopropylmalate does not interfere with the assay. The specific radioactivity of the substrate is determined by running the enzyme catalyzed reaction to completion in a stoppered tube equipped with a CO_2 trap (see

[6] R. H. Davis and F. J. de Serres, this series, Vol. 17, p. 79.

below) and measuring α-ketoisocaproate by the colorimetric assay described above and radioactive CO_2 by scintillation counting. The specific radioactivity of the β-isopropylmalate is diluted by addition of unlabeled β-isopropylmalate to 5×10^4 cpm per micromole.

Reagents

Tris-HCl buffer, 1.0 M, pH 8.0
KCl, 1.0 M
NAD, 10^{-2} M
$MnCl_2$, 10^{-2} M
^{14}C-Labeled β-isopropylmalate, about 5×10^4 cpm per micromole
Crude extract containing β-isopropylmalate dehydrogenase

Procedure

Add to a 12×75 mm tube on ice 0.3 ml of Tris-HCl buffer, 0.1 ml of KCl, 0.2 ml of NAD, 0.1 ml of $MnCl_2$, 7000 cpm of β-isopropylmalate, 0–0.1 ml of crude extract, and water to a total volume of 1 ml. A CO_2 trap containing 0.2 ml of $10\times$ hydroxide of Hyamine (New England Nuclear Corporation) is suspended in each tube before sealing the tube with an 11×17 mm serum stopper (Thomas Scientific). CO_2 traps are made by cutting disposable polypropylene medi-droppers (Fischer Scientific) to 3 cm in length. The sealed end is saved and used as the receptacle for a 1.8×0.75 cm piece of Whatman #1 filter paper, folded into an accordion shape. Because the sealed end of the trap is flared, the CO_2 trap can be suspended in the tube above the liquid. Alternatively, stoppers and center wells can be purchased from Kontes Glass Company, Vineland, NJ (items K-882310 and 882320). Sealed tubes are incubated at 37° for 30 min. The tubes are then rapidly chilled on ice and the reaction stopped by injection of 0.25 ml of 4 N HCl through the stopper with a 1-ml syringe fitted with an 18 gauge needle. After incubation for 45 min at room temperature with shaking to drive off CO_2, the traps are removed and the outer surface of the trap is rinsed with 95% ethanol and dried. The entire trap is then transferred to a scintillation vial, 5 ml of toluene containing 5 g of 2,5-diphenyloxazole and 0.22 g of p-bis[2-(5-phenyloxazolyl)] benzene per liter is added, and the radioactivity is measured by scintillation counting.

[28] Rapid Assay of Acetolactate Synthase in Permeabilized Bacteria

By JULIUS H. JACKSON

Principle

Frequently, in genetic analyses, it may be necessary to screen a large number of colonies for enzymatic activity or for measurable changes in an enzyme property. The acetolactate synthase (ALS, EC 4.1.3.18) reaction is the first pathway step common to biosynthesis of L-isoleucine and L-valine. The acetolactate synthase isozymes of *Escherichia coli* strain K12 display differential sensitivities to feedback inhibition by L-valine,[1–6] and both the maximum inhibition and the concentration of L-valine required for half-maximal inhibition can be altered by mutation.[6] This is a highly reproducible, quantitative method that can be used on small quantities of cells to measure specific activity, percent inhibition, and kinetic properties that relate to substrate and inhibitor concentrations. A major advantage to this method is that it can be routinely used on cells grown in 10 ml of culture medium. This method employs the cationic detergent cetyltri-methylammonium bromide (CETAB) to permeabilize cells for measurement of acetolactate synthase activity. Under the defined assay conditions, this permeabilized cell assay procedure gives 89% of the specific activity measured in crude extracts from sonically disrupted cells.[7]

Reagents

Potassium phosphate, pH 8.0, 1.0 M
$MgCl_2$, 100 mM
Flavin adenine dinucleotide (FAD), 20 μg/ml
Thiamin pyrophosphate (TPP), 1.0 mM
Sodium pyruvate, 200 mM
L-valine, 10 mM

[1] J. M. Blatt, W. J. Pledger, and H. E. Umbarger, *Biochem Biophys. Res. Commun.* **48**, 444 (1972).
[2] J. P. O'Neill and M. Freundlich, *Biochem. Biophys. Res. Commun.* **48**, 437 (1972).
[3] J. Guardiola, M. De Felice, and M. Iaccarino, *J. Bacteriol.* **120**, 536 (1974).
[4] M. De Felice, J. Guardiola, B. Esposito, and M. Iaccarino, *J. Bacteriol.* **120**, 1068 (1974).
[5] J. H. Jackson and E. K. Henderson, *J. Bacteriol.* **121**, 504 (1975).
[6] E. J. Davis, J. M. Blatt, E. K. Henderson, J. J. Whittaker, and J. H. Jackson, *Mol. Gen. Genet.* **156**, 239 (1977).
[7] J. M. Blatt and J. H. Jackson, *Biochim. Biophys. Acta* **526**, 267 (1978).

Cetyltrimethylammonium bromide (CETAB), 0.355 mg/ml
H_2SO_4, 50% (w/v)
Creatine hydrate, 0.5% (w/v)
α-Naphthol, 5.0% (w/v) in 4 N NaOH
ALS buffer (for one liter):

Potassium phosphate, pH 8.0, 1.0 M	50 ml
$MgCl_2$, 1.0 M	1 ml
Dithiothreitol (DTT)	77 mg
FAD	10 mg
TPP	48 mg
Distilled or deionized water	949 ml

Prepare ALS buffer fresh before each use, or store frozen ($-20°$) immediately after preparation. This buffer is not stable for more than 5 hr at room temperature.

Procedure

The procedure described was developed for optimal results with *E. coli* strain K12. It has also been used with *Salmonella typhimurium* strain LT2. It may be necessary to vary the concentration of CETAB to obtain optimal results with other genera.

1. Inoculate duplicate 10-ml batches of a desired growth medium for each strain to be assayed. This is conveniently done in 16 × 120 mm culture tubes. When a minimal salts medium is used with appropriate supplements, most strains can be grown overnight to a cell density limited by the initial concentration of glucose (0.02%, w/v). This is particularly desirable when handling a large number of strains, in order to allow all cultures to limit at the same cell density. Following 16- to 24-hr incubation at 37°, with aeration, add 0.1 ml of 50% (w/v) glucose to each tube. A tissue culture roller drum is the easiest way to accommodate aeration of large numbers of tubes, and less evaporation of culture medium occurs than when air is bubbled through the medium.

2. Use one tube of each duplicate culture to follow growth, and allow cultures to grow to mid-log phase before harvest. Carefully record the turbidity as precisely as possible for each culture, and harvest cells, from the second culture tube, by centrifugation at 1000 g for 10 min. Note: When large numbers of cultures are used, it is practical to follow growth by measuring turbidity in one culture and to harvest a second, identical culture.

3. Drain the pellet carefully by decanting the supernatant medium into a small, graduated cylinder, and record the supernatant volume. Resus-

pend the pellet in 0.8 ml of ALS buffer. From this step forward, maintain suspensions at $0-4°$

4. Prepare reaction cocktails as shown below, but take care to add $MgCl_2$ last, immediately before preequilibration of temperature for the reaction. Scale the cocktail volumes up, as needed, to accommodate increased sample numbers.

Reagents	Cocktail volume (ml)		
	1	2	3
Potassium phosphate	1.0	1.0	1.0
$MgCl_2$	1.0	1.0	1.0
FAD	1.0	1.0	1.0
TPP	1.0	1.0	1.0
Deionized water	3.0	1.0	—
Sodium pyruvate	—	2.0	2.0
L-Valine	—	—	1.0
CETAB	2.0	2.0	2.0
Total	9.0	9.0	9.0

5. Divide each suspension into three or four fractions with a volume of 0.1 ml each. The fractions should correspond with the numbered reaction cocktails described in step 4. Although these reaction cocktails consist of a blank (no pyruvate), one with pyruvate, and one with pyruvate and valine, any number beyond the first two may be constructed to accommodate specific experimental designs.

6. Preequilibrate cocktails and sample suspensions for at least 5 min at the appropriate reaction temperature (usually 37°). Add 0.9 ml of each numbered cocktail to each appropriately numbered sample suspension at carefully timed intervals.

7. Incubate for 10 min at the appropriate reaction temperature (usually 37°). Stop the reaction by addition of 0.1 ml of 50% (w/v) H_2SO_4 at carefully timed intervals to correspond to a precise incubation time of 10 min for each tube. The addition of sulfuric acid precipitates the protein to halt the reaction and acidifies the reaction mixture for subsequent acid-catalyzed decarboxylation of the reaction product, α-acetolactate, to form acetoin.

8. Incubate reaction mixtures at 37° for 30 min for decarboxylation.

The reaction mixtures may be stored at $-20°$, at this point or prior to decarboxylation, for later color development. Color development should be done within 3 to 5 days to obtain optimal results.

9. The method used to measure acetoin is a slight modification of the procedure described by Westerfeld.[8] The Grade III α-naphthol obtained from Sigma Chemical Co., St. Louis, MO, is suitable for use without recrystallization or sublimation under nitrogen. Add 1.0 ml of 0.5% (w/v) creatine hydrate to each reaction mixture and mix thoroughly. Add 1.0 ml of 5.0% (w/v) α-naphthol in 4 N NaOH to each tube. Incubate at room temperature, with shaking at 10-min intervals to provide aeration, for 1 hr.

10. Centrifuge to clarify cell debris from the reaction mixtures (10 min at 1000 rpm in Sorvall GLC-1).

11. Measure absorbance at 525 nm in a cuvette with a 1-cm light path.

12. Results may be expressed as units of specific activity where one unit of specific activity is defined as one nanomole of product formed per minute per milliliter of cell culture per unit of culture density.

Calculations

A general formula for calculation of specific activity in permeabilized cells is

$$S = (A/M)(V_s/V_c)(1 \times 1/V_r)(1/t)(1/V_t)(1/K) \tag{1}$$

where S is the specific activity; A, absorbance of reaction product; M, absorbance units per nanomole of product; V_s, volume (ml) in which total cells from culture are suspended; V_c, volume (ml) of V_s used for reaction; V_r, volume (ml) of reaction mixture used for color development; V_t, volume of culture supernatant; t, reaction time (min); K, units of culture turbidity or density at time of harvest. Under the conditions defined for this procedure, when $M = 0.0063$ absorbance units per nanomole, the expression for S reduces to

$$S = (CA)/(V_s K) \tag{2}$$

where $C = 127$ as long as the conditions do not change. The value for M was determined from a standard curve that related acetoin concentrations to absorbance at 525 nm. Acetoin concentrations were prepared from the crystalline dimer.

[8] W. W. Westerfeld, *J. Biol. Chem.* **161**, 495 (1945).

[29] Assay of Products of Acetolactate Synthase

By Natan Gollop, Ze'ev Barak and David M. Chipman

Acetolactate synthase (ALS, EC 4.1.3.18) can condense an "acetalde-hyde" moiety derived from pyruvate either with a second molecule of pyruvate to form 2-acetolactate (the first intermediate in the valine and leucine pathway), or with 2-ketobutyrate to form 2-aceto-2-hydroxybutyr-ate (on the isoleucine biosynthetic pathway).[1,2] The colorimetric assay of Bauerle et al.[3] for acetohydroxy acids is commonly used to follow the formation of acetolactate from pyruvate in the *absence* of 2-ketobutyrate. Acetohydroxybutyrate also yields the same color in this assay, with a color yield about 15% that of acetolactate (unpublished results). However, the second reaction obviously can not be followed independently of the first, and a method which differentiates between the two possible products is required to study this reaction. A bioassay for 2-aceto-2-hydroxybutyrate has been used,[4] but it is not suitable for application to quantitative studies. Shaw and Berg[5] followed the incorporation of 2-keto[1-^{14}C]butyrate into 2-aceto-2-hydroxybutyrate by determination of the $^{14}CO_2$ evolved upon acidification of the reaction mixture. The combined, tandem use of the

[1] H. E. Umbarger, *Annu. Rev. Biochem.* **47**, 533 (1978).
[2] H. E. Umbarger, *in* "Amino Acids: Biosynthesis and Genetic Regulation" (K. M. Her-mann and R. L. Somerville, eds), pp. 245–266. Addison-Wesley, Reading, Massachusetts, 1983.
[3] R. H. Bauerle, M. Freundlich, F. C. Stormer, and H. E. Umbarger, *Biochim. Biophys. Acta* **92**, 142 (1964).
[4] H. Grimminger and H. E. Umbarger, *J. Bacteriol.* **137**, 846 (1979).
[5] K. J. Shaw and C. M. Berg, *Anal. Biochem.* **105**, 101 (1980).

methods of Bauerle *et al.*[3] and of Shaw and Berg[5] could in principle allow quantitative study of the formation of the two ALS products, assuming one knows their relative color yields. However, such a double assay would be cumbersome.

We describe here a reliable method for the simultaneous quantitative assay of the two reactions catalyzed by ALS, with either crude or purified enzymes. This method[6] is based on gas–liquid chromatographic (GLC) analysis of the 2,3-diketones formed from the enzymatic products by oxidative decarboxylation.[7,8]

Assay Method

$$
\begin{array}{ccc}
\mathrm{CH_3} & & \mathrm{CH_3} \\
| & & | \\
\mathrm{C{=}O} & & \mathrm{C{=}O} \\
| & \xrightarrow[\ 80°,\ \mathrm{Fe\ salts}\ ]{[O_2]} & | \\
\mathrm{R{-}C{-}COO^-} & & \mathrm{R{-}C{=}O} \quad +\ CO_2 \\
| & & \text{2,3-Butanedione or} \\
\mathrm{OH} & & \text{2,3-pentanedione} \\
\end{array}
$$

Acetolactate or
acetohydroxybutyrate
R = CH$_3$ or CH$_2$CH$_3$

Principle

The enzymatically produced acetohydroxy acids are converted to the respective 2,3-diketones by oxidative decarboxylation in the presence of air.[7] Acetolactate yields 2,3-butanedione and acetohydroxybutyrate, 2,3-pentanedione. The effective oxidative decarboxylation of an acetohydroxy acid by atmospheric oxygen requires catalysis,[9] and an appropriate catalyst (e.g., iron salts) must be added when a purified enzyme has been used.[6] In the absence of appropriate catalysts a large fraction of the acetohydroxy acid undergoes decarboxylation to the hydroxyketone, which is resistant to oxidation. When analyzing the results of a reaction with a crude enzyme preparation, the added catalyst is often unnecessary.

The diketones resulting from oxidative decarboxylation are volatile, and are easily separated from the reaction mixture by transfer in a stream of air.[10] The GLC analysis of the diketones takes advantage of the particular sensitivity of an electron capture detector (ECD) toward vicinal diketones.[11] Hejgaard has proposed the use of direct head space GLC analysis.[8]

[6] N. Gollop, Z. Barak, and D. M. Chipman, *Anal. Biochem.* **160**, 323 (1987).
[7] P. Ronkainen, S. Brummer, and H. Suomalainen, *Anal. Biochem.* **34**, 101 (1970).
[8] J. Hejgaard, *Anal. Biochem.* **37**, 368 (1970).
[9] J. C. de Man, *Rec. Trav. Chim* **78**, 480 (1959).
[10] W. Postel and B. Meier, *Z. Lebensm.-Unters. Forsch.* **173**, 85 (1981).
[11] J. E. Lovelock, *Nature (London)* **189**, 729 (1961).

This method requires a precolumn and complex arrangements for back-flush of this column to avoid contamination of the detector. The method described here gives excellent results with much simpler equipment.

Reagents

Potassium phosphate buffer, pH 7.6
$MgCl_2$
Flavin adenine dinucleotide (FAD)
Thiamin pyrophosphate (TPP)
Sodium pyruvate
2-Ketobutyric acid (neutralized with NaOH)
2,3-Butanedione
2,3-Pentanedione
"Dried" Methanol
Phosphoric acid
$FeCl_3$
$FeSO_4$

Apparatus. The "air distillation" apparatus consists of a large (30×210 mm) test tube with a gas washing head and a 20×150 mm collecting test tube with an inlet tube reaching to within a few millimeters of the bottom, connected with Tygon tubing, as illustrated in Fig. 1. The system for the

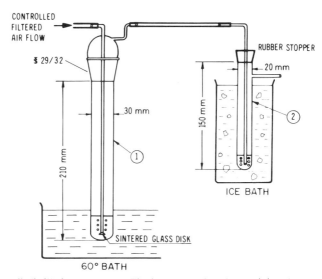

FIG. 1. Air distillation apparatus. The large test tube (1) containing the quenched enzymatic reaction mixture (pH ~4) with iron salts, with a total volume of ~10.5 ml, is held at 60°. The collecting tube (2) in an ice bath contains 2 ml dry methanol.

controlled supply of filtered air to 10 tubes can be simply constructed using a compressed air cylinder, filter, manifold of flexible PVC tubing and Y connectors, and individual needle valves (AS 2000, SMC, Tokyo, Japan or equivalent). Since there is virtually no pressure drop through the distillation apparatus, flow rates to each of the tubes are adjusted (using a Brooks Instrument or similar flow meter) before attaching the apparatus to the manifold.

A gas–liquid chromatograph equipped with a ^{63}Ni ECD and an appropriate recording integrator is required (we have used a Varian Model 3300 GLC with model 02-1972-01 ECD and a Spectra Physics SP4100 Computing Integrator *or* a Barspec Chromaset 401 PC-based Data Acquisition System). A column (⅛ inch × 1 m) packed with Chromasorb 102 80-100 Mesh (Johns-Manville) is used, with nitrogen as carrier gas. A column made with nickel tubing provides a very significant improvement in resolution compared with stainless steel.

Procedure

A standard enzymatic reaction mixture (3 ml) contains 10 mM MgCl$_2$, 20 μg/ml FAD, and 30 μg/ml TPP in 0.1 M potassium phosphate buffer (pH 7.6) and the substrates in the desired concentrations. Satisfactory kinetic results are obtained with pyruvate concentrations over the range of 0.1–25 mM and 2-ketobutyrate concentrations in the range of 0–25 mM. As the specificity of the ALS enzymes varies widely,[12,13] the ratio between the two substrates which will give detectable quantities of both products depends on the particular enzyme being assayed.

The reaction is initiated by addition of 0.005–0.05 enzyme units (one enzyme unit is defined as the quantity with catalyzes the formation of 1 μmol of acetolactate per minute in a standard reaction mixture, with 10 mM pyruvate only). After incubation at 37° for 6 min, the reaction is terminated by adding 0.12 ml of 2 M phosphoric acid to bring the pH to about 4. Such quenching is adequate to inactivate the bacterial enzymes, without interfering with the analysis. Note that if the enzymatic reaction mixture is quenched with strong acid, the acetohydroxy acids are decar-

---

[12] The specificity of a given ALS can be defined by the ratio: R = [(acetohydroxybutyrate formed)/(acetolactate formed)]/([2-ketobutyrate]/[pyruvate]). This ratio is essentially constant for each of the different isozymes of enteric bacteria we have examined, but varies from about 1–4 for ALS I to about 250 for ALS II. A constant R for a given enzyme is expected if 2-ketobutyrate and pyruvate compete for an intermediate on the enzyme formed irreversibly from the first pyruvate [e.g., as suggested by L. M. Abell, M. H. O'Leary, and J. V. Schloss, *Biochemistry* **24**, 3357 (1985)].

[13] Z. Barak, D. M. Chipman, and N. Gollop, *J. Bacteriol.* **169**, 3750 (1987).

boxylated to hydroxyketones without oxidation, and diketones are not formed. The enzymatic reaction can be carried out in a volume up to 10 ml with various buffers, so long as the amount of phosphoric acid added is adjusted so that the pH of the quenched reaction mixture is 3.7–4.3.

The quenched enzymatic reaction mixture is diluted to 10 ml with H_2O, and 0.15 ml each of 10 mM $FeCl_3$ and 10 mM $FeSO_4$ added. Stock solutions of $FeCl_3$ and $FeSO_4$ (10 mM) should be freshly made up every week. This mixture is incubated in a large (30 × 210 mm) test tube, tightly sealed with a rubber stopper, for 10 min at 80°, to convert the 2-aceto-2-hydroxy acids to the respective 2,3-diketones. The tube is cooled in ice for a few minutes (to prevent evaporative losses), and the stopper promptly replaced by the gas washing head of the apparatus. The "air distillation" is

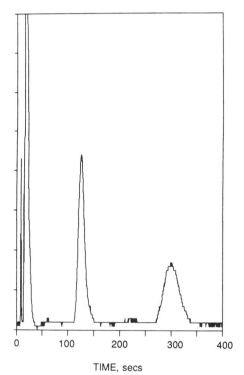

TIME, secs

FIG. 2. Gas chromatogram of products from enzymatic reaction of pyruvate (10 mM) and 2-ketobutyrate (0.05 mM) with 0.04 enzyme units of purified ALS II [J. V. Schloss, D. E. Van Dyk, J. F. Vastra, and R. M. Kutny, *Biochemistry* **24**, 4952 (1985)]. The standard buffer and cofactors were used, in a 5-ml reaction mixture, and the reaction was processed as described. The peaks are air (8 sec), methanol (18 sec), 2,3-butanedione (125 sec) and 2,3-pentanedione (295 sec).

carried out by passing filtered air through the quenched reaction mixture in the large test tube, held in a bath at 60°, and into the collecting tube containing 2 ml of dry methanol at 0°. After 25 min of air flow (180 ml/min through each tube), the solution in each collecting tube is brought to 4 ml by addition of cold, dry methanol, and transferred to a tightly capped polyethylene vial for storage at −20°. Each gas washing head and its associated delivery tubing is washed with ∼10 ml ethanol before reuse. Water interferes with the GLC analysis, and care should be taken to dry the apparatus, and to use dry methanol. Using a manifold with 10 air distillation apparatuses, it is possible to process 50 reaction mixtures in a working day. The chromatographic analysis of one group of samples can be carried out while processing the next.

Samples of 1 μl are analyzed by GLC, with the column at 150° and with a carrier gas flow rate of 25 ml/min. The injector and the detector are held at 150° and 250°, respectively. A chromatogram of the methanol solution of 2,3-diketones obtained after work-up of a typical enzymatic reaction is shown in Fig. 2. The peaks identified are air, methanol, 2,3-butanedione, and 2,3-pentanedione, in that order. When clean methanol is used in the collecting tube, no other significant peaks should be detected.

To obtain calibration curves, mixtures of known amounts of authentic 2,3-butanedione and 2,3-pentanedione are added to quenched reaction buffer and processed as described above. The calibration curves (Fig. 3) are used to calculate the amounts of products formed in enzymatic reactions.

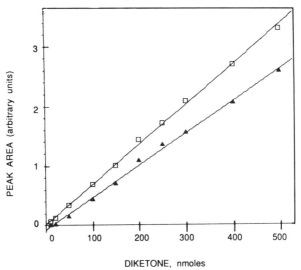

DIKETONE, nmoles

FIG. 3. Calibration curve. Equimolar samples of 2,3-butanedione (□) and 2,3-pentane-dione (▲), in the molar quantities indicated, were introduced into the large test tubes, and processed as described. Each point is the average integrator reading from two chromatographic injections.

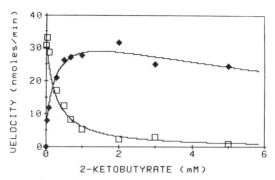

FIG. 4. Formation of 2-acetolactate (□) and 2-aceto-2-hydroxybutyrate (◆) as a function of 2-ketobutyrate concentration, in the presence of 15 mM pyruvate, by 0.033 U of purified ALS III (Z. Barak, J. M. Calvo, and J. V. Schloss, this volume [57]) under standard reaction conditions.

To be certain of the efficiency of conversion of the acetohydroxy acid product to the diketone, and of its transfer, a laboratory adopting this assay may want to compare it with the conventional colorimetric method.[3] When we followed the enzymatic production of acetolactate from pyruvate with time, we found excellent agreement between the two methods.[6] The correlation between the two assays had a weighted least-squares slope of 0.99 ± 0.05.

Other minor (and presumably nonphysiological) products of ALS-catalyzed reactions can also be analyzed using this method. For example, when the assay is applied to a reaction of *E. coli* ALS III with pyruvate and 2-ketopentanoate, 2,3-hexanedione, which is no doubt derived from the expected 2-aceto-2-hydroxypentanoate, is detected (retention time 11 min).

Comments

The outstanding characteristic of this method is its suitability for studying the specificity of purified ALS isozymes. Data of the sort illustrated in Fig. 4, in which both physiologically significant reactions are followed, are important for understanding the physiological control of branched-chain amino acid biosynthesis.[13] The method can also be used to investigate effects of inhibitors on specificity, and other aspects of the mechanism of action of the enzyme.

Acknowledgment

We are indebted to Ms. Batsheva Damri for excellent technical assistance. This work was supported in part by the U.S.–Israel Binational Science Foundation (Grant 8600205).

[30] Detection of the Acetolactate Synthase Isozymes I and III of Escherichia Coli K12

By Maurilio De Felice, Giovanna Griffo, Carmine T. Lago, Danila Limauro, and Ezio Ricca

The first pair of homologous reactions in the biosynthetic pathway leading to isoleucine, leucine, and valine in *Escherichia coli* K12 is catalyzed by two acetolactate synthase (ALS)[1] isozymes, termed I and III isozymes.[2,3] The study of the specific catalytic and regulatory role of each isozyme is difficult, because in standard assay conditions both isozymes contribute to α-aceto-α-hydroxy acid production. Some studies have been done with *E. coli* K12 strains in which one or the other isozyme is absent because of mutations in the structural genes.[4] This approach requires the preparation of specific strains, which is often very difficult. Furthermore, due to the partial specialization of isozymes I and III in the partition of carbon flow through the two sets of homologous reactions,[4,5] strains lacking one isozyme may have abnormal levels of the other one or of different enzymes in the pathway.

Assay Method

Principle

In this article we describe a simple method to assay each ALS isozyme in the presence of the other.

Isozyme I has its highest activity at pH 7.5; about 50% of this activity is retained at pH 6.6 and only about 5% at pH 9.3.[4] Isozyme III has its highest activity at pH 8.8; about 80% of this activity is retained at pH 9.3 and only about 2% at pH 6.6.[4] Furthermore, ALS I activity is reduced by 75% in the absence of FAD,[1] while ALS III activity is not influenced by this cofactor.[3,4] The amounts of the two isozymes can be selectively detected by performing the assay at pH 6.6 and 9.3 with and without FAD.

[1] Abbreviations: ALS, acetolactate synthase (EC 4.1.3.18); FAD, flavin adenine dinucleotide.

[2] M. De Felice, M. Levinthal, M. Iaccarino, and J. Guardiola, *Microbiol. Rev.* **43**, 42 (1979).

[3] F. La Cara and M. De Felice, *Biochem. Biophys. Res. Commun.* **91**, 319 (1979).

[4] M. De Felice, C. Squires, and M. Levinthal, *Biochim. Biophys. Acta* **541**, 9 (1978).

[5] M. De Felice, C. T. Lago, C. H. Squires, and J. M. Calvo, *Ann. Microbiol.* (Paris) **133A**, 251 (1982).

ALS I assay. When ALS activity is assayed at pH 6.6 in the presence of FAD, most of it is due to isozyme I, and contribution by isozyme III is negligible (normally below 5%) if the amounts of the two isozymes are comparable, or if isozyme I is in excess. This is confirmed by the observation that enzyme activity is reduced to one-fourth in the absence of FAD. If the amount of ALS III is much higher than that of ALS I, contribution of isozyme III to the ALS I assay may reach undesirable values, but it can be easily identified and subtracted, because in such a case the fraction of activity independent from FAD is higher than one-fourth. For example, if in a certain strain ALS-specific activity at pH 6.6 is 10.0 and 2.5 units in the presence and absence of FAD, respectively, we can assume that all activity is due to isozyme I; if it is 10.0 and 4.0 units, respectively, isozyme III is obviously interfering in the ALS I assay. In such a case, the actual ALS I activity can be calculated by subtracting the contribution of ALS III. In fact, if we indicate x to be the ALS I specific activity and y the ALS III activity to be subtracted, we have that

$$x + y = 10 \text{ units} \tag{1}$$

and

$$0.25\,x + y = 4 \text{ units} \tag{2}$$

This constitutes a simple system of two equations, from which the values of x and y are easily calculated as 8 and 2 units, respectively.

ALS III Assay. When ALS activity is assayed at pH 9.3, most of it is due to isozyme III, and contribution by isozyme I is negligible (below 5%) if the amounts of the two isozymes are comparable or if isozyme III is in excess. If the amount of ALS I is much higher than that of ALS III, contribution of isozyme I to the ALS III assay may reach undesirable values, but it can be identified and subtracted as above. For example, identical ALS-specific activities are found in the presence and in the absence of FAD in extracts of strains having comparable amounts of isozymes I and III or an excess of the latter. If we observe a reduction in activity in the absence of the cofactor, it is due to isozyme I. This contribution can be subtracted after calculating the activity of the two isozymes as indicated above for ALS I assay [(eq. (1) and (2)].

Preparation of Cell-Free Extracts

Bacteria grown under the desired conditions are harvested by centrifugation at 4°, washed twice with saline solution or with unsupplemented minimal medium, and either utilized immediately for ALS assay or frozen at −20°. Frozen cells can be kept for several weeks without significant loss

of activity. The cell pellet (50–200 mg) is suspended in 1 ml of $5 \times 10^{-3} M$ Tris-HCl buffer, pH 8.0. The suspension is sonicated at 4° (we normally sonicate twice for 45 sec with a 1-min interval for cooling, using an MSE sonic oscillator at its maximum intensity) and then centrifuged for 20 min at 25,000 g at 4°. Protein concentration of the extract ranges between 5 and 15 mg of protein per milliliter. In order to obtain reliable results, the ALS assay should be started soon after extraction, since isozymes I and III have different stabilities in extracts.

Assay Mixtures

The assay mixtures are the following 1-ml solutions.

A: 40 mM sodium pyruvate, 170 μM thiamin pyrosphosphate chloride, 10 mM MgSO$_4$, 100 mM potassium phosphate buffer, pH 6.6

B: Same as A plus 25 μM FAD

C: Same as A, with 100 mM Tris-HCl buffer, pH 9.3, in the place of potassium phosphate

D: Same as B with 100 mM Tris-HCl buffer, pH 9.3, in the place of potassium phosphate

Blanks containing the same components as in A–D, except for pyruvate are also prepared and marked A°, B°, C°, and D°.

Mixtures are normally prepared by diluting with water stock solutions of each component having 10-fold higher concentration.

Since the pH values of the various mixtures are critical, they should be measured carefully in the final mixtures.

Assay Reaction

ALS assay is performed with a procedure based on methods described by Umbarger and co-workers.[6,7] All mixtures are preincubated for 5 min at 37°, then an aliquot of extract containing from 50 to 500 μg of protein is added to each in order to start the ALS reaction. Reaction mixtures are incubated for 20 min at 37°, then the reaction is stopped by adding 0.1 ml of 50% (v/v) H$_2$SO$_4$ and samples are kept at 37° for 30 min. Acetolactate formed is assayed as acetoin according to Westerfeld.[8] 2.4 ml of freshly prepared coloring solution [containing 0.35 ml 5 N NaOH, 0.20 ml 5% (w/v) creatine, and 10 mg α-naphthol per milliliter of water solution] is added to each tube. After 30 min of incubation at 37°, samples are clarified

[6] R. H. Bauerle, M. Freundlich, F. C. Stormer, and H. E. Umbarger, *Biochim. Biophys. Acta* **92**, 142 (1964).

[7] F. C. Stormer and H. E. Umbarger, *Biochem. Biophys. Res. Commun.* **17**, 587 (1964).

[8] W. W. Westerfeld, *J. Biol. Chem.* **161**, 495 (1945).

with a brief centrifugation at room temperature and the A_{520} of each is read using A°, B°, C°, and D° as blanks. The activity of each sample is calculated as A_{520} per minute per milligram of protein. We suggest that specific activity is detected as a mean value of triplicate samples containing different aliquots of extracts with A_{520} ranging between 0.1 and 1.0. Outside this interval, values generally show poor precision and are not linear.

Standard Curve of Acetoin

Specific activity is generally reported in the literature as nanomoles of acetoin formed per minute per milligram of protein. Standard solutions of acetoin can be prepared by including various amounts of this compound (20 to 200 nmol) in solutions that are otherwise identical to A°, B°, C°, or D°. These samples are incubated, treated, and read as the assay samples after addition of extracts. The A_{520} values, plotted versus nanomoles, will lie on a straight line, which will be used as a standard curve of acetoin for the assay.

Remarks

Due to the different inactivation rates of ALS I and III during dialysis,[4] the assay described should be performed with nondialyzed extracts, in order to obtain reliable enzyme activities. This creates the problem that the pH profiles of the isozymes may be slightly influenced by growth conditions and concentration of extracts. Also, specific metabolic properties of strains under study may have some influence on pH profiles. In our experience, 6.6 and 9.3 are the most suitable pH values in most experiments, but in some cases highest accuracy of results may be obtained by using reaction mixtures buffered at slightly different pH values than those suggested above.

Acknowledgments

Research was supported by Ministero della Pubblica Instruzione, Progetto Finalizzato Ingegneria Genetica e Basi Molecolari delle Malattie Ereditarie of the C.N.R and a Grant from Enichem Agricoltura, Milano, Italy.

[31] Transport of Branched-Chain Amino Acids in Escherichia Coli

By TAMMY K. ANTONUCCI and DALE L. OXENDER

Introduction

Three primary systems serve to transport the branched-chain amino acids in *Escherichia coli.* The simplest system, LIV-II, is thought to consist of a single membrane component which has a low affinity for leucine, isoleucine, and valine. Two high-affinity transport systems also serve for the branched-chain amino acids. The high-affinity LIV-I system transports leucine, isoleucine, valine, threonine, and alanine and is composed of a periplasmic binding protein (LIV-BP) as well as three membrane components (LivH, LivM, and LivG). The high affinity LS system is specific for the D- and L-isomers of leucine and utilizes the same three membrane components of the LIV-I transport system and a distinct periplasmic binding protein (LS-BP). The genes for the two high-affinity transport systems have been cloned and sequenced.

In this chapter several methods used in characterizing prokaryotic transport systems will be presented. The transport assay to be described has been used to measure the rate of uptake of radiolabeled leucine by the bacterial cells. The rates of uptake determined by this assay reflect primarily the activities of the high-affinity LIV-I and LS transport systems. The binding proteins were isolated by the osmotic shock method, which is an adaptation of that published by Neu and Heppel.[1] This procedure results in the release of the periplasmic fluid and, consequently, all of the periplasmic components. In addition, methods are presented for the purification and crystallization of the periplasmic binding proteins from the LIV-I and the LS transport systems in *E. coli.*

Materials and Reagents

The calibrated dropper bottle used in the rapid transport assay was purchased from Microdels, Microbiological Associates, Rockville, MD. Amino acids and antibiotics were obtained from Sigma Co. except L-[^{3}H]leucine, which was from ICN Biomedicals, Inc. Hydroxylapatite was a product of Bio-Rad Laboratories and the 2-methyl-2,4-pentanediol was from Eastman Kodak Company. Vogel–Bonner (VB) minimal media[2]

---

[1] H. C. Neu and L. A. Heppel, *J. Biol. Chem.* **240**, 3685 (1965).

was used for transport assays. For all other protocols 3-N-morpholinopro-pane sulfonic acid (MOPS) minimal media[3] was used.

Methods

Transport Assay

Cells were grown overnight in 1-ml cultures shaking at 37° in VB media supplemented with nutrients as required. If the cells were harboring a plasmid, an antibiotic was included to select for retention of the plasmid. The overnight cultures were used to inoculate 4-ml cultures (1 : 10 dilution) which were grown to midlog phase ($OD_{420} = 0.4-0.7$). Each cell culture was grown in triplicate. The cells were washed and resuspended in cold, unsupplemented VB media so that their final OD_{420} was between 0.4 and 0.8. Tubes containing 0.4 ml of VB media plus L[^{3}H]leucine at a final concentration of 200 nM were equilibrated to 37°. After 5 min 0.4 ml of the cell culture (briefly equilibrated to 37°) was mixed with the tritiated solution for 10 sec, pipetted onto a hydrated 0.45 μm nitrocellulose filter, and immediately washed twice with 5 ml of a prewarmed 10 mM potassium phosphate buffer (pH 7.2) containing 0.1 mM MgSO$_4$. Each culture was assayed in triplicate. Medium without cells was assayed to determine the background. Radioactivity was determined using a Packard liquid scintillation counter and standard scintillation cocktail. Leucine uptake was calculated as nmol of leucine/min/mg dry cell weight in the following way. To determine cell quantity, an absorbance at 420 nm of 1.0 equaled 0.16 mg/ml dry weight. The specific activity of the tritiated leucine was calculated in cpm/nM by dividing the cpm of 10 μl of the leucine by the product of the volume counted and the concentration of the leucine.

An example of the results obtainable from this method is shown in Fig. 1. The values for derepressed transport phenotypes usually range from 0.3 to 0.6 nmol/min/mg dry cell weight, while the repressed values are rarely larger than 0.2 nmol/min/mg dry cell weight. The standard errors of these uptake values do not overlap with this method.

Rapid Transport Assay

A rapid alternate for the standard assay above was developed[4] for the large-scale assay of L-leucine transport phenotypes among progeny of

[2] H. J. Vogel and D. M. Bonner, J. Biol. Chem. 218, 97 (1956).
[3] F. C. Neidhardt, P. L. Bloch, and D. F. Smith, J. Bacteriol. 119, 736 (1974).
[4] J. J. Anderson, S. C. Quay, and D. L. Oxender, J. Bacteriol. 126, 80 (1976).

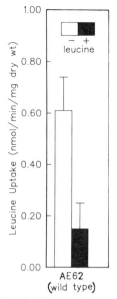

FIG. 1. Leucine transport in *E. coli*. The open bar represents the rate of leucine transport under derepressed conditions. The solid black bar represents the repressed rate of leucine transport.

genetic manipulations. VB minimal media was prepared with supplements as required except that 0.04% glucose was the carbon source and L-leucyl-L-alanine (15 μg/ml) was the source of leucine for both Leu$^+$ and Leu$^-$ strains. Two-milliliter volumes of this media were dispensed into test tubes, inoculated with small portions of recombinant colonies, and incubated overnight at 37° with shaking. Regardless of inoculum size or growth rate, all cultures ceased growth at the same density when glucose was exhausted. Glucose was then added back by a calibrated dropping bottle to 0.2%, and the tubes were incubated for 20 min to allow partial recovery. Chloramphenicol was then added similarly (150μg/ml), and the tubes were placed at 37° for assay. Cell samples (0.4 ml) were than added directly to 0.4 ml of unsupplemented VB media containing L-[³H]leucine at a final concentration of 0.75 μM. After 20 sec of incubation, 0.4 ml of the mixture was rapidly filtered, washed, and counted as above. The data obtained are reproducible and the range of transport phenotypes do not overlap under these conditions.

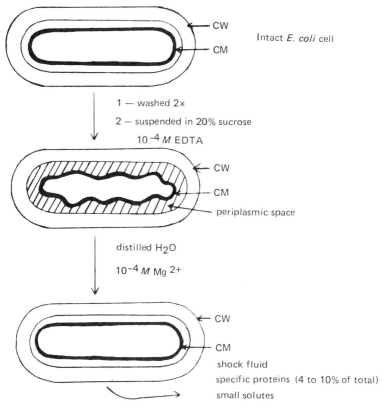

FIG. 2. Schematic diagram of the process of osmotic shock and its effect on the bacterial cell. From D. L. Oxender and S. Quay, *Ann. N.Y. Acad. Sci.* **264,** 358 (1976).

Osmotic Shock Procedure

A flow chart of this protocol is shown in Fig. 2. Cells were grown in 3 liters of MOPS minimal media to mid- or late-log phase, then washed 3 times in cold 0.01 M Tris-HCl, pH 7.0. The cells were resuspended in 80 parts (w/w) of room temperature 20% sucrose solution containing 0.1 mM EDTA. The cell suspension was stirred gently for 10 min, then centrifuged at 4° at 13,000 g for 10 min. The cell pellet was quickly resuspended in 80 parts ice-cold distilled water containing 0.1 mM MgCl$_2$. This suspension was gently stirred for 10 min and centrifuged as above. The supernatant fluid is called the "shock fluid" and contains approximately 4% of the cellular proteins, including the periplasmic binding proteins. Another os-

motic shock procedure has been developed for small (2 ml) volumes and yields adequate quantities for gel electrophoresis.[5]

Purification of the Binding Proteins

The first step in the purification of the binding proteins is to perform the large-scale osmotic shock procedure presented above. Three additional steps are employed in the following order.

DEAE-Cellulose Chromatography. Whatman DE-22 DEAE-cellulose was recycled according to the manufacturer's directions, equilibrated with 0.01 M potassium phosphate buffer, pH 6.9, washed with water, and stored at 4°. The supernatant fluid from the osmotic shock treatment was applied to a column of 2.2 cm diameter and 10 cm height at a flow rate of 5 to 20 ml per minute. The absorbance of the effluent at 280 nm was continuously monitored. After adding the sample, the column was washed with 0.02 M potassium phosphate buffer, pH 6.9, until two ultraviolet-absorbing peaks had been eluted. This usually required 180 to 200 ml of buffer. Frontal elution of the binding activity was accomplished with 0.15 M potassium phosphate buffer, pH 6.9. When 7-ml fractions were collected, the ultraviolet-absorbing material containing the binding activity was usually present in fractions 4 through 10. Fractions containing binding activity were pooled and dialyzed against distilled, deionized water for at least 4 hr and often overnight. A qualitative identification of the leucine-binding protein was made at various stages by layering an aliquot of each fraction over rabbit antisera. When binding protein was present, turbidity developed at the interface within a few minutes.

Hydroxylapatite Cellulose Chromatography. Hydroxylapatite was prepared by the method of Levin[6] and equilibrated with 1.0 mM potassium phosphate buffer, pH 6.2. Three grams of hydroxylapatite suspended in 25 to 30 ml of buffer were added slowly with stirring to 2 g of Whatman CF11 cellulose powder suspended in 15 ml of water. The mixture was stirred for 10 min, degassed for 5 min, and poured to form a column of 1 cm diameter and 16 cm height. The sample was applied to the column under a slight pressure in order to produce a flow rate of 0.5 ml per minutes. Elution was performed with a convex exponential gradient of pH and ionic strength. The first buffer container which fed directly into the column contained 125 ml of 0.04 M potassium phosphate buffer, pH 6.2. A reservoir containing 0.15 M potassium phosphate buffer, pH 6.9, was directed into the first buffer container so that its volume remained constant. When 2-ml fractions were collected, the ultraviolet-absorbing mate-

[5] J. J. Anderson, J. M. Wilson, and D. L. Oxender, *J. Bacteriol.* **140**, 351 (1979).
[6] O. Levin, this series, Vol. 5, p. 27.

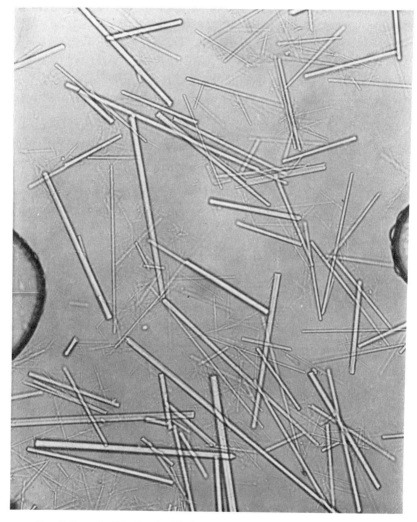

FIG. 3. Crystals of the leucine-binding protein (LS-BP). Magnification: × 252.

rial containing the binding activity was usually present in fractions 24 through 30. Fractions containing substantial levels of binding activity were pooled and dialyzed exhaustively against distilled deionized water. In order to store the protein, the dialyzed solution was concentrated to a small volume by lyophilization, divided into aliquots, and quickly frozen in an ethanol–dry ice bath for storage at −20°.

Crystallization. The binding proteins may be crystallized with the or-

ganic solvent 2-methyl-2,4-pentanediol as described below.[7,8] The dialyzed effluent from the hydroxylapatite cellulose column was lyophilized to dryness. The residue was dissolved in 2 mM potassium phosphate buffer, pH 6.9, and the final protein concentration was adjusted to 10 mg/ml; insoluble material was removed by centrifugation. Then 2-methyl-2,4-pentanediol was added slowly with mixing until a turbidity developed (50 to 55% diol). Distilled water was added until the turbidity nearly disappeared. Crystals appeared immediately as evidenced by schlieren patterns in the solution. By this method of crystallization, needlelike crystals could be obtained which were 0.1 mm in length. They did not appear to grow appreciably upon storage at −20°. Larger crystals were grown by allowing a 55 to 55% solution of diol to dialyze into a solution of the binding protein. By this method crystals were obtained after 10 to 12 hr; the largest were 0.5 mm in length and 0.1 mm in width (Fig. 3). The crystalline binding protein could be stored for at least 8 weeks at −20° without loss of activity. This method yielded crystals of the binding protein containing bound substrate.

An improved method for crystallization of the binding proteins with and without bound substrate has been developed by Saper and Quiocho.[9] In this method, crystals as large as 1.5 mm which diffracted to at least 2Å resolution were routinely grown.

Comments

The protocols presented in this article are applicable primarily to prokaryotic transport systems which utilize periplasmic components. It should be noted that, in the case of the transport assays, a physiological process is being measured so that interstrain variation is to be expected. Thus all samples should be assayed in triplicate. A quantitative difference is considered significant only if the standard errors of the ranges do not overlap.

[7] A. B. Pardee, *Science* **156**, 1627 (1967).
[8] W. R. Penrose, G. E. Nichoalds, J. R. Piperno, and D. L. Oxender, *J. Biol. Chem.* **243**, 5921 (1968).
[9] M. A. Saper and F. A. Quiocho, *J. Biol. Chem.* **258**, 11057 (1983).

[32] Transport of Branched-Chain Amino Acids and Their Corresponding 2-Keto Acids by Mammalian Cells

By MICHAEL S. KILBERG and MARY E. HANDLOGTEN

The catabolism of the branched-chain amino acids (BCAA) and their corresponding 2-keto acids (BCKA) involve a number of tissues. As a result, flows of the metabolites between various organs and their transport into and out of individual cell types represent important components of the overall metabolism of these nutrient molecules. The BCAA transaminase, responsible for the conversion of BCAA to BCKA, is present in relatively low levels in the liver in comparison to extrahepatic tissues such as muscle.[1,2] In contrast, the branched-chain 2-keto acid dehydrogenase activity, which catalyzes the oxidative decarboxylation of 2-keto acids, is present primarily in the liver.[2-4] Thus, the BCAA are transaminated in muscle to form the corresponding 2-keto acids which in turn are released into the bloodstream and removed from the circulation by the liver. As a result of the high levels of the transaminase activity in extrahepatic tissues, increased amounts of the BCKA are released from these tissues under metabolic conditions favoring tissue breakdown or proteolysis.

Branched-Chain 2-Keto Acid Transport

A number of studies have described 2-keto acid uptake by mitochondria[5-8] as well as the properties and regulation of the mitochondrial 2-keto acid dehydrogenase.[3,9,10] The mitochondrial carrier for monocarboxylates, including the BCKA has been reconstituted successfully, and partially purified.[11] Our laboratory,[12] as well as another,[13] has provided

[1] A. Ichihara and E. Koyama, *J. Biochem. (Tokyo)* **59**, 160 (1966).
[2] F. L. Shinnick and A. E. Harper, *Biochim. Biophys. Acta* **437**, 477 (1976).
[3] R. M. Wohlheuter and A. E. Harper, *J. Biol. Chem.* **245**, 2391 (1970).
[4] B. S. Khatra, R. K. Chawla, C. W. Sewell, and D. Rudman, *J. Clin. Invest.* **59**, 558 (1977).
[5] A. P. Halestrap, M. D. Brand, and R. M. Denton, *Biochim. Biophys. Acta* **367**, 102 (1974).
[6] E. Walajtys-Rode and J. R. Williamson, *J. Biol. Chem.* **255**, 413 (1980).
[7] T. B. Patel, P. P. Waymack, and M. S. Olson, *Arch. Biochem. Biophys.* **201**, 629 (1980).
[8] N. McKay and B. Robinson, *in* "Metabolism and Clinical Implications of Branched Chain Amino and Keto Acids" (M. Walser and J. R. Williamson, eds.) p. 55. Elsevier/North-Holland, New York, 1981.
[9] P. J. Parker and P. J. Randle, *FEBS Lett.* **90**, 183 (1978).
[10] D. J. Danner, S. K. Lemmon, and L. J. Elsas, *Biochem. Med.* **19**, 27 (1978).

evidence that the plasma membrane of rat hepatocytes contains specific transport systems for the removal of 2-keto acids from the bloodstream. A Na^+-dependent and a saturable, Na^+-independent carrier is present in rat hepatocytes held in primary culture.[12] Steele[14] has described a 2-keto acid transport system in brain tissue. The BCKA carriers in the brain and the liver also exhibited affinity for pyruvate, propionate, acetoacetate, and β-hydroxybutyrate. Although not tested in liver, the carrier in the brain did not exhibit high affinity for keto acids containing side-chain aromatic rings, e.g., phenylpyruvate. In contrast to the activity present in liver, [12,13] brain 2-keto acid transport was not blocked by α-cyanocinnamate, an inhibitor of mitochondrial 2-keto acid uptake.[5]

With respect to regulation of BCKA transport, starvation of rats resulted in a stimulation of 2-ketoisocaproate transport across the blood–brain barrier suggesting regulation of the carrier in that epithelial tissue.[14] Modulation of the Na^+-dependent 2-keto acid transport activity by either insulin treatment or substrate deprivation of cultured hepatocytes has been reported.[12] Demigne *et al.*[15] have shown that hepatic extraction of plasma BCKA is greater in rats fed a high-protein diet.

Branched-Chain Amino Acid Transport

The BCAA are transported into eukaryotic cells by a number of Na^+-dependent and Na^+-independent carriers with overlapping specificity. The System L carrier appears to mediate most of the BCAA transport in mammalian cells.[16] Since its original description in Ehrlich ascites tumor cells,[17] System L transport activity has been characterized in a number of tissues and cultured cells. Oxender and colleagues have described substrate-dependent regulation of System L in Chinese hamster ovary (CHO) cells.[16,18] They have studied a mutant CHO cell line (CHO-tsH1) that contains a temperature-sensitive leucine-tRNA ligase which renders the cell inefficient in forming leucyl-tRNA at marginally permissive temperatures.[19] Under these conditions, the mutant cells behave as though they are starved for leucine as a result of the decreased synthetase activity and

[11] K. A. Nalecz, R. Bolli, L. Wojtczak, and A. Azzi, *Biochim. Biophys. Acta* **851**, 29 (1986).

[12] M. B. Gwynn and M. S. Kilberg, *J. Biol. Chem.* **258**, 11524 (1983).

[13] K. A. Nalecz, A. B. Wojtczak, and L. Wojtczak, *Biochim. Biophys. Acta* **805**, 1 (1984).

[14] R. D. Steele, *Fed. Proc., Fed. Am. Soc. Exp. Biol.* **45**, 2060 (1986).

[15] C. Demigne, C. Remesy, and P. Fafournoux, *J. Nutr.* **116**, 2201 (1986).

[16] M. A. Shotwell, M. S. Kilberg, and D. L. Oxender, *Biochim. Biophys. Acta* **737**, 267 (1983).

[17] D. L. Oxender and H. N. Christensen, *J. Biol. Chem.* **238**, 3686 (1963).

[18] M. A. Shotwell and D. L. Oxender, *Trends Biochem. Sci.* **8**, 314 (1983).

[19] L. H. Thompson, J. L. Harkins, and C. P. Stanners, *Proc. Natl. Acad. Sci.* **70**, 3094 (1973).

consequently, leucine transport activity via System L increases 2- to 3-fold as a result of translational control.[20,21] Increased System L activity can also be observed in the parental CHO cell line if the cells are incubated in a culture medium containing limiting amounts of leucine.[22] Dantzig et al.[23] have isolated CHO cell mutants that contain altered System L activity following selection by resistance to melphalan toxicity; melphalan is a nitrogen mustard derivative of phenylalanine which is transported into cells via System L. The Na^+-independent transport of BCAA by isolated rat hepatocytes is mediated by multiple carriers.[24-26] Competitive inhibition and kinetic analyses of transport for a number of amino acids including the BCAA and a nonmetabolizable analog (2-aminobicyclo [2.2.1] heptane-2-carboxylic acid, BCH) have been used to delineate and characterize these systems. During the first several days of primary culture, one of these systems (System L1) exhibits a RNA- and protein synthesis-dependent increase in activity,[26] while transport by the second system (System L2) decreases. In contrast to freshly isolated adult hepatocytes that contain primarily System L2, hepatocytes isolated from rat fetuses or 10-day-old rats express System L1 predominantly and contain little or no detectable System L2 activity.[25] These results suggest the need for further studies on the ontogeny of Na^+-independent transport in liver tissue to determine the age dependence and mechanism by which the cells switch from expression of System L1 only to synthesis of both carriers with L2 predominating. Recently, another Na^+-independent carrier for the BCAA, System asc, has been characterized in red blood cells.[27,28] As suggested by the name, the substrate specificity of this transporter resembles that of the Na^+-dependent activity System ASC, which has been recognized for more than 20 years. Whether System asc is analogous to one of the systems in hepatocytes has not been determined.

Although considerable progress has been achieved during the last several years in characterizing Na^+-independent transport of branched-chain

[20] P. A. Moore, D. W. Jayme, and D. L. Oxender, J. Biol. Chem. 252, 3191 (1977).
[21] M. A. Shotwell, P. M. Mattes, D. W. Jayme, and D. L. Oxender, J. Biol. Chem. 257, 2974 (1982).
[22] A. Moreno, C. D. Lobaton, and D. L. Oxender, Biochim. Biophys. Acta 819, 271 (1985).
[23] A. H. Dantzig, M. Fairgrieve, C. W. Slayman, and E. A. Adelberg, Somatic Cell Mol. Genetics 10, 113 (1984).
[24] M. E. Handlogten, L. Weissbach, and M. S. Kilberg, Biochem. Biophys. Res. Commun. 104, 307 (1982).
[25] M. S. Kilberg, L. Weissbach, and E. F. Barber, in "Isolation, Characterization, and Use of Hepatocytes" (R. A. Harris and N. W. Cornell, eds.), pp. 227 Elsevier, New York, 1983.
[26] L. Weissbach, M. E. Handlogten, H. N. Christensen, and M. S. Kilberg, J. Biol. Chem. 257, 12006 (1982).
[27] D. A. Fincham, D. K. Mason, and J. D. Young, Biochem. J. 227, 13 (1985).
[28] J. V. Vadgama and H. N. Christensen, J. Biol. Chem. 260, 2912 (1985).

amino acids and their corresponding 2-keto acids, it is clear that additional research will be required to appreciate fully the impact of these systems on intermediary metabolism. It is with this thought in mind that we present the following protocols for the measurement of Na^+-independent transport of BCAA or BCKA in isolated hepatocytes. Most of these procedures can be easily adapted for use with a variety of other tissues or cell types.

Hepatocyte Isolation

Hepatocytes are prepared by a modification of the collagenase perfusion technique originally described by Berry and Friend.[29] Briefly, the rat is anesthetized and the inferior vena cava is cannulated. Immediately after the cannula has been secured, the hepatic artery and the portal vein are severed. The thoracic cavity is opened and the superior vena cava is clamped with a hemostat. The liver is perfused at a rate of 10 ml/min with a Ca^{2+}-free perfusion buffer (200 ml) containing 25 mM sodium phosphate, pH 7.4, 3.1 mM potassium chloride, 119 mM sodium chloride, 5.5 mM glucose. 0.1% bovine serum albumin (BSA), and 5 mg/liter phenol red. After the liver has been blanched with an initial 100 ml of the buffer, collagenase (Sigma Type 1) is added to the remaining 100 ml of perfusion buffer to a final concentration of 60–75 units collagenase per gram of body weight of the animal. The exact number of enzyme units per gram is determined experimentally for individual batches of collagenase.

Following the completion of the perfusion process, the liver is removed, the cells are dispersed with a large-bore Pasteur pipet, and then forced through a 75 μm nylon mesh to remove connective tissue and nondisassociated cells. The cells are washed repeatedly in ice-cold MEM (Gibco) culture medium supplemented to 25 mM NaHCO$_3$, 4 mM glutamine, 10 mg/liter streptomycin, 100 mg/liter penicillin, 28.4 mg/liter gentamycin, 0.23 mg/liter N-butyl-p-hydroxybenzoate, and 2 gram/liter BSA. Between washes, the cells are pelleted by centrifugation for 2 min at 100 g in a refrigerated clinical centrifuge. The supernatants are either discarded or saved for the isolation of nonparenchymal cells through the use of a Percoll gradient.[30] Following these washes, the hepatocytes are diluted to the appropriate volume with culture medium at 37° and placed in suspension or monolayer culture as described below.

For measurement of carrier-mediated Na^+-independent transport in either monolayer or suspension culture, the saturable uptake must be

[29] M. N. Berry and D. S. Friend, *J. Cell Biol.* **43**, 506 (1969).
[30] D. F. Gardner, M. S. Kilberg, M. M. Wolfe, J. E. McGuigan, and R.I. Misbin, *Am. J. Physiol.* **248**, G663 (1985).

separated from the nonsaturable diffusion. The saturable transport of a particular solute is monitored by measuring solute uptake in the presence and absence of a concentration of unlabeled substrate known to saturate the carrier. This concentration, typically 10 times the K_m value, must be determined experimentally by assaying the uptake of a constant concentration of radioactively labeled substrate in the presence of increasing concentrations of unlabeled substrate. The Na^+-independent transport that is carrier mediated is taken as the difference between the rates in the presence and absence of the excess unlabeled substrate. To test for exchange reactions or trans-stimulation, the cells are "loaded" by incubation in the presence or absence of an unlabeled substrate for a period of time that will allow equilibrium across the plasma membrane. The extracellular substrate is then removed by quickly rinsing cells in ice-cold buffer and the cells are transferred to the uptake solution containing the radioactively labeled substrate. The substrate used to load the cells can be the same one used to monitor uptake or conversely, different substrates can be employed to test for their reactivity with a particular carrier system. Carrier-mediated trans-stimulation will be observed as enhanced uptake by the cells preloaded with substrate.

Monolayer culture

Acid-soluble calfskin collagen (Sigma Type III) is weighed into an autoclaved beaker containing a stirring bar and placed under a UV light for several hours. A sterile solution of 0.5% acetic acid is used to suspend the collagen to a concentration of 1 mg/ml and the mixture is stirred at room temperature until dissolved. The solution is stored at $-20°C$ in 5-ml aliquots. To collagen-coat 24-well cluster trays, the 1 mg/ml stock is diluted 1 : 50 with sterile water just prior to use and 0.5 ml (10 μg collagen) is added to each well. The trays are incubated at room temperature overnight and the following day the liquid is aspirated from each well, leaving a collagen matrix attached to the bottom. These collagen-coated trays can be stored under sterile conditions at room temperature prior to use.

For monolayer cultures, the hepatocytes are diluted to 0.8×10^6 cells/ml in MEM culture medium at 37° supplemented as described above. The cells are plated onto collagen-coated 24-well cluster trays by adding 0.33 ml of the cell suspension to each well. The cells are then placed in a humidified CO_2-incubator containing an atmosphere of 5% CO_2/95% O_2. The use of serum for these cultures is optional; it does enhance the cell attachment efficiency, but it is not necessary if its use will complicate the experimental protocol.

For the study of Na^+-independent carriers, exchange reactions are an important consideration. To minimize trans-stimulation in cells that are to be assayed soon after initiating primary cultures, we have found that hepatocytes will attach and spread on collagen-coated dishes even when washed and cultured in serum-free Krebs-Ringer bicarbonate buffer lacking all nutrient supplements with the exception of BSA and glucose.[31] For cells maintained as monolayers, short-term culture (less than 24 hr) under these conditions has little or no effect on cell viability and eliminates the need for lengthy incubations designed to deplete the intracellular amino acid pools which can be significantly altered by culture in amino acid-enriched media. However, we find that the Krebs-Ringer bicarbonate buffer is not satisfactory for maintenance of cells in suspension culture (described below), as evidenced by significant loses in cell viability.

Substrates

All three of the BCAA are readily available in radioactive form from a number of commercial sources, but radiopurity should be checked by thin-layer or paper chromatography. A convenient solvent system for detection of radioactive contaminants of individual BCAA is n-butanol : acetic acid : water (2 : 1 : 1). Removal of $[^3H]$-H_2O can be achieved by drying an aliquot of labeled amino acid under nitrogen and then redissolving the residue in the appropriate buffer just prior to use.

Radioactively labeled 2-keto acids corresponding to leucine (2-ketoisocaproic acid), isoleucine (2-keto-3-methyl-n-valeric acid), and valine (2-ketoisovaleric acid) are also available commercially (Amersham Corp). However, they may be prepared from the amino acid by the method of Rüdiger et al.[32] Briefly, 100 μl of ^{14}C-labeled BCAA (1 mCi/ml) is added to 500 μl of 1 M Tris, pH 7.7, 100 μl (0.2 U) L-amino acid oxidase, 100 μl (200 U) of catalase. The solution is aerated with O_2 for 1 min and then incubated in a stoppered flask for 2 hr at 37°C. After the first hour, the aeration is repeated. At the end of the incubation, the entire mixture is passed over a 1×7 cm Dowex 50W($\times$8) column (H^+ form) and eluted with distilled water at a flow rate of 9 ml/hr. Fractions (0.5 ml) are collected and analyzed by thin-layer chromatography (n-butanol : acetic acid : H_2O, 12 : 3 : 5). The fractions containing significant amounts of purified 2-keto acid are pooled and taken to dryness under a stream of nitrogen. The yield of 2-keto acid is typically 70–80%.

[31] L. Weissbach and M. S. Kilberg, *J. Cell Physiol.* **121**, 133 (1984).
[32] H. W. Rüdiger, U. Langenbeck, and H. W. Goedde, *Biochem. J.* **126**, 445 (1972).

BCAA or BCKA Transport by Cells in Monolayer Culture

Transport of radioactively labeled BCAA or BCKA by cells cultured on 24-well cluster trays is assayed by a modification of the method described originally by Gazzola et al.[33] and reviewed more recently by Vadgama.[34] The lids to the 24-well cluster trays are modified so that they contain either a 12 × 75 mm plastic test tube (4-ml capacity) or a Beem (#00) embedding capsule (0.5-ml capacity) that is aligned over each well. Placing the bottom of the tray, to which the cells are attached, upside-down on either of the modified lids and then rapidly inverting the entire unit allows the transfer of either 2 ml (warm or cold washes) or 0.25 ml (uptake mixture) of medium to all 24 wells simultaneously. Just prior to the transport assays, the cultured cells are incubated in 2 ml of Na$^+$-free Krebs–Ringer phosphate buffer at 37° for two 10-min periods. These incubations serve to minimize trans-stimulation by partially depleting the cells of amino acids or 2-keto acids, and to rinse the cells free of extracellular Na$^+$. After removal of the second rinse by inverting the tray over a dishpan, the transport assay is initiated by transferring 0.25 ml of the uptake medium to all 24 wells through the use of the lids modified with the Beem embedding capsules. The uptake solution is prepared by adding radioactively labeled test substrate at a known concentration (plus unlabeled substrate where appropriate as described above) to Na$^+$-free Krebs–Ringer phosphate buffer, pH 7.4. For many radioactively labeled substrates the contribution of the labeled form to the total concentration used is negligible, but for compounds of low specific radioactivity this must be calculated and compensation made during addition of unlabeled substrate.

For the preparation of Na$^+$-free buffers, a 10× stock solution of choline phosphate is prepared by heating a 500 mM solution of choline bicarbonate in the presence of 250 mM phosphoric acid at 100° for at least 1 hr (continue to heat until no CO_2 generation is visible) to produce $Chol_2HPO_4$. This stock solution is then titrated to the appropriate pH with HCl. After diluting this stock solution 10-fold during the preparation of the 25 mM buffer mixture, the pH should be readjusted to 7.4. Thus, the final buffer composition is 25 mM choline phosphate, pH 7.4, 119 mM choline chloride, 5.9 mM potassium chloride, 1.2 mM magnesium sulfate, 1.2 mM potassium bicarbonate, 5.5 mM glucose, and 0.5 mM calcium chloride.

Transport is terminated by inverting the cluster tray over a dishpan to discard the uptake mixture and then rapidly rinsing the cells several times with ice-cold Na$^+$-free Krebs–Ringer phosphate buffer. The cells are solu-

[33] G. C. Gazzola, V. Dall'Asta, R. Franchi-Gazzola, and M. F. White, *Anal. Biochem.* **115**, 368 (1981).
[34] J. Vadgama, this series, Vol. 171

bilized by adding 0.2 ml of 0.2% SDS in 0.2 N sodium hydroxide to each well. Following a 30-min incubation at room temperature, 0.1-ml aliquots of the cell extract are transferred to scintillation vials for determination of radioactivity following neutralization with 0.1 ml of 0.2 N HCl. The remaining 0.1 ml of cell extract is left in each well and used to assay the total cellular protein content by a modification of the Lowry procedure.[26] To each well, 600 μl of a copper reagent (0.58 mM EDTA copper disodium salt, 189 mM sodium carbonate, 100 mM sodium hydroxide, and 1% SDS) is added. After incubating for 10 min at room temperature, 60 μl of diluted phenol reagent (1 : 1 dilution with water just prior to use) is added, the mixture is incubated for 30 min, and then the absorbance at 500 or 750 nm is determined. The data are expressed as picomoles of substrate transported per milligram protein per unit time. Through the use of alternate assays or conversion factors, the data can also be expressed per cell number or per intracellular water rather than protein content. Intracellular water content can be easily measured by the distribution of a solute which is not subject to active transport such as 3-O-methylglucose.[35]

BCAA or BCKA Transport by Cells in Suspension

For suspension cultures, the cells are diluted to 20×10^6 cells per milliliter in Eagle's minimum essential medium supplemented with amino acids and pyruvate as described by Seglen et al.[36] and containing 10% fetal bovine serum. The cell suspension is placed in a 500-ml roller bottle and rotated at 2.5 rpm through the use of a Wheaton 2-place roller bottle apparatus that is small enough to fit into a 37° humidified CO_2-incubator. To measure transport of a test substrate by cells maintained in suspension culture, a modification of the method of McGivan et al.[37] is used. Just prior to the assay, 8 ml of the suspended cells is removed from the roller bottle, separated from the culture medium by centrifugation at 100 g for 3 min, washed twice in Na^+-free Krebs–Ringer phosphate buffer, and then resuspended to 20×10^6 cells per milliliter. Assay tubes (2-ml microcentrifuge tubes) are prepared prior to the experiment by layering, in order from the bottom, 0.2 ml of perchloric acid (20% v/v), 1 ml of n-butyl phthalate, and 0.5 ml of Na^+-free Krebs–Ringer phosphate buffer containing the radioactively labeled amino acid or 2-keto acid. A multichannel adjustable

[35] Kletzien, R. F., Pariza, M. W., Becker, J. E., and Potter, V. R., Anal. Biochem. **68**, 537 (1975).
[36] P. O. Seglen, A. E. Solheim, B. Grinde, P. B. Gordon, P. E. Schwarze, R. Gjessing and A. Poli, Ann. N.Y. Acad. Sci. **349**, 1 (1980).
[37] J. D. McGivan, N. M. Bradford, and J. Mendes-Mourao, FEBS Lett. **80**, 380 (1977).

pipet is used to add simultaneously 50-μl aliquots of cells (approximately 1×10^6 cells) to the top of 6 assay tubes, equilibrated previously in a 37° waterbath. Transport is terminated by a 30-sec centrifugation (15,000 g) in a microcentrifuge. The density of the oil layer (n-butyl phthalate) may be modified through the use of alternate oils so as to be compatible with the density of any particular cell type under study. The top two layers (uptake buffer and oil layer) are discarded by aspiration and the residual radioactivity is removed from the side of the tube with a paper wick. A 0.1-ml sample of the perchloric acid extract is placed in a scintillation vial for the determination of radioactivity. The precipitated protein can be solubilized in 0.2 N NaOH containing 0.2% SDS and an aliquot used to determine the total protein content. As above, the results can be expressed as picomoles of substrate transported per milligram protein per unit time or related to other cell-associated parameters.

[33] Leucine-tRNA Ligase Complexes

By Arnold Hampel and Richard Tritz

The aminoacyl-tRNA synthetases are a group of 20 different enzymatic activities which esterify amino acids to cognate tRNA in the initial step of protein synthesis. As early as 1964,[1] these enzymes were suggested to be components of high-molecular-weight complexes in mammalian systems. The first description of the existence of the complexes themselves was made by this laboratory in 1973[2] — other earlier reports have been retracted. After characterizing the nature of these complexes, we showed that the likely subcellular role for these multicomponent high-molecular-weight synthetase complexes is utilization of extracellular amino acids for protein synthesis immediately upon transport and before these amino acids have equilibrated with the internal pool. In contrast, the low-molecular-weight forms of the enzymes preferentially utilize the amino acids from the intracellular pool for protein synthesis.[3]

This article describes a method for the preparation of the largest aminoacyl-tRNA synthetase (aaRS) complex found to date. The complex from mammalian cells is 30S and contains 10 aminoacyl-tRNA synthetase ac-

[1] H. Hird, E. McLean, and H. N. Munro, *Biochim. Biophys. Acta* **87**, 219 (1964).

[2] A. E. Hampel and M. D. Enger, *J. Mol. Biol.* **79**, 285 (1973).

[3] A. Hampel, A. Mansukhani, and T. Condon, *Fed Proc., Fed. Am. Soc. Exp. Biol.* **43**, 2991 (1984).

tivities. Enzymatic activities found in the 30S complex are specific for the amino acids isoleucine, leucine, threonine, aspartic acid, glutamine, glutamic acid, arginine, lysine, methionine, and proline while the aaRS for the branched-chain amino acid valine is found in a 19S complex and a 9S low-molecular-weight form.[4-6]

Through the use of mutants and revertants from the model Chinese hamster ovary cell (CHO) system we have shown properties of leucine-tRNA ligase, EC 6.1.1.4 (leuRL) in the complex correlate with defined phenotypic properties. LeuRL exists in normal CHO cells in three distinct forms having sedimentation coefficients of 30S, 20S, and 8S, respectively. The temperature-sensitive mutant *ts*H1[7] contains only the low-molecular-weight 8S form of leuRL which has an *in vitro* thermolability three times greater than wild-type 8S.[8] Thus, this is a case where the leuRL is not found in the complex due to an enzymatic alteration. We have characterized another CHO leuRL mutant, *ts*025C1, which is temperature sensitive for leuRL but has a perfectly normal 8S form of the enzyme.[9]

Revertants of the *ts*H1 mutant fall into two classes: (1) transport revertants and (2) enzymatic revertants. Transport revertants show a stable 2- to 3-fold enhancement in the activity of transport system L, the transport system for the branched-chain amino acids. This leads to increased transport of leucine, thereby increasing the leucine pools and overcoming the leuRL defect.[10]

Enzymatic revertants showed a correlation between extent of reversion and degree of ability to reform the complex,[5,8] with an inverse correlation between extent of reversion and internal leucine pool sizes. From these studies, it was possible to correlate the internal leucine pool sizes with the amount of leuRL in the higher-molecular-weight 20S and 30S complexes and formulate a model for function of the high-molecular-weight complex. In this model, it is proposed leuRL in the complex utilizes amino acid for protein synthesis immediately upon transport before it can equilibrate with the internal pool, while the low-molecular-weight 8S form of leuRL utilizes amino acid only from the internal pool.[3]

Other properties of the leuRL complex are a K_m much less than that of

[4] P. Ritter, M. D. Enger, and A. Hampel, *in* "Onco-Developmental Gene Expression" (H. Fishman and S. Sell, eds.), p. 47. Academic Press, New York, 1976.

[5] E. Pahuski, M. Klekamp, T. Condon, and A. E. Hampel, *J. Cell. Physiol.* **114**, 82 (1983).

[6] P. Ritter, M. D. Enger, and A. Hampel, *Biochim. Biophys. Acta* **562**, 377 (1979).

[7] L. H. Thompson, J. Harkins, and C. P. Stanners, *Proc. Natl. Acad. Sci. U.S.A.* **70**, 3094 (1973).

[8] M. Klekamp, E. Pahuski, and A. Hampel, *Somatic Cell Genet.* **7**, 725 (1981).

[9] A. Mansukhani, T. Condon, A. Hampel, and D. Oxender, *Biochem. Genet.* **22**, 349 (1984).

[10] M. Shotwell, E. Collarini, A. Mansukhani, A. Hampel, and D. Oxender, *J. Biol. Chem.* **258**, 8183 (1983).

the low-molecular-weight 8S form and a thermostability twice that of the 8S form. Levels of complex decreased in leu-G$_1$ arrested cells, showing a correlation between cell state and the level of complex.[11] Thus, data support the argument that such a complex as described herein does indeed exist and plays an important cellular role.

To date, this largest aaRS complex is only poorly characterized biochemically. The complex has been difficult to work with as it falls apart on columns due to their large absorptive surface, dissociates also when salt levels exceed 0.15 M, changes composition at low salt levels due to formation of nonspecific aggregates, and changes composition when precipitated, even though the precipitated form is very stable. Consequently, methods for studying this complex are very limited at this time.

Methodologies are described in this article for study of this complex which use gentle methods for resolution with salt concentrations near 0.15 M and without precipitation steps. The very gentle method of sucrose gradient centrifugation has been ideal for this purpose. Its utility has been proved in preparing such labile high-molecular-weight structures as polysomes.[12]

Leucine-tRNA Ligase Assay

The assay for detecting leuRL is a simple, rapid, and highly accurate procedure developed in our laboratory.[13] The assay determines the amount of leu-tRNA produced in the reaction:

$$tRNA + ATP + leucine \longrightarrow leu\text{-}tRNA + PP_i + AMP$$

Radioactive leucine is used and the reaction is stopped with acid-soaked filter paper disks. Wash conditions are used which precipitate the esterified amino acid on the disk and wash off the nonesterified amino acid. The disks are then simply dried and counted to determine the amount of leu-tRNA product of the enzyme assay.

The final reaction mixture contains 10 mM Tris-HCl, pH 8.6, at 25°; 0.5 mM EDTA adjusted to pH 7.0 with KOH; 5 mM ATP adjusted to pH 7.0 with KOH; 0.34 mM CTP; 10 μM each of the 19 amino acids except leucine; 0.1 mM dithiothreitol; 8 mM MgCl$_2$; 0.25–0.5 mg/ml rat liver tRNA; and 10 μM L-(^{3}H)-leucine (10,000 Ci/mol). Mechanically, the reactions are carried out by placing drops of reaction mixture on Saran wrap on a slide warmer at 37°. The drop size is 50 μl minus the volume of leuRL to

---

[11] M. D. Enger, P. O. Ritter, and A. E. Hampel, *Biochemistry* **17**, 2435 (1978).
[12] H. Bloemendal, E. Benedetti, and W. Bont, this series, Vol. 30, p. 313.
[13] A. Hampel, M. D. Enger, and P. O. Ritter, this series, Vol. 59, p. 229.

be added. The Saran wrap is placed on a slide warmer for 1 min prior to the addition of leuRL. Limiting amounts of enzyme are added such that the reaction is linear with time. Normally, the reaction is run for 6 min before being terminated by placing a previously 10% trichloroacetic acid (TCA) soaked and dried filter paper disk (Whatman 3 MM, 1.5 cm) on the reaction drop. The disk can be labeled with a pencil and not impair the synthetase reaction. Five minutes later, the disks are placed in ice-cold 10% TCA for up to 2 hr without affecting the results. After completion of the set of assays all reaction disks are washed together three times in 10% TCA (5 ml/disk) and three times in 5% TCA (4 min for each wash). Disks are then rinsed two times in ether and dried under a heat lamp. The TCA-insoluble radioactivity on each disk corresponding to radioactive leu-tRNA is determined by liquid scintillation counting in 5 ml scintillation fluid (5 g PPO and 0.3 g dimethyl-POPOP per liter of toluene).

Preparation of LeuRL Complexes

Cells used were Chinese hamster ovary cells grown in α-MEM medium containing 10% calf serum.[5] Cells were grown in suspension culture to a density of 5×10^5 per milliliter; collected by placing in ice and adding 250 ml frozen 0.25 M sucrose ice cubes per liter of culture, centrifuged 1,000 g for 10 min at 4°, washed once with 0.25 M sucrose, resuspended (10^8 cells/0.9 ml) in buffer A (100 mM KCl, 1 mM MgCl$_2$, 0.1 mM dithiothreitol, 10 mM Tris-HCl, pH 7.5, at 25°), and stored at $-90°$.

Cells (10^8) were thawed in a 37° water bath, then lysed for 30 min by the addition of Nonidet P-40 to 1%. Nuclei and cell debris were removed by centrifugation (1,000 g, 30 min). The supernatant was carefully layered on 12 ml 10–30% w/v sucrose gradients in 10 mM KCl, 1 mM MgCl$_2$, 0.1 mM dithiothreitol, and 10 mM Tris-HCl, pH 7.5, at 25°. Gradients were centrifuged in a Beckman SW27 rotor at 97,000 g at 4° for 18 hr, fractionated, and assayed for leuRL activity as above. Sedimentation coefficients were calculated by a modification of the computer program of Dingman[14] and protein determined by the method of Bradford.[15] Three activity peaks with sedimentation coefficients at 8S, 20S, and 30S are seen.

Leucine Michaelis Constants

Peak tubes from sucrose gradients were assayed for leuRL activity over a range of leucine concentration of 2, 4, 8, 12, 16, and 20 μM. Assay times

[14] W. Dingman, *Anal. Biochem.* **49**, 124 (1972).
[15] M. Bradford, *Anal. Biochem.* **72**, 248 (1976).

TABLE I
PROPERTIES OF LEURL COMPLEXES

Cell type	Sedimentation form	K_m for Leu (μM)
Wild type	8S	16
	20S	8
	30S	9
tsH1	8S	71
ts025C1	8S	16

of 0.5, 1.0, and 1.5 min gave linear product formation. Double-reciprocal plots were analyzed by linear regression and the K_m values calculated by standard statistical kinetic analysis[16] (Table I).

Note in the wild type that the complexes (20S and 30S) have K_m values smaller than the 8S form. The mutant tsH1, which contains an altered leuRS, has a K_m value over 4 times larger than its corresponding wild-type 8S form. The leuRL mutant ts025C1 has a K_m identical to the wild-type 8S, supporting the notion that it is not altered in any way except that the large-molecular-weight complexes are missing in this mutant.

Determination of Temperature Sensitivity

It is necessary to have a reliable and reproducible temperature-sensitivity assay in order to identify the functional role of the complexes and characterize the leuRL activity in temperature-sensitive mutants. Two problems have been encountered in the past which produce artifacts. The first is the inherent lability of leuRL at low protein concentrations. This is overcome by adjusting all solutions to the same protein concentration (10 mg/ml) using bovine serum albumin. The second problem is the tendency of isolated leuRL to contain the aminoacyl-adenylate intermediate at varying levels. The leuRL is greatly stabilized by the aminoacyl-adenylate intermediate, making it possible to carry out reproducible thermolability studies.[17] The addition of tRNA (3 mg/ml) effectively removes the intermediate from the catalytic sites, resulting in reproducible results with linear kinetics of inactivation.

The assay is carried out on peak tubes of the various forms of the enzyme obtained from sucrose gradients. The protein concentration is adjusted to 10 mg/ml with bovine serum albumin, tRNA (3 mg/ml) is

---

[16] G. N. Wilkinson, *Biochem. J.* **80**, 324 (1961).

[17] L. Haars, A. Hampel, and L. H. Thompson, *Biochim. Biophys. Acta* **454**, 493 (1976).

added, and the tubes are heated at 40.5° for 0, 2, 4, 6, 8, 10, 12, and 14 min. Ten-microliter aliquots are removed and assayed for leuRL activity as described. Typical results of these assays are given in Table II.

Note that the high-molecular-weight forms of the enzyme are more thermostable than the low-molecular-weight forms.[8] The temperature-sensitive mutant tsH1 has no high-molecular-weight forms of the enzyme but does exhibit a temperature-sensitive low-molecular-weight form.[18] The mutant ts025C1 is unique as it shows a thermosensitive phenotype but has a thermostable 8S form of the leuRS enzyme. Thermolability was shown due to the absence of high-molecular-weight forms of the enzyme[9] which are known to have greater thermostability. Thus, this CHO cell line contains a temperature-sensitive leuRL with a mutation not in leuRL itself but rather in some other location which affects the integrity of the leuRL complex.

Summary

The methodologies described in this chapter allow the reproducible preparation of native high-molecular-weight synthetase complexes of leuRL. These complexes have the ability to preferentially utilize extracellular leucine immediately upon transport and are likely the forms of the enzyme most important in the utilization of leucine for protein synthesis.

[18] A. Hampel, P. O. Ritter, and M. D. Enger, *Nature (London)* **276,** 844 (1978).

Section III

Enzymes

[34] Purification of Branched-Chain-Amino-Acid Aminotransferase from Pig Heart

By T. K. KORPELA

Branched-chain-amino-acid aminotransferase (BCAT; EC 2.6.1.42) from mammals is a protein with a molecular mass of 75,000 Da; it contains one molecule of pyridoxal phosphate and is probably composed of two subunits.[1,2] The coenzyme is tightly bound and cannot be resolved without irreversible loss of the catalytic activity.[3] Mammalian BCAT displays maximal activity in the presence of thiols[4] and is inhibited by di- and polyvalent metal cations[4,5] and by mono- and dicarboxylic acids.[6] The optimum pH is 8.0–8.7.[1,2,7]

Distribution of BCAT in organs of mammals such as rat and pig is well studied. The highest activities are found in stomach and pancreas. Skeletal muscles also contain relatively high levels of BCAT, while, in contrast to other transaminases, only a low activity is detected in liver.[8,9]

Isoenzymes of BCAT are found in mammalian tissues but their exact structural differences and physiological roles are not yet clear. BCAT exists in both cytosolic and mitochondrial fractions of several tissues, and these fractions usually contain about equal activity.[9,10] These forms of BCAT differ by their K_m values and isoelectric points.[9,11]

In addition to the cytosolic and mitochondrial isoenzymes, three other subforms of BCAT, termed I, II, and III, have been identified. The most conspicuous difference between them is their elution behavior during DEAE-cellulose chromatography: they elute from the anion exchanger with 0.02, 0.18, and 0.2 M phosphate buffer, respectively.[2,12] BCAT I is present in most tissues of rat while BCAT III is found in larger amounts in

[1] R. T. Taylor and W. T. Jenkins, *J. Biol. Chem.* **241,** 4396 (1966).
[2] K. Aki, A. Yokojima, and A. Ichihara, *J. Biol. Chem.* **65,** 539 (1969).
[3] E. E. Snell and S. J. DiMari, *in* "The Enzymes" (P. D. Boyer, ed.), Vols. 1–2, p. 335. Academic Press, New York, 1970.
[4] R. T. Taylor and W. T. Jenkins, *J. Biol. Chem.* **241,** 4406 (1966).
[5] R. Odessey and A. L. Goldberg, *Biochem. J.* **178,** 475 (1979).
[6] R. T. Taylor, V. Shakespeare, and W. T. Jenkins, *J. Biol. Chem.* **245,** 4880 (1970).
[7] K. Aki, K. Ogawa, and A. Ichihara, *Biochim. Biophys. Acta* **159,** 276 (1968).
[8] A. Ichihara, C. Noda, and M. Goto, *Biochem. Biophys. Res. Commun.* **67,** 1313 (1975).
[9] A. Ichihara and E. Koyama, *J. Biochem. (Tokyo)* **59,** 160 (1966).
[10] A. Ichihara, *Ann. N.Y. Acad. Sci.* **259,** 347 (1975).
[11] K. Aki, K. Ogawa, A. Shirai, and A. Ichihara, *J. Biochem. (Tokyo)* **62,** 610 (1967).
[12] A. Ichihara, C. Noda, and K. Ogawa, *Adv. Enzym. Regul.* **11,** 155 (1972).

METHODS IN ENZYMOLOGY, VOL. 166

brain, ovary, and placenta.[2,13,14] Pig liver and kidney[7] and many human tissues[15] contain BCAT III. The forms I and III are found in both cytosolic and mitochondrial fractions.

BCAT II acts only on leucine and does not transaminate other branched-chain amino acids. It is not inhibited by antisera produced against BCAT I or III.[2,12] BCAT II was originally found in rodent liver[7] and was subsequently purified from rat liver mitochondria.[16] The mitochondrial BCAT II also transaminates methionine. Thus, this enzyme may represent a side activity of other transaminases and its classification as a BCAT is uncertain.[17]

In this chapter we report a method of purification of pig heart cytosolic BCAT I to near homogeneity.[18] It seems probable that the method is also applicable to cytosolic BCAT I from other mammals and tissues other than heart. The first steps of the purification are adapted from the work of Taylor and Jenkins.[1,19] The subsequent stages involve exclusion, anion exchange, and affinity chromatography.

Assay Methods

The method of assay of BCAT is a modified version from the literature.[19-21] The substrate solution contains 30 mM L-leucine, 30 mM 2-ketoglutarate, 15 μM pyridoxal phosphate in 80 mM Tris-HCl buffer, pH 8.0. Centrifuge tubes of about 10-ml volume (resistant to toluene), which can be tightly capped, are required. Pipet 0.4 ml of the substrate solution into tubes and warm to 37°. Add 0.1 ml of enzyme sample and incubate at 37° for 30 min. Stop the reaction with 0.1 ml of 5 M H$_2$SO$_4$. Mix and allow to cool to room temperature. Then add 1 ml of 15 mM solution of 2,4-dinitrophenylhydrazine in 2 M HCl. After 5 min, apply 2.5 ml of toluene to extract the hydrazone of 2-ketoisocaproic acid. Stopper the tubes and mix with an end-over-end stirrer for 2 min with a speed of about 100 rpm.

[13] K. Ogawa, A. Yokojima, and A. Ichihara, *J. Biochem. (Tokyo)* **68,** 901 (1970).
[14] A. Ichihara, Y. Yamasaki, H. Musuji, and J. Sato, in "Isozymes" (C. L. Markert, ed.), Vol. 3, p. 875. Academic Press, New York, 1975.
[15] M. Goto, H. Shinno, and A. Ichihara, *Gann* **68,** 663 (1977).
[16] T. Ikeda, Y. Konishi, and A. Ichihara, *Biochim. Biophys. Acta* **445,** 622 (1976).
[17] A. Ichihara, in "Transaminases" (P. Christen and D. E. Metzler, eds.), p. 437. Wiley, New York, 1984.
[18] T. Korpela and R. Saarinen, *J. Chromatogr.* **318,** 333 (1985).
[19] W. T. Jenkins and R. T. Taylor, this series, Vol. 17, p. 802.
[20] R. Raunio, *Acta Chem. Scand.* **23,** 1168 (1969).
[21] K. Aki and A. Ichihara, this series, Vol. 17, p. 807.

Separate the layers by centrifugation (2 min, 2000 g) and measure the absorbance of the toluene phase at 350 nm. Take care not to transfer any 2-oxoglutarate hydrazone from the walls of tubes or from the interphase of the layers. This may occasionally cause high values.

For each assay series, make blank tubes using water in place of the sample. Standard solutions are made from 2-ketoisocaproate (1 – 5 mM) in water (pH adjusted to neutral). They can be stored in proper portions in a freezer. In routine series, use a 2.5 mM standard as a control. It should yield an absorbance of about 0.86.

The enzyme activity is expressed as micromoles of 2-ketoisocaproate formed/minute at 37° by dividing the number of absorbance units obtained in 30 min by a factor of 103.

Protein in pooled fractions of the chromatographic runs can be measured by the Coomassie Brilliant Blue method[22] with bovine serum albumin as the standard.

Preparation of the Affinity Gel

The synthesis involves linking of D-cycloserine to cyanogen bromide (CNBr)-activated Sepharose 4B gel. Suitable techniques for the activation can be found elsewhere in this series.[23] The author has no experience with the properties of gels activated by other than the "titration method." Use 0.15 g of CNBr (dissolved in acetonitrile) per gram of suction-dried moist gel. If BCAT from 2 kg of starting material is chromatographed in one batch, about 400 ml of the gel should be prepared.

Immediately after the activation procedure, pour the suction-dried gel into an equal volume of ice-cold 0.25 M D-cycloserine solution (pH adjusted to 9.5 with NaOH). The degree of the ligand content of the gel is increased if the activated gel is finally washed on the funnel with the ligand solution. Allow the reaction to proceed overnight at 4 – 8° with gentle shaking or with a nondestructive magnetic stirrer[24] which does not disintegrate the gel particles. Wash the gel with water (5 liters), 1 M NaCl (5 liters), water (2 liters), and finally with 25 mM sodium phosphate, pH 6.0, supplemented with 0.1 M NaCl, 1 mM mercaptoethanol, and 0.5 mM ethylenediaminetetraacetic acid (EDTA) (2 liters). Store the gel at 4°.

The ligand content bound to agarose should be 10 μmol or more per gram of moist gel. It can be tested as follows: incubate 2 g of the gel with 2 ml of concentrated HCl for 12 hr at 95° and measure the serine that

[22] M. Bradford, *Anal. Biochem.* **72,** 248 (1976).
[23] I. Parikh, S. March, and P. Cuatrecasas, this series, Vol. 34, p. 77.
[24] T. Korpela, *Lab. Pract.* **34,** 111 (1985).

appears in the solution (cycloserine decomposes to serine on such treatment). Analysis can be by amino acid analyzer or paper chromatography. Standards are made from cycloserine and Sepharose 4B gel.

Purification Procedure

Extraction and Heat Treatment

The following steps should be done without interruption until the dialysis. Avoid metallic vessels at all steps.

Cool fresh pig hearts in ice and remove any visible fat and the auricles to yield 2 kg of the substance. Treat it twice with a conventional meat grinder precooled in ice. Suspend the minced meat in 2 liters of ice-cold solution containing 50 mM sodium caproate and 5 mM EDTA (pH 6.0). Transfer the homogenate to a kettle of about 6 liters and sink it in a water bath at 70°. Gently stir the content. When its temperature reaches 40°, add 10 ml of 1 M 2-ketoglutarate (pH adjusted to 6.0). Continue the stirring until the temperature is 62° (it should take about 10 min). Keep the homogenate at 62° for 20 min and thereafter cool it in an ice-water bath to 5–10°. Pour the suspension onto a large Büchner funnel covered with cheesecloth or into a cheesecloth sock and collect the percolating fluid in a cold room. Centrifuge it at 10,000 g for 10 min and filter the supernatant through a glass wool pad to remove floating lipids.

$(NH_4)_2SO_4$ Precipitation and Dialysis

The main purpose of this step is to lower the volume of the supernatant. Add calculated amount of solid $(NH_4)_2SO_4$ at 4° in portions to stirred supernatant to achieve 50% saturation. Check the pH (6.0) in 1:30 dilutions between the additions. After dissolving, allow the precipitation to take place on an ice-bath for 30 min. Longer times drastically lower the recovery of BCAT. Centrifuge the precipitate at 4° (10,000 g, 10 min) and dissolve the pellet in a minimum (30–50 ml) of 20 mM Tris (pH adjusted with caproic acid to 6.0) supplemented with 2 mM EDTA. Again centrifuge the opalescent solution. Eventually dialyze overnight against 2 liters of the above buffer. If the last centrifugation is omitted, the dialyzate tends to turn gelatinous. The enzyme activity remains unaltered but the subsequent purification steps are tedious.

Exclusion Chromatography

Use a column about 110 cm long with an inner diameter of 3.4 cm filled with Ultrogel AcA 44 gel (LKB Products). Equilibrate it with 2–3

bed volumes of 20 mM Tris-caproate (pH 6.0) supplemented with 2 mM EDTA. Perform the chromatography in the cold room, preferably in two lots (the sample volume should be below 5% of the gel bed). Elute with the equilibration buffer with a flow rate of 50–70 ml/hr. Pool the fractions containing 10% or more of the maximum activity of BCAT. Concentrate the pooled fractions with ultrafiltration using an Amicon PM 30 membrane.

The recovery of BCAT in all fractions is nearly complete in this step. The yield is only dependent on the number of fractions pooled.

DEAE-Sepharose CL-6B

Dialyze the BCAT preparation overnight against 2 liters of starting buffer (50 mM Tris-caproate, pH 8.5, and 5 mM EDTA). Thoroughly wash about 400 ml of DEAE-Sepharose CL-6B (Pharmacia) on a funnel with the starting buffer to preequilibrate the gel. Pack the gel into a column that is about 50 cm long and further equilibrate the gel. Finally, ascertain that the pH is exactly the same at the inlet and outlet of the column. Apply the sample and continue the elution (150 ml/hr) with a linear gradient from A to B (650 ml each), where A is the starting buffer and B is the same as A but supplemented with 0.5 M NaCl.

Concentrate the fractions containing BCAT with ultrafiltration and dialyze overnight in the cold room against 2 liters of 50 mM sodium caproate, pH 6.0, containing a 5 mM EDTA with one change of the buffer.

Affinity Chromatography

Pack 400 ml of cycloserine-agarose into a 50-cm-long column and equilibrate it at 2–4° with 2–3 bed volumes of sodium phosphate, pH 6.0, containing 0.1 M NaCl, 1 mM mercaptoethanol, and 0.5 mM EDTA. The low temperature is essential for the chromatography. Apply the sample and elute with the same buffer. Collect fractions into test tubes into which sodium caproate and EDTA are pipetted to bring the fractions to 5 mM and 1 mM caproate and EDTA, respectively. These ingredients can also be added to the elution buffer. Wash the gel before the next run with 1 M NaCl. The gel can be stored for months at 2–4° in 1 M NaCl supplemented with 5 mM D-cycloserine.

Concentrate the fractions containing BCAT to 1–2 ml and store at 2°. The enzyme retains 75% of its activity for 4 months.[18] The enzyme preparation can be frozen only if thiols are removed from the solution.[11] A summary of the purification procedure is shown in Table I.

TABLE I
PURIFICATION OF BRANCHED-CHAIN-AMINO-ACID AMINOTRANSFERASE
FROM 2 KG OF PIG HEARTS

Procedure	Volume (ml)	Total activity (units)	Protein (mg/ml)	Specific activity (units/mg)	Yield (%)	Purification
Extraction	4000	515	12.3	0.011	—	—
Heat treatment	1940	1184	6.8	0.090	100	1
$(NH_4)_2SO_4$ precipitation	43	1019	53.7	0.44	86	5
Dialysis	60	845	31.6	0.45	71	5
Ultrogel AcA 44	149	693	2.0	2.33	58	26
DEAE-Sepharose CL-6B	68	543	0.9	8.86	46	98
Cycloserine-agarose	240	325	0.03	45.2	27	501

Concluding Remarks

The method described is shorter and apparently more reproducible than the earlier ones. The specific activity of the product corresponds to that resulting from the technique of Taylor and Jenkins[1] and the purification coefficient is the same. Provided the preparation was homogeneous, 2 kg of pig hearts produces about 7 mg of the enzyme with a yield of 27% (Table I). Purification of BCAT from crude extracts with cycloserine-agarose has not been tested but may be useful for this purpose because the same gel can be used to purify alanine aminotransferase from human serum[25] and the latter enzyme and BCAT elute similarly on the gel.[18,26]

BCAT, alanine aminotransferase, and aspartate aminotransferase remain active after the heat treatment. If cycloserine-agarose is employed for crude extracts, BCAT and alanine aminotransferase will probably not be separated in the chromatography[18,26] while aspartate aminotransferase will coelute with the bulk protein.[25] In the procedure described, alanine aminotransferase separates at least in part on the gel filtration step and finally on the anion exchanger.[19] In our laboratory cycloserine-agarose has proved applicable also for the purification of alanine racemase and BCAT from *Escherichia coli.*

[25] T. Korpela, *J. Chromatogr.* **143**, 519 (1977).
[26] T. Korpela, A. Hinkkanen, and R. Raunio, *J. Solid-Phase Biochem.* **1**, 216 (1976).

[35] Pancreatic Branched-Chain-Amino-Acid Aminotransferase

By RYO KIDO

Branched-chain L-amino acids
(valine, leucine, and isoleucine)
+ 2-ketoglutarate $\rightleftharpoons$ branched-chain 2-keto acids + L-glutamate
(2-ketoisovalerate, 2-ketoisocaproate, and 2-keto-3-methylvalerate)

Assay Method[1,2]

Principle

The branched-chain 2-keto acid formed is converted to the 2,4-dinitrophenylhydrazone, which is selectively extracted with toluene. The hydrazone is then transferred to sodium carbonate solution and the color developed by addition of NaOH is measured at 440 nm.[1,3]

Reagents

Branched-chain L-amino acid, 80 mM in potassium phosphate buffer, 0.2 M, pH 8.0
2-Ketoglutarate, 0.16 M
Pyridoxal 5′-phosphate, 1.6 mM
Trichloroacetic acid, 25% (w/v)
2,4-Dinitrophenylhydrazine, 0.3% in 2 M HCl (w/v)
Toluene
HCl, 0.5 M
Sodium carbonate, 10% (w/v)
NaOH, 1.0 M

Procedure

The reaction mixture contains in a total volume of 1.2 ml: 0.6 ml of L-amino acid, 0.075 ml of 2-ketoglutarate, 0.03 ml of pyridoxal 5′-phosphate and an appropriate amount of enzyme. The reaction is initiated, after preincubation for 5 min at 37°, by addition of 2-ketoglutarate; it is

---

[1] See also R. T. Taylor and W. Jenkins, this series, Vol. 17a, p. 802; K. Aki and A. Ichihara, this series, Vol. 17a, pp. 807, 811, and 814.
[2] T. K. Korpela, this volume, [34].
[3] R. T. Taylor and W. T. Jenkins, *J. Biol. Chem.* **241**, 4391 (1966).

stopped by addition of 0.25 ml of trichloroacetic acid. For the blank, 2-ketoglutarate is added after termination of the reaction. Then 2 ml of 2,4-dinitrophenylhydrazine is added and the incubation continued for 5 min at 37°. The 2,4-dinitrophenylhydrazone of the branched-chain 2-keto acid formed is extracted with 5 ml of toluene by shaking vigorously for 2 min. The phases are separated by centrifugation (3 min at 3000 rpm); the aqueous phase is carefully removed with a syringe. The organic phase is washed with 5 ml of 0.5 M HCl, separated by centrifugation as above, and 2 ml of the upper phase is transferred to a 12-ml conical centrifuge tube. Care must be taken not to transfer any 2-oxoglutarate hydrazone from the walls or interface. The 2-keto acid hydrazone is extracted from the toluene with 2 ml of 10% sodium carbonate by shaking vigorously. After centrifugation for 2 min, 1.0 ml of the carbonate phase is transferred to a test tube and 1 ml of 1.0 M NaOH is added. The absorbance at 440 nm is read 5 min later. One unit of activity was defined as the amount of enzyme which catalyzed conversion of 1 μmol of branched-chain amino acid to the corresponding oxo acid per min at 37°. Protein was measured by the method of Lowry et al.[4]

Purification Procedure for Branched-Chain-Amino-Acid Aminotransferase from Rat Pancreas

Considerable activity of branched-chain-amino-acid aminotransferase is found in rat pancreas.[5] The following method[6] for purification of the enzyme from this tissue is based in part on procedures described by others for various tissues.[1]

All manipulations were carried out at 0°–4°. About 120 rat pancreases were homogenized in 5 volumes of 5 mM potassium phosphate, (KP$_i$), pH 7.5, in a Waring blender and then sonicated at 10 kHz for 4 min. After centrifugation of 105,000 g, the resulting supernatant was added to 200 ml of DEAE-Sepharose equilibrated with 5 mM KP$_i$, pH 7.5, and gently stirred for 30 min at 4°. The resulting DEAE-Sepharose suspension was centrifuged at 3,000 g for 15 min. The precipitate was resuspended in 1 liter of 5 mM KP$_i$, pH 7.5, and packed into a column (5 × 10 cm). The column was washed with 300 ml of 5 mM KP$_i$, pH 7.5, before the elution by a linear gradient prepared from 1 liter of this buffer and 1 liter of 0.25 M

[4] O. H. Lowry, N. J. Rosebrough, A. L. Farr, and R. J. Randall, J. Biol. Chem. 193, 265 (1951); see also E. Layne, this series, Vol. 3, p. 447.

[5] A. Ichihara, C. Noda, and M. Goto, Biochem. Biophys. Res. Commun. 67, 1313 (1975).

[6] M. Makino, Y. Minatogawa, E. Okuno, and R. Kido, Comp. Biochem. Physiol. 77B, 175 (1984).

TABLE I
PURIFICATION OF BRANCHED-CHAIN-AMINO-ACID AMINOTRANSFERASE FROM
RAT PANCREAS

Fraction	Total protein (mg)	Specific activity for leucine (units/mg protein)	Purification (-fold)	Recovery (%)
Homogenate	7540	0.363	1	100
Supernatant	2330	1.07	3	91
DEAE-Sepha-rose	209	6.57	18	50
Hydroxylapatite	37	18.5	51	25
Sephadex G-150	27	21.8	50	23

KP_i, pH 7.5. Fractions (20 ml) were collected and the enzyme was eluted as a single peak from the column at the buffer concentration of 0.15 M. The enzyme solution was applied to a hydroxylapatite column (2.1 × 5 cm) equilibrated with 5 mM KP_i, pH 7.5. The column was washed with 50 ml of 5 mM KP_i, pH 7.5, containing 0.2 mM pyridoxal 5'-phosphate, 1 mM 2-mercaptoethanol, and 1 mM 2-ketoglutarate. The enzyme was eluted with a linear gradient of 5 mM potassium phosphate to 0.25 M KP_i at the same pH and with the supplements given above. After active fractions were concentrated by ultrafiltration with a Diaflo apparatus, the enzyme solution (3.7 ml) was applied to a Sephadex G-150 column (2.5 × 100 cm) equilibrated with 50 mM KP_i, pH 7.5, supplemented as above. The enzyme was eluted with the same buffer from the column. The active fractions were collected and concentrated by ultrafiltration. A summary of the purification procedure is given in Table I.

Properties of Branched-Chain-Amino-Acid Aminotransferase Purified from Rat Pancreas

Homogeneity. The purified enzyme was shown to be a single protein on SDS-polyacrylamide disc. gel electrophoresis.

Heat Stability. The enzyme was labile to heat. Activity was completely lost by heating at 65° for 1 min. About 30% was lost within 1 min at 50° and about 70% within 10 min.

pH Optimum. The enzyme gave a pH optimum between 7.5 and 8.0.

Molecular Weight and Isoelectric Point. The molecular weight was estimated as 68,000, 79,000, and 35,000 by Sephadex G-150 gel filtration,

sucrose density gradient centrifugation, and SDS-polyacrylamide disc. gel electrophoresis, respectively. The enzyme appears to consist of two identical subunits. The isoelectric point was pH 5.6.

Kinetic Properties. The absolute K_m values for leucine aminotransferase (L-leucine:2-oxoglutarate aminotransferase) activity were 11.1 mM for leucine, 22.2 mM for isoleucine, and 143 mM for valine with 2-ketoglutarate as the amino group acceptor. The K_m for 2-ketoglutarate varied with the amino acid: 4 mM with leucine, 14.2 mM with isoleucine, and 12.5 mM with valine.

Substrate Specificity. Activity was greatest with leucine. Isoleucine and valine gave activities 74 and 49%, respectively, of that observed with leucine.

Inhibitors. The enzyme was inhibited by semicarbazine, hydroxylamine, and hydrazine but not by isonicotinic acid hydrazide or KCN. *p*-Chloromercuribenzoate (5 mM) inhibited activity completely. 2-Mercaptoethanol protected against inactivation by sulfhydryl reagents.

Purification Procedure for Branched-Chain-Amino-Acid Aminotransferases from Human Pancreas

Two types of branched-chain-amino-acid aminotransferases with properties very much like those of hog heart and brain[1,7] have been isolated from human pancreas.[6] Their properties, however, are quite different from those of the rat pancreas enzyme described above. Like the rat pancreas, the human pancreas is a rich source of branched-chain-amino-acid aminotransferase activity.[6] The following method is based in part on procedures described by others for various tissues.[1]

All procedures were carried out at $0°-4°$. About 60 g of pancreas were obtained at autopsy, 4–5 hr post mortem. The homogenization, sonication, and centrifugation were the same as described above for the rat pancreas. The resulting supernatant solution was quickly warmed to 60° in a 100° water bath and maintained at this temperature with constant stirring for 1 min and then quickly cooled on ice. After centrifugation at 10,000 g for 30 min, solid $(NH_4)_2SO_4$ was added to the supernatant, with gentle stirring, to 30% saturation. After 30 min, the precipitate was removed by centrifugation at 10,000 g for 30 min and discarded. Additional $(NH_4)_2SO_4$ was added to the supernatant to 75% saturation, and after 30 min the resulting precipitate was collected by centrifugation at 10,000 g for 30 min, dissolved in 5 mM KP_i, pH 7.5, and dialyzed against the same buffer overnight. The dialyzed solution was centrifuged at 10,000 g for 30

[7] M. Goto, H. Shinno, and A. Ichihara, *Gann* **68**, 663 (1977).

min and the supernatant was applied to a column (3.3 × 8 cm) of DEAE-Sepharose equilibrated with the same buffer. The column was washed with 500 ml of this buffer followed by a linear gradient (1 liter) from 5 mM to 0.3 M KP$_i$, pH 7.5. Fractions of 20 ml were collected. Two activity peaks for branched-chain-amino-acid aminotransferase were obtained from the DEAE-Sepharose column. The first enzyme eluted was designated enzyme I; the slower one enzyme III. Enzymes I and III were eluted from the DEAE-Sepharose column at buffer concentrations of 0.02 and 0.23 M, respectively. Enzyme I and enzyme III were separated and purified individually as follows. Each enzyme solution separated by DEAE-Sepharose column was diluted to adjust the KP$_i$ concentration to about 5 mM and then concentrated by ultrafiltration. Enzyme solutions were than applied to hydroxylapatite columns (2.0 × 8 cm) equilibrated with 5 mM KP$_i$, pH 7.5, supplemented as above with pyridoxal 5′-phosphate, 2-mercaptoethanol, and 2-ketoglutarate. The column was washed with 100 ml of the same buffer and elution of the enzyme accomplished with a linear gradient (500 ml) from 5 mM to 0.25 M KP$_i$, pH 7.5. Fractions of 20 ml were collected. The active fractions were pooled and concentrated by ultrafiltration and the enzyme solution applied to a Sephadex G-150 gel filtration. A single and symmetric activity peak was obtained with each enzyme preparation. Enzymes I and III were purified 47- and 973-fold, respectively (see Table II).

Properties of Branched-Chain-Amino-Acid Aminotransferases from Human Pancreas

Purified enzymes I and III were not inactivated by heating at 65° for 10 min. Enzyme I had a pH optimum of 9.0, enzyme III of 8.0. Molecular weights were estimated by Sephadex G-150 to be 80,000 for enzyme I and 90,000 for enzyme III. Isoelectric points were pH 6.6 and 4.2, respectively. The absolute K_m values (mM) for enzyme I with 2-ketoglutarate as amino group acceptor were 25 for leucine, 10.3 for isoleucine, and 30.8 for valine. K_m values (mM), for 2-ketoglutarate were 7.6 with leucine, 4.8 with isoleucine, and 5.0 with valine. Those of enzyme III with 2-oxoglutarate as the amino group acceptor were 6.8 for leucine, 3.0 for isoleucine, and 8.3 for valine. K_m values (mM) for 2-ketoglutarate with enzyme III were 1.0 with leucine, 1.1 with isoleucine, and 1.7 with valine. Both enzymes gave greatest activity with isoleucine followed by about 60 and 30% as much activity with leucine and valine, respectively. Enzymes I and III were equally inhibited by the carbonyl reagents hydroxylamine, hydrazine, and KCN but not by isonicotinic acid and semicarbazide. p-Chloromercuribenzoate completely inhibited enzyme I, but only partially inhibited en-

TABLE II
PURIFICATION OF BRANCHED-CHAIN-AMINO-ACID AMINOTRANSFERASE FROM
HUMAN PANCREAS

Fraction	Total protein (mg.	Specific activity (units/mg protein)	Purification (-fold)	Recovery (%)
Homogenate	4520	0.082	1	100
Supernatant	1930	0.150	2	75
Heat treatment	1840	0.220	3	67
$(NH_4)_2SO_4$	169	1.03	13	47
Enzyme I				
DEAE-				
Sepharose	37.2	1.58	19	16
Hydroxylapa-				
tite	7.78	1.05	13	2.3
Sephadex				
G-150	1.98	3.85	47	2.1
Enzyme III				
DEAE-				
Sepharose	3.37	10.63	130	9.6
Hydroxylapa-				
tite	0.80	40.3	491	8.7
Sephadex				
G-150	0.27	98.0	973	5.7

zyme III. Inactivation by sulfhydryl reagents was prevented by 2-mercaptoethanol.

Purification Procedure for Branched-Chain-Amino-Acid Aminotransferases from Dog Pancreas

Two types of branched-chain-amino-acid aminotransferases have been isolated from dog pancreas. Their properties were very much like those of the human pancreas enzymes.[6]

The same procedures described above for human pancreatic enzyme were used to purify dog pancreatic enzyme. However, dog pancreatic enzymes I and III were eluted from the DEAE-Sepharose column at buffer concentrations of 0.06 and 0.11 M, respectively. Enzymes I and III were purified 89-fold and 602-fold, respectively (see Table III).

Properties of Branched-Chain-Amino-Acid Aminotransferase from Dog Pancreas

Purified Enzymes I and III stained as single protein bands on SDS-polyacrylamide disc. gel electrophoresis. Neither enzyme was inactivated by

TABLE III
PURIFICATION OF BRANCHED-CHAIN-AMINO-ACID AMINOTRANSFERASE FROM DOG
PANCREAS

Fraction	Total protein (mg)	Specific activity (units/mg protein)	Purification (-fold)	Recovery (%)
Homogenate	4290	0.123	1	100
Supernatant	1260	0.337	3	80
Heat treatment	577	0.842	7	92
$(NH_4)_2SO_2$	140	2.25	18	60
Enzyme I				
DEAE-Sepharose	32.8	1.70	14	11
Hydroxylapatite	17.6	3.03	25	10
Sephadex G-150	4.2	10.92	89	8.7
Enzyme III				
DEAE-Sepharose	9.4	6.46	53	11
Hydroxylapatite	0.89	55.0	447	9.3
Sephadex G-150	1.21	74.0	602	2.9

heating at 65° for 10 min. Enzyme I had a pH optimum of 9.0, enzyme III between 8.5 and 9.5. Molecular weights for enzyme I and II were estimated as 80,000 and 96,000, respectively, by Sephadex G-150 chromatography, and as 39,000 and 43,000, respectively, by SDS-polyacrylamide disc. gel electrophoresis. Isoelectric points were pH 5.6 and pH 4.6, respectively. The absolute K_m values (mM) of enzyme I with 2-ketoglutarate as amino group acceptor were 3.6 for leucine, 22.2 for isoleucine, and 7.4 for valine. K_m values for 2-ketoglutarate were 2.1 with leucine, 10.0 with isoleucine, and 2.7 with valine. K_m values of enzyme III with 2-ketoglutarate as amino group acceptor were 2.3 for leucine, 4.2 for isoleucine, and 3.0 for valine. K_m values for 2-ketoglutarate were 1.4 with leucine, 1.0 with isoleucine, and 1.7 with valine. Enzyme I had greatest activity with isoleucine whereas enzyme III showed nearly equal activity for leucine and isoleucine and about half as much activity with valine. Both enzymes were inhibited by hydroxylamine and hydrazine but not isonicotinic acid hydrazide and KCN. Enzyme III was somewhat sensitive to semicarbazide; enzyme I was not. Enzyme I was completely inhibited by p-chloromercuribenzoate whereas enzyme III was less sensitive. Inactivation by sulfhydryl reagents was prevented by 2-mercaptoethanol.

[36] Isolation and Characterization of Leucine Dehydrogenase from Bacillus subtilis

By Geoffrey Livesey and Patricia Lund

Leucine dehydrogenase [L-leucine:NAD$^+$ oxidoreductase (deaminating), EC.1.4.1.9] is associated with several bacterial species and has been isolated to purity from *Bacillus subtilis*,[1] *Bacillus sphaericus*,[2] *Bacillus cereus*,[3] *Bacillus natto*,[4] and *Bacillus stearothermophilus*.[5] Induction of the enzyme occurs in some strains of these organisms when grown in a medium rich in one or more of the branched-chain amino acids. The appearance of the enzyme during the growth and sporulation of *Bacillus* species is often accompanied by the related enzyme, alanine dehydrogenase, which is usually the major dehydrogenase contaminant. Alanine dehydrogenase may also be inducible so the concentration of alanine in the microbial growth medium should be kept to a minimum. Of several *B. subtilis* investigated, we chose strain 8577 from the National Collection of Industrial Bacteria, U.K., because it yielded leucine dehydrogenase in largest quantity, of highest specific activity in the crude cell extract, and the highest leucine-to-alanine dehydrogenase activity. Procedures for the purification of leucine dehydrogenase differ from one laboratory to another and choice is partly determined by the microbial source, the quantity to be produced, and the intended use of the enzyme. The method described here was developed for the preparation of small batches of enzyme for estimation of the branched-chain amino acids and their keto acids in biological extracts.[6] The object of the isolation procedure is therefore a high yield of leucine dehydrogenase free of interfering enzyme activities rather than high purity as defined by physical criteria.

[1] J. Hermier, J. M. Lebeault, and C. Zévaco, *Bull. Soc. Chim. Biol.* **52**, 1089 (1970).
[2] T. Ohshima, H. Misono, and K. Soda, *J. Biol. Chem.* **253**, 5719 (1978).
[3] H. Schütte, W. Hummel, H. Tsai, and M.-R. Kula, *Appl. Microbiol. Biotechnol.* **22**, 306 (1985).
[4] K. Matsui, Y. Kameda, T. Atsusaka, M. Higaki, and K. Sasaki, *Chem. Pharmacol. Bull.* **25**, 761 (1977).
[5] T. Ohshima, S. Nagata, and K. Soda, *Arch. Microbiol.* **141**, 407 (1985).
[6] G. Livesey and P. Lund, *Biochem. J.* **188**, 705 (1980).

Assay Method

Reagents

Buffer solution: 0.1 M Na_2CO_3 containing 2 mM ethylenediaminetetraacetic acid (EDTA), pH 10.0
Leucine solution: 200 mM L-leucine
NAD^+ solution: 2% (w/v) NAD^+
Enzyme solution: Leucine dehydrogenase should be diluted to approximately 0.5 U/ml with buffer solution

Assay Procedure

The enzyme catalyzes the oxidative deamination of L-leucine, which is followed by measuring the increase in absorption at 340 nm as NAD^+ is reduced.[6]

$$(CH_3)_2CHCH_2CH(NH_2)CO_2H + NAD^+ + H_2O \rightleftharpoons$$
L-Leucine

$$(CH_3)_2CHCH_2COCO_2H + NADH + NH_4^+$$
4-Methyl-2-ketovaleric acid

The reaction cuvette (glass, 10 mm) contains, at 25°: 1.0 ml buffer solution, 0.1 ml leucine solution, 0.1 ml NAD^+ solution, and 0.78 ml of water. The reaction is started with 20 μl of the appropriately diluted enzyme solution and the rate is followed for 2 min. The reference blank contains water in place of substrate solution.

Definition of Units

The reaction is stoichiometric, 1 mol of NADH (coefficient of absorption 6.22 A per μmol) is produced per mole of L-leucine oxidatively deaminated. One unit of leucine dehydrogenase is the quantity of enzyme that catalyzes the production of 1.0 μmol NADH per minute at 25° in the oxidative deamination assay when 10 mM L-leucine is present in the cuvette.

Comments on the Procedure

The assay of enzyme suspended in $(NH_4)_2SO_4$ solution may give low values since the reaction is reversible and ammonium ions inhibit the reaction. The reverse reaction (reductive amination) can also be used to assay activity by measuring the decrease in absorbance at 340 nm[6] in cuvettes (glass, 10 mm) at 25° containing 1.0 ml of 0.75 M $NH_4Cl/$ NH_4OH, pH 9.5, containing 2 mM EDTA, 0.1 ml of 0.3% (w/v) NADH,

0.1 ml of 50 mM sodium 4-methyl-2-ketovalerate, 0.78 ml of water, and 20 μl of appropriately diluted enzyme (0.1 U/ml). This reductive amination assay is much more sensitive than the oxidative deamination assay but is more expensive.

Production of Leucine Dehydrogenase

Reagents

Growth medium I: 3% (w/v) dextrose peptone broth
Growth medium II: 26 g K_2HPO_4, 11.14 g KH_2PO_4, 4.0 g trisodium citrate · H_2O, 0.8 g $MgSO_4 \cdot 7H_2O$, 40 g L-leucine, 4.0 g L-isoleucine, 4.0 g L-valine, 20 g Oxoid powdered yeast extract, 40 μl silicone MS antifoam A dissolved and diluted to 4 liters
Cells: *B. subtilis* 8577, approximately 50 mg wet weight
Buffer solution: 0.05 M Tris-HCl, pH 8.0, containing 0.01% (v/v) 2-mercaptoethanol

Procedure

Growth is started from 50 mg of the cells in 100 ml of autoclaved medium I, at 37° in a 1-liter Roux bottle. After 24 hr the medium, containing between 0.5 and 1.0 g wet weight cells, is added to 4 liters of autoclaved medium II in a 5-liter aspirator at 37°. The medium is aerated at a rate of approximately 0.5 liters/min through a 5 μm Millipore filter. Cells are harvested by centrifugation and washed twice by resuspending in the buffer solution. The cell yield is usually 20–25 g wet weight. Inadequate aeration markedly decreases cell yields. The cells can be stored frozen at $-20°$ for up to 2 weeks but repeated freeze/thawing can cause loss of leucine dehydrogenase activity.

Purification of the Enzyme

The procedure involves lysis of the cells with lysozyme, precipitation of nucleic acid with $MnSO_4$, and separation of proteins by $(NH_4)_2SO_4$ fractionation and chromatography on calcium phosphate gel, DEAE-cellulose, and Sephadex G-150. All procedures are performed in the presence of 0.01% (v/v) 2-mercaptoethanol and at 0–4° unless otherwise stated.

Crude Cell Extract. Cells, 50 g wet weight, are suspended in 200 ml of 0.05 M Tris-HCl, pH 8.5, at 37°. After mixing with 100 mg lysozyme the incubation is continued for 30 min. The supernatant after centrifugation at 30,000 g for 30 min is the crude extract.

Ammonium Sulfate Fractionation. To the crude extract is added 0.05

volume of 1 M MnSO$_4$ and the precipitate produced is removed by centrifugation at 30,000 g for 10 min. To the supernatant is added powdered (NH$_4$)$_2$SO$_4$ to give a 45% saturated solution.[7] The precipitate formed after 1 hr is removed (30,000 g, 10 min) and (NH$_4$)$_2$SO$_4$ increased to 70% saturation.[7] After 1 hr the precipitate containing the enzyme is collected by centrifugation and dissolved in 50 ml of 0.01 M KH$_2$PO$_4$ adjusted to pH 7.4 with NaOH. The solution is dialyzed for 16 hr against 5 liters of the same buffer and the residue removed by centrifugation to give a clear (NH$_4$)$_2$SO$_4$ extract.

Calcium Phosphate Gel Chromatography. The calcium phosphate gel is prepared as described in a previous volume of this series[8] and is suitable for use immediately. The gel is poured into a Büchner funnel to create a 10 cm diameter, 4 cm depth bed supported on Whatman No. 1 filter paper. Chromatography is at 20–25° after washing the gel bed with 200 ml of 0.01 M K$_2$HPO$_4$ adjusted to pH 7.4 with NaOH. The (NH$_4$)$_2$SO$_4$ extract is filtered on the bed, which is then washed with increasing concentrations of the phosphate buffer: 100 ml each of 0.01 M, 0.015 M, 0.02 M, and 200 ml of 0.08 M. The highest activities of leucine dehydrogenase are contained in the latest fractions. Alanine dehydrogenase is separated and appears in the 0.015 and 0.02 M fractions. Brown pigmented material elutes between the two enzyme peaks. The fractions with the highest activities of leucine dehydrogenase are pooled and dialyzed for 16 hr against 50 volumes of 0.01 M K$_2$HPO$_4$/KH$_2$PO$_4$, pH 7.2. The residue is clarified by centrifugation to give the calcium phosphate gel extract.

DEAE-Cellulose Chromatography. A 10 cm diameter, 5 cm depth bed of DEAE-cellulose, equilibrated with 0.01 M K$_2$HPO$_4$/KH$_2$PO$_4$, pH 7.2, is prepared in a Büchner funnel supported by Whatman No. 1 filter paper at 20–25°. The calcium phosphate gel extract is filtered on the bed, which is then washed with 100 ml of the phosphate buffer. Protein is eluted successively with 100 ml volumes of the buffer containing 0.1, 0.15, 0.2 and 0.3 M NaCl. The leucine dehydrogenase elutes with buffer containing 0.3 M NaCl. The fractions with the highest activity of the enzyme are pooled, made 80% saturated with (NH$_4$)$_2$SO$_4$[7], and, after 1 hr, centrifuged to collect the leucine dehydrogenase in the pellet. The pelleted protein is dissolved in 10 ml of 0.05 M Tris-HCl, pH 9.0, and clarified by centrifugation.

Sephadex G-150 Chromatography. The DEAE-cellulose fraction is filtered through a Sephadex G-150 (120 mesh) column (1.6 cm diameter × 100 cm) equilibrated with 0.05 M Tris-HCl, pH 9.0, at 0–4° and eluted at a rate of 30 ml/hr. The enzyme elutes shortly after the void volume and is

[7] A. A. Green and W. L. Hughes, this series, Vol. 1, p. 67.
[8] M. Koike and M. Hamada, this series, Vol. 22, p. 339.

TABLE I
PURIFICATION OF LEUCINE DEHYDROGENASE

Extract	Units recovered	Specific activity[a]	Percentage recovery
Crude cell	705	0.18	100
Ammonium sulfate	564	0.43	80
Calcium phosphate gel	451	0.71	64
DEAE-cellulose	311	4.78	44
Sephadex G-150	259	15.23	37

[a] The specific activity is expressed in units per milligram protein. The protein is assayed by the Lowry procedure with bovine serum albumin standard.

concentrated from pooled fractions with highest enzyme activity by addition of powdered $(NH_4)_2SO_4$ to give an 80% saturated solution,[7] followed by centrifugation. The pellet is resuspended in 75% saturated $(NH_4)_2SO_4$ at a concentration of 50 units/ml of suspension, to give the Sephadex G-150 fraction.

Comments on the Purification

Removal of nucleic acids from the crude cell extract with too high a concentration of $MnSO_4$ causes precipitation of the leucine dehydrogenase.[9] Dialysis of the enzyme against too low a concentration of the phosphate buffer, after, for example, elution from the calcium phosphate gel, results in a loss of enzyme activity. A complete loss of activity occurs during dialysis against distilled water.[10] Increasing the size of either the calcium phosphate gel bed or the DEAE-cellulose bed to accommodate more enzyme and eluting with larger volumes of buffer solutions can result in poor resolution and poor recovery of the enzyme.

The purification of leucine dehydrogenase is summarized in Table I. Other NAD-linked dehydrogenases associated with *B. subtilis* reacting with glutamate, lactate, and malate are of low activity in the crude cell extract and absent from the Sephadex G-150 fraction. The purification procedure removes 99% of the alanine dehydrogenase activity.

Enzymological Properties

Substrate Specificity. In the oxidative deamination assay, the leucine dehydrogenase preparation reacts with L-leucine, L-valine, L-isoleucine,

[9] G. Livesey, unpublished observations (1979).
[10] P. Lund and G. Baverel, *Biochem. J.* **174**, 1079 (1978).

L-norleucine, L-norvaline, and L-methionine with relative velocities of 100, 92, 65, 40, 7, and 0.18 respectively. In the reductive amination assay, the leucine dehydrogenase reacts with 4-methyl-2-ketovalerate (ketoleucine), 3-methyl-2-ketobutyrate (ketovaline), 3-methyl-2-ketovalerate (ketoisoleucine), and 3-methylthio-2-ketobutyrate (ketomethionine) with relative velocities of 100, 113, 109, and 32, respectively. NAD(H) but not NADP(H) reacts. NH_4^+ is essential for the oxidation of NADH in the presence of leucine dehydrogenase and the reactive keto acids. The following substrates show no oxidative deamination: L-glutamate, L-glutamine, L-aspartate, L-citrulline, L-asparagine, L-histidine, L-threonine, L-ethionine, L-ornithine, L-proline, L-hydroxyproline, L-tryptophan, L-arginine, L-phenylalanine, L-cysteine, L-homocysteine, L-serine, L-tyrosine, L-lysine, glycine, L-lactate, L-malate. The following substrates show no reductive amination: glyoxylate, acetoacetate, phenylpyruvate. Pyruvate reacts in the reductive amination assay and L-alanine reacts in the oxidative deamination assay at velocities determined by the degree of contamination of the leucine dehydrogenase with alanine dehydrogenase. The purification procedure can remove 99% of alanine dehydrogenase activity, giving a relative velocity of 0.23 compared to 100 for L-leucine.

Kinetic Properties. Michaelis constants at pH 10.0 for oxidative deamination in the presence of 1.5 mM NAD$^+$ at 25° are 1.33 mM for L-leucine, 2.4 mM for L-isoleucine, 3.4 mM for L-valine, and 25 mM for L-methionine and, at pH 9.5 in the reductive amination assay in the presence of 0.2 mM NADH and 0.2 M NH_4^+ at 25°, are 0.6 mM for 4-methyl-2-ketovalerate, 0.9 mM for 3-methyl-2-ketovalerate, 3.3 mM for 3-methyl-2-ketobutyrate, and 6.7 mM for 4-methylthio-2-ketobutyrate.

pH Optimum and Inhibitors. With L-leucine, optimum oxidative deamination is at pH 11.0,[9] and with 4-methyl-2-ketovalerate optimum reductive amination is at pH 9.5.[9] Leucine is a competitive inhibitor ($K_i = 1$ mM) of the reductive amination of 4-methylthio-2-ketobutyrate (ketomethionine). EDTA (2 mM) is without effect on activity of leucine dehydrogenase. NH_4^+ and Mn^{2+} are inhibitors of the oxidative deamination reaction.[9] High concentrations of Mn^{2+} cause the leucine dehydrogenase to precipitate.[9]

Comments on the Properties of Leucine Dehydrogenase

The properties of the leucine dehydrogenase from *B. subtilis* 8577 are similar to the enzyme from *B. cereus, B. sphaericus, B. natto,* and *B. stearothermophilus* and other strains of *B. subtilis.*[1–5,11] Between these species, optimum pH values for the oxidative deamination of L-leucine

---

[11] M. W. Zink and B. D. Sanwal, this series, Vol. 17, p. 799.

range between 10.3 and 11.5 and for reductive amination of 4-methyl-2-ketovalerate range between 9.0 and 9.5. The range of K_m values between species are 1.0 to 6.3 mM for L-leucine, 1.3 to 5.2 mM for L-isoleucine, 1.7 to 20.0 mM for L-valine, 0.3 to 3.3 mM for 4-methyl-2-ketovalerate, 0.9 to 2.2 mM for 3-methyl-2-ketovalerate, and 1.7 to 3.3 mM for 3-methyl-2-ketovalerate. Methionine reacts with leucine dehydrogenase from *B. sphaericus* and *B. cereus* as with the enzyme from *B. subtilis* 8577.

The equilibrium constant for leucine dehydrogenase[1-5] is in favor of branched-chain amino acid formation. The value has been determined[11] for the enzyme from *B. cereus* to be

$$K'_{eq.} = \frac{(4\text{-Methyl-2-ketovalerate}^-)(H^+)(NADH)(NH_4^+)}{(Leucine)(H_2O)(NAD^+)}$$
$$= (11.1 \pm 1) \times 10^{-14}$$

The molecular weight of leucine dehydrogenase from *B. subtilis* 8577 has not been determined; that from *B. subtilis* SJ2 is reported to be 230,000[1] compared with a range of values from other species of between 245,000 and 360,000, consisting of subunits of molecular weight about 39,000.[2-5]

The effects of *p*-chloromercuribenzoate and $HgCl_2$ on the activity of leucine dehydrogenase from *B. subtilis* 8577 have not been tested. Both inhibit the enzyme from *B. subtilis* SJ2 and other species from which leucine dehydrogenase has been isolated.

[37] Purification and Assay of α-Ketoisocaproate Dioxygenase from Rat Liver

By P. J. Sabourin and L. L. Bieber

The normal catabolic route for metabolism of the keto acid of leucine (α-ketoisocaproic acid) is via mitochondrial oxidative decarboxylation by the branched-chain α-keto acid dehydrogenase to form a branched-chain acyl-CoA ester and CO_2. However, in liver an alternate metabolic pathway can occur in the cytosol compartment via α-ketoisocaproate dioxygenase. This little investigated, unstable enzyme catalyzes the following reaction:

$$(CH_3)_2\text{-CH-CH}_2\text{-COCO}_2H \xrightarrow[Fe^{2+}]{O_2} (CH_3)_2\text{-C(OH)-CH}_2\text{-CO}_2H + CO_2$$

The rat liver enzyme has been partially purified and characterized.[1-5] The product, β-hydroxyisovaleric acid, is frequently elevated in urine of individuals with certain types of organic acid acidemias[6-12] and apparently is a detoxification product.

Assay of α-Ketoisocaproate Dioxygenase

Decarboxylation Assay

The α-ketoisocaproate dioxygenase activity can be measured by monitoring the release of $^{14}CO_2$ from α-keto[1-^{14}C]isocaproate or α-keto-γ-methio[1-^{14}C]butyrate.[3,4] The activity may also be measured polarographically by monitoring oxygen consumption in the presence of substrate.[4] [1-^{14}C]-Labeled α-keto acids are synthesized enzymatically, using L-amino-acid oxidase, from the corresponding carboxyl ^{14}C-labeled L-amino acids according to the method of Rüdiger *et al.*[13] Solutions of 1-^{14}C-labeled α-keto acids are then adjusted to the desired concentration with the addition of unlabeled α-keto acid (Sigma Chemical Co., St. Louis, MO).

Assays for the determination of decarboxylating activity are performed in 1.5 cm diameter culture tubes, which are tightly stoppered with a serum cap. A plastic cup (Kontes) hanging from the serum cap contains 0.2 ml Hyamine (1 M methylbenzethonium hydroxide in methanol, Sigma) to trap $^{14}CO_2$ released during the reaction. The assay stock solution A contains: 0.32 M Tris base, 0.32 M maleate, pH adjusted to 6.5 with NaOH. Stock solution B contains 16 mM FeSO$_4$, 8 mM ascorbic acid, and 16 mM

[1] R. M. Wohlhueter and A. E. Harper, *J. Biol. Chem.* **245**, 2391 (1970).

[2] W. D. Grant and J. L. Connelly, *Fed. Proc., Fed. Am. Soc. Exp. Biol.* **33**, 1570 (Abstr.) (1974).

[3] P. J. Sabourin and L. L. Bieber, *Arch. Biochem. Biophys.* **206**, 132 (1981).

[4] P. J. Sabourin and L. L. Bieber, *J. Biol. Chem.* **257**, 7460 (1982).

[5] P. J. Sabourin and L. L. Bieber, *J. Biol. Chem.* **257**, 7468 (1982).

[6] D. E. Hene and K. Tanaka, *Pediatr. Res.* **18**, 508 (1984).

[7] C. Jacobs, L. Sweetman, W. L. Nyhan, and S. Packman, *J. Inheret. Metab. Dis.* **7**, 15 (1984).

[8] G. Steen and L. Ransnas, *Acta Neurol. Scand.* **68**, 231 (1983).

[9] Y. Shigematsu, M. Sudu, Y. Momoi, Y. Inoue, Y. Suzuki, and J. Kameyama, *Pediatr. Res.* **16**, 771 (1982).

[10] L. Sweetman, *J. Inheret. Metab. Dis.* **4**, 53 (1981).

[11] B. M. Charles, G. Hosking, A. Green, R. Pollitt, and K. Bartlett, *Lancet* **2**, 118 (1979).

[12] T. Kuhara, T. Shinka, M. Matsuo, and I. Matsumoto, *Clin. Chem. Acta* **123**, 101 (1982).

[13] H. W. Rüdiger, U. Langenbeck, and H. W. Goedde, *Biochem. J.* **126**, 445 (1972).

dithiothreitol. This solution (cofactor mix) should be prepared fresh daily. Addition of 25 μl of stock solution B to 0.25 ml of A, 0.10 ml enzyme, and 25 μl 1-^{14}C-labeled α-keto acid gives the desired final concentration of FeSO$_4$, ascorbic acid, and dithiothreitol (1.0 mM, 0.5 mM, and 1.0 mM, respectively) in a volume of 0.4 ml. To obtain maximal reaction rates, stock solution B should be kept on ice and not added to the reaction mixture more than 50 min before the addition of enzyme. The enzyme sample, 100 μl, is added last and the reaction mixture preincubated for 1 hr at 25° with all components except the 1-^{14}C-labeled α-keto acid. Then 25 μl of 16 mM 1-^{14}C-labeled α-keto acid (approximately 100 dpm/nmol) is added to the reaction mixture (final concentration of 1.0 mM) to start the reaction, and the tube is quickly stoppered with the hanging cup previously filled with 0.2 ml of Hyamine.

The reaction is incubated in a 25° shaking water bath. The reaction rate is linear for at least 60 min. To terminate the reaction, 0.2 ml of 20% trichloroacetic acid is injected through the serum cap and an additional 1 hr with shaking allowed for collection of ^{14}CO$_2$. The Hyamine cup is placed in a scintillation vial containing 10 ml of scintillation fluid.[14] Specific activity is determined by releasing all of the ^{14}CO$_2$ from the 1-^{14}C-labeled α-keto acid using ceric sulfate.[15]

Oxygen Consumption Assay

The α-ketoisocaproate dioxygenase can be assayed by polarographic measurement of oxygen consumption. The reaction mixture contains 0.2 M Tris, 0.2 M maleate, pH 6.5, and 0.5 ml of the enzyme preparation in a final volume of 3.0 ml. A basal rate of O$_2$ consumption is established at 25° before the α-keto acid is added (final concentration of 1 mM) and the rate of oxygen consumption determined. Oxygen solubility in the reaction mixture is determined by the method of Robinson and Cooper.[16] Ferrous sulfate, ascorbate, and dithiothreitol can not be included in O$_2$ consumption assays due to the rapid autooxidation of the ferrous iron. Therefore, this assay procedure does not give optimal reaction rates. It is a useful method, however, for screening of α-keto acids (which are not ^{14}C-labeled) as potential substrates. Ammonium sulfate (1.5 M) doubles the rate of O$_2$ consumption, but a correction must be made for the decreased solubility of O$_2$ in the assay solution containing ammonium sulfate.[3]

[14] W. A. Johnson and J. L. Connelly, *Biochemistry* **11**, 1967 (1972).
[15] A. Meister, *J. Biol. Chem.* **197**, 309 (1952).
[16] J. Robinson and J. M. Cooper, *Anal. Biochem.* **33**, 390 (1970).

Tissue Sources of α-Ketoisocaproate Acid Dioxygenase

α-Ketoisocaproate, but not α-ketoisovalerate or α-keto-β-methylvalerate (the keto acid analogs of valine and isoleucine, respectively) is decarboxylated by cytosolic preparations from liver and kidney of mouse, rat, rabbit, guinea pig, and beef and from the liver only of chicken. This decarboxylase activity does not require added CoASH or NAD$^+$ and is presumed to be due to the α-ketoisocaproate dioxygenase.[2–4] In rat liver the dioxygenase occurs in the soluble cellular fraction.[3,4] No activity is found in rat brain, heart, muscle, or pancreas using assay conditions optimized for the liver dioxygenase. The cytosolic decarboxylation of ketoisocaproate apparently increases when high circulating levels of the branched-chain amino acids are present, e.g., high dietary protein[17] or diabetes.[18]

Purification of α-Ketoisocaproate Dioxygenase

Purification of the dioxygenase can be monitored both by decarboxylation of α-keto[1-^{14}C]isocaproate or by measuring α-ketoisocaproate stimulated O_2 consumption in the presence of respiratory chain inhibitors. All purification steps are done at 4° unless noted otherwise. The dioxygenase preparations can be stored at any step of the purification at −80°, but are very unstable at 4° and −20°.

Preparation of Extract

For large-scale preparation, approximately 200 Sprague-Dawley rats (150–200 g) are fasted for 24 hr, decapitated, and the livers removed and homogenized with a Potter-Elvehjem glass homogenizer in 5–10 volumes of 0.25 *M* sucrose. The homogenate is centrifuged for 12 min at 500 *g* and 12 min at 10,000 *g* and the supernatant fluid saved. It is made 1% in 2-propanol, which stabilizes the enzyme activity during the initial purification steps. The 10,000 *g* supernatant fractions can be stored at −80° for at least 2 years with little loss of enzyme activity. *Protease inhibitors were not tested, but could be beneficial.* The combination of 1 m*M* phenylmethylsulfonyl fluoride and 0.5 mg/ml soybean trypsin inhibitor did not stabilize the enzyme.

[17] J. L. Dixon and A. E. Harper, *Fed. Proc., Fed. Am. Soc. Exp. Biol.* **40,** 900 (1981).
[18] M. E. May, V. S. Mancusi, and M. G. Buse, *Fed. Proc., Fed. Am. Soc. Exp. Biol.* **38,** 1028 (1979).

$(NH_4)_2SO_4$ Fractionation

The 10,000 g supernatant is made to 45% of saturation with powdered ammonium sulfate, stirred for 30 min, and centrifuged for 30 min at 10,000 g. The supernatant is then brought to 75% of saturation with the addition of powdered $(NH_4)_2SO_4$, stirred for 30 min, and centrifuged for 30 min at 10,000 g. The pellet is suspended in 1.5 liter of 20 mM Tris-HCl, pH 7.8, 1% 2-propanol (buffer A), and dialyzed against this solution to remove the ammonium sulfate. The dialyzate is centrifuged for 15 min at 8,000 g to remove precipitated material.

DEAE-Cellulose Chromatography

The dialyzate is applied to a 4.8 × 38 cm DEAE-cellulose (Whatman DE-52) column which has been equilibrated with buffer A. The column is washed with 2.5 liters of buffer A and the activity eluted with a linear gradient of 0–0.2 M NaCl in buffer A. The activity elutes at approximately 0.06 M NaCl.[4] Fractions containing appreciable activity are pooled and concentrated to 250 ml using an Amicon PM10 ultrafiltration membrane ("concentrated DEAE pool").

Although 2-propanol initially stabilized the dioxygenase, at this point in the purification the enzyme is very unstable except when stored at −80°. The addition of 5% monothioglycerol stabilized the dioxygenase throughout the remainder of the purification.

Initially, during subsequent column chromatography steps, large losses of activity were encountered. For example, during passage over a Sephacryl S-200 column, only 10% of the applied activity was recovered. This loss is caused by tight binding of the enzyme to the column. The "high affinity" sites on the column material can be masked using a procedure similar to that employed by Kaufman and Fisher[19] for the purification of phenylalanine monooxygenase from rat liver. The column is pretreated with crude cytosolic preparations. For this procedure the protein containing fractions from the first DEAE column of the purification, which did not contain dioxygenase activity, were pooled and used to pretreat other columns. The pooled fractions are referred to as "DEAE side fractions."

Phenyl-Sepharose Chromatography

A 4.0 × 40 cm column of phenyl-Sepharose CL-4B (Pharmacia) is prepared by washing the column with 1.5 liters of buffer A, followed by 1.0 liter of buffer A containing 5% monothioglycerol and 2.5 M NaCl. Then

[19] S. Kaufman and D. B. Fisher, *J. Biol. Chem.* **245,** 4745 (1970).

1.8 liters of DEAE side fractions containing 2.5 M NaCl is applied to the column which is washed with 2 liters of buffer A to remove protein that is not tightly bound. The column is then equilibrated with 1.5 liters of buffer A containing 2.5 M NaCl and 5% monothioglycerol.

Sephacryl S-200 Chromatography

A 4.8 × 82 cm Sephacryl S-200 column is prepared by washing the column with 3 liters of buffer A and application of 1.5 liters of DEAE side fractions. Then the column is washed well with 0.1 M NaCl in buffer A to remove extra protein. The column is then equilibrated by washing with 2 liters of 0.1 M NaCl plus 5% monothioglycerol in buffer A.

The concentrated DEAE pool is adjusted to 5% monothioglycerol and 2.5 M NaCl by slowly adding these compounds while stirring. Half of this mixture is applied to the phenyl-Sepharose column and the column washed with 1.5 liters of buffer A containing 2.5 M NaCl and 5% mono-thioglycerol. The dioxygenase is then eluted with 2 liters of linear gradient of 2.5–0 M NaCl in buffer A containing 5% monothioglycerol. Fractions containing α-ketoisocaproate dioxygenase are pooled and concentrated to 20 ml.

The concentrate is then applied to the 4.8 × 82 cm Sephacryl S-200 column prepared as described above. This column is eluted with buffer A containing 0.1 M NaCl and 5% monothioglycerol at a flow rate of 14 ml/hr. Fractions with high specific activity dioxygenase are pooled and stored at −80°. This purified preparation has a specific activity of about 130 nmol of α-keto[1-^{14}C]isocaproate decarboxylated/min/mg protein. The remaining concentrated DEAE pool enzyme can be stored at −80° for future purification.

Monothioglycerol Removal

Since monothioglycerol at concentrations greater than 0.6% in the assay reaction mixture inhibits the dioxygenase, only small aliquots of preparations containing 5% monothioglycerol can be assayed accurately. To remove monothioglycerol, 0.5 ml of the enzyme preparation is passed over a 0.75 × 14 cm BioGel P-6 (50–100 mesh) column which was pre-treated with DEAE side fractions and equilibrated with buffer A containing 0.1 M NaCl. Protein is monitored by adding 50 μl of Coomassie Blue reagent[20] to 10 μl of each fraction and checking visually for blue color. To monitor monothioglycerol, 1 ml of 0.3 mM DTNB (5,5'-dithiobis-2-nitro-benzoic acid) in 0.25 M glycylglycine, pH 8.2, is added to the same assay

[20] M. M. Bradford, *Anal. Biochem.* **72**, 248 (1976).

tubes and yellow color checked visually. Fractions containing the dioxygenase, but not monothioglycerol, are pooled and stored at −80°. The purified enzyme is stable at −80° for at least 1 week in the absence of monothioglycerol.

Properties of α-Ketoisocaproate Dioxygenase

Stability

The dioxygenase is stable at any stage of purification for at least several weeks at −80°. However, as the enzyme is purified, activity is rapidly lost at 4°. Interestingly, activity is less stable at −20° and +20° than at 4° (see Fig. 1). 2-Propanol at 1% stabilizes the activity during the initial purification steps; however, with purification the enzyme becomes quite unstable despite this addition. Monothioglycerol (5%) completely stabilizes the purified enzyme (Table I). Dithiothreitol at 1 or 5 mM and in the presence or absence of 5% glycerol did not stabilize the enzyme. In the presence of 5% monothioglycerol, the purified dioxygenase retained 90% activity for 30 days at 4°, but the activity was completely lost by 70 days.

Cofactor Requirements

The cytosolic dioxygenase is activated by the addition of ferrous iron, a reducing agent (such as ascorbate), and dithiothreitol.[3] In the presence of ascorbate and dithiothreitol, ferric iron can replace ferrous iron, presumably due to the reduction of the ferric to the ferrous form. About 25% of optimal activity is obtained in the absence of any added iron, presumably

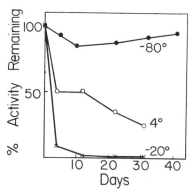

FIG. 1. Stability of the α-ketoisocaproate dioxygenase in an 80-fold purified preparation. Aliquots were stored at the temperatures indicated. Activity was measured by the decarboxylation assay. Frozen samples were thawed only once, then discarded. An activity of 100% was 42 nmol/min/mg protein for −80° experiments and 31.7 nmol/min/mg protein for 4° and −20° experiments. Values are means of duplicate assays.

TABLE I
STABILITY OF α-KETOISOCAPROATE DIOXYGENASE IN THE PRESENCE OF
MONOTHIOGLYCEROL, DITHIOTHREITOL, OR GLYCEROL[a]

	α-Ketoisocaproate dioxygenase activity (nmol $^{14}CO_2$ produced/ min/mg protein)		Days of storage at 4°	Percentage activity remaining
Addition	Initial	Final		
None	26.4	11.6	6	44
1 mM Dithiothreitol	25.5	9.48	6	37
5 mM Dithiothreitol	25.0	7.93	6	32
1% Monothioglycerol	24.3	7.13	6	29
5% Monothioglycerol	20.9	21.1	6	101
5% Glycerol + 1 mM dithiothreitol	21.7	11.6	8	53
5% Glycerol + 5 mM dithiothreitol	21.1	6.43	8	30

[a] Aliquots (0.15 ml) of the purified enzyme were incubated at 4° with the additions shown. All aliquots were adjusted to the same final volume (0.17 ml). Activity was measured immediately (Initial activity) and after 6 or 8 days storage at 4° (Final activity). Results are mean of 2 replicate assays. Each assay contained 29 μg protein of sample.

due to contaminating iron in the buffer. This residual activity is completely abolished by o-phenanthroline, an iron chelator. Several other divalent metals were tested but they did not substitute for iron. Maximal stimulation is obtained at 0.5 to 2.0 mM FeSO$_4$ in the presence of 0.5 mM ascorbic acid and 1 mM dithiothreitol (Fig. 2). FeSO$_4$ is inhibitory at concentrations above 5 mM. Whereas ascorbate stimulates the dioxygenase activity greater than 2-fold at 0.05 mM FeSO$_4$, there is no effect of added ascorbate at a FeSO$_4$ concentration of 2.0 mM. The stimulatory effect of ascorbate, therefore, may be solely due to its capacity to keep iron in the reduced, ferrous state. Other reducing agents including NADH and NADPH can replace ascorbate.

Substrate Specificity

A variety of α-keto acids have been tested as possible substrates for the dioxygenase by measuring O$_2$ consumption by rat liver cytosolic preparations. Only α-ketoisocaproate and α-keto-γ-methiobutyrate are appreciably oxidized by this enzyme. The product of the reaction, using α-ketoisocaproic acid as the substrate, has been identified as β-hyroxyisovaleric acid.[4,5] The product of the α-keto-γ-methiobutyrate oxidation has not been identified, but 3-hydroxy-3-methylthiopropionate should be expected. The enzyme has a much greater affinity for α-ketoisocaproate than α-keto-γ-

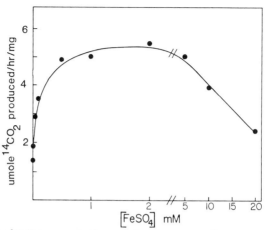

FIG. 2. Effect of $FeSO_4$ concentration on α-ketoisocaproate dioxygenase activity. Activity was measured by the decarboxylation assay, except for variation of the $FeSO_4$ concentration. Each assay contained 10 μl (12 μg protein) of the enzyme. Values are means of duplicate assays.

methiobutyrate ($K_m \times 0.32$ mM versus 1.9 mM); however, the maximal velocity of the enzyme is greater with α-keto-γ-methiobutyrate (247 versus 130 nmol/min/mg). During purification, the ratio of α-ketoisocaproate to α-keto-γ-methiobutyrate decarboxylating activities remains constant, indicating that a single enzyme is responsible for both activities. α-Ketoisovalerate, α-keto-β-methylvalerate, pyruvate, α-ketoglutarate, β-phenyl pyruvate, α-ketobutyrate, and α-ketononanoate have been tested and are not substrates.

Other Properties

The subunit molecular weight of the dioxygenase as determined by SDS-polyacrylamide gel electrophoresis is 46,000. A molecular weight of 51,000 was determined by Sephacryl S-200 chromatography with nondenaturing conditions. The enzyme therefore appears to be a monomer.

Optimal activity is obtained at pH 6.5 using 0.1 M Tris plus 0.1 M maleate buffer. The use of Tris-HCl, MOPS (morpholinopropanesulfonic acid), MES [2-(N-morpholino)ethanesulfonic acid] or BisTris-propane {1,3-bis[tris(hydroxymethyl)methylamino]propane} as assay buffers resulted in less than 30% of the activity seen in the presence of the Tris-maleate buffer (Table II). Very little activity can be measured in the presence of phosphate buffer. Optimal activation is obtained at 0.1 to 0.2 M maleate, but concentrations of maleate above 0.4 M are inhibitory. The concentration of Tris base (between 50 and 200 mM) does not affect the activity. Activation by maleate is not due to ionic strength since 0 to 0.4 M

TABLE II
EFFECT OF VARIOUS BUFFERS ON α-
KETOISOCAPROATE DIOXYGENASE ACTIVITY[a]

Assay buffer	α-Ketoisocaproate dioxygenase activity (μmol/hr/mg protein)
0.2 M Tris, 0.2 M maleate	5.93
0.1 M Tris, 0.1 M maleate	4.40
50 mM Tris, 50 mM maleate	2.96
20 mM Tris, 20 mM maleate	2.31
50 mM sodium phosphate	0.46
50 mM MES	1.90
50 mM BisTris-propane	1.54

[a] Activity was assayed by the decarboxylation assay, except the buffer was varied as shown above; pH was 6.5 for all determinations. Each assay contained 10 μl (12 μg protein) of purified enzyme. The pH of Tris-maleate buffers was adjusted with NaOH. Values are mean of duplicate assays.

sodium chloride did not change enzyme activity. Maleate may activate the dioxygenase by forming an iron chelate which is favorable for the catalytic reaction. Other iron chelators (o-phenanthroline, EDTA, and ADP) are very inhibitory. Thus, the dioxygenase either prefers the iron–maleate complex or maleate activates this enzyme by some other mechanism.

Mechanism

Little is known about the mechanism of this enzyme. Isovaleric acid is not an intermediate in the oxidation of α-ketoisocaproate to β-hydroxyisovaleric acid.[4,5] One O_2 molecule is consumed for each CO_2 molecule released from the α-keto acid. Incorporation of ^{18}O into both the carboxyl group and the β-hydroxyl group of β-hydroxyisovaleric acid during incubation of the enzyme with α-ketoisocaproate indicates both oxygen atoms of the product come from molecular oxygen. The mechanism proposed by Hamilton[21] for dioxygenases which require an α-keto acid as a cosubstrate would be consistent with the properties of the dioxygenase. An intramolecular reaction mechanism may be involved with the α-keto group of the substrate itself.

[21] G. A. Hamilton, *in* "Molecular Mechanisms of Oxygen Activation" (O. Hayaishi, ed.) p. 405. Academic Press, New York, 1974.

[38] Isolation of Branched-Chain α-Keto Acid Dehydrogenase as Active Complex from Bovine Liver

By DEAN J. DANNER and SUE C. HEFFELFINGER

Assay Method

$$R-COCO_2H \xrightarrow{\text{CoASH, NAD}} R-COSCoA + NADH + H^+ + CO_2$$

The reaction of this multienzyme complex[1] can be followed by quantifying any of the reaction products. Beginning with 1-^{14}C-labeled keto acid, evolved [14]CO_2 is trapped by a base and radioactivity measured by liquid scintillation. Using U-^{14}C-labeled keto acid, reaction products are separated by thin-layer chromatography and the products migrating with a mobility of the acyl-CoA compounds are quantified by liquid scintillation. NADH production is quantified by absorption change with time at 340 nm using a recording spectrophotometer.

Reagents. Stock solutions of cofactors and substrates are prepared either in distilled water or 30 mM potassium phosphate buffer and stored at −20°.

2 mM thiamin pyrophosphate (TPP)
2 mM MgCl_2
5 mM CoASH
10 mM NAD
10 mg/ml bovine serum albumin (BSA)
30 mM potassium phosphate, pH 7.5, stored at room temperature
50 mM dithiothreitol (DTT) made fresh daily

Reaction Mixture.[2] Assay conditions for each reaction product contained reactants at a final concentration of 0.2 mM TPP, 0.2 mM MgCl_2, 0.5 mM CoASH, 1.0 mM NAD, 5 mM DTT, 100 μg BSA, 0.1 mM keto acid, and 5–50 μg enzyme protein in 30 mM potassium phosphate buffer, pH 7.5.

Procedures

CO_2 Production.[3] A 250-μl reaction volume containing 1-^{14}C-labeled

[1] See also R. Paxton, this volume [41]; F. H. Pettit and L. J. Reed, this volume [40]; K. G. Cook and S. J. Yeaman, this volume [39].
[2] S. C. Heffelfinger, E. T. Sewell, and D. J. Danner, *Biochemistry* **22**, 5519 (1983).
[3] D. J. Danner, S. K. Lemmon, and L. J. Elsas, *Biochem. Med.* **19**, 27 (1978).

keto acid as substrate was placed in a Teflon tube closed with a serum cap and fitted tightly in a scintillation vial containing 500 μl Hyamine hydroxide. Holes in the sides of the Teflon tube are blocked by the neck of the scintillation vial. After incubation at 37° for varying times, the reaction is terminated with 250 μl of 1 M sodium acetate, pH 3.5. Liberated CO_2 is allowed to collect into the base solution by pushing the Teflon tube further into the vial exposing the holes. After 20 min of distillation, the Teflon tube is removed and scintillation fluid added for quantifying the $^{14}CO_2$.

Acyl-CoA Production.[4] Using a 250-μl reaction mixture and U-^{14}C-labeled keto acid as substrate, reactions are run as described above. A 20- to 50-μl aliquot is then spotted on a 20 × 20 cm thin-layer plate (Silica gel 1B-F, Baker Chemical Co.) with 50 μg of carrier acyl-CoA. The chromatograph is developed for 6.5 hr in butanol : acetic acid : water (6 : 2 : 2). Acyl-CoA (R_f 0.45) is identified by spraying the dried plate with 0.05% (w/v) of o-diaminobenzene in 20% trichloroacetic acid (TCA) and viewing under UV light. Identified spots are scraped into scintillation vials using scrapings from nonreactive portions of the plate as background. Quantification is by liquid scintillation. Vials are dark adapted for several hours prior to counting.

NADH Production.[2] Using a 500-μl reaction volume, NADH production is followed with time by change in absorption at 340 nm. Incubation is at 37°. This reaction monitors total complex activity.

Definition of Activity. A unit of activity is that amount of enzyme which will catalyze the conversion of 1 μmol of keto acid per minute under defined conditions. Specific activity is expressed as 1 μmol of product formed per minute per milligram of enzyme protein.

Purification

Mitochondria Isolation

Fresh bovine liver is cut into cubes and the tissue covered with 0.25 M sucrose buffered to pH 7.2 with 1 mM potassium phosphate containing 0.1 mM ethylenediaminetetraacetic acid (EDTA). The pH is adjusted to 8.5 with NaOH and after 4 hr at 4° readjusted to pH 8.0. After standing for 12–16 hr at 4°, the tissue is drained of fluid by filtering through cheesecloth and minced with a meat grinder. The wet weight of the mince is determined. One liter of mince is mixed with an equal volume of buffered sucrose and the pH adjusted to 7.0 before homogenization with a blender for 45 sec at low speed. All procedures are done at 4°. This homogenate is

[4] D. J. Danner, S. K. Lemmon, and L. J. Elsas, *Arch. Biochem. Biophys.* **202**, 23 (1980).

diluted 4-fold with buffered sucrose and centrifuged at 800 g for 10 min. Mitochondria are pelleted from the resulting supernatant using a TZ-28 flow-through rotor in a Sorvall RC5B centrifuge. Rotor speed is set for 13,000 g and a flow rate of 80–120 ml/min. Mitochondria are suspended in a volume of 0.25 M sucrose buffered to pH 7.2 with 5 mM potassium phosphate equal to two times the original weight of the mince. Centrifugation is repeated as above. This mitochondrial pellet is suspended in one volume of 6.25 mM sucrose, 1.25 mM potassium phosphate, pH 7.2, and 0.1 mM EDTA and centrifugation repeated. The final wash for these mitochondria is in one volume of 20 mM potassium phosphate, pH 6.5. The final mitochondrial pellet is suspended in a minimal volume of 20 mM potassium phosphate buffer, pH 6.5, and shell frozen in a 4-liter Erlenmeyer flask using a dry ice/acetone bath and stored at $-70°$. It can be stored up to 6 months. Five kilograms of liver will yield 300–400 g wet weight of mitochondria. This procedure is a modification from Roche and Cate.[5]

BCKD Complex Purification

Mitochondria are thawed in cold running water and the slurry made 50 mM with NaCl and 1% with bovine serum. Centrifugation at 32,000 g for 30 min removes extraneous protein. All procedures are done at 4°. The supernatant containing the BCKD complex is made 0.2 mM with TPP, 1% with bovine serum, 10 mM with $MgCl_2$, and the pH adjusted to 7.5 with 10 M KOH. While stirring at 4°, the solution is made 2% with a 50% solution of polyethylene glycol 8000 (PEG) and stirring continued for 10 min. Precipitated protein is removed by centrifugation at 24,000 g for 15 min and the supernatant made 4% with PEG. After stirring for 10 min at 4°, the precipitated BCKD complex is removed by centrifugation at 24,000 g for 15 min. This pellet is suspended in 50 mM potassium phosphate, pH 7.5, containing 0.2 mM TPP, 5 mM DTT, 1 mM $MgCl_2$, and 1% bovine serum, made 0.2 M with NaCl and left at 4° overnight. Extraneous protein is removed by centrifugation at 40,000 g for 20 min and the supernatant made 1% with PEG. Again nonspecific protein is removed by centrifugation at 24,000 g for 15 min. The BCKD complex is precipitated from the supernatant by adjusting the PEG concentration to 3% and pelleted by centrifugation at 24,000 g for 15 min. The pellet is suspended in a minimal volume of 50 mM potassium phosphate, pH 7.5, containing 0.2 mM TPP, 5 mM DTT, and 1 mM $MgCl_2$ and applied to a Sepharose 6B column (1.5 $\times$ 25 cm) equilibrated with this same buffer. Fractions are eluted with this same buffer without $MgCl_2$. Fractions containing protein with BCKD activity, as judged by NAD^+ reduction, are pooled and pel-

[5] T. E. Roche and R. L. Cate, *Arch. Biochem. Biophys.* **183**, 664 (1977).

TABLE I
PURIFICATION OF BOVINE LIVER BRANCHED-CHAIN
α-KETOACID DEHYDROGENASE COMPLEX

Fraction	Protein[a]	Specific activity[b]	Total activity[c]
Mitochondrial extract	7200	30	216
4% PEG	272	710	193
2% PEG/high salt	36	2260	82
Sepharose 6B	24	2000	48
Clarified final complex	11	5500	62

[a] Milligrams as determined by Bio-Rad Coomassie Blue dye binding.
[b] Nanomoles NADH/mg protein/min with α-ketoisovalerate as substrate.
[c] Micromoles NADH/min with α-ketoisovalerate as substrate.

leted by centrifugation at 109,000 g for 3 hr in a Beckman Ti 70.1 rotor. The final BCKD complex is suspended in 50 mM potassium phosphate, pH 7.5, containing 0.2 mM TPP, 2 mM DTT, 1 mM MgCl$_2$, 1 mM NAD, and 0.1 mM EDTA and the solution clarified by centrifugation in a microfuge. Aliquots of the final preparation are stored at $-70°$. Data from a typical isolation are summarized in Table I.

Properties

Resolution of the BCKD complex by electrophoresis under reducing conditions in a linear gradient polyacrylamide gel (10–16%) containing sodium lauryl sulfate[6] reveals four proteins by Coomassie Blue staining. Strongly evident is the lipoamide dehydrogenase at M_{12} 55,000 (Fig. 1). This is in contrast to other preparations of BCKD described in this volume. Three other proteins are also present (M_r, 52,000, 46,500, and 37,500). These correspond to the transacylase and α- and β-subunits of the decarboxylase, respectively.[2]

Catalytically, the specific activity varies between 5 and 8 μmol of NADH/min/mg protein for different preparations with α-ketoisovalerate as substrate. Exogenous lipoamide dehydrogenase is not required, and addition of the enzyme does not increase the specific activity of this BCKD complex. Relative activity for the three branched-chain keto acid sub-

6 U. K. Laemmli, *Nature (London)* **227**, 680 (1970).

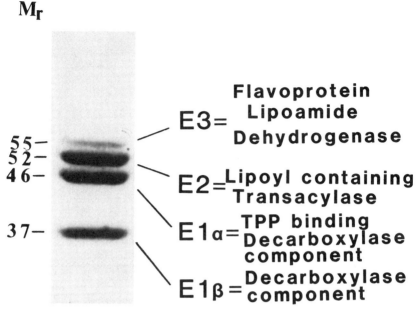

FIG. 1. Purified bovine liver BCKD complex resolved by electrophoresis in a linear 10–16% polyacrylamide gel in the presence of sodium lauryl sulfate and 2-mercaptoethanol. The proteins are stained with Coomassie Blue.

strates is $1.5:1.1:1$ for α-ketoisovalerate:α-ketoisocaproate:α-keto-β-methylvalerate, respectively, with essentially no activity toward α-ketoglutarate or pyruvate. Activity remains stable for at least 2 years at $-70°$. Reaction products, CO_2, acyl-CoA, and NADH, are produced in a $1:1:1$ stoichiometry. The complex also contains the endogenous kinase activity to phosphorylate the α-subunit and inactivate the complex. No phosphatase activity is endogenous with the complex as prepared.

[39] Purification, Resolution, and Reconstitution of Branched-Chain 2-Keto Acid Dehydrogenase Complex from Bovine Kidney

By Kenneth G. Cook and Stephen J. Yeaman

Branched-chain 2-keto acid dehydrogenase complex (BCKADC) is a multienzyme complex consisting of three enzyme components.[1] These are an acyltransferase (E2), which forms a central core around which are arranged the 2-keto acid dehydrogenase (E1) and dihydrolipoamide dehydrogenase (E3) components. The enzymes in the complex act in sequence to catalyze the oxidation of the 2-keto acid substrates derived from transamination of the essential branched-chain amino acids: leucine, isoleucine, and valine.[2] There is also evidence that the complex catalyzes the oxidative decarboxylation of 2-ketobutyrate[3,4] and 4-methylthio-2-ketobutyrate,[4] implicating it in the catabolism of threonine and methionine, two additional essential amino acids. The activity of the complex is regulated by reversible phosphorylation of the α-subunit of the E1 component.[5]

Assay

BCKADC activity is measured at 30° in a 1-ml cuvette by monitoring the formation of NADH at 340 nm.[2] The reaction mixture contains 50 mM potassium phosphate (pH 8.0), 0.2 mM thiamin pyrophosphate, 1.0 mM MgCl$_2$, 2.5 mM NAD$^+$, 0.13 mM CoA, 2.6 mM cysteine hydrochloride, 2 mM 2-keto acid (sodium salt), 10 μg/ml E3 and 1–5 μg/ml BCKADC. The final pH is approximately 7.4. One unit of enzyme catalyzes the formation of 1 μmol of NADH/min.[2] E2 activity is measured in the presence of excess E3 (5 μg) and E1 (3 μg). E1 activity is measured in the presence of excess E3 (5 μg) and E2 (3 μg).

[1] See also D. J. Danner and S. C. Heffelfinger, this volume [38]; R. Paxton, this volume [41]; F. H. Pettit and L. J. Reed, this volume [40].

[2] F. H. Pettit, S. J. Yeaman, and L. J. Reed, *Proc. Natl. Acad. Sci. U.S.A.* **75**, 4881 (1978).

[3] R. Paxton, P. W. D. Scislowski, E. J. Davis, and R. A. Harris, *Biochem. J.* **234**, 295 (1986).

[4] S. M. A. Jones and S. J. Yeaman, *Biochem. J.* **237**, 621 (1986).

[5] P. J. Randle, H. R. Fatania, and K. S. Lau, *Mol. Aspects Cell. Regul.* **3**, 1 (1984).

Purification of BCKADC

Buffers and Reagents

Buffer A: 50 mM potassium phosphate, 2 mM ethylenediaminetetraacetic acid (EDTA), 3% (w/v) Triton X-100, 1 mM benzamidine, 1 mM phenylmethylsulfonyl fluoride (PMSF), 0.1 mM dithiothreitol (DTT), pH 7.5

Buffer B: 30 mM potassium phosphate, 0.1 mM EDTA, 0.1 mM ethylene glycol bis(β-aminoethyl ether)-N,N^1–tetraacetic acid (EGTA), 1 mM benzamidine, 1 mM PMSF, 1 mM DTT, pH 7.3

Buffer C: 30 mM sodium phosphate, 0.1 mM EDTA, 0.1 mM EGTA, 1 mM benzamidine, 1 mM PMSF, 1 mM DTT, pH 7.3

Buffer D: 50 mM sodium phosphate, 0.1 mM EDTA, 0.1 mM EGTA, 5 mM DTT, pH 7.2

Buffer E: 50 mM Tris-HCl, 0.1 mM EDTA, 0.1 mM EGTA, 1 mM benzamidine, 1 mM PMSF, 1 mM DTT, pH 7.3

Polyethylene glycol 6000 (PEG) is used as a stock solution of 35% (w/v)

Spheroidal hydroxylapatite, Triton X-100, and polyethylene glycol 6000 are from British Drug Houses; antifoam A concentrate and 2-keto acids from Sigma Chemical Co.; and E3 (diaphorase) from Boehringer Mannheim, and Ultrogel AcA 34 from LKB.

Procedure

Ox kidney is obtained within 10 min of slaughter and transported on ice to the laboratory. All subsequent operations are carried out at 4° unless otherwise stated. Cortex is removed ($\sim$ 220 g/kidney), finely chopped, and then homogenized in a Waring Blendor for 2 min at low speed in 2 volumes buffer A containing antifoam A (0.5 ml/liter). After homogenization, an equal volume of buffer A is added and the homogenate centrifuged at 30,000 g for 15 min. The resulting supernatant (first supernatant) is adjusted to pH 7.15 with 10% acetic acid if necessary and 0.12 volume polyethylene glycol (35% w/v) is added. The mixture is stirred for 30 min, and the resulting precipitate collected by centrifugation at 30,000 g for 15 min. This precipitate is resuspended in buffer B to approximately one-third the initial volume, resuspension being completed by treating the sample with a Polytron PT10S at setting 3 for 20 sec. The mixture is stirred for 30 min at room temperature (PEG 1) and then clarified by centrifugation at 30,000 g for 30 min (PEG 1 clarified). This first precipitation step is crucial as it removes most of the 2-ketoglutarate dehydrogenase and pyruvate

dehydrogenase complexes, the remainder of these complexes being removed at subsequent stages of the purification procedure. The clarified solution is made to 1 mM MgCl$_2$ by addition of a 1 M solution. Polyethylene glycol, 0.055 volume (35% w/v), is then added and the solution stirred for 20 min. Centrifugation is then carried out at 30,000 g for 15 min, the resulting precipitate discarded, and a further 0.065 volume of polyethylene glycol is added to the supernatant. After stirring for 30 min, the precipitate is again collected by centrifugation at 30,000 g for 15 min. The precipitate is resuspended in buffer B (one-twentieth initial volume), homogenized, stirred, and clarified (PEG 2 clarified) as above.

The clarified material is applied to spheroidal hydroxylapatite (2.6 × 8 cm) preequilibrated in buffer B. The column is washed extensively with buffer B containing 150 mM potassium phosphate, either overnight or until the absorbance at 280 nm of the eluent decreases to zero. BCKADC is then eluted using buffer B containing 300 mM potassium phosphate. Fractions containing enzyme activity are pooled and further purified and concentrated by ultracentrifugation at 180,000 g for 3 hr at 4°. The final pellet is carefully resuspended in buffer C and any insoluble material removed by centrifugation at 10,000 g for 2 min. A convenient method of storage for long periods (at least 2 months without any detectable loss in enzyme activity) is at −20° following dialysis into buffer C containing 50% (w/v) glycerol.

The whole procedure can be completed in 1.5 working days. All centrifugation steps (except the 180,000 g spin) can be carried out conveniently in one 6 × 300 ml rotor (e.g., Beckman JA-14). The procedure can be scaled up severalfold without any significant increase in preparation time. If larger quantities of tissue are being processed, it is convenient to use a larger capacity rotor and lower gravitational forces (e.g., Beckman JA-10 at 17,700 g) to obtain the first supernatant and PEG 1 fractions. No change in the subsequent purification steps are necessary. The final material is essentially homogeneous as judged by SDS-polyacrylamide gel electrophoresis. In agreement with Pettit et al.[2] and Fatania et al.,[6] the complex is devoid of the E3 subunit; however, the sample does contain endogenous BCKADC kinase activity. A summary of the purification is shown in Table I. Using four fresh bovine kidneys (800–1000 g of cortex), 100–150 units of purified enzyme can be obtained. Frozen tissue can also be used; however, this results in a lower yield of approximately 50 units.

The specific activity of the final material, using 3-methyl-2-ketobutyrate as substrate, ranges from 4.0–6.0 U/mg. The purified complex is active against all five known substrates, namely 3-methyl-2-ketobutyrate, 4-

[6] H. R. Fatania, K. S. Lau, and P. J. Randle, *FEBS Lett.* **132**, 285 (1981).

TABLE I
PURIFICATION OF BRANCHED-CHAIN 2-KETO ACID DEHYDROGENASE COMPLEX FROM
BOVINE KIDNEY CORTEX[a]

Fraction	Volume (ml)	Protein (mg)	PDC (U)	KGDC (U)	BCKADC (U)	Specific activity BCKADC (U/mg)	yield (%)
					Total activity		
First supernatant	4,340	130,200	3,340	8,542	ND[b]	—	—
PEG 1 clarified	1,480	10,360	144	468	360	0.035	100
2% supernatant	1,480	8,140	45	446	330	0.041	92
PEG 2 clarified	250	2,188	—	10	260	0.12	72
Hydroxylapatite	136	260	—	—	132	0.51	37
180,000 g pellet	3.6	32	—	—	129	4.03	36

[a] Values are taken from a preparation using four bovine kidneys as starting material. Pyruvate dehydrogenase complex (PDC) and 2-ketoglutarate dehydrogenase complex (KGDC) are assayed as for BCKADC, in the presence of the appropriate substrate and in the absence of exogenous E3.

[b] ND, Not determined; the NADH-production assay is unsuitable for determination of BCKADC activity in crude extracts containing high concentrations of Triton X-100.

methyl-2-ketopentanoate, 3-methyl-2-ketopentanoate, 2-ketobutyrate, and 4-methylthio-2-ketobutyrate. At saturating concentrations of substrate (2 mM) the ratio of activity against each substrate is 1.0 : 0.7 : 0.4 : 0.5 : 0.3. The purified BCKADC also has some activity against pyruvate (activity ratio of 0.06) but is completely devoid of activity against 2-ketoglutarate.

Resolution of BCKADC into Active Components

The purified complex consists of E1, E2, and an intrinsic kinase (E3 being lost on purification). This complex can then be resolved into E1 and E2–kinase subcomplex by gel filtration on Ultrogel AcA 34 at pH 7.2 in the presence of 1 M NaCl. Purified complex (5–10 mg) is preincubated for 15 min at room temperature in buffer D containing 2 M NaCl. The sample (400 μl) is then applied to AcA 34 (1.5 cm × 42 cm) equilibrated at 4° in buffer D containing 1 M NaCl. Chromatography is carried out at 4°, at a flow rate of 8.0 ml/hr and approximately 2.0 ml fractions are collected. E2 has an M_r in excess of 1 × 10^6, and is eluted in the void volume, whereas E1 is included in the matrix, being eluted at K_{av} = 0.28. Analysis by SDS-polyacrylamide gel electrophoresis of both pooled components (Fig. 1) shows the absence of any major impurities and confirms that E2 is a single polypeptide of M_r 52,000 and that E1 contains two nonidentical polypeptides of M_r 46,000 (α-subunit) and M_r 35,000 (β-subunit).[2] After elution, each component can be concentrated by (NH$_4$)$_2$SO$_4$ precipitation,

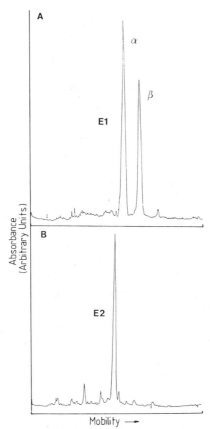

FIG. 1. Densitometric scans of SDS-polyacrylamide gels of the purified components of
BCKADC. Samples are analyzed using a Tris-glycine buffer system [U. K. Laemmli and M.
Favre, *J. Mol. Biol.* **80,** 575 (1973)] with 10% (w/v) acrylamide in the running gel and 5% in
the stacking gel. The gel is stained using Coomassie Brilliant Blue R. (A) E1 and (B) E2.

E2 being collected at 15–25% and E1 at 35–50% saturation. After concentration and dialysis into buffer E containing 50% glycerol, the components can be stored at −20° for at least 2 months (longest period tested)
without appreciable loss of enzyme activity.

An alternative method for resolution utilizes FPLC (Pharmacia) gel
filtration on a Superose 6 Prep Grade Column (16 × 500 mm). Sample
preparation and conditions of chromatography are the same as for soft gel
chromatography; however, the procedure is carried out at room temperature at a flow rate of 1.0 ml/min. Greater resolution is achieved in minutes
as compared to hours on conventional soft gel chromatography. Due to the
higher exclusion limit of this matrix, E2 is included into the gel. A smaller
peak eluting at the void volume consists mainly of aggregated impurities.

Reconstitution of the Components into Active BCKADC

In experiments using the overall activity of the complex as a monitor of reconstitution, it is found that reassociation is an extremely rapid process. Complex activity can be detected immediately (less than 3 sec) after the addition of the separate components into a cuvette containing all the essential substrates and cofactors required for activity. With a ratio of 1.2 μg of E1 to 1.0 μg of E2 (with 5 μg of E3), a specific activity of 8.5 U/mg can be achieved. This compares well with the specific activity obtained for the purified complex. By using excess E2 (3 μg) to assay E1 activity and excess E1 (5 μg) to assay E2, specific activities of approximately 25.0 and 20.0 U/mg can be obtained for E1 and E2, respectively. This also compares well with values expected from calculations using the specific activity of the isolated complex. These results suggest that essentially all the activity of the original complex can be regained by reconstitution of the component enzymes.

The purified E1 component can also reconstitute with phosphorylated inactive BCKAD complex to regain all the initial activity of the complex. Addition of E1 directly into an assay cuvette containing phosphorylated complex allows immediate regain of enzyme activity, as indicated by the formation of NADH. Dissociation and reconstitution of E1 with the E2 core may occur naturally in the mitochondria, as it has been shown that rat liver and kidney mitochondria contain excess E1 component which is found free in the high-speed supernatant of the mitochondrial extracts.[7,8] The physiological significance of this additional E1 component is not clear, as it has been shown that it is still susceptible to phosphorylation and inactivation *in vitro* by the BCKAD kinase.[8]

Reconstitution of the components of BCKADC can also be monitored using HPLC gel filtration chromatography on TSK G4000 SWG (7.5 × 300 mm) equilibrated in buffer D. Chromatography is carried out at room temperature at a flow rate of 0.5 ml/min. Chromatography of reconstituted components (50 μg E1 and 50 μg E2 in 25 μl buffer D) shows that E1 and E2 reassociate to form a complex which elutes before the individual E1 and E2 components, indicating that it possesses a higher apparent molecular weight than native E2.

Acknowledgments

Work in the authors' laboratory was supported by grants from the Medical Research Council, U.K.. S.J.Y. is a Lister Institute Research Fellow.

[7] H. R. Fatania, K. S. Lau, and P. J. Randle, *FEBS Lett.* **147**, 35 (1982).
[8] S. J. Yeaman, K. G. Cook, R. W. Boyd, and R. Lawson, *FEBS Lett.* **172**, 38 (1984).

[40] Branched-Chain α-Keto Acid Dehydrogenase Complex from Bovine Kidney

By FLORA H. PETTIT and LESTER J. REED

$$RCOCO_2H + CoA + NAD^+ \rightarrow RCO\text{-}CoA + CO_2 + NADH + H^+$$

In eukaryotic cells the branched-chain α-keto acid dehydrogenase (BCKDH)[1] complex is located in mitochondria, within the inner membrane-matrix compartment. The complex is composed of multiple copies of three major component enzymes: branched-chain α-keto acid dehydrogenase (E1), dihydrolipoamide acyltransferase (E2), and dihydrolipoamide dehydrogenase (E3). These three enzymes, acting in sequence, catalyze the oxidative decarboxylation of α-ketoisovaleric, α-ketoisocaproic, and α-keto-β-methylvaleric acids, which are derived from the amino acids valine, leucine, and isoleucine, respectively.

Assay Method

Principle

Activity of this multienzyme complex is assayed spectrophotometrically by measurement of the initial rate of NADH production.[2] Because E3 dissociates from the complex during its purification, it is necessary to add E3 to the assay mixture to obtain maximal rates.

Reagents. Dilution buffer: BCKDH complex is diluted in 50 mM sodium 2-(N-morpholino)propane sulfonate (MOPS), pH 7.0, 1 mM MgCl$_2$, 0.1% bovine serum albumin.

Procedure

The assay mixture contains 50 mM potassium phosphate, pH 8.0, 0.2 mM thiamin diphosphate, 1.0 mM MgCl$_2$, 2.5 mM NAD$^+$, 0.13 mM CoA, 0.32 mM dithiothreitol, 2.0 mM sodium α-ketoisovalerate, 4 units of dihydrolipoamide dehydrogenase (Sigma Chemical Co., St. Louis, MO, type VI), and enzyme complex in a final volume of 1.0 ml. The pH of the solution is 7.4. After equilibration of the cuvette to 30°, the reaction is

---

[1] See also D. J. Danner and S. C. Heffelfinger, this volume [38]; R. Paxton, this volume [41]; K. G. Cook and S. J. Yeaman, this volume [39].

[2] F. H. Pettit, S. J. Yeaman, and L. J. Reed, *Proc. Natl. Acad. Sci. U.S.A.* **75,** 4881 (1978).

started by addition of enzyme complex, and the increase in absorbance at 340 nm is followed with a recording spectrophotometer.

Definition of Unit

One unit of BCKDH complex activity is the amount of enzyme which catalyzes the production of 1.0 μmol of NADH per minute in the standard assay. Protein is determined by the biuret method[3] with crystalline bovine serum as standard.

Purification Procedure

In the procedure described below, care is taken to remove or inhibit endogenous proteases. A lysosome-enriched fraction is carefully removed during preparation of mitochondria, and the mitochondrial fraction is washed extensively to remove proteases released from ruptured lysosomes. In addition, the serine protease inactivator phenylmethylsulfonyl fluoride (PMSF), the thiol protease inactivator benzyloxycarbonylphenylalanyla-lanine diazomethylketone[4] (Z-Phe-AlaCHN$_2$), and rabbit serum (a source of protease inhibitors, particularly α_2-macroglobulin) are added in the early steps of the purification procedure.

The purification procedure is modified from that of Pettit *et al.*[2] It was developed originally[5] to obtain highly purified preparations of the pyruvate dehydrogenase complex and the α-ketoglutarate dehydrogenase complex, as well as partially purified preparations of pyruvate dehydrogenase phosphatase and pyruvate dehydrogenase kinase. All operations are carried out at $2°-5°$.

Preparation of Mitochondrial Extract

Bovine kidney mitochondria are prepared and washed as described by Pettit and Reed.[6] The washed mitochondrial paste is suspended in 20 mM potassium phosphate, pH 6.5, containing 0.1 mM PMSF and 0.01 mM Z-Phe-AlaCHN$_2$, and the mixture is shell-frozen and thawed once. The thawed suspension is made 50 mM in NaCl, 1% by volume of rabbit serum (Pel-Freez Biologicals, Rogers, AR, type 2) is added, and the mixture is

[3] A. G. Gornall, C. J. Bardawill, and M. M. David, *J. Biol. Chem.* **177,** 751 (1949).
[4] H. Watanabe, G. D. J. Green, and E. Shaw, *Biochem. Biophys. Res. Commun.* **89,** 1354 (1979).
[5] T. C. Linn, J. W. Pelley, F. H. Pettit, F. Hucho, D. D. Randall, and L. J. Reed, *Arch. Biochem. Biophys.* **148,** 327 (1972).
[6] F. H. Pettit and L. J. Reed, this series, Vol. 89, p. 376.

centrifuged at 30,000 g for 30 min in a Beckman type J-14 rotor. The supernatant fluid is decanted carefully; the residue is discarded.

Isoelectric Precipitation of BCKDH Complex

The mitochondrial extract is made 0.2 mM in thiamin diphosphate and 10 mM in MgCl$_2$, and 1% by volume of rabbit serum is added. The pH is carefully lowered to 6.18 by dropwise addition, with stirring, of 10% acetic acid. The mixture is stirred for 20 min, and the precipitate is collected by centrifugation at 30,000 g for 10 min. The precipitate contains essentially all of the BCKDH complex activity. The supernatant fluid contains the pyruvate and α-ketoglutarate dehydrogenase complexes, which are purified further as described elsewhere.[6]

Ultracentrifugation

The precipitate from the previous step is suspended, by means of a glass/Teflon homogenizer, in 15 ml of 100 mM potassium phosphate, pH 7.0, containing 2 mM dithiothreitol and 0.1 mM ethylenediaminetetraacetic acid (EDTA). The suspension is allowed to stand overnight and then is centrifuged at 40,000 g for 15 min. The pellet is discarded, and the supernatant fluid is centrifuged at 105,000 g for 3.5 hr. The faintly yellow pellet is redissolved in 100 mM potassium phosphate, pH 7.0, containing 2 mM dithiothreitol and 0.1 mM EDTA. The solution is clarified by centrifugation at 40,000 g for 15 min. A summary of the purification is presented in Table I.

The enzyme complex is stable for several months at −50°, but it is sensitive to freezing and thawing.

TABLE I

PURIFICATION OF BCKDH COMPLEX FROM BOVINE KIDNEY[a]

Step	Volume (ml)	Protein (mg)	Specific activity[b]	Recovery (%)
Mitochondrial extract	3,000	42,066	0.034	(100)
Isoelectric precipitation	15	314	4.5	99
Ultracentrifugation	15	176	8.8	108

[a] From about 37.4 kg of bovine kidney.
[b] Micromoles of NADH produced per minute per milligram of protein in the presence of excess dihydrolipoamide dehydrogenase.

Properties

The enzyme complex is near homogeneous as judged by analytical ultracentrifugation and by electrophoresis on a 2% agarose slab gel at pH 7.0.[2] The sedimentation coefficient ($s_{20,w}$) is about 40S. The sodium dodecyl sulfate-polyacrylamide gel electrophoresis pattern shows three major bands corresponding to E2 and the E1α and E1β subunits. The apparent molecular weights are 52,000 (E2), 46,000 (E1α), and 35,000 (E1β).

As usually isolated from kidney mitochondria the BCKDH complex is deficient in, if not completely free of, dihydrolipoamide dehydrogenase (E3). That E3 dissociates more readily from the BCKDH complex than from the pyruvate dehydrogenase complex is confirmed by measurement of dissociation constants (K_D). The K_D value for the E3/BCKDH complex is 55 nM,[7] whereas K_D for the E3/pyruvate dehydrogenase complex is 3 nM.[8]

E2 has a molecular weight of about 1,250,000 and consists of 24 apparently identical subunits arranged in a cube.[2,9] The isolated BCKDH complex contains about 24 E1α chains and 24 E1β chains that are attached to the E2 core by noncovalent bonds. The BCKDH complex prepared by the procedure described above exhibits little, if any, BCKDH kinase activity, because the kinase is apparently unstable below pH 7.0.[10] A procedure to obtain, from bovine kidney mitochondria, near homogeneous BCKDH complex that contains active BCKDH kinase is described elsewhere.[11]

Apparent K_m values are 40, 50, and 37 μM for α-ketoisovalerate, α-ketoisocaproate, and α-keto-β-methylvalerate, respectively.[2] Pyruvate and α-ketobutyrate (but not α-ketoglutarate) are also oxidized by the BCKDH complex, at rates about 20 and 50%, respectively, of the rate observed with α-ketoisovalerate. The apparent K_m values are 1 mM for pyruvate and 56 μM for α-ketobutyrate. Isobutyryl-CoA and isovaleryl-CoA, products of the oxidation of α-ketoisovalerate and α-ketoisocaproate, respectively, inhibit the activity of the BCKDH complex about 48 and 56%, respectively, at a concentration of 0.2 mM. At a concentration of 0.1 mM, NADH inhibits the activity of the complex about 44%.

[7] T.-L. Wu and L. J. Reed, unpublished observations.
[8] T.-L. Wu and L. J. Reed, *Biochemistry* **23**, 221 (1984).
[9] F. H. Pettit and L. J. Reed, unpublished observations.
[10] Z. Damuni, M. L. Merryfield, J. S. Humphreys, and L. J. Reed, *Proc. Natl. Acad. Sci. U.S.A.* **81**, 4335 (1984).
[11] Z. Damuni and L. J. Reed, this volume [42].

[41] Branched-Chain α-Keto Acid Dehydrogenase and Its Kinase from Rabbit Liver and Heart

By RALPH PAXTON

Branched-chain α-keto acid dehydrogenase (BCKDH),[1] an intramitochondrial multienzyme complex, catalyzes the oxidative decarboxylation of numerous α-keto acids (α-ketoisovalerate, α-ketoisocaproate, α-keto-β-methylvalerate, α-ketobutyrate, α-ketoγ-thiomethylbutyrate, α-ketocaproate, α-ketovalerate, and pyruvate).[2-5] The complex is composed of three different subunits (E1α and E1β subunit of branched-chain α-keto acid decarboxylase, EC 1.2.4.4; E2 or dihydrolipoyl acyltransferase, no EC number; and E3 or dihydrolipoamide reductase, EC 1.8.1.4).[2] The complex is subject to phosphorylation (inactivation) on the E1α subunit by a copurifying kinase[2] (no EC number) and dephosphorylation (activation) by an intramitochondrial phosphatase[6,7] and by a broad-specificity phosphatase.[3,8]

Assay Methods

BCKDH Activity

A spectrophotometric assay accurately determines BCKDH activity with mitochondrial extracts, partially purified and highly purified BCKDH preparations. However, a radiochemical assay of the decarboxylation of 1-^{14}C-labeled α-keto acid is needed for accurate determinations of BCKDH activity in whole-tissue homogenates.[9] The spectrophotometric assay is based on the reduction of NAD[+].[2] The 1-ml assay volume is composed of 30 mM K$_2$HPO$_4$/KH$_2$PO$_4$ (KP$_i$), pH 7.5 at 20°, 3 mM MgCl$_2$, 0.4 mM thiamin pyrophosphate (TPP), 3 mM NAD$^+$, 0.1% (v/v)

---

[1] See also D. J. Danner and S. C. Heffelfinger, this volume [38]; K. G. Cook and S. J. Yeaman, this volume [39]; F. H. Pettit and L. J. Reed, this volume [40].
[2] R. Paxton and R. A. Harris, *J. Biol. Chem.* **257**, 14433 (1982).
[3] R. Paxton and R. A. Harris, *Arch. Biochem. Biophys.* **231**, 48 (1984).
[4] R. Paxton, P. W. D. Scislowski, E. J. Davis, and R. A. Harris, *Biochem. J.* **234**, 295 (1986).
[5] G. Livesey and P. Lund, *Biochem. Soc. Trans.* **8**, 540 (1980).
[6] Z. Damuni, M. L. Merryfield, J. S. Humphreys, and L. J. Reed, *Proc. Natl. Acad. Sci. U.S.A.* **81**, 4335 (1984).
[7] R. Paxton and R. A. Harris, *Fed. Proc., Fed. Am. Soc. Exp. Biol.* **43**, 2283 (1984).
[8] R. A. Harris, R. Paxton, and R. A. Parker, *Biochem. Biophys. Res. Commun.* **107**, 1497 (1982).
[9] G. W. Goodwin, B. Zhang, R. Paxton, and R. A. Harris, this volume [23].

Triton X-100, 2 mM dithiothreitol (DTT), 2 units of dihydrolipoamide reductase from porcine heart (Sigma Chem. Co., 10–15 units/mg protein), and the source of BCKDH activity. The reaction is initiated with 0.05 ml of 20 mM α-ketoisovalerate and the increase in optical density at 340 mm is recorded. With all BCKDH preparations but the highly purified enzyme, the assay mixture is incubated in the spectrophotometer for 2–3 min until the change in optical density (usually increasing) becomes zero before addition of substrate. With very crude preparations it becomes necessary to have a second simultaneous assay where, after waiting several minutes, this cuvette receives the same volume of water as the reaction cuvette received of substrate. All reagents, with the exception of the enzymes, and the chamber housing the cuvettes are maintained at 30°. Routinely, a 2-fold concentrated assay cocktail, without MgCl$_2$ and dihydrolipoamide reductase, can be stored frozen at −20° in 10-ml portions for several weeks without any noticeable deleterious effects on BCKDH activity. On thawing, 1 M MgCl$_2$ can be added to this 2-fold concentrated cocktail to achieve 6 mM. Dihydrolipoamide reductase, which is readily lost upon purification of the complex,[2] is added to the assay to ensure that this enzyme is not rate limiting for BCKDH activity.

BCKDH Kinase

The copurifying BCKDH kinase activity can be determined in two ways: by BCKDH activity, and by acid-stable incorporation of phosphate (P$_i$) from [γ-^{32}P]ATP into the E1α subunit of BCKDH.[2,10] In crude or partially purified BCKDH preparations, both procedures can be limited by the presence of ATPase activity which produces an unknown concentration of ATP and ADP (competitive inhibitor),[2] and by endogenous phosphatase activity. The P$_i$ incorporation assay is further limited by the incorporation of P$_i$ into other phosphorylated proteins. Nevertheless, both assays can be used with highly purified preparations of BCKDH kinase by determining the lack of ATPase and phosphatase activity in the preparation, and by determining that the P$_i$ incorporation, from [γ-^{32}P]ATP, is solely in the E1α subunit by autoradiography after polyacrylamide gel electrophoresis in the presence of sodium dodecyl sulfate (SDS-PAGE).[2,10] The BCKDH kinase assay is conducted, whether by measuring lost of BCKDH activity (see above) or by P$_i$ incorporation in 30 mM potassium N-2-hydroxyethylpiperazine-N'-2-ethane sulfonate (HEPES–KOH), 1.5 mM MgCl$_2$, 2 mM DTT, and 0.2 mM ethylene glycol bis(β-aminoethyl ether)-N,N,N',N'-tetraacetic acid (EGTA; all at pH 7.35 at 20°). The assay is routinely done at 1 mg/ml of BCKDH kinase complex in the presence of

[10] R. Paxton, M. Kuntz, and R. A. Harris, *Arch. Biochem. Biophys.* **244**, 187 (1986).

0.4 mM [γ-^{32}P]ATP (specific activity of 600–800 cpm/pmol). With either BCKDH kinase assay a time course must be determined to ensure linearity. If effectors (e.g., α-keto acids) of the kinase are to be evaluated by changes in BCKDH activity, then the influence on enzyme activity in the absence of ATP must also be determined, since several effectors of the kinase (e.g., α-ketoisocaproate, α-keto-β-methylvalerate, and α-ketoisovalerate) can cause a time-dependent inactivation of BCKDH activity.[2,11] The P$_i$ incorporation assay may also give misleading results since there are two phosphorylation sites, only one of which appears responsible for inactivation.[9,12] While certain BCKDH kinase inhibitors (e.g., α-chloroisocaproate) diminish phosphorylation of both sites, they almost completely block phosphorylation of site 2.[10] These effects on different phosphorylation sites can be determined by reversed-phase HPLC analysis of the tryptic phosphopeptides from BCKDH[10] or by SDS-PAGE where mobility of E1α subunit is altered depending on the site and degree of phosphorylation.[13] Due to the limitation of activity and P$_i$ incorporation measurements, both parameters should be analyzed simultaneously for a complete understanding of the relationship between activity and phosphorylation.[10]

The assay for acid-stable, protein-bound P$_i$ is as follows.[10] Remove a portion of the above BCKDH kinase assay mixture and place on the crown of a folded 2 × 2 cm piece of slightly moist Whatman 3MM filter paper previously soaked in 24% (w/v) trichloroacetic acid, 2 mM K$_4$(PO$_4$)$_2$, 0.2 M H$_3$PO$_4$, and 1 mM ATP. The papers are immediately added to 0.5 liter of stirring ice-cold 10% (w/v) trichloroacetic acid, 2.5 mM K$_4$(PO$_4$)$_2$, and 1 mM ATP. After all of the papers have been added, including papers that have been spotted with the same assay mixture but without enzyme (i.e., the assay blank), continue to stir on ice for 15 min. Stirring should be vigorous enough to circulate the wash but not so vigorous as to swirl all of the papers. After 15 min, decant the wash and add another 0.5 liter of ice-cold wash. Mix this for 15 min, decant, add 0.5 liter of room temperature 5% (w/v) trichloroacetic acid, and mix for 15 min. Repeat. Decant the final wash, put papers in about 80 ml of ether for 5–10 min, remove papers and dry. Place papers in 5 ml of scintillation fluid and determine the radioactivity by liquid scintillation counting. The specific activity of the ATP is determined by removing a known amount of the [γ-^{32}P]ATP mixture and diluting with 0.5 mM ATP in a 100-ml volumetric flask. A

[11] R. A. Harris, R. Paxton, and A. A. DePaoli-Roach, *J. Biol. Chem.* **257**, 13915 (1982).
[12] K. G. Cook, A. P. Bradford, S. J. Yeaman, A. Aitken, J. M. Fearnley, and J. E. Walker, *Eur. J. Biochem.* **145**, 587 (1984).
[13] M. J. Kuntz, R. Paxton, Y. Shimomura, G. W. Goodwin, and R. A. Harris, *Biochem. Soc. Trans.* **14**, 1077 (1986).

known volume is removed, spotted on the filter paper (as above except not treated), directly dried, and radioactivity determined.

Purification of Rabbit Liver BCKDH Kinase Complex

Homogenization

Frozen ($-70°$) rabbit liver (0.6 kg; see Table I), from Pel-Freez Biologicals or obtained locally, is homogenized until a uniformly smooth, icy mixture is achieved with a large (3.6 liter) Waring blender in about 2 volumes to weight (v/w) of ice-cold 50 mM KP$_i$ (pH 7.5 at 20°), 2 mM DTT, 0.2 mM TPP, 2% (v/v) Triton X-100, 2 mM EDTA, 2% (v/v) bovine serum, 0.1 mM N^α-p-tosyl-L-lysine chloromethyl ketone (TLCK), 0.5 μM leupeptin, 0.5 μM pepstatin A, 1 μg/ml aprotonin (lyophilized powder from bovine lung; Sigma Chemical Co.), 0.1 mg/ml trypsin inhibitor from turkey egg white, and, immediately before homogenization, 0.1 mM phenylmethylsulfonyl fluoride (PMSF). Unless noted, all solution and operations are done at 1° to 4°. Portions (about 0.3 liter) of this tissue extract are further homogenized with a Polytron for 30 sec at setting #4 and centrifuged at 9,000 rpm (14,000 g) in a GS-3 rotor (Sorvall) for 30 min. The supernatant is filtered through 4 layers of cheesecloth. The pellet is extracted with 0.5 (v/w) as above. The pH of the combined supernatant is adjusted to 7.5 with 2 M Tris.

TABLE I
PURIFICATION OF RABBIT LIVER BCKDH

Purification stage	Total activity (units)	Total protein (mg)	Specific activity (units/mg)	Yield (%)
Crude[b]	313	118,000	0.0026	100
PEG pellet	274	22,500	0.012	88
Supernatant[b]	260	15,000	0.017	83
PEG pellet	172	2,530	0.068	55
Hydroxylapatite	164	500	0.33	52
PEG pellet	144	90	1.6	46
Hydroxylapatite	100	50	2.0	32
High-speed centrifugation pellet	85	10	8.6	27

[a] Specific activity given as μmol/min/mg protein at 30° with 1 mM α-ketoisovalerate as substrate. One unit equals 1 μmol/min.

[b] Refers to protein and activity of supernatant after centrifugation.

Polyethylene Glycol (PEG) Fractionation

The supernatant is made to 2% (w/v) in PEG 8000 by slow addition with constant mixing of the appropriate volume of cold 50% (w/v) PEG. This mixture is gently stirred for 30 min and centrifuged in a GS-3 rotor at 9,000 rpm for 20 min. The supernatant is made to 5% (w/v) PEG (i.e., assuming the recovered supernatant is 2% in PEG and the addition of PEG will expand the volume by the volume of the addition) and centrifuged as above. The 2 to 5% PEG pellet is taken up with a motor-driven Potter–Elvehjem homogenizer to a volume of 0.7 liter of buffer B [50 mM KP$_i$, 2 mM DTT, 0.1 mM TPP, and 0.1% (v/v) Triton X-100; pH 7.5 @ 20°].

Clarification by Centrifugation

The suspended pellet is centrifuged in a 45 Ti rotor (Beckman) at 40,000 rpm (125,000 g) for 25 min. The supernatant is decanted until the appearance of turbid material. The pellet is again suspended to a volume of 0.7 liter of buffer B and centrifuged as above. This is repeated one more time. All supernatants are combined.

PEG Fractionation

The combined supernatant is made to 5% (w/v) PEG as given above. The pellet is suspended with a motor-driven Potter-Elvehjem homogenizer in about 0.15 liter of 50 mM HEPES-KOH, 2 mM DTT, and 0.1% (v/v) Triton X-100 (pH 7.5 at 20°). This mixture is clarified by centrifugation in an SS-34 (Sorvall) rotor at 16,000 rpm (31,000 g) for 15 min. The pellet is extracted in 25 ml, and centrifuged as above. This is repeated one additional time.

Phosphatase Treatment

The above mixture is made to 2 mM MgCl$_2$, 0.5 μM leupeptin, 0.5 μM pepstatin A, 0.1 mM TLCK, 0.1 mM PMSF, and about 1 ml (10 to 15 mg protein) of broad-specificity phosphoprotein phosphatase[3,8,10] is added. This mixture is incubated at 30° with constant monitoring of BCKDH activity. Usually, by 30 min, no further increase in activity is obtained; however, with liver it is rare to see any increase in activity. This material, which usually becomes slightly turbid, is clarified by centrifugation in the SS-34 rotor as above. The pellet is extracted and centrifuged two additional times with buffer B. The supernatant is made to the conductivity of buffer B by adding 1 M KP$_i$ (pH 7.5 at 20°) and the pH adjusted to 7.5.

Hydroxylapatite Chromatography

The sample is adsorbed to a column (5 × 21 cm) of hydroxylapatite (BioGel HTP; Bio-Rad) equilibrated in buffer B. The column is washed with about 0.3 liter of buffer B followed by a 1.5 liter linear gradient of 0.05 to 0.55 M KP$_i$ in buffer B. BCKDH activity usually elutes between 0.26 and 0.28 M KP$_i$.

PEG Fractionation

The pooled fractions are made to 5% PEG as above. The pellet is extracted in about 40 ml of buffer B and clarified by centrifugation in an SS-34 rotor at 16,000 rpm for 15 min. The resulting pellet is extracted (about 10 ml of buffer B) and centrifuged as above. This is repeated one more time.

Hydroxylapatite Chromatography

The above is applied to a smaller (1.5 × 12 cm) column of hydroxylapatite equilibrated in buffer B. The column is washed with 0.1 liter of buffer B and eluted with a 0.25 liter linear gradient from 0.05 to 0.5 M KP$_i$ in buffer B.

High-Speed Centrifugation

The pooled activity is centrifuged at 50,000 rpm (226,000 g) for 120 min in a 50 Ti (Beckman) rotor. The pellet is taken up with a 5-ml Potter-Elvehjem homogenizer in a volume of storage buffer [50 mM HEPES-KOH, 5 mM DTT, 0.05 mM TPP, 0.05 mM EDTA, 15% (v/v) glycerol, and 0.1% (v/v) Triton X-100 with a pH of 7.5 at 20°] to give about 40–50 units/ml. This is based on the total activity of the pooled fractions from the previous chromatography. This mixture is clarified by centrifugation at 16,000 rpm in an SS-34 rotor for 15 min. The pellet is extracted with 0.5 volume of the volume used above and centrifuged. Protein concentration is determined by a modified Lowry procedure.[2]

The overall purification (Table I) generally gives a specific activity of 7–9 units/mg protein with a yield of about 25 to 40%. About 2 kg of liver can be processed in the manner given with no modifications. The overall procedure takes about 3 days. If more tissue is to be processed, the pellet from the second PEG fractionation (usually obtained in about 8 hr) can be stored on ice. An equivalent amount of tissue can be processed the next day, combined with the previous material, treated with phosphatase, and applied to the large hydroxylapatite column.

Purification of BCKDH Kinase Complex from Rabbit Heart and Other Tissues

The same procedure as outlined above has been used[10] to isolate the BCKDH kinase complex from rabbit heart, skeletal muscle, brain, kidney cortex, rat heart, and bovine kidney. The procedure described for liver is generally applicable for all of these tissues; however, there is variability between the tissues regarding the solubility of the PEG pellets. Consequently, it is important to sometimes use a different number of clarification steps, depending on the turbidity of the sample or how much of the BCKDH activity is pelleted. With all of these tissues but kidney, BCKDH activity is generally not measurable until after the second PEG precipitation. Phosphatase treatment becomes of paramount importance for all of these tissues in order to obtain a reliable determination of the true activity of the complex. Certainly, complete activation and dephosphorylation of the complex, regardless of the tissue, is important to determine the relationship between activity and phosphorylation.[10]

Properties of Isolated BCKDH Kinase Complex

Molecular Weight Estimate

The isolated complex, regardless of the tissue, apparently lacks the dihydrolipoamide reductase since, with purification, BCKDH activity becomes dependent on exogenously added reductase and the highly purified BCKDH kinase complex lacks the subunits of this enzyme.[2] The estimated molecular weight of the rabbit liver complex is greater than 2×10^6 based on gel filtration (Sepharose CL-2B).[2] The relative molecular weights of the E1α, E1β, and E2 subunits are 46,800, 38,000, and 51,000, respectively.[2] The relative molecular weight of these subunits are the same for the complex isolated from all of the above tissues.[10] There is a molar ratio of about 1 : 1 : 1 for these subunits with the rabbit liver complex.

BCKDH Kinase

The BCKDH kinase has an optimum pH of about 7.5, with maximum activity at this pH in HEPES (100%) relative to KP_i (60%), imidazole (55%), and 3-[N-morpholino]ethane sulfonate (85%). Maximum activity is with 1.5 mM $MgCl_2$. Inhibition occurs at higher concentrations. Half-maximum stimulation by $MgCl_2$ is at 25 μM. The presence of 0.2 mM EGTA stimulates rabbit liver BCKDH kinase activity and may partially be due to Ca^{2+} chelation, since this divalent cation slightly inhibits kinase

activity. The apparent K_m for ATP is 25 μM and ADP is a competitive inhibitor of ATP with an apparent K_i of 130 μM. The rabbit liver BCKDH kinase is inhibited by several compounds[2,3,11,14]: α-ketoisocaproate, α-ketoisovalerate, α-keto-β-methylvalerate, dichloroacetate, clofibric acid, pyruvate, phenylpyruvate, phenyl acetate, p-hydroxyphenylacetate, phenyllactate, p-hydroxyphenylpyruvate, α-chloroisocaproate, α-ketocaproate, α-ketovalerate, α-ketoadipate, α-ketobutyrate, acetoacetyl-CoA, methylmalonyl-CoA, octanoate, NADP$^+$, heparin, isovaleryl-CoA, isobutyryl-CoA, and malonyl-CoA. The kinase phosphorylates two different serines separated by nine amino acids within the E1α subunit at about equal rates and to about the same extent (0.75 mol P_i/site) with total phosphorylation of about 1.5 mol P_i/α subunit.[10] With BCKDH kinase complex isolated from all of the tissues mentioned above, there is a linear relationship between total phosphorylation and degree of inactivation up to about 1 mol P_i/α subunit and 95% inactivation. Beyond this level of inactivation, there is generally greater phosphorylation than inactivation. Inactivation appears, however, to be dependent on phosphorylation of only one site. The specificity of the kinase toward one site or the other, while apparently equal, can be altered by the presence of a kinase inhibitor (e.g., α-chloroisocaproate).[10] The activity of the BCKDH kinase, as isolated with the above procedure, is apparently tissue dependent with heart having the highest activity and liver the lowest.[10] However, these differences may also be dependent on the relative difficulty of purification of the complex. The principal difficulty resides in the number of clarification steps needed at the different stages, with liver requiring many more than heart. Whether this difference explains the differences in kinase activity has not been established.

Acknowledgments

This work was supported in part by grants from U.S. Public Health Service (AM19259 and DK39263), Established Investigatorship and Grant-In-Aid from The American Heart Association, and the Grace M. Showalter Residuary Trust.

[14] R. Paxton and R. A. Harris, *Arch. Biochem. Biophys.* **231**, 58 (1984).

[42] Branched-Chain α-Keto Acid Dehydrogenase Phosphatase and Its Inhibitor Protein from Bovine Kidney

By ZAHI DAMUNI and LESTER J. REED

The mammalian branched-chain α-keto acid dehydrogenase (BCKDH) complex is located in mitochondria, within the inner membrane-matrix compartment. Phosphorylation and concomitant inactivation of the BCKDH (E1) component of the multienzyme complex are catalyzed by a kinase that copurifies with the complex.[1-5] Dephosphorylation and concomitant reactivation of the complex are catalyzed by a specific phosphatase[5] that is only loosely associated with the complex. The activity of BCKDH phosphatase is regulated by a potent heat- and acid-stable protein inhibitor.[6]

Assay Methods

BCKDH Phosphatase

Principle. Assay of BCKDH phosphatase activity is based on measurement of the initial rate of release of ^{32}P-labeled phosphoryl groups from enzyme complex that has been phosphorylated with [γ-^{32}P]ATP.[6] To ensure that rates of dephosphorylation are linear with respect to time and the amount of phosphatase, the extent of dephosphorylation of substrate is restricted to ≤20%.

Reagents. Dilution buffer: BCKDH phosphatase is diluted in 50 mM imidazole chloride (Sigma Chemical Co., St. Louis, MO, grade III), pH 7.3, 10% glycerol, 0.1 mM ethylenediaminetetraacetic acid (EDTA), 0.1 mM ethylene glycol bis(β-aminoethyl ether)-N, N'-tetraacetic acid (EGTA), 0.1 mM phenylmethylsulfonyl fluoride (PMSF), 1 mM benzamidine, 1 mM dithiothreitol (buffer B) containing bovine serum albumin (1 mg/ml).

Preparation of Substrate. ^{32}P-Labeled, inactive BCKDH complex is

---

[1] H. R. Fatania, K. S. Lau, and P. J. Randle, *FEBS Lett.* **132**, 285 (1981).

[2] R. Odessey, *Biochem. J.* **204**, 353 (1982).

[3] R. Paxton and R. A. Harris, *J. Biol. Chem.* **257**, 14433 (1982).

[4] R. Lawson, K. G. Cook, and S. J. Yeaman, *FEBS Lett.* **157**, 54 (1983).

[5] Z. Damuni, M. L. Merryfield, J. S. Humphreys, and L. J. Reed, *Proc. Natl. Acad. Sci. U.S.A.* **81**, 4335 (1984).

[6] Z. Damuni, J. S. Humphreys, and L. J. Reed, *Proc. Natl. Acad. Sci. U.S.A.* **83**, 285 (1986).

prepared by incubating for 15 min at 30° a solution containing 5–10 mg of BCKDH complex (prepared as described below), buffer B, 1 mM PMSF, 2 mM MgCl$_2$, and 0.1 mM [γ-^{32}P]ATP (2–4 × 10^6 cpm/nmol, New England Nuclear, Boston, MA) in a final volume of 0.5 ml. The solution is filtered through a column (10 × 1.5 cm) of fine Sephadex G-50 equilibrated and developed with buffer B. The phosphorylated BCKDH complex contains 4–6 nmol of ^{32}P-labeled phosphoryl groups per milligram of protein. All the radioactivity is localized in the E1α subunit.[7] The preparations are free of endogenous phosphatase activity.

Procedure. Prior to assay, BCKDH phosphatase is diluted with buffer B containing bovine serum albumin (1 mg/ml). The diluted enzyme (30 μl) and additions (10 μl) as noted in the text are incubated for 5 min at 30° in a plastic microcentrifuge tube. The reaction is initiated with 10 μl of ^{32}P-labeled BCKDH complex (2.5 mg/ml; 1.2–2.4 × 10^7 cpm/mg). After 5 min at 30°, 0.1 ml of 10% trichloroacetic acid is added. The mixture is centrifuged at 10,000 rpm for 2 min in an Eppendorf microcentrifuge. A 0.12-ml aliquot of the supernatant fluid is transferred to a plastic microcentrifuge tube containing 1 ml of scintillant (Amersham, Arlington Heights, IL). The tube is placed in a plastic vial and radioactivity is determined.

Definition of Unit. One unit of BCKDH phosphatase activity is the amount of enzyme which catalyzes the release of 1.0 nmol of [^{32}P]P$_i$ per minute in the standard assay. Protein is determined as described by Bradford.[8]

BCKDH Phosphatase Inhibitor Protein

Principle. Assay of inhibitor protein activity is based on measurement of BCKDH phosphatase activity, as described above, in the absence and presence of inhibitor protein. In the standard assay, inhibition is proportional to the amount of inhibitor protein up to ~60%.

Procedure. The reaction mixture contains 0.001–0.003 unit of BCKDH phosphatase in 30 μl of buffer B containing bovine serum albumin (1 mg/ml), and inhibitor protein in 10 μl of the same buffer. The solution is incubated for 5 min at 30° in a plastic microcentrifuge tube. The reaction is initiated with 10 μl of ^{32}P-labeled BCKDH complex (2.5 mg/ml) and the assay is continued as described above for BCKDH phosphatase. One control tube lacks inhibitor protein, and another lacks both inhibitor protein and BCKDH phosphatase.

---

[7] Z. Damuni and L. J. Reed, unpublished observations.
[8] M. M. Bradford, *Anal. Biochem.* **72**, 248 (1976).

Definition of Unit. One unit of inhibitor protein is the amount of protein that inhibits 1 unit of BCKDH phosphatase by 50% in the standard assay.

Purification of BCKDH Phosphatase

Buffers Used

Buffer A: 0.1 M potassium phosphate, pH 7.3, 0.5 mM PMSF, 1 mM benzamidine, 1 mM EDTA, 1 mM EGTA,[9] 1 mM dithiothreitol

Buffer B: 50 mM imidazole chloride, pH 7.3, 10% glycerol, 0.1 mM EDTA, 0.1 mM EGTA, 0.1 mM PMSF, 1 mM benzamidine, 1 mM dithiothreitol

Buffer C: 50 mM imidazole chloride, pH 5.0, 1 mM EDTA, 1 mM EGTA, 0.1 mM PMSF, 1 mM benzamidine, 0.01 mM benzyl-oxycarbonylphenylalanylalanine diazomethyl ketone (Z-Phe-AlaCHN₂), a thiol protease inhibitor,[10] 1 mM dithiothreitol, 30% ethylene glycol, 0.03% Brij 35

Buffer D: 50 mM imidazole chloride, pH 7.3, 1 mM EDTA, 1 mM EGTA, 0.1 mM PMSF, 1 mM benzamidine, 0.01 mM Z-Phe-AlaCHN₂, 1 mM dithiothreitol, 30% ethylene glycol, 0.03% Brij 35

Buffer E: 50 mM potassium phosphate, pH 7.3, 1 mM EDTA, 1 mM EGTA, 0.1 mM PMSF, 1 mM benzamidine, 0.01 mM Z-Phe-AlaCHN₂, 1 mM dithiothreitol, 30% ethylene glycol, 0.03% Brij 35

Purification Procedure

The purification procedure provides highly purified preparations of BCKDH phosphatase and near homogeneous preparations of BCKDH phosphatase inhibitor protein, BCKDH complex, and pyruvate dehydrogenase complex. The procedure is modified from that of Damuni *et al.*[5,6] Unless specified otherwise, all operations are performed at $2° - 5°$.

Step 1. Preparation of Mitochondrial Extract. Bovine kidney mitochondria are prepared and washed as described by Pettit and Reed,[11] with the following modifications. The mitochondria are washed once with 0.25 M sucrose, 10 mM potassium phosphate, pH 7.6, 0.1 mM EDTA; once with 14 mM 2-mercaptoethanol, 0.1 mM PMSF, 0.01 mM Z-Phe-AlaCHN₂, 1 mM benzamidine; and twice with 50 mM phosphate buffer, pH 7.1. The

[9] In recent runs, EGTA has been omitted from all buffers, without noticeable effect on the results.

[10] H. Watanabe, G. D. J. Green, and E. Shaw, *Biochem. Biophys. Res. Commun.* **89,** 1354 (1979).

[11] F. H. Pettit and L. J. Reed, this series, Vol. 89, p. 376.

washed mitochondrial paste is suspended in 50 mM phosphate buffer, pH 7.1, containing 0.1 mM PMSF, 0.01 mM Z-Phe-AlaCHN$_2$, and 1 mM benzamidine. The mixture is shell-frozen and thawed once. The thawed suspension is diluted with 0.6 volume of a solution containing 50 mM phosphate, pH 7.3, 0.1 mM PMSF, 0.01 mM Z-Phe-AlaCHN$_2$, 1 mM benzamidine, 0.1 mM EDTA, 0.1 mM EGTA, and 1 mM dithiothreitol. NaCl is added to a final concentration of 100 mM, the pH is adjusted to 7.3 with dilute KOH, and the mixture is stirred for 20 min. The mixture is centrifuged at 30,000 g for 30 min in a Beckman JA-14 rotor, and the pellets are discarded.

Step 2. Separation of BCKDH Complex from BCKDH Phosphatase and Its Inhibitor Protein. To the mitochondrial extract is added, with stirring, 0.08 volume of 50% (w/v) aqueous polyethylene glycol 8000 (J. T. Baker Co., Phillipsburg, PA). After 20 min, the precipitate is collected by centrifugation at 30,000 g for 15 min. This precipitate (A) contains essentially all of the BCKDH complex activity and the pyruvate dehydrogenase complex activity. Precipitate A is processed further as described below. The supernatant fluid contains BCKDH phosphatase and its inhibitor protein.

Step 3. Separation of BCKDH Phosphatase from Its Inhibitor Protein. The supernatant fluid from step 2 is diluted with 1 volume of a solution containing 0.1 mM EDTA, 0.1 mM EGTA, 0.1 mM PMSF, 1 mM benzamidine, and 1 mM dithiothreitol. This solution is passed at a flow rate of 5–10 liters/hr through a column (7 × 14.5 cm) of DEAE-cellulose (Whatman DE-52) equilibrated with buffer B in a Büchner funnel with a fritted disk. The effluent contains essentially all of the BCKDH phosphatase activity. The column is washed with about 20 liters of buffer B containing 0.1 M NaCl, and then with 200 ml of buffer B containing 0.3 M NaCl. Inhibitor protein is eluted with an additional 400–500 ml of buffer B containing 0.3 M NaCl.

Step 4. Polyethylene Glycol Precipitation. The effluent from the DEAE-cellulose column is adjusted to pH 7.3 by dropwise addition, with stirring, of 10% acetic acid. To this solution is added, with stirring, 0.18 volume of 50% polyethylene glycol. After 45 min, the precipitate is collected by centrifugation at 30,000 g for 15 min. This precipitate is suspended, by means of a glass/Teflon homogenizer, in 400 ml of buffer B containing 1 mM PMSF and 0.01 mM Z-Phe-AlaCHN$_2$. The mixture is clarified by centrifugation at 30,000 g for 15 min.

Step 5. First ADP-Sepharose Chromatography. The supernatant fluid from step 4 is applied to a column (2.5 × 5 cm) of N^6-(6-aminohexyl)-ADP-Sepharose (P-L Biochemicals, Piscataway, NJ, type 2) equilibrated with buffer B. The column is washed with buffer B until the absorbance of the effluent at 280 nm is less than 0.02 (~1.5 liters). The column is washed

TABLE I
PURIFICATION OF BCKDH PHOSPHATASE[a]

Step	Volume (ml)	Protein (mg)	Specific activity[b]	Recovery (%)
Mitochondrial extract	5,550	84,095	0.02[c]	100
DEAE-cellulose effluent	11,300	38,257	0.06	91
Polyethylene glycol precipitate	400	20,500	0.1	81
First ADP-Sepharose	43	69	13.5	55
Second ADP-Sepharose	23	2.8	158	26

[a] Approximately 22 kg of kidney cortex (≈ 2.5 kg of mitochondria, wet weight) are used. Mitochondria from ~ 11 kg of cortex are processed through the polyethylene glycol precipitation, and the active fractions from two runs are combined prior to the first ADP-Sepharose chromatography.

[b] Nanomoles of [^{32}P]P_i released per minute per milligram of protein.

[c] BCKDH phosphatase activity in mitochondrial extracts is determined after gel filtration on Sephadex G-50 to remove inorganic orthophosphate and other low-molecular-weight inhibitors, e.g., ATP, ADP. However, the activity of the phosphatase in the gel-filtered extracts is increased up to 4-fold by dilution,[6] due to the presence of inhibitor protein. Therefore, specific activity of BCKDH phosphatase in the gel-filtered extracts is determined at the highest possible dilution and the extent of dephosphorylation of substrate is restricted to 1–3%.

consecutively with 2 liters of buffer B containing 0.06 M NaCl and 500 ml of buffer B. BCKDH phosphatase is eluted with buffer B containing 10 mM MgCl$_2$.

Step 6. Second ADP-Sepharose Chromatography. The active fractions from step 5 are pooled, made 12 mM with respect to EDTA, and the solution is applied to a column (2.5 × 5.5 cm) of ADP-Sepharose equilibrated with buffer B. The column is washed with buffer B until the absorbance of the effluent approaches 0 (~ 500 ml). The column is washed with 1.5 liters of buffer B containing 1 mM ATP. BCKDH phosphatase is eluted with buffer B containing 10 mM ATP. A summary of the purification is presented in Table I.

Properties

Despite the extensive purification, the preparations of BCKDH phosphatase are not homogeneous. The phosphatase has an M_r of about 460,000 as estimated by gel permeation chromatography. This phosphatase is essentially inactive ($<0.1\%$) with ^{32}P-labeled phosphorylase a as substrate, but it shows some activity with ^{32}P-labeled pyruvate dehydrogenase complex, i.e., about 10% of the activity observed with ^{32}P-labeled BCKDH complex.

The activity of BCKDH phosphatase is not affected by protein phosphatase inhibitor-1 and inhibitor-2 at concentrations up to 50 and 300 nM, respectively.[12] This phosphatase is inhibited noncompetitively by the inhibitor protein described below.[6] The inhibition constant (K_i) is about 0.13 nM.

In contrast to pyruvate dehydrogenase phosphatase, which requires Mg^{2+} or Mn^{2+} and is markedly stimulated by Ca^{2+}, BCKDH phosphatase is active in the absence of divalent cations.[5] BCKDH phosphatase is inhibited by nucleoside tri- and diphosphates. Half-maximal inhibition is observed at the following concentrations (μM): GTP, 60; GDP, 200; ATP, 200; ADP, 400; UTP, 100; UDP, 250; CTP, 250; CDP, 400. These inhibitions are reversed completely by 2 mM Mg^{2+}. GTP is replaceable by guanosine 5'-(β,γ-imido)triphosphate. GMP, AMP, UMP, NAD^+, and NADH have little effect, if any, on BCKDH phosphatase activity at concentrations up to 1 mM. Heparin shows half-maximal inhibition at 2 μg/ml. CoA and various acyl-CoA compounds exhibit half-maximal inhibition at 150–300 μM. These inhibitions are not reversed by 2 mM Mg^{2+}. BCKDH phosphatase activity is stimulated 1.5- to 3-fold by protamine (3.6 μg/ml), poly(L-lysine) (3.6 μg/ml), poly(L-arginine) (3.6 μg/ml), and histone H3 (36 μg/ml). Spermine and spermidine are inactive at concentrations up to 2 mM. It is possible that there is a nucleotide-binding regulatory site or sites on BCKDH phosphatase and that its activity is modulated by the energy status of mitochondria. It is also possible that protamine and other basic polypeptides mimic an as-yet-unidentified regulator of the phosphatase activity.

Purification of Inhibitor Protein

Purification Procedure

Step 7. DEAE-Cellulose Chromatography at pH 5.0. The inhibitor protein preparation from step 3 is diluted with 2 volumes of a solution containing 1 mM EDTA, 1 mM EGTA, 0.1 mM PMSF, 1 mM benzamidine, 0.01 mM Z-Phe-AlaCHN$_2$, 1 mM dithiothreitol. The solution is applied to a column (2.5 × 12 cm) of DEAE-cellulose equilibrated with buffer B. The column is washed with 6 liters of buffer C[13] containing

---

[12] Z. Damuni, H. Y. L. Tung, and L. J. Reed, *Biochem. Biophys. Res. Commun.* **133**, 878 (1985).

[13] Inclusion of 30% ethylene glycol and 0.03% Brij 35 in buffers C, D, and E is necessary to stabilize the inhibitor protein in dilute solutions, particularly in the late stages of the purification.

0.05 M NaCl and is then developed with a 500-ml salt gradient from 0.05–0.5 M NaCl. The inhibitor protein is eluted at ~0.3 M NaCl.

Step 8. DEAE-Cellulose Chromatography at pH 7.3. The active fractions from step 7 are pooled and diluted with 3 volumes of buffer D. The solution is applied to a column (2.5 × 6 cm) of DEAE-cellulose equilibrated with buffer D. The column is washed with 400 ml of buffer D containing 0.05 M NaCl and developed with a 280-ml salt gradient from 0.05–0.5 M NaCl. The inhibitor protein is eluted at ~0.3 M NaCl.

Step 9. Hydroxylapatite Chromatography. The active fractions from step 8 are pooled and diluted with 1 volume of buffer D. The solution is applied to a column (2.5 × 3 cm) of hydroxylapatite (Calbiochem, San Diego, CA, high-resolution grade) equilibrated with buffer E. The column is washed with buffer E until the absorbance of the effluent at 280 and 260 nm approaches zero (about 1 liter). Inhibitor protein is eluted stepwise with buffer E containing 0.15 M potassium phosphate, pH 7.3 (~430 ml) and then with buffer E containing 0.3 M potassium phosphate (~260 ml). The activity is divided equally between the 0.15 and 0.3 M potassium phosphate eluates.[14]

Step 10. Sephacryl S-200 Chromatography. The 0.3 M phosphate eluate from step 9 is diluted with 8 volumes of buffer D and concentrated by using a small column (1 × 2.5 cm) of DEAE-cellulose equilibrated with buffer D. The column is washed with about 50 ml of buffer D, and the inhibitor protein is eluted with buffer D containing 0.4 M NaCl. The concentrated inhibitor protein (~1.5 ml) is applied to a column (2 × 90 cm) of Sephacryl S-200 equilibrated and developed with buffer E containing 0.2 M NaCl. The active fractions are combined and concentrated about 8-fold, prior to protein determination, using a small column of DEAE-cellulose as described above. A summary of the purification is presented in Table II.

Properties

The highly purified (> 100,000-fold) preparations of inhibitor protein show a major silver staining band of M_r ~36,000, which appears to be a closely spaced doublet, on SDS-polyacrylamide gel electrophoresis.[6] By gel-permeation chromatography on a calibrated column of Sephadex G-100, the apparent M_r of the inhibitor protein is estimated to be ~33,000. The activity of inhibitor protein is not affected by heating the protein at 90° for 5 min or by keeping the inhibitor at pH 2.0 for 1 hr at 30°.

---

[14] Purification through step 9 is effective in removing most of the low-molecular-weight proteins and nucleic acids.

TABLE II
PURIFICATION OF INHIBITOR PROTEIN[a]

Step	Volume (ml)	Protein (mg)	Specific activity[b]	Recovery (%)
Mitochondrial extract	5,550	84,095	0.19[c]	100
DEAE-cellulose, pH 7.3	865	4,481	4.2	118
DEAE-cellulose, pH 5.0	110	59	119	44
DEAE-cellulose, pH 7.3	25	20.3	212	27
Hydroxylapatite	260	7.2	272	12
Sephacryl S-200	8	0.029[d]	22,860	4

[a] From 22 kg of kidney cortex.
[b] Units per milligram of protein.
[c] Inhibitor protein activity in mitochondrial extracts is determined after heat treatment (90° for 5 min) and gel filtration on Sephadex G-50.
[d] Protein is determined after concentrating the solution about 8-fold as described in the text.

Inhibition of BCKDH phosphatase by inhibitor protein is noncompetitive. The apparent K_i, determined from a Henderson plot, is 0.13 nM.[6] This protein inhibitor is specific for BCKDH phosphatase. The inhibitor does not affect the activity of protein phosphatase-1, protein phosphatase-2, or pyruvate dehydrogenase phosphatase at concentrations up to 10 nM.[12] The inhibitor protein at 60 ng/ml has no effect on the activity of BCKDH complex or its endogenous kinase.[6]

Inhibitor protein activity is reversed by Mg^{2+}.[6] Half-maximal reversal is obtained at ~1.3 mM, near physiological concentration. Reversal of inhibition is also obtained with the polyamine spermine, with 50% reversal at ~0.1 mM. Basic polypeptides such as protamine sulfate, poly(L-lysine), and poly(L-arginine) also reverse the inhibition, with 50% reversal at ~36 μg/ml.

Purification of BCKDH Complex

To obtain BCKDH complex that contains active BCKDH kinase, it is necessary to maintain a pH above 7.0, i.e., 7.0–7.3, in all operations and to include protease inhibitors in all buffers. BCKDH complex activity is assayed spectrophotometrically by measurement of NADH production in the presence of excess dihydrolipoamide dehydrogenase and with α-ketoisovalerate as substrate.[15]

[15] F. H. Pettit, S. J. Yeaman, and L. J. Reed, *Proc. Natl. Acad. Sci. U.S.A.* **75**, 4881 (1978).

Purification Procedure

Polyethylene glycol precipitate A from step 2 is suspended, by means of a glass/Teflon homogenizer, in 250 ml of buffer A containing 0.01 mM Z-Phe-AlaCHN$_2$. The suspension is clarified by centrifugation at 30,000 g for 20 min. The supernatant fluid is diluted with buffer A to 4 mg of protein per milliliter, and 0.09 volume of 25% polyethylene glycol is added, with stirring. After 20 min, this mixture is centrifuged at 30,000 g for 15 min, and the precipitate is discarded. To the supernatant fluid is added 0.5 mM PMSF and 0.01 mM Z-Phe-AlaCHN$_2$ (final concentrations). The solution is warmed to 20°, and 0.06 volume of 25% polyethylene glycol is added, with stirring. After 20 min, the mixture is centrifuged at 30,000 g and 20° for 15 min. The precipitate is discarded. To the supernatant fluid is added again 0.5 mM PMSF and 0.01 mM Z-Phe-AlaCHN$_2$ (final concentrations), and the solution is cooled to 5° and stirred for 20 min. The mixture is centrifuged at 30,000 g for 20 min. The precipitate contains 80–90% of the BCKDH complex activity, and the supernatant fluid contains 80–90% of the pyruvate dehydrogenase complex activity. The pyruvate dehydrogenase complex is purified further as described[6] to obtain near homogeneous preparations of this complex.

The polyethylene glycol precipitate is extracted with about 5 ml of buffer A containing 0.01 mM Z-Phe-AlaCHN$_2$. After clarification, the solution is applied to a column (2.6 × 6 cm) of hydroxylapatite equilibrated with buffer A containing 0.03 M potassium phosphate, pH 7.3. The column is washed with 1.5 liters of the same buffer and then with 1.5 liters of buffer A containing 0.15 M potassium phosphate. BCKDH complex is eluted with buffer A containing 0.35 M potassium phosphate. The active fractions are combined and centrifuged at 144,000 g for 4 hr in a Beckman type 40 rotor to sediment the BCKDH complex. The pellets are dissolved in buffer B. The purified BCKDH complex has a specific activity of 8–13 μmol of NADH produced per minute per milligram of protein, and the overall recovery is 20–50%. These preparations are apparently homogeneous, as shown by SDS-polyacrylamide gel electrophoresis.

[43] Isolation and Properties of the Branched-Chain 2-Keto Acid and Pyruvate Dehydrogenase Multienzyme Complex from *Bacillus subtilis*

By RICHARD N. PERHAM and PETER N. LOWE

The oxidative decarboxylation of 2-keto acids in almost all organisms is brought about by multienzyme complexes of unusual complexity and sophistication of mechanism.[1-3] Separate multienzyme complexes which convert pyruvate into acetyl-CoA, 2-oxoglutarate into succinyl-CoA, and the branched-chain 2-keto acids (as part of the catabolism of valine, leucine, and isoleucine) into the corresponding acyl-CoA, have been purified and studied from numerous sources. In general, these complexes coexist in any given organism and are reasonably specific for their substrates. For example, the distinct branched-chain 2-keto acid dehydrogenase complexes of ox kidney[4] and *Pseudomonas aeruginosa*[5] have a low activity with pyruvate as substrate and do not act on 2-oxoglutarate.

Bacillus subtilis, unlike *Escherichia coli,* has a requirement for branched-chain fatty acids, such as isovalerate (3-methylbutyrate) and isobutyrate (2-methylpropionate), to act as starter units for the synthesis of branched-chain lipids for the cell membrane.[6] The branched-chain 2-keto acid dehydrogenase complex is therefore essential for normal growth of *B. subtilis,* whereas *E. coli* has no such enzyme and cannot grow on branched-chain amino acids, e.g., leucine. Similarly, the production of acetyl-CoA by the pyruvate dehydrogenase reaction is essential for normal growth of *B. subtilis.*[7-10]

During a study of wild-type and mutant forms of the pyruvate dehydrogenase multienzyme complex of *B. subtilis,*[10] it became apparent that

[1] L. J. Reed, *Acc. Chem. Res.* **7**, 40 (1974).

[2] R. N. Perham, *in* "Frontiers in Biochemical and Biophysical Studies of Proteins and Membranes" (T.-Y. Liu, S. Sakakibara, A. N. Schechter, K. Yagi, and K. Yasunobu, eds.), p. 249. Elsevier, New York, 1983.

[3] R. N. Perham, L. C. Packman, and S. E. Radford, *Biochem. Soc. Symp.* **54**, 67 (1987).

[4] K. G. Cook, A. P. Bradford, and S. J. Yeaman, *Biochem. J.* **225**, 731 (1985).

[5] V. McCully, G. Burns, and J. R. Sokatch, *Biochem. J.* **233**, 737 (1986).

[6] T. Kaneda, *Can. J. Microbiol.* **19**, 87 (1973).

[7] W. Klofat, G. Picciolo, E. W. Chappelle, and E. Freese, *J. Biol. Chem.* **244**, 3270 (1969).

[8] U. Fortnagel and E. Freese, *J. Bacteriol.* **95**, 1431 (1968).

[9] E. Freese and U. Fortnagel, *J. Bacteriol.* **99**, 745 (1969).

[10] J. A. Hodgson, P. N. Lowe, and R. N. Perham, *Biochem. J.* **211**, 463 (1983).

the pyruvate and branched-chain 2-keto acid dehydrogenase activities of this organism copurified. All the structural and genetic evidence points to these two complexes being one and the same, which is an economical solution to the need to supply two different sets of essential metabolites.[11] It is possible that the two complexes are also identical in *Bacillus stearothermophilus.*[11]

This article describes the purification of the pyruvate-branched chain 2-keto acid dehydrogenase complex of *B. subtilis,* based on the simple method described by Hodgson *et al.*[10] Some of its characteristic molecular and catalytic properties[10-12] are summarized.

Assays for 2-Keto Acid Dehydrogenase Multienzyme Complexes and Constituent Enzyme Activities

Overall Complex Activity

Principle. Reduction of NAD+ in the presence of 2-keto acid and CoASH, and the cofactor thiamin pyrophosphate, is monitored continuously in a spectrophotometer at 340 nm.

Materials. Solution A contains 200 mg of NAD+, 1 ml of 105 mM $MgCl_2$, 1 ml of 21 mM thiamin pyrophosphate, and 5.2 ml of 1 M KH_2PO_4/K_2HPO_4, pH 8.0, and is made up with water to a final volume of 100 ml. It can be stored for several months at $-20°$.

Solution B contains 0.13 mM coenzyme A dissolved in a solution of 130 mM cysteine hydrochloride. The cysteine acts as a preservative to keep the coenzyme A in the reduced form. Solution B is stored in small aliquots at $-20°$ for later use.

Solution C consists of 100 mM sodium 2-oxoisovalerate, 100 mM sodium 2-oxoglutarate, or 100 mM sodium pyruvate dissolved in water. It may be stored for several weeks at $-20°$.

Method. A stock assay solution containing 9.5 ml of solution A, 0.2 ml of solution B, and 0.2 ml of solution C is mixed shortly before use and warmed to 30°. Samples (0.99 ml) are placed in cuvettes kept at 30° and the reaction is initiated by the addition of enzyme (0.02–0.5 units). The reaction is followed at 340 nm. Rates are calculated using $\epsilon_{340} = 6,220$ liter mol^{-1} cm^{-1}. One unit of enzyme activity is that amount of enzyme which causes a rate of reduction of NAD+ of 1 μmol/min under these conditions.

Comments. The rate of increase of A_{340} is usually linear except in crude extracts where the initial rise in absorbance is followed by a decrease,

[11] P. N. Lowe, J. A. Hodgson, and R. N. Perham, *Biochem. J.* **215,** 133 (1983).
[12] P. N. Lowe, J. A. Hodgson, and R. N. Perham, *Biochem. J.* **225,** 249 (1985).

presumably caused by contaminating enzyme activities that oxidize NADH. Thus, activity measurements in a crude extract can only be approximate, and are likely to be underestimates.

Pyruvate dehydrogenase, 2-oxoglutarate dehydrogenase, or 2-oxoisovalerate dehydrogenase complex activities can be measured independently by using the three different formulations of solution C containing the relevant substrate.

Dihydrolipoamide Dehydrogenase Activity

Reduction of NAD$^+$ by dihydrolipoamide dehydrogenase is assayed as described by Danson et al.[13] DL-Dihydrolipoamide is prepared by the reduction of lipoamide with NaBH$_4$.[14] The assay mixture consists of 0.95 ml solution A (see above), 0.02 ml of 20 mM dihydrolipoamide dissolved in ethanol, and 0.03 ml of sample or water. Production of NADH is monitored at 30° at 340 nm.

2-Keto Acid Decarboxylase Activity

Branched-chain 2-keto acid or pyruvate decarboxylase (E1) activities are assayed by the reduction of 2,6-dichlorophenolindophenol in the presence of the relevant 2-keto acid, as described by Lowe et al.[15] Enzyme is added to 0.98 ml of assay mixture containing 0.1 M potassium phosphate, pH 7.0, 100μM dichlorophenolindophenol, 0.2 mM thiamin pyrophosphate, and 1 mM MgCl$_2$ at 30°. The reaction is started by addition of 20 μl of solution C (see above). The decrease in A_{600} is followed.

Protein Determination

In cell-free extracts, protein is conveniently determined by the Biuret method.[16] As purification proceeds, it becomes more accurate and convenient to determine protein by precipitation with trichloroacetic acid, followed by a modified Lowry procedure.[17]

Purification of Branched-Chain 2-Keto Acid Dehydrogenase Multienzyme Complex

Materials. Lysozyme hydrochloride (egg white, grade VI) and ribonuclease A (ox pancreas, Type 1-AS) are obtained from Sigma Chemical Co.

[13] M. J. Danson, E. A. Hooper, and R. N. Perham, *Biochemistry* **175**, 193 (1978).
[14] L. J. Reed, F. R. Leach, and M. Koike, *J. Biol. Chem.* **232**, 123 (1958).
[15] P. N. Lowe, F. J. Leeper, and R. N. Perham, *Biochemistry* **22**, 150 (1983).
[16] A. G. Gornall, C. J. Bardawill, and M. M. David, *J. Biol. Chem.* **177**, 751 (1949).
[17] G. L. Peterson, *Anal. Biochem.* **83**, 346 (1977).

Deoxyribonuclease I (bovine pancreas, grade II) is obtained from Boehringer Mannheim.

Growth of Bacteria

Bacillus subtilis strain 168 (auxotrophic marker *trp*C2) is subcultured on tryptose-blood agar base (Difco). Cells are grown in L-broth [10 g of Bacto-tryptone (Difco), 5 g yeast extract (Difco), and 5 g of NaCl per liter of distilled water]. Prewarmed medium (1 liter in 2-liter conical flasks) is inoculated to A_{650} of 0.05 with an exponentially growing culture of the bacterium, and incubated at 37° with vigorous aeration until the culture reaches late exponential phase (A_{650} of 1.0–1.5). The grown cultures are chilled in ice and harvested within 3 hr by centrifugation at 4° at 8000 g for 15 min. The cell paste is stored at −20°. Approximately 3 g of wet cell paste per liter of medium is obtained. It is important that cultures should not be allowed to reach stationary phase as a low yield of 2-keto acid dehydrogenase complex will be obtained, probably caused by damage due to endogenous proteinases.

All subsequent operations are performed at 4°, unless otherwise stated. The complex purification can be followed by assaying for either pyruvate or branched-chain 2-keto acid dehydrogenase complex activity. The pyruvate dehydrogenase complex activity is the larger and therefore in some ways easier to measure, but 2-oxoisovalerate can be used as substrate perfectly well.

Step 1. Preparation of Cell-Free Extract

Bacillus subtilis cell paste is suspended (1 ml of buffer added to 0.5 g of cell paste) in 50 mM sodium phosphate buffer, pH 7.0, containing 5 mM EDTA, 0.15 mM phenylmethylsulfonyl fluoride, lysozyme (6 mg/ml), and deoxyribonuclease I (5 μg/ml), and stirred for 2 hr. Centrifugation for 15 min at 15,000 g generates a large pellet and a yellow supernatant that contains no pyruvate dehydrogenase complex activity. The pellet is washed by suspending it in 50 mM sodium phosphate buffer, pH 7.0, containing 5 mM EDTA and 0.15 mM phenylmethylsulfonyl fluoride and recovered by centrifuging again. Both supernatants are discarded.

The pellet is resuspended in the starting volume of 50 mM sodium phosphate buffer, pH 7.0, containing 5 mM EDTA, 0.15 mM phenylmethylsulfonyl fluoride, and 5 μg of deoxyribonuclease I/ml. The suspension is kept in an ice bath while being sonicated for periods of 30 sec with a Dawe Soniprobe type 7532A at an output of 80–100 W. Intervals of 5 min are allowed between periods of sonication to ensure that the temperature of the suspension does not rise above 10°. After each sonication, a sample

(0.1 ml) of the suspension is centrifuged in an Eppendorf microcentrifuge at 10,000 g for 2 min. The supernatant is assayed for pyruvate dehydrogenase complex activity. When this reaches its maximum value (after two to five sonication periods), the whole extract is centrifuged for 1 hr at 50,000 g (Beckman SW27 rotor, 20,000 rpm). The yellow cell-free extract containing pyruvate (or branched-chain) dehydrogenase complex activity is retained and the small pellet is discarded.

Step 2. Ribonuclease Treatment

This step removes a contaminant with a high A_{260}/A_{280} ratio, probably 70S ribosomes. Ribonuclease (1 mg/ml in water) is boiled for 7 min to ensure destruction of any contaminating proteinase activity and then cooled at 4° for 12 hr. It is added to the enzyme sample at a final concentration of 32 μg/ml. The mixture is kept at 10–15° for 70 min, during which time a fine white precipitate forms. The precipitate is removed by centrifugation for 25 min at 30,000 g (Beckman 42.1 rotor at 20,000 rpm).

Step 3. Ultracentrifugation

The supernatant is diluted to 5–10 mg of protein/ml and layered on top of a solution of sucrose (12.5% w/v) that itself is resting on a cushion of sucrose (70% w/w) about 10 mm deep at the bottom of a centrifuge tube. All solutions are made up in 50 mM sodium phosphate buffer/5 mM EDTA, pH 7.0. Centrifugation is performed for 5 hr at 90,000 g (Beckman SW27 rotor, 26,000 rpm) or for 3–4 hr at 150,000 g (Beckman 42.1 rotor, 40,000 rpm). The brown protein band formed at the interface between the 12.5 and 70% sucrose layers is drawn off. It contains > 90% of the pyruvate (or 2-oxoisovalerate) dehydrogenase complex activity originally loaded. (The supernatant is retained if it is desired to purify free dihydrolipoamide dehydrogenase; see below.) Sucrose is removed from the enzyme sample by dialysis for 2 hr against 50 mM sodium phosphate buffer/5 mM EDTA, pH 7.0.

Step 4. Gel Filtration

The enzyme complex is next loaded on to a column of Sepharose CL-2B, bed volume approximately 200 times the sample volume, and eluted with 20 mM sodium phosphate buffer/2 mM EDTA, pH 7.0. An opalescent contaminant, excluded from the gel matrix, emerges first and is followed by the 2-keto acid dehydrogenase complex activity, which emerges approximately halfway between V_o and V_t. Fractions containing pyruvate (or 2-oxoisovalerate) dehydrogenase complex activity are pooled

and concentrated by ultracentrifugation for 3 hr at 150,000 g (Beckman 42.1 rotor, 40,000 rpm).

Step 5. Ammonium Sulfate Fractionation

The protein concentration is adjusted to 10 mg/ml by adding 50 mM sodium phosphate buffer, pH 7.0, containing 5 mM EDTA. The enzyme solution is stirred while solid $(NH_4)_2SO_4$ is gradually added to 45% saturation (277 g/liter) and the pH is maintained at 7.0 by adding 0.75 M NaOH. The mixture is stirred for a further 30 min and any white precipitate is then removed by centrifugation for 15 min at 12,000 g (Sorvall SS34 rotor, 10,000 rpm). The supernatant retains almost all the pyruvate (or 2-oxoisovalerate) dehydrogenase complex activity.

The saturation of $(NH_4)_2SO_4$ in the supernatant is further increased by three increments of 5% each (33 g/liter), as above. After each increase the suspension is stirred for 30 min and the precipitate collected by centrifugation. When the supernatants are assayed for pyruvate dehydrogenase complex activity, it is generally found that 85–90% of the enzyme activity has been precipitated in two of the 5% additions. The two appropriate precipitates are combined and resuspended in the minimum volume of 50 mM sodium phosphate buffer, pH 7.0, containing 5 mM EDTA. Resuspension is immediate, yielding a clear yellow solution with >80% recovery of enzymatic activity. The solution is dialyzed overnight against 2 × 1 liter of the same buffer.

For small-scale preparations, when less than 10 ml of cell-free extract is available, $(NH_4)_2SO_4$ is added as a saturated solution in 50 mM sodium phosphate buffer/5 mM EDTA, adjusted to pH 7.0.

The enzyme solution is frozen rapidly at −70°. It can then be stored at −20° for several months without loss of activity. A typical purification procedure is summarized in Table I.

Purification of Dihydrolipoamide Dehydrogenase from *Bacillus subtilis*

Dihydrolipoamide dehydrogenase can readily be purified from the supernatants obtained from the ultracentrifugation through 12.5% sucrose (step 3 above) used to purify the multienzyme complex. Some 20–50% of the total dihydrolipoamide dehydrogenase activity in the cell-free extract appears to be uncomplexed (cf. Ref. 18) and is found there.

The dihydrolipoamide dehydrogenase is concentrated from this solution by addition of $(NH_4)_2SO_4$ to 75% saturation. The yellow precipitate is

[18] C. J. Lusty and T. P. Singer, *J. Biol. Chem.* **239**, 3733 (1964).

TABLE I
PURIFICATION OF PYRUVATE BRANCHED-CHAIN 2-KETO ACID DEHYDROGENASE COMPLEX FROM WILD-TYPE B. subtilis[a]

Step	PDHC (units)	PDHC (units/mg)	OIVDHC (units)	OIVDHC (units/mg)	Ratio PDHC/OIVDHC	E3 (units/mg)	OGDHC (units)	Ratio Pyr-E1/OIV-E1
1. Cell-free extract	1220	0.30	77	0.019	15.9	0.80	57	1.41
3. Ultracentrifugation pellet	1240	1.50	99	0.12	12.5	3.2	52	1.28
4. Sepharose-2B pool	885	6.4	60	0.46	13.9	10.3	0	1.25
5. (NH$_4$)$_2$SO$_4$ precipitate	656	10.0	43	0.70	14.3	14.1	0	1.18
Yield	54%		56%					

[a] PDHC, pyruvate dehydrogenase complex; OIVDHC, 2-oxoisovalerate dehydrogenase complex; OGDHC, 2-oxoglutarate dehydrogenase complex; Pyr-E1, activity of E1 component (reduction of 2,6-dichlorophenolindiphenol) in the presence of pyruvate; OIV-E1, activity of E1 component (reduction of 2,6-dichlorophenolindophenol) in the presence of 2-oxoisovalerate.

collected by centrifugation for 15 min at 14,000 *g* and dissolved in 0.05 the initial volume of 20 m*M* sodium phosphate buffer/5 m*M* EDTA, pH 7.0. After dialysis against the same buffer, the enzyme solution is treated with trypsin (final concentration 0.1 mg/ml) for 3 hr at 22°. The resistance of dihydrolipoamide dehydrogenase to proteolysis has been used previously to facilitate its purification from *E. coli*[19] and *B. stearothermophilus*.[20] The trypsin treatment leads to no loss of E3 activity; the trypsin is then inhibited by adding soybean trypsin inhibitor (final concentration 70 μg/ml).

The trypsin-treated sample (1.2 ml) is fractionated on a column of Sephacryl S-200 (37 cm × 1 cm) equilibrated with the same buffer, and the dihydrolipoamide dehydrogenase activity is eluted as a symmetrical peak (V_o = 5 ml, V_e = 13 ml) at a position consistent with an M_r of about 110,000. Fractions containing the enzyme activity are pooled and concentrated by $(NH_4)_2SO_4$ precipitation as above, except that the enzyme is redissolved in and dialyzed against 10 m*M* potassium phosphate buffer, pH 7.0.

At this stage, the preparation contains one major band (apparent M_r ~ 55,000) and several minor bands when analyzed by SDS-polyacrylamide gel electrophoresis. Contaminating protein can be removed by passing the enzyme preparation through a column of spheroidal hydroxylapatite (2 cm × 1.5 cm) equilibrated with 10 m*M* potassium phosphate buffer, pH 7.0. The dihydrolipoamide dehydrogenase activity is not retained by the column and emerges free of other proteins. The specific catalytic activity of the purified enzyme is 125 units/mg of protein.

Properties of the Pyruvate-Branched Chain 2-Keto Acid
Dehydrogenase Multienzyme Complex

Ultracentrifugation

The value of $s°_{20,w}$ for the *B. subtilis* complex is estimated to be 73S by analytical ultracentrifugation.[10] At higher protein concentrations (≥ 2mg/ ml) there is also a second, smaller peak that sediments more rapidly ($s°_{20,w}$ = 106S). This probably represents a dimer of the major peak, as observed also for the pyruvate dehydrogenase complexes from *E. coli*[21] and *B. stearothermophilus*.[22]

[19] J. P. Brown and R. N. Perham, *FEBS Lett.* **26**, 221 (1972).
[20] C. E. Henderson, R. N. Perham, and J. T. Finch, *Cell* **17**, 85 (1979).
[21] M. J. Danson, G. Hale, P. Johnson, R. N. Perham, J. Smith, and P. Spragg, *J. Mol. Biol.* **129**, 603 (1979).
[22] C. E. Henderson and R. N. Perham, *Biochem. J.* **189**, 161 (1980).

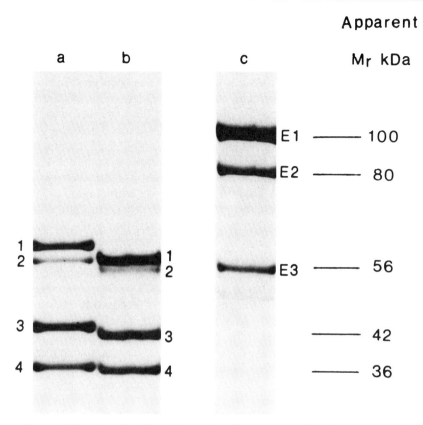

FIG. 1. SDS-polyacrylamide gel electrophoresis of pyruvate dehydrogenase complexes from (a) *B. subtilis,* (b) *B. stearothermophilus,* and (c) *E. coli.* Samples of purified complexes were analyzed by electrophoresis in phosphate-buffered SDS/7.5% (w/v) polyacrylamide gel. Reproduced by permission from Hodgson *et al.*[10] Copyright © 1983 by The Biochemical Society, London.

Subunit Composition

The purified enzyme shows four protein bands when subjected to SDS-polyacrylamide gel electrophoresis (Fig. 1). The apparent M_r values for the *B. subtilis* polypeptide chains, estimated on phosphate-buffered 7.5% polyacrylamide gels, are band 1, 59,500; band 2, 55,000; band 3, 42,500; band 4, 36,000. On a Tris/glycine-buffered system (12.5 or 15% polyacrylamide), somewhat different M_r values are obtained: 66,500, 63,500, 42,500, and 38,000 for bands 1–4, respectively. The banding pattern is similar to, but not identical with, that observed for the *B. stearothermophilus* enzyme[22] and is quite different from the three-band

pattern of the pyruvate dehydrogenase complex of *E. coli* (Fig. 1), the apparent M_r values for which are approximately 100,000, 83,000, and 56,000 (see Ref. 21 and references therein). The apparent M_r estimated for the E2 chain of the *E. coli* complex is known to be anomalously high,[23] its true value (based on the DNA sequence of its structural gene[24]) being 67,000. This is due, at least in part, to the curious amino acid sequence and domain structure of this polypeptide chain.[3,25] It is likely that the apparent M_r of the *B. subtilis* E2 chain is also overestimated, but to a lesser extent (perhaps 10%).

In the electron microscope, negatively stained images of the *B. subtilis* complex resemble those of the *B. stearothermophilus* enzyme,[10] indicative of the icosahedral symmetry reported for the E2 core of the latter complex.[20,22]

Catalytic Properties and Role of Subunits

Band 1 can be selectively labeled with N-ethyl[2,3-^{14}C]maleimide in the presence of pyruvate and absence of CoA (Fig. 2) and is therefore thought to be the E2 (lipoate acyltransferase) chain.[10] Band 2 has the same electrophoretic mobility as purified dihydrolipoamide dehydrogenase and is therefore identified as the E3 chain.[10]

Bands 3 and 4 are assigned to E1α and E1β chains, by analogy with the mammalian 2-keto acid dehydrogenase complexes.[1] This view is supported by the observation that selective proteolysis of band 3 (E1α) is accompanied by an inhibition of E1 (pyruvate or branched-chain 2-keto acid decarboxylase) activity.[12] Moreover, the branched-chain 2-keto acid dehydrogenase complex from the *B. subtilis* *ace* mutant 61141R[10] loses the subunits that correspond with bands 3 and 4 during purification, and simultaneously loses its overall complex and E1 activities. Evidence on the exact roles of bands 3 and 4 in catalyzing E1 activity is lacking.

The *B. subtilis* branched-chain 2-keto acid dehydrogenase complex catalyzes the oxidative decarboxylation of the three branched-chain 2-keto acids, 2-oxoisovalerate, 4-methyl-2-oxopentanoate, and 3-methyl-2-oxopentanoate, with activities in the proportions 1 : 0.07 : 0.09 (each substrate 2 mM). The K_m for 2-oxoisovalerate is 1.3 mM, with 2.5 mM NAD$^+$ and 0.13 mM CoA. The same enzyme complex also catalyzes the oxidative

[23] D. M. Bleile, P. Munk, R. M. Oliver, and L. J. Reed, *Proc. Natl. Acad. Sci. U.S.A.* **76,** 4385 (1979).
[24] P. E. Stephens, M. G. Darlison, H. M. Lewis, and J. R. Guest, *Eur. J. Biochem.* **133,** 481 (1983).
[25] J. R. Guest, H. M. Lewis, L. D. Graham, L. C. Packman, and R. N. Perham, *J. Mol. Biol.* **185,** 743 (1985).

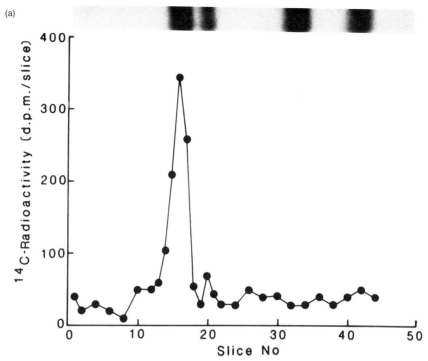

FIG. 2. Assignment of catalytic activities to bands 1 and 2 of pyruvate-branched chain 2-keto acid dehydrogenase complex from *B. subtilis*. (a) The complex was treated with *N*-ethyl[2,3-^{14}C]maleimide in the presence of pyruvate at pH 7. The inhibited enzyme was subjected to SDS-polyacrylamide gel electrophoresis and the gel was sliced and counted for ^{14}C radioactivity. (b) The complex and dihydrolipoamide dehydrogenase were subjected to SDS-polyacrylamide gel electrophoresis and the Coomassie Blue-stained gels were scanned at 550 nm. Track 1, enzyme complex; track 2, mixture of complex and dihydrolipoamide dehydrogenase; track 3, dihydrolipoamide dehydrogenase. Reproduced by permission from Hosgson *et al.*[10] Copyright © 1983 by The Biochemical Society, London.

decarboxylation of pyruvate with $K_m = 0.33$ mM. No reaction is observed with 2-oxoglutarate.[11]

Comparison with Other 2-Keto Acid Dehydrogenase Complexes

The *B. subtilis* pyruvate-branched chain 2-keto acid dehydrogenase complex closely resembles the pyruvate dehydrogenase complex from *B. stearothermophilus*, which can also catalyze the oxidative decarboxylation of branched-chain 2-keto acids[11] and the pyruvate dehydrogenase complexes from the mitochondria of eukaryotes, such as ox heart and kidney[1]

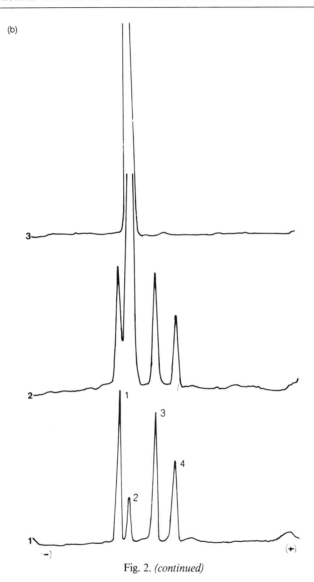

Fig. 2. *(continued)*

and yeast.[26] All these enzyme complexes consist of multiple copies of four types of subunit, based on an E2 core comprising 60 E2 chains arranged with icosahedral symmetry. However, the various complexes differ in their

[26] E. E. Keha, H. Ronft, and G.-B. Kresze, *FEBS Lett.* **145,** 289 (1982).

sensitivities to the proteinases trypsin and chymotrypsin, the *B. subtilis* enzyme resembling more closely the pyruvate dehydrogenase complex of ox heart in this regard.[12] The possibility that a fifth and additional component, protein X, may have a part to play in the mammalian pyruvate dehydrogenase complexes is currently under serious study.[27,28]

The pyruvate and 2-oxoglutarate dehydrogenase complexes of *E. coli* and the 2-oxoglutarate and branched-chain 2-keto acid dehydrogenase complexes of mammalian mitochondria are of a different structure, each being composed of multiple copies of three different types of subunit and based on an E2 core consisting of 24 polypeptide chains arranged with octahedral symmetry.[1,21,29,30] *E. coli* does not possess a branched-chain 2-keto acid dehydrogenase complex activity, but there is substantial sequence homology between the E2 chains of the *E. coli* and *B. stearothermophilus* pyruvate dehydrogenase complexes.[31]

Acknowledgments

Work in the authors' laboratory was supported by research grants from the Science and Engineering Research Council, U.K. We thank Dr J. A. Hodgson for his valuable collaboration.

[27] J. A. Hodgson, O. G. De Marcucci, and J. G. Lindsay, *Eur. J. Biochem.*, **158**, 595 (1986).
[28] M. Rahmatullah, J. M. Jilka, G. A. Radke, and T. E. Roche, *J. Biol. Chem.* **261**, 6515 (1986).
[29] R. N. Perham and G. C. K. Roberts, *Biochem. J.* **199**, 733 (1981).
[30] L. J. Reed, Z. Damuni, and M. L. Merryfield, *Curr. Top. Cell. Regul.* **27**, 41 (1985).
[31] L. C. Packman, A. Borges, and R. N. Perham, *Biochem. J.* **252**, 79 (1988).

[44] Purification of Branched-Chain Keto Acid Dehydrogenase and Lipoamide Dehydrogenase-Valine from *Pseudomonas*

By JOHN R. SOKATCH

Pseudomonas putida and *Pseudomonas aeruginosa* have been used extensively to study catabolic pathways because of the wide variety of compounds used by these organisms as sources of carbon.[1] Both species possess inducible pathways for catabolism of branched-chain amino acids which are induced by growth on these amino acids and repressed by

[1] R. Y. Stanier, N. J. Palleroni, and M. Doudoroff, *J. Gen. Microbiol.* **43**, 159 (1966).
[2] P. J. Sykes, G. Burns, J. Menard, K. Hatter, and J. R. Sokatch, *J. Bacteriol.*, **169**, 1619 (1987).

ammonium ion and glucose.[2] Branched-chain keto acid dehydrogenase is regulated independently of the other enzymes of these pathways, which facilitates its role as a common enzyme in the catabolism of valine, isoleucine, and leucine.[3] The actual inducers of branched-chain keto acid dehydrogenase are the branched-chain keto acids.[4]

Branched-chain keto acid dehydrogenase complexes from *P. putida* and *P. aeruginosa* are similar in structure and function. The complex from *P. putida* is composed of four polypeptides, with molecular weights of 37,000, 39,000, 47,000, and 49,000.[5] The 47,000 and 49,000 peptides have been identified as the E2 and E3 subunits, respectively.[5,6] The 37,000 and 39,000 peptides are assumed to be the E1 units. The corresponding molecular weights of the *P. aeruginosa* complex are 36,000, 42,000, 49,000, and 50,000.[6]

Both species are unique in that the E3 subunit is a specific lipoamide dehydrogenase, LPD-valine, which is induced during growth on branched-chain amino acids.[6,7] It appears that the sole function of LPD-Val is to serve as the E3 subunit of branched-chain keto acid dehydrogenase in pseudomonads. A second lipoamide dehydrogenase, LPD-Glc, which is produced during growth on either glucose or valine, serves as the E3 subunit of 2-ketoglutarate deyhrogenase,[7] the L-factor of the glycine oxidation complex,[8] and possibly pyruvate dehydrogenase, although there is no direct evidence of this. Branched-chain keto acid dehydrogenase of pseudomonads is regulated by L-valine, which increases the affinity of the E1 subunit for branched-chain keto acids.[5] There is no evidence of phosphorylation/dephosphorylation of the complex. The complex and LPD-Val can be isolated from either species relatively easily; however, E2 can more easily be isolated from the *P. aeruginosa* complex.[6]

Growth of Bacteria

Bacterial Strains

Pseudomonas putida strain PpG2 comes from the laboratory of I. C. Gunsalus, where it was originally designated strain PUG2.[9] It can be obtained from any of the laboratories currently using this strain. *Pseudo-*

[3] V. P. Marshall and J. R. Sokatch, *J. Bacteriol.* **110**, 1073 (1972).

[4] R. R. Martin, V. P. Marshall, J. R. Sokatch, and L. Unger, *J. Bacteriol.* **115**, 198 (1973).

[5] J. R. Sokatch, V. McCully, and C. M. Roberts, *J. Bacteriol.* **148**, 647 (1981).

[6] V. McCully, P. G. Burns, and J. R. Sokatch, *Biochem. J.* **233**, 737 (1986).

[7] J. R. Sokatch, V. McCully, J. Gebrosky, and D. J. Sokatch, *J. Bacteriol.* **148**, 639 (1981).

[8] J. R. Sokatch and G. Burns, *Arch. Biochem. Biophys.* **228**, 660 (1984).

[9] A. M. Chakrabarty, C. F. Gunsalus, and I. C. Gunsalus, *Proc. Natl. Acad. Sci. U.S.A.* **60**, 168 (1968).

monas aeruginosa strain PAO can be obtained from the American Type Culture Collection as strain number 156292.

Both organisms can be grown in synthetic media with branched-chain amino acids as the sole carbon and energy source. When valine is the carbon source, some isoleucine must be present in order to reverse the inhibitory effect of valine on acetolactate synthase which shuts down the synthesis of isoleucine. Branched-chain keto acid dehydrogenase is not produced in rich media such as L-broth or $2 \times$ YT medium.

Media

Valine/isoleucine medium[10]: 3 g L-valine, 1 g L-isoleucine, 100 ml Basal G, 990 ml water. Sterilize then add 10 ml Salts S

The compositions of the salt solutions are

Basal G: 43.5 g K_2HPO_4, 17 g KH_2PO_4, 21.4 g NH_4Cl, water to 1 liter

Salts S: 39.44 g $MgSO_4 \cdot 7H_2O$, 5.58 g $MnSO_4 \cdot 2H_2O$, 1.11 g $FeSO_4 \cdot 7H_2O$, 0.48 g $Na_2MoO_4 \cdot 2H_2O$, 0.33 g $CaCl_2$, 0.12 g NaCl, 1.0 g ascorbic acid and water to 1 liter

L-Broth[11]: 10 g tryptone, 5 g yeast extract, 5 g NaCl, 1 g glucose, and water to 1 liter.

All solutions can be sterilized by autoclaving. Stock suspensions of *Psesudomonas* are prepared by growing the organism overnight with shaking in 5 ml of L-broth. The culture is harvested by centrifuging for 10 min at top speed in a table-top clinical centrifuge, and the pellet is resuspended in 5 ml of physiological saline. This suspension is viable for up to 6 months when kept at 5°. A 5-ml L-broth culture is inoculated with 0.05 ml of the physiological saline suspension, and is grown overnight. The L-broth culture is used to inoculate 100 ml of valine/isoleucine medium, which is also grown overnight. The 100-ml culture is used to inoculate 1 liter of valine/isoleucine medium. *Pseudomonas putida* is grown at 30° with aeration; however, the organism is sensitive to shearing forces and lyses in fermentors with propellors. It can be grown satisfactorily in shake flasks or in fermentors such as the Labline/SMS Hi-Density Fermentor, which aerates by rotating the growth chamber. *Pseudomonas aeruginosa* is also grown with aeration, but at 37°. Hourly samples are taken, diluted 1:5 with physiological saline, and read at 660 nm in a cuvette with a 1 cm light path. When the optical density reaches 0.5–0.6, the culture is harvested and frozen at −15°. The yield is about 8 g of cell paste per liter of medium.

[10] J. R. Sokatch, V. McCully, J. Gebrosky-Sahm, and M. Reyes-Maguire, *J. Bacteriol.* **153**, 969 (1983).
[11] E. S. Lennox, *Virology* **1**, 190 (1955).

Enzyme Assays

Branched-Chain Keto Acid Dehydrogenase

The reaction catalyzed by the complex is

$$RCOCOOH + NAD^+ + CoASH \rightarrow RCO\text{-}SCoA + CO_2 + NADH + H^+$$

The reaction is followed by recording the change in absorbance at 340 nm due to reduction of NAD^+.

Reagents

   1 M potassium phosphate, pH 7.0
   0.1 M ethylenediaminetetraacetic acid (EDTA)
   0.1 M L-valine
   0.1 M NAD^+
   0.01 M coenzyme A in 0.02 M dithiothreitol
   0.021 M thiamin pyrophosphate
   0.1 M magnesium chloride
   0.1 M 2-ketoisovalerate
   Purified LPD-Val, 2–300 units/ml or heat-treated Sepharose CL-4B
     fraction (see Purification of LPD-Val)

Procedure. The reaction mixture contains 100 μl phosphate buffer, pH 7.0, 20 μl NAD^+, 10 μl coenzyme A plus dithiothreitol, 10 μl thiamin pyrophosphate, 10 μl magnesium chloride, 50 μl L-valine, 0.42 units of purified lipoamide dehydrogenase, and 0.008–0.032 units of branched-chain keto acid dehydrogenase. The heat-treated Sepharose CL-4B fraction from the purification of LPD-Val (Table II) can be substituted for purified LPD-Val. This fraction contains E2 and LPD-Val as well as some LPD-Glc; however, it is satisfactory for assay of column fractions. Enough water is added to make the final volume 0.96 ml and the absorbance is read at 340 nm for 1 min to correct for any endogenous reaction. The reaction is started with 40 μl of 0.1 M 2-ketoisovalerate and the absorbance read at 340 nm. The assay is linear with initial absorbance changes of up to 0.2 per minute. One unit is the amount of enzyme which produces 1 μmol of NADH per minute.

It is also possible to measure the activity of individual E1 and E2 subunits. These assays are described in Refs. 6 and 12.

LPD-Val

LPD-Val is the specific E3 subunit for branched-chain keto acid dehydrogenase of *P. putida* and *P. aeruginosa.* This subunit can be assayed

---

[12] P. J. Sykes, J. Menard, V. McCully, and J. R. Sokatch, *J. Bacteriol.* **162,** 203 (1985).

independently of the complex by virtue of its ability to reduce or oxidize lipoamide. The reaction catalyzed by LPD-Val, like other lipoamide dehydrogenases, is stimulated by the presence of NAD$^+$.[7] The assay is based on the oxidation of NADH to NAD$^+$ with the resulting decrease in absorbance at 340 nm:

$$\text{Lipoamide} + \text{NADH} + \text{H}^+ \rightarrow \text{dihydrolipoamide} + \text{NAD}^+$$

Reagents

1 M potassium phosphate, pH 7.0
0.01 M NAD$^+$
0.01 M NADH in 10 mM potassium phosphate, pH 7.9
0.05 M DL-lipoamide dissolved in ethanol or acetone

Procedure. Add to a 1-ml cuvette, 150 μl potassium phosphate buffer, 30 μl NAD$^+$, 10 μl NADH, 60 μl lipoamide, and enough water to bring the volume to 1 ml after the enzyme has been added. The initial absorbance is read at 340 nm against a water blank and should be 0.7–0.8. The reaction is started with enough enzyme to give a change in absorbance of between 0.02 and 0.15 per minute. A control cuvette should be run with an equivalent amount of ethanol since crude extracts of *P. putida* contain an alcohol dehydrogenase which reduces NAD$^+$. This problem can be overcome by using acetone as the solvent, but this etches the sides of plastic cuvettes. Specific activity is defined as 1 μmol NADH oxidized per minute per milligram protein. If the change in absorbance exceeds 0.15 OD per minute, the error in measurement of total complex becomes too large for accuracy.

Purification of the Complex

Buffers

A: 50 mM potassium phosphate, pH 7.0, 1 mM L-valine, 1 mM EDTA, 0.5 mM dithiothreitol, 0.5 mM thiamin pyrophosphate
B: 5 mM potassium phosphate, pH 7.0, 1 mM L-valine, and 0.5 mM dithiothreitol

Procedure

A typical purification procedure is shown in Table I. The preparation described here begins with 30–50 g of cell paste. The cells are resuspended in 5 ml of buffer A per gram of cell paste and the suspension is divided between two 150-ml beakers. Each beaker is chilled in an ice–salt bath and

TABLE I
PURIFICATION OF BRANCHED-CHAIN KETO ACID DEHYDROGENASE FROM
Pseudomonas aeruginosa PAO

Fraction	Total protein[a]	Total units		Specific activity	
		Alone	+LPD-Val[b]	Alone	+LPD-Val[b]
90,000 g supernatant	1485	24	190	0.016	0.128
176,000 g pellet	293	66	227	0.225	0.775
Sepharose CL-4B column	273	173	310	0.633	1.14
DEAE-Sepharose CL-6B column	41	45	136	1.10	3.32
BioGel HT column	9.2	26	100	2.83	10.9

[a] From 32 g of cells.
[b] The assays reported were supplemented with the heat-treated Sepharose CL-4B fraction obtained in the purification of LPD-Val.

the cells treated with sonic oscillation. As an example, we use a Heat Systems oscillator, model W 225R set at 7 which provides an output of about 30 using the regular tip. After 3 min, the beaker is chilled while the other batch of cells is being treated. The temperature should not be allowed to rise above 10°. The total period of sonication is about 1.2 min per gram of cells.

The broken cell suspension is centrifuged for 1 hr at 90,000 g in an ultracentrifuge at 4°. This step greatly reduces membrane-bound enzymes which interfere with the enzyme assay by oxidizing NADH in the absence of keto acid substrates. The supernatant solution is then centrifuged at 176,000 g for 3 hr at 4°, which brings down the complex. Both the pellet and the supernatant from this step should be assayed to ensure that at least 80% of the complex is removed from the supernatant enzyme solution. Resuspend the pellet in the smallest volume of buffer A possible, usually 25–30 ml, and add to a column of Sepharose CL-4B, 2.5 × 60 cm. The complex is eluted with buffer A at a flow rate of 0.5 ml per minute and fractions of 4 ml are collected. The complex begins to elute at about tube 40. Tubes containing the bulk of the activity are collected and pooled.

The pooled fractions from the Sepharose CL-4B column are loaded onto a column of DEAE-Sepharose CL-6B, 2.5 × 20 cm., equilibrated

with buffer A. The column is washed with 100 ml of buffer A plus 100 mM sodium chloride, and branched-chain keto acid dehydrogenase is eluted with a sodium chloride gradient. The receiving flask contains buffer A plus 100 mM sodium chloride and the reservoir flask contains 350 ml buffer A plus 350 mM sodium chloride. Fractions of 6 ml are collected and branched-chain keto acid dehydrogenase is eluted beginning at about 200 mM sodium chloride. There is also a significant amount of branched-chain keto acid dehydrogenase which comes off the column immediately. These fractions are frequently turbid and appear to contain cell wall fragments from which it is not possible to isolate branched-chain keto acid dehydrogenase. Occasionally, not all enzyme is eluted before the buffer runs out, in which case the remaining enzyme can be eluted with an additional 100 ml buffer A plus 350 mM sodium chloride. Frequently, more than one peak of activity is obtained, apparently due to complex composed of different proportions of subunits.

All active fractions are pooled, dialyzed against several changes of buffer B, and then applied to a column of hydroxylapatite, 1.5 × 7 cm, equilibrated in buffer B. BioGel HT (Bio-Rad) provides an acceptable flow rate, about 0.3 ml/min and good reproducibility. Branched-chain keto acid dehydrogenase is eluted with a linear gradient of phosphate ion. The gradient is produced by use of 150 ml of buffer B in the mixing flask and 150 ml of a solution of 200 mM potassium phosphate, pH 7.0, 1 mM L-valine, and 0.5 mM dithiothreitol in the reservoir flask. During the loading step, 10-ml fractions are collected, but after the gradient is started 3-ml fractions are collected. Branched-chain keto acid dehydrogenase starts to elute at about 100 mM potassium phosphate. The active fractions are pooled, dialyzed against buffer A, and assayed. The specific activity ranges from 6–10 μmol NADH produced per minute per milligram protein. The subunits can be separated on 7.5% polyacrylamide-SDS gels; however, the E2 subunit tends to aggregate, and therefore the sample should be boiled in 5% SDS plus 0.5% 2-mercaptoethanol and electrophoresed in 0.5% SDS to reduce the number of bands to four with the correct molecular weights.

The purified complex is active with 2-ketoisovalerate, 2-ketoisocaproate, 2-keto-3-methylvalerate, and 2-ketobutyrate. There is almost no activity with pyruvate, although the E1 subunit binds pyruvate,[6] and none with 2-ketoglutarate. The pH optimum is 7.0. It is fairly easy to demonstrate a partial dependence on thiamin pyrophosphate, magnesium chloride, and coenzyme A.[5,6] In order to demonstrate a dependence of L-valine, it is necessary to pass the complex over a short column of Sephadex G-25 in order to remove small molecules.

Purification of LPD-Val

Buffers

C: 50 m*M* potassium phosphate, pH 7.0, 1 m*M* EDTA, and 0.5 m*M* dithiothreitol

D: 10 m*M* potassium phosphate, pH 7.0, 1 m*M* EDTA, and 0.5 m*M* dithiothreitol

Procedure

A typical purification of LPD-Val from *P. aeruginosa* is shown in Table II. This preparation began with 20 g of cells, but the column sizes listed will accommodate preparations beginning with up to 50 g of cells. The first three steps through chromatography on Sepharose CL-4B are identical to the purification of branched-chain keto acid dehydrogenase. The pool from the Sepharose CL-4B column is heated at 65° for 5 min in a rotary evaporator in order to ensure even heating. The heated extract is chilled in an ice bath and, if turbid, is centrifuged at 45,000 g for 30 min at 5°C. It is then applied to a DEAE-Sepharose CL-4B column, 2.5 × 60 cm, equilibrated with buffer A. LPD-Val is eluted with a gradient of 0–300 m*M* sodium chloride. The mixing flask contains 250 ml of buffer C and the reservoir flask contains 250 ml of buffer C plus 300 m*M* sodium chloride. Fractions of 6 ml are collected and LPD-Val elutes beginning at about 100 m*M* sodium chloride. LPD-Glc elutes at about 150 m*M* sodium chloride. Fractions containing LPD-Val are pooled and dialyzed against 10 m*M* potassium phosphate, pH 7.0, 1 m*M* EDTA, and 5 m*M* 2-mercaptoethanol. The dialyzed pool is then added to a column of Affi-Gel Blue (Bio-Rad), 1.5 × 17 cm, which has been equilibrated in 10 m*M* potassium phosphate buffer, pH 7.0, 1 m*M* EDTA, and 0.5 m*M* dithiothreitol. The

TABLE II

PURIFICATION OF LPD-VAL FROM *Pseudomonas aeruginosa* PAO

Fraction	Total protein[a]	Total units	Specific activity
90,000 g supernatant	870	1047	1.20
176,000 g pellet	227	323	1.42
Sepharose CL-4B column	77	455	5.90
Heat-treated Sepharose CL-4B fraction	72	435	6.04
DEAE-Sepharose CL-6B Sepharose	7.2	367	51
Affi-Gel Blue column	3.7	252	68

[a] From 20 g of cells.

column is washed with 50 ml of the same buffer and LPD-Val is eluted with a gradient of 0–500 mM sodium chloride. The gradient is produced with 150 ml of buffer D in the mixing flask and 150 ml of buffer D plus 500 mM sodium chloride in the reservoir. Fractions of 3 ml are collected and LPD-Val appears at about 200 mM sodium chloride. The active fractions are pooled and dialyzed against buffer C.

The only demonstrated function for LPD-Val is its role as the E3 subunit of branched-chain keto acid dehydrogenase in *Pseudomonas*.[6-8] LPD-Val from *P. putida* and *P. aeruginosa* has an absorption maximum at 460 nm as opposed to 455 for LPD-Glc and other lipoamide dehydrogenases.[6,7] Antibodies to LPD-Val are specific for this protein and antibodies against LPD-Glc are specific for LPD-Glc.

[45] Cloning of Genes for Branched-Chain Keto Acid Dehydrogenase in *Pseudomonas putida*

By P. J. SYKES and JOHN R. SOKATCH

Pseudomonas putida is a metabolically versatile organism which has been used to study a variety of catabolic pathways. The following procedure has been used for cloning the structural genes for subunits of branched-chain keto acid dehydrogenase (BCKAD) which are located on the *P. putida* chromosome. The vector used is pKT230 (Fig. 1), a broad host-range plasmid which will transform *P. putida* and *Escherichia coli* and which contains structural genes for kanamycin and streptomycin resistance *(Km^{r}Sm^r).*[1] The cloning strategy utilizes the recombinant plasmid to transform mutants of *P. putida* which are unable to produce one or more of the complex subunits. Such mutants are unable to grow in media with branched-chain amino acids as the sole carbon source.[2] A recombinant plasmid pSS1-1 (Fig. 1) containing structural genes for BCKAD was obtained in this fashion and, when used to transform these mutants, complementation of the genetic lesion occurs and transformants acquire the ability to metabolize branched-chain amino acids.[3] Transformants are easily distinguished from revertants since transformants are *Km^{r}Sm^s* while

[1] M. Bagdasarian, R. Lurz, B. Ruckert, F. C. H. Franklin, M. M. Bagdasarian, J. Frey, and K. N. Timmis, *Gene* **16**, 237 (1981).

[2] P. J. Sykes, J. Menard, V. McCully, and J. R. Sokatch, *J. Bacteriol.* **162**, 203 (1985).

[3] P. J. Sykes, P. G. Burns, J. Menard, K. Hatter, and J. R. Sokatch, *J. Bacteriol.*, **169**, 1619 (1987).

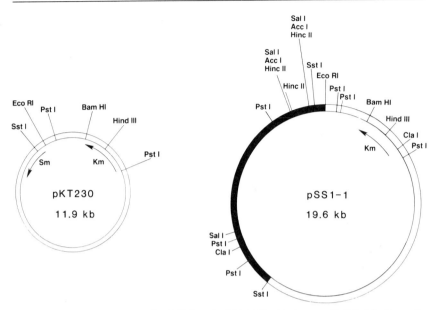

FIG. 1. Restriction maps of pKT230 and pSS1-1. The black area of pSS1-1 is *P. putida* chromosomal DNA.

revertants are *Km^{r}Sm^r*. When pSS1-1 is used to transform *E. coli,* good expression of BCKAD structural genes is obtained. *Escherichia coli* does not metabolize branched-chain amino acids; therefore detection of BCKAD in *E. coli* is good evidence that the structural genes have been cloned. Since *E. coli* does not produce many of the catabolic enzymes produced by *P. putida,* the method appears to have general applicability to the study of catabolic pathways of *P. putida.*

Bacterial Strains and Plasmids

Pseudomonas putida PpG2, originally obtained from I. C. Gunsalus, was used as the wild-type strain.

Pseudomonas putida PRS2003 containing the broad-host range vector pKT230[1] was obtained from Mark Shanley and L. N. Ornston, Department of Biology, Yale University.

Media

L-broth:[4] 10 g Tryptone, 5 g yeast extract, 5 g NaCl, and 1 g glucose per liter of water

---

[4] E. S. Lennox, *Virology* **1,** 190 (1955).

Valine/isoleucine medium:[5] 3 g L-valine, 1 g L-isoleucine, 100 ml Basal G, and 990 ml water. The medium is sterilized, then 10 ml Salts S is added

GASV medium:[5] 1.8 g D-glucose, 0.16 g sodium acetate, 0.32 g sodium succinate, 3 g L-valine, 1 g L-isoleucine, 100 ml Basal G, 10 ml Salts S, and water to 1 liter

Isobutyrate medium: 2.2 ml isobutyric acid neutralized with 1 N sodium hydroxide to pH 7.0, 100 ml Basal G, 10 ml Salts S, and water to 1 liter

The above media are solidified by adding 15 g agar per liter of medium. The compositions of the salt solutions are[6]

Basal G: 43.5 g K_2HPO_4, 17 g KH_2PO_4, 21.4 g NH_4Cl, water to 1 liter
Salts S: 39.44 g $MgSO_4 \cdot 7H_2O$, 5.58 g $MnSO_4 \cdot 2H_2O$, 1.11 g $FeSO_4 \cdot 7H_2O$, 0.48 g $Na_2MoO_4 \cdot H_2O$, 0.33 g $CaCl_2$, 0.12 g NaCl, 1.0 g ascorbic acid, water to 1 liter

Buffers

Buffer 1: 10 mM 3-(N-morpholino)propanesulfonic acid (MOPS), pH 7.0, 10 mM RbCl, 100 mM $MgCl_2$

Buffer 2: 100 mM MOPS, pH 6.5, 10 mM RbCl, 100 mM $CaCl_2$

TBE buffer: 0.89 M Tris, 0.08 M boric acid, 0.002 M ethylenediaminetetraacetic acid (EDTA), pH 8.0

TE buffer: 10 mM Tris, 1 mM EDTA, pH 8.0

High-salt buffer: 100 mM NaCl, 50 mM Tris, pH 7.5, 10 mM $MgCl_2$, 1 mM dithiothreitol

Ligase buffer: 500 mM Tris, pH 7.4, 100 mM $MgCl_2$, 10 mM dithiothreitol, 10 mM ATP

Low-salt buffer: 10 mM Tris, pH 7.5, 10 mM $MgCl_2$, 1 mM dithiothreitol

Sucrose solution: 15% sucrose, 50 mM EDTA, 50 mM Tris, pH 8.5

Growth Conditions

Pseudomonas putida is grown with shaking at 30°. BCKAD mutants are grown in GASV medium since they cannot grow in valine/isoleucine medium.[5] GASV medium contains L-valine and L-isoleucine which are

[5] J. R. Sokatch, V. McCully, J. Gebrosky-Sahm, and M. Reyes-Maguire, *J. Bacteriol.* **153**, 969 (1983).
[6] L. A. Jacobson, R. C. Bartholomous, and I. C. Gunsalus, *Biochem. Biophys. Res. Commun.* **24**, 955 (1966).

TABLE I
Pseudomonas putida BRANCHED-CHAIN KETO
ACID DEHYDROGENASE MUTANTS

Strain	Genotype[a]	Source
JS287	lpdV	b
JS113	bkdA	c
JS326	bkdAB lpdV	c
JS161	bkdAB lpdV	c

[a] bkdA, E1 subunit of BCKAD; bkdB, E2 subunit of BCKAD; and lpdV, LPD-Val (E3 subunit).
[b] From Sokatch et al.[5]
[c] From Sykes et al.[2]

converted to their respective keto acids, the inducers of BCKAD.[7] When antibiotic supplements are added, the final concentrations are 1.0 mg/ml of streptomycin (Sm) and 100 μg/ml of kanamycin (Km).

Isolation of Mutants

The method for isolation of BCKAD mutants of P. putida is described in detail in Ref. 2. Selection of BCKAD mutants is made by picking colonies which are unable to use valine as a carbon source, but can grow on isobutyrate agar. When the mutagenized cell suspension is plated on valine/isoleucine medium supplemented with 1 mM glucose, cells unable to use L-valine as a carbon source appear as tiny colonies whose growth is limited by the amount of glucose present. The tiny colonies are picked and those which grow on isobutyrate agar but not on valine agar are kept as putative BCKAD mutants. Some of these mutants have been characterized and shown to lack specific subunit enzyme activities of the BCKAD complex. The mutants employed in cloning the BCKAD genes are listed in Table I.

Preparation of extracts and assays for BCKAD have been described previously.[8] The E1 and E2 assays are described in detail in McCully et al.[9]

Cloning Protocol

Pseudomonas putida DNA is shot-gun cloned into the broad-host range plasmid vector pKT230 (see Fig. 2). As P. putida is unusually resistant to

[7] R. R. Martin, V. D. Marshall, J. R. Sokatch, and L. Unger, J. Bacteriol. 115, 198 (1973).
[8] J. R. Sokatch, V. McCully, and C. M. Roberts, J. Bacteriol. 148, 647 (1981).
[9] V. McCully, G. Burns, and J. R. Sokatch, Biochem. J. 233, 737 (1986).

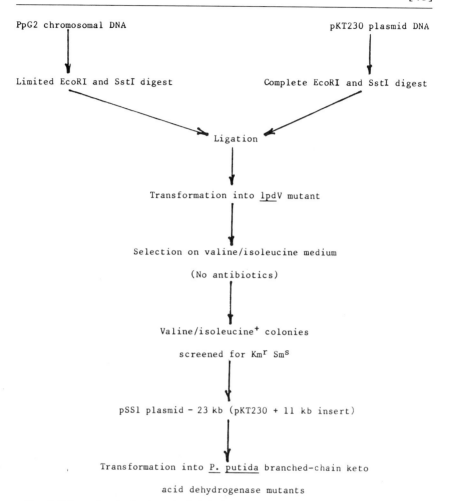

FIG. 2. Scheme for cloning branched-chain keto acid dehydrogenase genes in *P. putida.*

Sm and spontaneous Km resistance is rare in strain PpG2, it is convenient to clone into the Sm resistance gene and use Km resistance to select for pKT230 transformants. *Pseudomonas putida* chromosomal DNA is isolated, partially digested with the restriction endonucleases *Eco*RI and *Sst*I, and ligated into pKT230 which has been cut to completion with the same enzymes. The ligation mix is used to transform *P. putida* mutant JS287 deficient in the E3 (LPD-Val) subunit of the BCKAD complex. BCKAD mutants cannot utilize branched-chain amino acids as a carbon source and hence will not grow on valine/isoleucine agar unless the recombinant plasmid contains a *P. putida* chromosomal insert capable of complement-

ing the mutants. The transformants are selected which are Km^rSm^s and able to grow on valine/isoleucine medium.

Isolation of *Pseudomonas putida* Chromosomal DNA

Many of the general methods for isolation of DNA and cloning have been described in Maniatis *et al.*[10] and in Volumes 65, 68, 100 and 101 of this series. A 200-ml L-broth culture inoculated from a single colony of *P. putida* is grown overnight with shaking at 30°. The cells are centrifuged for 15 min at 5800 *g*, the supernatant discarded, and the pellet resuspended in 20 ml sucrose solution and centrifuged again. *Pseudomonas putida* is sensitive to high concentrations of EDTA, and therefore care should be taken when pouring off the supernatant as the cells may not form a discrete solid pellet. The cells are resuspended in 20 ml of sucrose solution and 6 ml of freshly made lysozyme, 6 mg/ml, in sucrose solution is added. After gentle mixing, the sample is incubated at room temperature for 30 min and then 1.5 ml of 20% SDS is added and mixed by inverting the tube gently. The sample is extracted three times with an equal volume of phenol/chloroform.[10] After each extraction, the sample is centrifuged for 20 min in polypropylene tubes at 16,000 *g* and 20°. Phenol is removed by extraction with an equal volume of chloroform[10] and centrifuging as above. Add enough 2.5 *M* sodium acetate, pH 5.3, to give a final concentration of 0.25 *M* and add 2 volumes of cold ethanol to precipitate the DNA. The mixture is kept at 4° overnight. DNA is centrifuged at 4° and 17,000 *g* for 30 min and the pellet resuspended in TE buffer, pH 8.0. The *P. putida* chromosome may be further purified by means of an ethidium bromide–cesium chloride gradient [refractive index (RI) = 1.393]. Ethidium bromide is extracted from the DNA solution with water-saturated *n*-butanol. The solution is dialyzed twice against 2 liters of TE buffer, pH 8.0, and DNA precipitated with ethanol as above and resuspended in TE buffer, pH 8.0. However, *P. putida* PpG2 is free of plasmids; therefore the cesium chloride gradient is usually omitted and DNA from alcohol precipitation used directly. The resuspended pellet is treated with 20 μl of RNase (10 mg/ml) and incubated at 37° for 30 min. The solution is treated with phenol/chloroform and precipitated with sodium acetate and ethanol. Four to 6 mg of purified DNA is obtained from a 200-ml culture.

Isolation of Plasmid DNA

pKT230 is isolated from an 800-ml overnight culture of *P. putida* (pKT230) in L-broth plus Km in a manner similar to isolation of chromo-

[10] T. Maniatis, E. F. Fritsch, and J. Sambrook, "Molecular Cloning: A Laboratory Manual." Cold Spring Harbor Lab. Cold Spring Harbor, New York, 1982.

somal DNA. After the SDS step, enough 5 M NaCl is added to give a final concentration of 1.1 M NaCl in order to precipitate chromosomal DNA. After 30 min on ice the solution is centrifuged at 4° and 47,000 g for 30 min. The supernatant containing pKT230 is removed and extracted three times with phenol/chloroform, washed with chloroform, and precipitated with ethanol as described above. The pKT230 band is isolated from an ethidium bromide–cesium chloride gradient. The DNA precipitate is resuspended in 5.25 ml TE buffer, transferred to a 13.5 ml ultracentrifuge tube, and 6 g of cesium chloride is added, mixed by inversion, and covered with mineral oil. Then 0.77 ml of ethidium bromide (10 mg/ml) is added and the tube filled with mineral oil. The tube is sealed and centrifuged for 36–48 hr at 20° and 48,000 rpm in the Ti 75 or equivalent fixed-angle rotor.

Miniplasmid Preparations

Miniplasmid preparations are used to determine whether chromosomal DNA has been inserted into pKT230 by restriction enzyme analysis. Miniplasmid preparations are made using the method of Holmes and Quigley[11] by boiling for 55 sec, and this proved to be the best for *P. putida*. The pellets appeared blobby after boiling but could still be removed with a sterile toothpick. The preparations are extracted twice with an equal volume of phenol/chloroform washed with chloroform and precipitated with ethanol. Restriction digests of the plasmid preparations are carried out in the presence of RNase (1 μl of a 50 mg/ml stock solution). The size of the insert is identified by viewing the samples with UV after electrophoresis in 0.8% agarose, that was stained with ethidium bromide and run in TBE buffer at 80 V until the bromphenol blue tracking dye reached the end of the gel.

Restriction Endonuclease Digestion

Double restriction enzyme digests are carried out on both pKT230 and *P. putida* chromosomal DNA to reduce possible religation of the pKT230 plasmid. Preliminary experiments should be carried out to determine the appropriate times for restriction endonuclease digestion of *P. putida* chromosomal DNA with *Eco*RI and *Sst*I which would give a range of DNA fragment lengths of 5–15 kb.

Pseudomonas putida chromosomal DNA (3 μg) is incubated with 5 units *Eco*RI for 1 hr at 37° and then for 20 min at 68° in a total reaction

[11] D. S. Holmes and M. Quigley, *Anal. Biochem.* **114**, 193 (1981).

volume of 50 μl containing 5 μl of 10X high-salt buffer. The DNA is extracted twice with phenol, washed once with chloroform, and precipitated with 2 volumes of ethanol in a final concentration of 0.25 M sodium acetate. The precipitate is resuspended in 50 μl of TE buffer, pH 8.0. Twenty five microliters of this DNA is then incubated with 5 units SstI for 15 min at 37° and 20 min at 68°, in low-salt buffer.

Complete cutting of pKT230 (1.5 μg) with EcoRI and SstI is achieved consecutively in a total volume of 50 μl as above with incubations of 1 hr at 37° per digestion. The DNA is extracted twice with phenol, washed with chloroform, and precipitated with ethanol between restriction digests.

Ligation

EcoRI–SstI restriction digests of chromosomal DNA (9 μl) and plasmid pKT230 (2 μl) are mixed together, extracted twice with phenol, washed with chloroform, and precipitated with ethanol, and the precipitate washed again with cold ethanol. The precipitate is resuspended in 50 μl distilled water and 5.5 μl of 10X ligase buffer and 2 units of T4 ligase are added. Ligation is carried out overnight in a 14° water bath.

Transformation

The rubidium chloride transformation method of Bagdasarian and Timmis[12] proved to be the most efficient transformation method. Of an overnight L-broth culture of a P. putida BCKAD mutant (Table I) 1.2 ml is used to inoculate 24 ml L-broth. The cells are incubated at 30° with shaking until the OD$_{660}$ = 0.6–0.8. The cells are chilled at 0° in an ice-bath and then pelleted in a clinical centrifuge for 8 min at 4°. The cells are resuspended in 20 ml cold buffer 1. Cells are again pelleted at 4°, resuspended in 20 ml cold buffer 2, and incubated for 30 min at 0°. The pellet is resuspended in 2 ml cold buffer 2 and 0.2 ml of the suspension is mixed with plasmid DNA or ligation mix (5 μl) and incubated at 0° for 45 min. The transformation mix is heat-shocked by immersing the tube in a 42° water bath for 1 min and then diluted with 1 ml L-broth and incubated at 30° for 90 min with shaking.

Selection

The transformation mixture is washed in 4 ml 0.8% NaCl to prevent subsequent growth due to carry over of L-broth, resuspended in 1 ml 0.8%

---

[12] M. Bagdasarian and K. N. Timmis, *Curr. Top. Microbiol. Immunol.* **96,** 47 (1981).

NaCl, and then 0.1 ml is plated directly onto valine/isoleucine minimal agar. BCKAD mutants will occasionally revert to growth on valine/isoleucine agar, but selection directly onto valine/isoleucine agar supplemented with Km is too harsh an initial selection pressure since no transformants were found when the transformation mixture was plated directly onto this medium. After 3 days, the few colonies that appear on valine/isoleucine agar are picked onto fresh L-agar plates. After approximately 8 hr growth, the colonies are replica-plated onto valine/isoleucine agar, valine/isoleucine agar plus Km, L-agar plus Km, and L-agar plus Sm. *Pseudomonas putida* is sensitive to 90 μg/ml Km but quite high concentrations of the Sm (1 mg/ml) are required to achieve clear-cut sensitivity. About half of the transformants should contain pKT230 plasmids with inserts based on antibiotic resistance *(Km^{r}Sm^s)*. Miniplasmid preparations using *Eco*RI and *Sst*I to digest plasmid DNA are used to show that the plasmid contains an insert of chromosomal DNA of the correct size.

Very few of the *Km^{r}Sm^s* transformants will contain the correct structural gene assuming that *Eco*RI and *Sst*I cut randomly. pKT230 is stable in *P. putida,* but pSS1-1 is rapidly lost when cultured in the absence of kanamycin; therefore strains carrying recombinant plasmids should be maintained in the presence of kanamycin.

Recombinant plasmids can be subcloned into pKT230 using appropriate restriction enzymes. pSS1-1 was obtained by a limited *Sal*I digest of the first recombinant plasmid followed by ligation. However, the number of restriction sites in pKT230 suitable for cloning is somewhat limited, and it is advisable to subclone into an expression vector such as pUC[13] provided detection systems such as the use of specific antibody or enzyme assay are available. One advantage of subcloning into pUC is that the host, *E. coli,* does not metabolize branched-chain amino acids, and therefore there is no background to contend with.

Expression of pSS1 in *Pseudomonas putida*

Pseudomonas putida PpG2 transformed with pSS1-1 expressed 3–5 times the amount of BCKAD compared with wild-type strain PpG2 alone. Extracts of strain JS287 which had been transformed with pSS1-1 reacted with antibody to LPD-Val, whereas extracts of JS287 which had been transformed with pKT230 alone did not. In all cases expression is constitutive.

---

[13] C. Yanisch-Perron, J. Vieira, and J. Messing, *Gene* **33**, 103 (1985).

Expression of pSS1-1 in *Escherichia coli*

pKT230 isolated from *E. coli* does not transform *P. putida* efficiently. However, pKT230 isolated from *P. putida* will transform both *P. putida* and *E. coli* HB101[14] efficiently. For this reason, pKT230 used in the cloning procedure should be isolated from *P. putida*.

Extracts of *E. coli* transformed with pSS1-1 have about the same specific activity of BCKAD as *P. putida* PpG2, react with specific antibody to *P. putida* Lpd-Val, and complement extracts of *P. putida* BCKAD mutants, restoring activity of the BCKAD complex. Although pSS1-1 encodes the structural genes for the three subunits of BCKAD, *E. coli* minicells transformed with pSS1-1 produce anomalous proteins, possibly fusion products, when carrying pSS1-1. Therefore, care should be taken in interpretation of gene products unless supporting information is available. However, the correct number of polypeptides with the correct molecular weights are produced when plasmid preparations constructed from pUC18 are used as DNA templates in an *in vitro* prokaryotic translation kit.[3]

Discussion

Although the methods used here to clone the branched-chain keto acid dehydrogenase genes in *P. putida* are based on general recombinant DNA procedures, there are a number of unique features. The plasmid pKT230, although quite large, proved a convenient and stable vector for cloning procedures, allowing expression of the cloned BCKAD genes in *P. putida* and *E. coli* without further manipulation of the cloned insert. Restriction mapping of pSS1-1 is more difficult than, for example, if pBR322 is used as a plasmid vector because of the limited knowledge of the pKT230 restriction map.

The transformation method of Bagdasarian and Timmis[12] results in efficient transformation in both *P. putida* and *E. coli*, and the direct selection method for branched-chain keto acid dehydrogenase genes on valine/isoleucine minimal agar is a simple and convenient procedure to identify the proper recombinant plasmid.

---

[14] H. W. Boyer and D. Roulland-Dussoix, *J. Mol. Biol.* **41**, 459 (1969).

[46] 2-Methyl Branched-Chain Acyl-CoA Dehydrogenase from Rat Liver

By Yasuyuki Ikeda and Kay Tanaka

Rat 2-methyl branched-chain acyl-CoA dehydrogenase is a mitochondrial flavoprotein that catalyzes the third step in the oxidative metabolism of isoleucine and valine, as illustrated in Fig. 1.[1,2] The enzyme removes one hydrogen each from C-2 and C-3 of (S)-2-methylbutyryl-CoA and isobutyryl-CoA in a stereospecific manner,[2,3] resulting in the formation of tiglyl-CoA (2-methylcrotonyl-CoA) and methacrylyl-CoA, respectively. In these reactions, a second flavoprotein, electron-transfer flavoprotein (ETF),[1,2] is required as a natural electron acceptor. In in vitro assay, ETF may be replaced with phenazine methosulfate (PMS). This enzyme was not known until recently.

Isobutyryl-CoA and 2-methylbutyryl-CoA, like isovaleryl-CoA, were previously assumed to be dehydrogenated by short-chain acyl-CoA dehydrogenase. In fact, Green, Beinert, and co-workers previously stated that bovine short-chain acyl-CoA dehydrogenase dehydrogenated isobutyryl-CoA,[4,5] but in retrospect, their enzyme preparation was probably contaminated with 2-methyl branched-chain acyl-CoA dehydrogenase. While systematically studying enzymes which dehydrogenate various acyl-coenzyme A esters with a straight- or branched-acyl chain, Ikeda and Tanaka found that isobutyryl-CoA and 2-methylbutyryl-CoA were not dehydrogenated by any one of the four previously known acyl-CoA dehydrogenases, including short-chain acyl-CoA and isovaleryl-CoA dehydrogenases. Subsequently, for the first time, Ikeda and Tanaka identified[1] and purified[2] a previously unknown enzyme, which dehydrogenated these substrates, from rat liver mitochondria. Since this enzyme acts on the substrates with a methyl group at C-2, this enzyme was designated as 2-methyl branched-chain acyl-CoA dehydrogenase.

[1] Y. Ikeda, C. Dabrowski, and K. Tanaka, J. Biol. Chem. 258, 1066 (1983).
[2] Y. Ikeda and K. Tanaka, J. Biol. Chem. 258, 9477 (1983).
[3] K. Tanaka, J. J. O'Shea, G. Finocchiaro, Y. Ikeda, D. J. Aberhart, and P. K. Ghoshal, Biochim. Biophys. Acta, in press.
[4] D. E. Green, S. Mii, H. R. Mahler, and R. M. Bock, J. Biol. Chem. 206, 1 (1954).
[5] H. Beinert, this series, Vol. 5, p. 546.

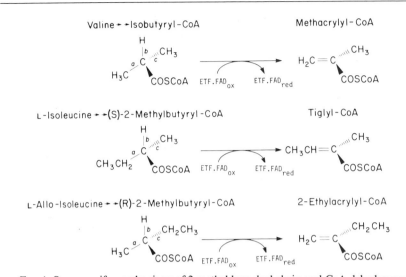

FIG. 1. Stereospecific mechanisms of 2-methyl branched-chain acyl-CoA dehydrogenase reaction with isobutyryl-CoA, (S)-2-methylbutyryl-CoA, and (R)-2-methylbutyryl-CoA as substrate. From Ikeda and Tanaka.[2]

Assay Method

Principle

In routine assay, the activity of this enzyme is determined using PMS as intermediary electron acceptor and 2,6-dichlorophenolindophenol (DCIP) as terminal acceptor indicator. The reduction of DCIP is followed spectrophotometrically.

Reagents

PMS, 33 mM, stored at 4° in a brown bottle as soon as prepared

Flavin adenine dinucleotide (FAD), 1 mM, stored at 4° in a brown bottle

DCIP, 0.123 mM, in 0.25 M potassium phosphate, pH 8.0

Substrate: isobutyryl-CoA and (S)-2-methylbutyryl-CoA, 2 mM solution in 2 mM acetate buffer, pH 5.0, stored frozen. The former is available from commercial sources (Pharmacia, PL-Biochemicals), but the latter must be prepared as described below

Preparation of (R)- and (S)-2-Methylbutyryl-CoA

These compounds are not commercially available, and are synthesized in our laboratory. Only (RS)-2-methylbutyric acid is available from com-

mercial sources. The commercial *(RS)*-2-methylbutyric acid contains 15% isovaleric acid as an impurity. *(S)*-2-Methylbutyric acid is conveniently prepared in a highly pure form from L-isoleucine (Sigma) using snake venom L-amino-acid oxidase *(Crotalus adamanteus)* from Sigma. L-Amino-acid oxidase (200 mg) is dissolved in 2 ml of water and is dialyzed against 3 liters of water at 4°. Precipitate is removed by centrifugation. The supernatant is used as L-amino-acid oxidase solution. L-Isoleucine (200 mg) is dissolved in 20 ml of 100 mM sodium phosphate buffer, pH 7.4, and is mixed with the L-amino-acid oxidase solution. The mixture is incubated at 37° for 24 hr with a constant, gentle bubbling of water-saturated oxygen stream. After incubation, 2 ml of 3 N perchloric acid (PCA) is added to the reaction mixture and is vigorously mixed. The mixture is allowed to stand for 20 min on an ice bath in order to precipitate the protein. After centrifugation, 1.4 ml of 2 N K$_2$CO$_3$ is added to the supernatant solution and the mixture is again centrifuged to remove PCA as precipitate. After acidification with 5 M phosphoric acid, S-2-methylbutyric acid is isolated from the supernatant solution by steam distillation using a Markham still. The distillate is titrated with 0.1 N NaOH to neutrality and then lyophilized. The purity of the *(S)*-2-methylbutyric acid is 99.2% by gas chromatographic analysis, and it is free of isovaleric acid by mass spectrometric analysis. *(R)*-2-Methylbutyric acid is also prepared from L-alloisoleucine (Sigma) according to the method described above, and the purity is 99%. Coenzyme A thioesters of *(S)*- and *(R)*-2-methylbutyric acids are synthesized by the mixed anhydride synthesis.[6]

Procedure

　　Into a 1 cm light path semimicrocuvette, the following reagents are added: 150 μl of the enzyme, 50 μl of FAD, 50 μl of PMS, and 200 μl of DCIP buffer solution, pH 8.0. The reaction is started by the injection of 50 μl of isobutyryl-CoA or *(S)*-2-methylbutyryl-CoA into the cuvette. The final volume is 500 μl. Bleaching of DCIP is followed at 600 nm at 32° for 1 min using Beckman model 3600 double-beam spectrophotometer. Enzyme activity is expressed as micromoles or nanomoles of DCIP reduced per milligram of enzyme (or per milliliter of enzyme solution) per minute. The extinction coefficient of DCIP (21 mM^{-1} cm^{-1}) at 600 nm is used as the basis for computation of the amount of DCIP reduced.

　　The enzyme activity can also be measured using an appropriate amount of ETF (1–15 nmol of bound FAD) as an intermediary electron acceptor, replacing PMS.[2] The method is otherwise the same as described above.

[6] T. Wieland and L. Rueff, *Angew Chem., Int. Ed. Engl.* **65**, 186 (1953).

Purification Procedure

The entire purification is summarized in Table I. In the third and fourth steps, activities to dehydrogenate various acyl-CoAs are spectrophotometrically monitored in numerous sequential fractions eluting from column chromatographies: these acyl-CoA substrates include (S)-2-methylbutyryl-CoA, isobutyryl-CoA, isovaleryl-CoA, n-butyryl-CoA, n-octanoyl-CoA, and palmitoyl-CoA. All purification procedures are carried out at 4°.

First Three Steps of Purification

Procedures used for the preparation of mitochondria, solubilization, ammonium sulfate fractionation, and DEAE-Sephadex A-50 column chromatography are identical to those described for the purification of isovaleryl-CoA dehydrogenase (see [47], this volume). In DEAE-Sephadex chromatography (step 3), 2-methyl branched-chain acyl-CoA dehydrogenase is coeluted together with short-chain acyl-CoA, medium-chain acyl-CoA, and long-chain acyl-CoA dehydrogenases with 0.35 M NaCl, but it is partially separated from isovaleryl-CoA dehydrogenase (see Fig. 1 of Ref. 7). The latter is eluted with 0.4 M NaCl. The fractions 60–93 (fraction B), containing 2-methyl branched-chain acyl-CoA dehydrogenase, are pooled and dialyzed against 10 mM potassium phosphate buffer, pH 7.5. This fraction does not contain a significant amount of isovaleryl-CoA dehydrogenase.

Hydroxylapatite Column Chromatography (Step 4)

The dialyzate (1.5 g of protein in 110 ml) from step 3 is applied onto two hydroxylapatite columns (4.6 × 20 cm) equilibrated with 10 mM potassium phosphate buffer, pH 7.5. The column is washed with 2 liters of the equilibration buffer. Proteins are eluted with a linear gradient formed with 1 liter each of the equilibration buffer and 0.33 M potassium phosphate buffer, ph 7.5. As shown in Fig. 2, 2-methyl branched-chain acyl-CoA dehydrogenase, dehydrogenating both (S)-2-methylbutyryl-CoA and isobutyryl-CoA, is eluted with 0.2 M potassium phosphate buffer as a single peak. At this stage, the peak overlaps with the descending slope of short-chain acyl-CoA dehydrogenase and the ascending slopes of long-chain acyl-CoA dehydrogenase but it is completely devoid of medium-chain acyl-CoA dehydrogenase. Fractions 64 to 83 (fraction E) are pooled and concentrated using Amicon Diaflow membrane PM30, and are dialyzed against 10 mM potassium phosphate buffer containing 0.5 mM EDTA, pH 8.0.

TABLE I
Purification Steps for 2-Methyl Branched-Chain Acyl-CoA Dehydrogenases from Rat Liver Mitochondria[a]

Purification	Total protein (mg)	Total activity (nmol/min)		Specific activity (nmol/min/mg)		Purification (-fold)		Yield (%)	
		iC_4CoA	S-2-meC_4CoA	iC_4CoA	S-2-meC_4CoA	iC_4CoA	S-2-meC_4CoA	iC_4CoA	S-2-meC_4CoA
1. Sonication supernatant	35,000	—[b]	—[b]	—	—	—	—	—	—
2. Ammonium sulfate (40–80%)	14,400	—[b]	—[b]	—	—	—	—	—	—
3. DEAE-Sephadex A-50	1,500	40,400	40,500	26.9	27.0	1	1	100	100
4. Hydroxylapatite	108	12,900	13,600	119	126	4.4	4.7	32	34
5. Matrex Gel Blue A	5.0	6,200	6,500	1,240	1,300	46	48	15	16
6. Agarose–hexane-CoA	0.8	1,570	1,710	1,963	2,138	73	79	3.9	4.2
7. BioGel A-0.5m	0.42	985	1,040	2,345	2,476	87	92	2.4	2.6

[a] From Ikeda and Tanaka.[2] iC$_4$CoA, isobutyryl-CoA; S-2-meC$_4$CoA, (S)-2-methylbutyryl-CoA.

[b] Activities could not be determined due to interference of the dye-reduction assay by nonspecific reductants.

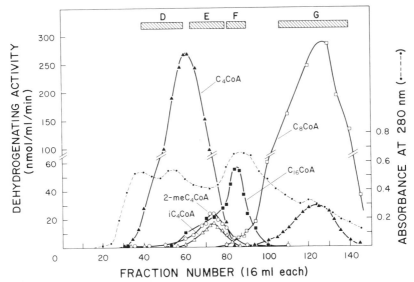

FIG. 2. Elution profile of 2-methyl branched-chain acyl-CoA dehydrogenase from hydroxylapatite column chromatography. The sample preparation (750 mg of protein, 55 ml) was applied to a hydroxylapatite column (4.5 × 20 cm). Fractions were assayed using the dye-reduction method. O, (S)-2-methylbutyryl-CoA; △, isobutyryl-CoA; ▲, n-butyryl-CoA; □, n-octanoyl-CoA; ■, palmitoyl-CoA. The hatched rectangles (D–G) indicates the fractions combined. From Ikeda et al.[1]

Matrex Gel Blue A Column Chromatography (Step 5)

The dialyzate (108 mg of protein in 25 ml) from step 4 is applied to a Matrex Gel Blue A column (1.5 × 8 cm) equilibrated with 10 mM potassium phosphate buffer, pH 8.0, containing 10% glycerol and 0.5 mM EDTA. The column is washed with 200 ml of the equilibration buffer, and then washed with the same buffer containing 0.35 M NaCl. Proteins are eluted with a linear gradient formed with 90 ml each of 10 mM phosphate buffer containing 10% glycerol, 0.5 mM EDTA, and 0.35 M NaCl, and the same buffer containing 0.4 M NaCl and 7 mM FAD. As shown in Fig. 3, 2-methyl branched-chain acyl-CoA dehydrogenase is eluted as a sharp single peak at 1–2 mM FAD, whereas only a small amount of short-chain acyl-CoA dehydrogenase is eluted in this region; no significant activity of long-chain acyl-CoA dehydrogenase is detectable. Most of the short-chain and long-chain acyl-CoA dehydrogenases are eluted as very broad peaks at FAD concentrations higher than 3 mM. When the column is further eluted with 10 mM phosphate buffer, pH 8.0, containing 0.8 M NaCl and 3 mM FAD, additional amounts of short-chain and long-chain acyl-CoA dehy-

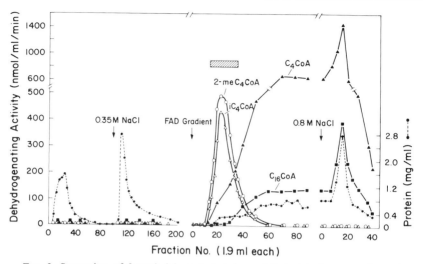

FIG. 3. Separation of 2-methyl branched-chain acyl-CoA dehydrogenase from short-chain acyl-CoA and long-chain acyl-CoA dehydrogenases on Matrex Gel Blue A column chromatography. The crude 2-methyl branched-chain acyl-CoA dehydrogenase preparation (108 mg of protein in 25 ml) obtained from step 4 was applied to a Matrex Gel Blue A column (1.5 × 8 cm) equilibrated with 10 mM potassium phosphate buffer, pH 8.0, containing 0.5 mM EDTA and 10% glycerol. The column was first washed with the starting buffer and then washed with the same buffer containing 0.35 M NaCl. The adsorbed proteins were eluted with a linear gradient of 10 mM potassium phosphate buffer containing 0.35 M NaCl and the same buffer containing 0.4 M NaCl and 7 mM FAD. The column was finally washed with 10 mM potassium phosphate buffer containing 0.8 M NaCl and 3 mM FAD. The hatched rectangle indicates the fractions combined. 2-meC$_4$CoA, 2-methylbutyryl-CoA; iC$_4$CoA, isobutyryl-CoA; C$_4$CoA, n-butyryl-CoA; C$_{16}$CoA, palmitoyl-CoA. From Ikeda and Tanaka.[2]

drogenases are eluted as a sharp peak. This affinity chromatography results in a 10-fold enrichment of 2-methyl branched-chain acyl-CoA dehydrogenase specific activity. Fractions 18 to 35, containing 2-methyl branched-chain acyl-CoA dehydrogenase, are pooled and concentrated. The sample is dialyzed against 10 mM potassium phosphate buffer, pH 8.0, containing 10% glycerol and 0.5 mM EDTA.

Agarose-Hexane-CoA Column Chromatography (Step 6)

The dialyzate (5 mg of protein in 7 ml) from step 5 is applied to an agarose-hexane-CoA (Pharmacia-PL Biochemicals, Type 1) column (1 × 7 cm) equilibrated with 10 mM potassium phosphate buffer, pH 8.0, containing 10% glycerol and 0.5 mM EDTA. The column is washed with 60 ml of the same buffer. Proteins are eluted with a linear gradient formed with 40 ml each of the equilibration buffer and the same buffer containing

0.5 M NaCl. This affinity chromatography is effective in separating 2-methyl branched-chain acyl-CoA dehydrogenase from a small amount of residual short-chain acyl-CoA dehydrogenase. 2-Methyl branched-chain acyl-CoA dehydrogenase is eluted with 0.2 M NaCl as a single peak, while short-chain acyl-CoA dehydrogenase is eluted with 0.06 M NaCl (Fig. 4).[2] The fractions containing 2-methyl branched-chain acyl-CoA dehydrogenase are pooled and concentrated.

BioGel A-0.5m Gel Filtration (Step 7)

The concentrated sample (0.8 mg of protein in 1.5 ml) from step 6 is applied to a BioGel A-0.5m column (1.5 × 100 cm) equilibrated with 10 mM potassium phosphate buffer, pH 8.0, containing 10% glycerol and 0.2 mM EDTA. The column is eluted with the same buffer at a flow rate of 0.16 ml per min and 1.6 ml of each is fractionated. As in all other steps, the

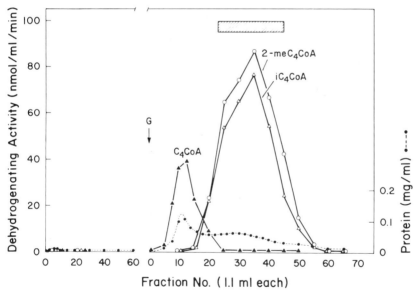

FIG. 4. Separation of 2-methyl branched-chain acyl-CoA dehydrogenase from the residual butyryl-CoA dehydrogenase by agarose-hexane-CoA column chromatography. The sample (5.0 mg of protein in 7.3 ml) obtained from Matrex Gel Blue A chromatography was applied to an agarose-hexane-CoA column (1 × 7 cm) equilibrated with 10 mM potassium phosphate buffer, pH 8.0, containing 0.5 mM EDTA and 10% glycerol. The adsorbed proteins were eluted with a linear gradient of the starting buffer and same buffer (pH 8.0) containing 0.5 M NaCl. The hatched rectangle indicates the fractions combined. Abbreviations of substrates are the same as in Fig. 2. G indicates the point at which the gradient was started. From Ikeda and Tanaka.[2]

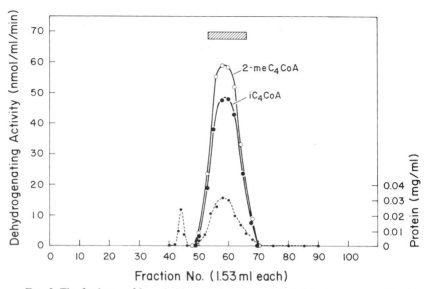

Fraction No. (1.53 ml each)

FIG. 5. The final step of 2-methyl branched-chain acyl-CoA dehydrogenase purification by BioGel A-0.5m column chromatography. The sample (0.8 mg of protein in 1.5 ml) obtained from step 6 was applied to a BioGel A-0.5m column (1.5 × 100 cm) equilibrated with 10 mM potassium phosphate buffer, pH 8.0, containing 0.2 mM NaCl. The hatched rectangle indicates the fractions combined as the final preparation. From Ikeda and Tanaka.[2]

activities to dehydrogenate both *(S)*-2-methylbutyryl-CoA and isobutyryl-CoA are eluted as a single peak in fractions 50 to 70 (Fig. 5), indicating that a single protein catalyzes these two reactions. These fractions are combined and concentrated. This represents the final preparations. When the purified enzyme is stored in 50% glycerol at −20°, it is stable for at least 6 months. After the entire seven purification steps, 0.42 mg of 2-methyl branched-chain acyl-CoA dehydrogenase is obtained. The preparation shows a single protein band on SDS-polyacrylamide gel electrophoresis, both in the absence and presence of 2-mercaptoethanol (Fig. 6, band C). It also gives a single band on a 5% gel by polyacrylamide gel electrophoresis without SDS (Fig. 6, E).

Properties of the Enzyme

Properties of 2-methyl branched-chain acyl-CoA dehydrogenase are summarized in comparison to those of short-chain acyl-CoA and isovaleryl-CoA dehydrogenases in Table II of [47] in this volume.[7]

---

[7] Y. Ikeda and K. Tanaka, this volume [47].

SDS - PAGE   **PAGE without SDS**

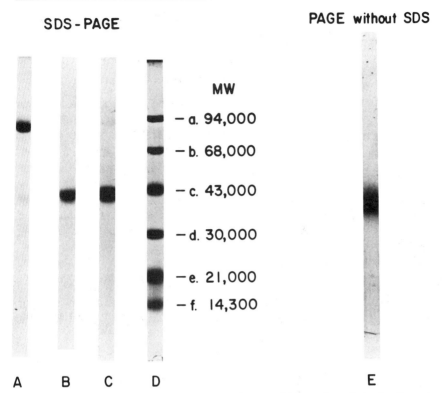

MW

– a. 94,000

– b. 68,000

– c. 43,000

– d. 30,000

– e. 21,000

– f. 14,300

A   B   C   D   E

FIG. 6. Polyacrylamide gel electrophoresis, with and without sodium dodecyl sulfate, of the purified 2-methyl branched-chain acyl-CoA dehydrogenase. The purified enzyme (10 μg of protein each) was applied to SDS-PAGE as follows: A, without 2-mercaptoethanol (2-ME) and without boiling the enzyme preparation; B, without 2-mercaptoethanol but with boiling; C, with 2-mercaptoethanol and with boiling. D represents the authentic proteins as calibration standards. The purified enzyme protein (10 μg of protein) was also subjected to 5.0% polyacrylamide gel without SDS, as shown in E. The migration is from top to bottom. From Ikeda and Tanaka.[2]

Molecular Weight and Subunit Structure

The molecular weight of the native enzyme is estimated to be 170,000 by gel filtration on BioGel A-0.5m chromatography. Analysis of the subunit by SDS-PAGE (Fig. 6, B and C) indicates that 2-methyl branched-chain acyl-CoA dehydrogenase is composed of four subunits of equal size with a molecular weight of 41,500. When the pure enzyme is subjected to SDS-PAGE without boiling, the enzyme gives an apparent single protein band with a molecular weight of 85,000, both in the presence and absence of 2-mercaptoethanol, indicating a dimeric form of the protein (Fig. 6, A).

These data suggest that the binding forces between four subunits are not equal, and that the force between two subunits in a dimer is stronger than that which binds two dimers.

Isoelectric Point, Amino Acid Composition, and Absorbance Coefficient

In isoelectric focusing, the enzyme is recovered in a broad peak with an isoelectric point of 5.5 ± 0.2.[1] Amino acid composition of the enzyme, as expressed in number of residues per subunit, is as follows: Asx, 29; Thr, 19; Ser, 40; Glx, 54; Pro, 12; Gly, 67; Ala, 38; ½Cys, 5; Val, 23; Met, 5; Ile, 19; Leu, 26; Tyr, 7; Phe, 16; His, 17; Lys, 20; Arg, 11; Trp, 2. The values are based on a subunit molecular weight of 41,500. The subunit molecular weight of the enzyme is calculated to be 42,400 from the amino acid composition. The specific volume of the enzyme is 0.72. The absorbance coefficient is estimated to be 12.5 and 10 mM phosphate buffer, pH 7.8, with the protein concentration determined by the microbiuret method.

Absorption and Fluroescence Spectra, and Prosthetic Group

The visible and ultraviolet spectra of the purified enzyme are shown in Fig. 7. The major absorption maxima are found at 275, 340, and 435 nm.

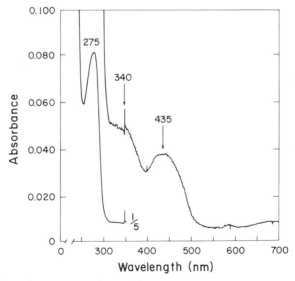

FIG. 7. Absorption spectrum of the purified 2-methyl branched-chain acyl-CoA dehydrogenase. The enzyme preparation ($A_{280\,nm}$ = 0.393, $A_{450\,nm}$ = 0.038) in 0.01 M potassium phosphate buffer, pH 7.8, at 25°, was used for the determination. The absorbance was scanned from 700 to 300 nm on the 0 to 0.1 scale and from 300 to 220 nm on the 0 to 0.5 scale. From Ikeda and Tanaka.[2]

The ratios of absorbance at 275, 340, and 435 are 10.3 : 1.3 : 1.0. The fluorescence emission of the enzyme excited at 450 nm shows a peak at 520 nm, as in the case of authentic FAD, and its fluorescence intensity is 28% of the equivalent amount of authentic FAD. The excitation spectrum of the enzyme fluorescence emitted at 530 nm shows three maxima at 370, 445, and 460 nm, as in the case of authentic FAD. Its fluorescence intensity is 25% of authentic FAD, indicating quenching due to FAD–protein interaction. These spectroscopic studies indicate that 2-methyl branched-chain acyl-CoA dehydrogenase contains FAD as a prosthetic group. The FAD content was 0.6 mol of FAD per mole of subunit, as calculated from these observed values. The absorption coefficient of FAD ($E_{450} = 11.3$ mM^{-1}) and the estimate of 41,500 for the subunit molecular weight are used for this calculation. The FAD content was probably underestimated due to a partial loss of FAD during the purification process. The relatively labile nature of FAD binding in this enzyme is also indicated by the observation that the activities of the crude and the purified preparation are enhanced 1.5-fold by the addition of 0.1 mM FAD. Thus, the FAD content of 2-methyl branched-chain acyl-CoA dehydrogenase is estimated to be originally 1 mol of FAD per mole of subunit in the native form.

Substrate Specificity, Reaction Products, and Kinetic Analysis

The optimum pH is 8.0 using *(S)*-2-methylbutyryl-CoA or isobutyryl-CoA as a substrate. The enzyme exhibits its highest activity with either *(S)*-2-methylbutyryl-CoA or isobutyryl-CoA as substrate. Activities for these two substrates are similar (Table II). When *(R)*-2-methylbutyryl-CoA is used as a substrate, its activity is 22% of that observed with *(S)*-2-methylbutyryl-CoA. The activity is extremely low or not detectable when the following compounds are used as substrate: isovaleryl-CoA, propionyl-CoA, *n*-butyryl-CoA, *n*-valeryl-CoA, *n*-hexanoyl-CoA, *n*-octanoyl-CoA, palmitoyl-CoA, glutaryl-CoA, and sarcosine. The V_{max} and K_m values for isobutyryl-CoA are 2.0 μmol per minute per milligram and 89 μM, respectively, and the V_{max} and K_m values for *(S)*-2-methylbutyryl-CoA are 2.2 μmol per minute per milligram and 20 μM, respectively.

The reaction products of the purified enzyme with *(S)*-2-methylbutyryl-CoA and isobutyryl-CoA are identified as tiglyl-CoA and methacryl-CoA, respectively, by detection of their hydrolysis products using chromatographic analysis.[2] When *(R)*-2-methylbutyryl-CoA is used as a substrate, ethylacrylyl-CoA is produced as the reaction product,[2] indicating that the two enantiomers of 2-methylbutyryl-CoA are stereospecifically dehydrogenated. Likewise, isobutyryl-CoA is dehydrogenated by removal of one hydrogen each from (pro-2S)-methyl and α-methine, as shown in an experiment using (2S)-[3-^{13}C]isobutyryl-CoA as substrate.[3] Thus, the de-

TABLE II
SUBSTRATE SPECIFICITY AND K_M OF THE PURIFIED
2-METHYL BRANCHED-CHAIN ACYL-CoA
DEHYDROGENASE[a]

Substrate	Specific activity (nmol DCIP reduced $min^{-1} mg^{-1}$)	K_m (μM)
Isobutyryl-CoA	2024[b]	89
(S)-2-Methylbutyryl-CoA	2200[b]	20
(R)-2-Methylbutyryl-CoA	484	—
Isovaleryl-CoA	0	—
Propionyl-CoA	0	—
n-Butyryl-CoA	110	—
n-Valeryl-CoA	0	—
n-Hexanoyl-CoA	0	—
Palmitoyl-CoA	0	—
Glutaryl-CoA	0	—
Sarcosine	0	—

[a] The dehydrogenating activity was determined by the dye-reduction assay in the presence of 100 μM FAD. The substrate concentration was 100 μM, except for isobutyryl-CoA which was used at 200 μM. From Ikeda and Tanaka.[2]

[b] Represents V_{max}.

hydrogenation of (S)- and (R)-2-methylbutyryl-CoAs, and isobutyryl-CoA follows the identical stereochemical course (Fig. 1). 2-(Propyl)valeryl-CoA (Valproyl-CoA) is dehydrogenated by this enzyme but not by any other acyl-CoA dehydrogenases (Y. Ikeda, J. G. Arnez, and K. Tanaka, unpublished observations).

2-Methyl branched-chain acyl-CoA dehydrogenase activity with (S)-2-methylbutyryl-CoA and that with isobutyryl-CoA are both competitively inhibited by tiglyl-CoA, the reaction product from (S)-2-methylbutyryl-CoA. The K_i values for tiglyl-CoA are 7 μM when (S)-2-methylbutyryl-CoA is the substrate and 3 μM when isobutyryl-CoA is the substrate. This product inhibition is highly specific, and the enzyme activities are not significantly inhibited by 3-methylcrotonyl-CoA.

Requirement for ETF as Natural Electron Acceptor

In this experiment, ETF purified from rat liver mitochondria[1,2,8] is used as an intermediary electron acceptor in place of PMS in the spectrophoto-

[8] Y. Ikeda and K. Tanaka, *J. Biol. Chem.* **258,** 1077 (1983).

metric assay as described under Assay Method. ETF serves as a natural electron acceptor for 2-methyl branched-chain CoA dehydrogenase with either (S)-2-methylbutyryl-CoA or isobutyryl-CoA as substrate. The specific activity with (S)-2-methylbutyryl-CoA in the presence of 3.4 μM ETF is 0.37 μmol per minute per milligram and that of isobutyryl-CoA with 6.6 μM ETF is 0.39 μmol per minute per milligram. These values are approximately 20% of the corresponding values obtained using PMS. When isobutyryl-CoA is the substrate, twice as much ETF is required as that required with (S)-2-methylbutyryl-CoA to obtain an approximately equal specific activity.

Inhibitors

The purified enzyme is inhibited by several sulfhydryl reagents. Essentially identical inhibitory effects are observed with either substrate. The activity is severely inhibited by 0.1 mM *p*-hydroxymercuribenzoate. *N*-Ethylmaleimide (2 mM) and methylmercury iodide (0.1 mM) inhibit the enzyme activities by 70 and 50%, respectively. Iodoacetamide (2 mM) does not significantly inhibit the activity. The enzyme activities are also completely inhibited by heavy metals such as Hg^{2+} (0.1 mM), Cu^{2+} (0.1 mM), and Ag^+ (0.1 mM), which are also known to affect the thiol groups in proteins. These results suggest that 2-methyl branched-chain acyl-CoA dehydrogenase contains an essential cysteine residue, as in the case of three straight-chain acyl-CoA dehydrogenases.[9] This cysteine residue may be the base that obstructs C-2 protons.[9]

Immunochemical Properties

Antibody directed against 2-methyl branched-chain acyl-CoA dehydrogenase has not been raised, since this enzyme was not purified in an amount sufficient for this purpose. However, 2-methyl branched-chain acyl-CoA dehydrogenase is immunologically distinct from the four other acyl-CoA dehydrogenases and the final enzyme preparation is not contaminated by any other acyl-CoA dehydrogenases. This is indicated by the following experiments. In both the immunoprecipitation and Ouchterlony double-diffusion experiments, the purified 2-methyl branched-chain acyl-CoA dehydrogenase (10 μg of protein) exhibits no cross-reactivity with any of the antisera, each raised against rat isovaleryl-CoA-,[8] short-chain acyl-CoA-,[10] medium-chain acyl-CoA-,[10] or long-chain acyl-CoA dehydrogenase.[10]

---

[9] K. O. Ikeda, Y. Ikeda, and K. Tanaka, *J. Biol. Chem.* **260**, 1338 (1985).
[10] Y. Ikeda, K. O. Ikeda, and K. Tanaka, *J. Biol. Chem.* **260**, 1311 (1985).

[47] Isovaleryl-Coa Dehydrogenase from Rat Liver

By YASUYUKI IKEDA and KAY TANAKA

Mammalian isovaleryl-coa dehydrogenase (EC 1.3.99.10) is a mito-chondrial flavoprotein that catalyzes the third step of the leucine oxidative metabolism.[1-3]

$$\text{ETF} \cdot \text{FAD}_{ox} \quad \text{ETF} \cdot \text{FAD}_{red}$$

$$\text{E} \cdot \text{FAD}_{ox} + \text{CH}_3(\text{CH}_3)-\text{CH}-\text{CH}_2-\text{COSCoA} \rightleftharpoons \text{E} \cdot \text{FAD}_{ox}$$
$$+ \text{CH}_3(\text{CH}_3)-\text{C}=\text{CH}-\text{COSCoA}$$

The enzyme removes one hydrogen each at C-2 and C-3 of isovaleryl-coa in a stereospecific manner producing 3-methylcrotonyl-coa.[3,4] This reaction requires a second flavoprotein, electron-transfer flavoprotein (ETF), as an electron acceptor.[3]

Historically, it has long been known that in the first two steps the three branched-chain amino acids (leucine, isoleucine, and valine) are metabolized in an analogous manner, via the corresponding 2-keto acids, resulting in the formation of isovaleryl-coa, (S)-2-methylbutyryl-coa, and isobutyryl-coa, respectively.[5] However, the identities of the enzymes which catalyze the dehydrogenation of these branched-chain acyl-coas were not known until recently. Previously, three acyl-coa dehydrogenases were known. These are butyryl-coa (short-chain acyl-coa),[6] general acyl-coa (medium-chain acyl-coa),[7] and palmitoyl-coa (long-chain acyl-coa) dehydrogenases.[8] These mitochondrial enzymes catalyze the 2,3-dehydrogenation of straight-chain acyl-coas with varying chain length. Butyryl-coa, octanoyl-coa, and palmitoyl-coa, respectively, are their optimal substrates. Initially, the branched-chain acyl-coas were believed to be dehydrogenated by short-chain acyl-coa dehydrogenase.[6,9] However, the

[1] C. Noda, W. J. Rhead, and K. Tanaka, Proc. Natl. Acad. Sci. U.S.A. 77, 2646 (1980).
[2] Y. Ikeda, C. Dabrowski, and K. Tanaka, J. Biol. Chem. 258, 1066 (1983).
[3] Y. Ikeda and K. Tanaka, J. Biol. Chem. 258, 1077 (1983).
[4] D. J. Aberhart, G. Finocchiaro, Y. Ikeda, and K. Tanaka, Bioorg. Chem. 14, 170 (1986).
[5] K. Tanaka and L. E. Rosenberg, in "Metabolic Basis of Inherited Disease" (J. B. Stanbury, J. B. Wyngaarden, D. S. Fredrickson, J. L. Goldstein, and M. S. Brown, eds.), 5th Ed., p. 440. McGraw-Hill, New York, 1982.
[6] D. E. Green, S. Mii, H. R. Mahler, and R. M. Bock, J. Biol. Chem. 206, 1 (1954).
[7] F. L. Crane, S. Mii, J. G. Hauge, D. E. Green, and H. Beinert, J. Biol. Chem. 218, 701 (1956).
[8] J. G. Hauge, F. L. Crane, and H. Beinert, J. Biol. Chem. 219, 727 (1956).
[9] H. Beinert, this series, Vol. 5, p. 546.

existence of a specific isovaleryl-CoA dehydrogenase was suggested in 1966 by Tanaka et al.,[10-12] from the biochemical observations on patients with isovaleric acidemia, an inborn error of leucine metabolism. Isovaleric acid[10] and its glycine conjugate[11] accumulate alone in the blood and urine, respectively, of the patients with this disease. Other short-chain fatty acids, including n-butyric, isobutyric, and n-hexanoic acids, did not accumulate. This observation suggested that isovaleryl-CoA is not dehydrogenated by short-chain acyl-CoA dehydrogenase. However, the enzymological identification of isovaleryl-CoA dehydrogenase was accomplished only in 1980.[1] The purification of this enzyme from rat liver mitochondria was subsequently achieved in 1983.[3] The delay in the biochemical characterization of this enzyme was mainly due to its low abundance in mammalian tissues.

The procedure, outlined here, made it possible to purify isovaleryl-CoA dehydrogenase in a homogeneous form from rat liver. Monitoring of activities to dehydrogenate various acyl-CoA substrates in sequential fractions at each column chromatography steps was necessary to accomplish purification. The use of Matrex Gel Blue A with elution with FAD/NaCl double gradient was particularly useful, achieving eightfold enrichment of isovaleryl-CoA dehydrogenase and its complete separation from short-chain acyl-CoA dehydrogenase. In the process of purifying isovaleryl-CoA dehydrogenase, another previously unknown enzyme, 2-methyl branched-chain acyl-CoA dehydrogenase, which dehydrogenates acyl-CoA having an alkyl branching at the α-position, was identified[2,13] (see chapter [46] in this volume). Human isovaleryl-CoA dehydrogenase has recently been purified from cadaver liver,[14] using procedures similar to those used for the purification of rat isovaleryl-CoA dehydrogenase.

Assay Method

Tritium Release Assay

Principle. When isovaleryl-CoA dehydrogenase is reacted with [2,3-³H]isovaleryl-CoA, the tritium that is taken up by the enzyme exchanges with protons in water, forming tritiated water.[15] In this assay, the activity of isovaleryl-CoA dehydrogenase is determined by quantifying the

[10] K. Tanaka, M. A. Budd, M. L. Efron, and K. J. Isselbacher, *Proc. Natl. Acad. Sci. U.S.A.* **56,** 236 (1966).

[11] K. Tanaka and K. J. Isselbacher, *J. Biol. Chem.* **242,** 2966 (1967).

[12] W. J. Rhead and K. Tanaka, *Proc. Natl. Acad. Sci. U.S.A.* **77,** 580 (1980).

[13] Y. Ikeda and K. Tanaka, *J. Biol. Chem.* **258,** 9477 (1983).

[14] G. Finocchiaro, M. Ito, and K. Tanaka, *J. Biol. Chem.* **262,** 7982 (1987).

[15] W. J. Rhead, C. L. Hall, K. Tanaka, *J. Biol. Chem.* **256,** 1616 (1981).

amount of tritium released from [2,3-³H]isovaleryl-CoA as tritium water.[15] This method has an advantage of being free from interference by nonspecific reductants, so that it can be utilized for assays in crude preparations.

Reagents

Potassium phosphate, 0.5 M, pH 8.0

Phenazine methosulfate (PMS), 33 mM, store at 4° in a brown bottle as soon as prepared

Flavin adenine dinucleotide (FAD), 1 mM, store at 4° in a brown bottle

Isovaleryl-CoA, 2.5 mM (10 mCi per 10 mmol), in 2 mM acetate buffer, pH 5.0, store frozen

Preparation of [2,3-³H]isovaleryl-CoA. [2,3-³H]Isovaleric acid (10 mCi/mmol) was synthesized by New England Nuclear (Boston, MA). The coenzyme A ester of [2,3-³H]isovaleric acid was synthesized by a modified mixed anhydride synthesis.[16] The synthesized [2,3-³H]isovaleryl-CoA was purified by ascending paper chromatography using ethanol:0.1 M potassium acetate (1:1), pH 4.5, as a developing solvent.

Assay Procedure. Into a disposable culture tube (12 × 75 mm) the following reagents are added: 20 μl of potassium phosphate buffer, pH 8.0, 10 μl of PMS, 10 μl of FAD, and 10 μl of isovaleryl-CoA. The reaction is started by addition of 50 μl of enzyme solution to the test tube. The test tube is incubated at 37° for 30 min. The reaction is stopped by addition of 10 μl of 0.5 N HCl into the reaction mixture and cooling to 0°. The whole reaction mixture is applied on to a column (0.5 × 30 mm: actually a Pasteur pipet) packed with anion-exchange resin (AGI-X8), and the resin was washed with 3 ml of distilled water. Tritiated water formed is recovered in the wash while the labeled substrate is bound to the resin. Aliquot (2 ml) of the eluate is transferred into a vial and mixed with 15 ml of Biofluor (New England Nuclear). The sample is then counted in a scintillation counter.

Spectrophotometric Assay

Principle. The reduction of 2,6-dichlorophenolindophenol (DCIP) is followed spectrophotometrically using PMS as an intermediary electron acceptor. This method is subject to interference by nonspecific reductants and thioesterases, and applicable only for assaying the activity of partially purified specimens from the steps after ammonium sulfate fractionation.

---

[16] T. Wieland and L. Rueff, *Angew Chem. Int. Ed. Engl.* **65**, 186 (1953).

Reagents

PMS, 33 mM
FAD, 1 mM
DCIP, 0.123 mM, in 0.25 M potassium phosphate, pH 8.0, store at 4°
Substrate: Isovaleryl-CoA or other acyl-CoAs, 1 mM, in 2 mM acetate
buffer, pH 5.0, store frozen

Preparation of Isovaleryl-CoA. Isovaleryl-CoA is commercially available, but commercial samples mostly contain 15% 2-methylbutyryl-CoA as a contaminant. It is recommended to synthesize isovaleryl-CoA from pure isovaleric acid and coenzyme A. Isovaleric acid must be analyzed by gas chromatography/mass spectrometry[17] before use in the synthesis since it sometimes also contains 15% 2-methylbutyric acid. The coenzyme A thioester synthesis is carried out using pure isovaleric acid (Eastman) by the mixed anhydride synthesis.[16] The thioester is purified by paper chromatography as described above.

Procedure. Into a 1 cm light path semimicrocuvette, the following reagents are added: 150 μl of the enzyme, 50 μl of FAD, 50 μl of PMS, and 200 μl of DCIP buffer solution, pH 8.0. The reaction is started by the injection of 50 μl of isovaleryl-CoA into the cuvette, making the final volume 500 μl. Bleaching of DCIP is followed at 600 nm wavelength at 32° for 1 min using a Beckman model 3600 double-beam spectrophotometer. The use of a microprocessor-controlled spectrophotometer such as Beckman DU7 makes the assay faster and more accurate. Enzyme activity is expressed as micromoles or nanomoles of DCIP reduced per milligram of protein or milliliter of enzyme solution per minute. The extinction coefficient of DCIP (21 mM^{-1} cm^{-1}) at 600 nm is used as the basis for computation of the amount of DCIP reduced.

The enzyme activity can also be measured using an appropriate amount of ETF (1–15 nmol of bound FAD) as an intermediary electron acceptor, replacing PMS.[3] The method is essentially the same as the assay method described above.

Purification Procedure

The entire purification procedure is summarized in Table I. The tritium release assay is used to determine the activity of isovaleryl-CoA dehydrogenase in crude preparations such as disrupted mitochondria, their sonicated supernatant, and ammonium sulfate fractions. Spectrophotome-

[17] K. Tanaka and G. M. Yu, *Clin. Chim. Acta* **175**, 151 (1973).

TABLE I
PURIFICATION STEPS FOR ISOVALERYL-CoA DEHYDROGENASE FROM RAT LIVER
MITOCHONDRIA[a]

Step	Total protein (mg)	Total activity (nmol/min)	Specific activity (nmol/min/mg)	Purifi- cation (-fold)	Yield (%)
Mitochondria	55,620	155,200	2.8	1	100
1. Sonication supernatant	34,400	131,900	3.5	1.3	85
2. Ammonium sulfate (40–80%)	17,070	109,500	4.2	1.5	71
3. DEAE-Sephadex A-50	1,160	60,740	52.4	18.7	39
4. Hydroxylapatite	137	38,600	282	100	25
5. Matrex Gel Blue A	8.7	20,000	2,300	821	13
6. BioGel A-0.5m	3.0	8,040	2,680	957	5

[a] From Ikeda and Tanaka.[3]

tric assay is more convenient for monitoring acyl-CoA dehydrogenase activities in fractions from column chromatographic steps.

Preparation of Mitochondria, Solubilization, and Ammonium Sulfate Fractionation

In a typical experiment, 100 adult male Charles River CD rats, each weighing 200 to 280 g, are sacrificed by decapitation. The livers are removed and perfused with saline to remove blood. Mitochondria are prepared by the method of de Duve *et al.*[18] The isolated mitochondrial pellets (55.6 g of protein), suspended in 10 mM potassium buffer (pH 8.0) with 1 mM ethylenediaminetetraacetic acid (EDTA), are sonicated 10 times for 2 min with 1-min rest intervals at 0° using a Branson sonifier. The sonicated mitochondria are centrifuged at 105,000 g for 60 min (step 1). The supernatant is fractionated by the ammonium sulfate precipitation (40–80%) (step 2). This fraction contains most of the activity of all of the acyl-CoA dehydrogenases. This fraction is dialyzed against 10 mM potassium phosphate buffer containing 0.5 mM EDTA, pH 7.5.

DEAE-Sephadex A-50 Column Chromatography (Step 3)

The dialyzed sample preparation (17 g of protein) is applied to four DEAE-Sephadex A-50 columns (4.5 × 20 cm) equilibrated with 10 mM potassium phosphate buffer containing 0.5 mM EDTA, pH 7.5. All of the acyl-CoA dehydrogenase activities are adsorbed onto the resin. After wash-

[18] C. de Duve, R. Wattiaux, and P. Baudhuin, *Adv. Enzymol.* **24**, 291 (1962).

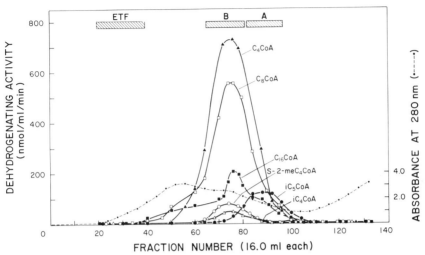

FIG. 1. DEAE-Sephadex A-50 column chromatography of the ammonium sulfate precipitate fraction. The enzyme solution (3310 mg of protein, 250 ml) from the ammonium sulfate fractionation step (done in the presence of 0.5 mM EDTA) applied to a DEAE-Sephadex A-50 column (4.5 × 20 cm) equilibrated with 10 mM potassium phosphate buffer (pH 8.0) containing 0.5 mM EDTA as described in Experimental Procedures. Fractions were assayed for various acyl-CoA dehydrogenating activities using the dye-reduction assay. ●, Isovaleryl-CoA; ○, (S)-2-methylbutyryl-CoA; △, isobutyryl-CoA; ▲, n-butyryl-CoA; □, n-octanoyl-CoA; ■, palmitoyl-CoA. Tubes 83 to 98 were pooled as preparation A. Tubes 65 to 82 were pooled as preparation B. Tubes 20 to 40 were pooled as the electron transfer flavoprotein preparation (ETF). From Ikeda et al.[2]

ing with 5 liters of the equilibration buffer, proteins are eluted by a linear gradient formed with 1 liter each of the equilibration buffer and the same buffer containing 0.6 M NaCl. Elution is done at a flow rate of 2 ml/min. In this procedure, isovaleryl-CoA dehydrogenase is partially separated from four other acyl-CoA dehydrogenases (Fig. 1). It is eluted with 0.4 M NaCl, while the four other acyl-CoA dehydrogenases are eluted together with 0.35 M NaCl. The fractions which contain isovaleryl-CoA dehydrogenase are pooled (fraction A) and dialyzed against 10 mM potassium phosphate buffer, pH 7.5.

Hydroxylapatite Column Chromatography (Step 4)

Hydroxylapatite, prepared in our laboratory according to the method of Tiselius et al.,[19] is mixed with an equal volume of cellulose powder (CF11, Whatmann) and packed into a column (4.5 × 17 cm). Hydroxyl-

[19] A. Tiselius, S. Hjerten, and O. Levin, *Arch. Biochem. Biophys.* **65**, 132 (1956).

apatite obtained from commercial sources does not give as good a resolution as the home-made preparation. The column is equilibrated with 10 mM potassium phosphate buffer, pH 7.5. The dialyzate (1.2 g of protein) from step 3 is applied on two hydroxylapatite columns. Each column is washed with 2 liters of the equilibration buffer. Proteins are eluted with a linear gradient formed with 1 liter each of the equilibration buffer and 0.33 M potassium phosphate buffer, pH 7.5. As shown in Fig. 2, isovaleryl-CoA dehydrogenase is coeluted with residual short-chain acyl-CoA dehydrogenase with 0.2 M phosphate buffer as a sharp single peak. They are completely separated from medium-chain acyl-CoA dehydrogenase, and almost separated from long-chain and 2-methyl branched-chain acyl-CoA dehydrogenases. Fractions 40–64 are pooled (fraction C) and concentrated using Amicon diaflow membrane PM-30, and are dialyzed against 10 mM potassium phosphate buffer containing 0.5 mM EDTA, pH 8.0.

Matrix Gel Blue A Column Chromatography (Step 5)

The dialyzate from step 4 is applied to a Matrex Gel Blue A column (1 × 17 cm) equilibrated with 10 mM potassium phosphate buffer con-

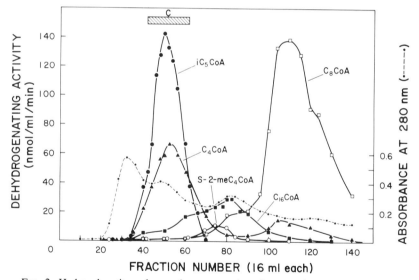

FIG. 2. Hydroxylapatite column chromatography of preparation A. The solution of preparation A (770 mg of protein, 60 ml) was applied to a hydroxylapatite column (4.5 × 15 cm) as described in Experimental Procedures. Fractions were assayed for various dehydrogenating activities using the dye-reduction method. ●, Isovaleryl-CoA; ○, (S)-2-methylbutyryl-CoA; ▲, n-butyryl-CoA; □, n-octanoyl-CoA; ■, palmitoyl-CoA. Tubes 42 to 62 were combined as preparation C. From Ikeda et al.[2]

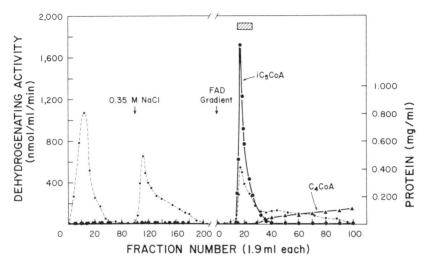

FRACTION NUMBER (1.9 ml each)

FIG. 3. Separation of isovaleryl-CoA dehydrogenase from butyryl-CoA dehydrogenase on Matrex Gel Blue A column chromatography. The sample solution (138 mg of protein in 45 ml) was applied to a Matrex Gel Blue A column (1 × 17 cm) equilibrated with 10 mM potassium phosphate buffer (pH 8.0) containing 0.5 mM EDTA. The column was first washed with the same buffer and then further washed with the same buffer containing 0.35 M NaCl. The adsorbed proteins were eluted with a linear gradient of 10 mM KPO$_4$ buffer containing 0.35 mM NaCl and the same buffer containing a 0.40 M NaCl and 8 mM FAD. The hatched rectangle indicates the fractions combined. C$_4$CoA, n-butyryl-CoA; iC$_5$CoA, isovaleryl-CoA. From Ikeda and Tanaka.[3]

taining 0.5 mM EDTA, pH 8.0. The column is washed with the same buffer until the absorbance at 280 nm returns to the baseline, and further washed with the same phosphate buffer containing 0.35 M NaCl. Proteins are eluted with a linear gradient formed with 90 ml each of 10 mM phosphate buffer containing 0.35 M NaCl and the same buffer containing 0.4 M NaCl and 8 mM FAD. As shown in Fig. 3, neither isovaleryl-CoA dehydrogenase nor short-chain acyl-CoA dehydrogenase is eluted with 10 mM phosphate buffer containing 0.35 M NaCl alone. However, this procedure effectively removes proteins other than these acyl-CoA dehydrogenases. Isovaleryl-CoA dehydrogenase is eluted as a sharp peak at 1–3 mM FAD, while no significant amount of short-chain acyl-CoA dehydrogenase is eluted in this region. Short-chain acyl-CoA dehydrogenase is eluted as a very broad peak in the presence of FAD at concentrations higher than 3 mM. The affinity chromatography results in an 8-fold enrichment of isovaleryl-CoA dehydrogenase specific activity. The resulting preparation is completely devoid of any other acyl-CoA dehydrogenases. Fractions 14 through 25, containing isovaleryl-CoA dehydrogenase, are pooled and concentrated.

BioGel A-0.5m Gel Filtration (Step 6)

The concentrated sample (8.7 mg of protein in 1.5 ml) from step 5 is applied to a BioGel A-0.5m column (1.5 × 100 cm) equilibrated with 10 mM phosphate buffer containing 0.2 M NaCl, pH 8.0. The column is eluted with the same buffer at a flow rate of 0.16 ml per min, and 1.6 ml each is collected in each fraction. Isovaleryl-CoA dehydrogenase is eluted as a sharp peak in fractions 51 to 70 (Fig. 4). These fractions are combined and concentrated. This represents the final preparation. When the purified enzyme is stored in 50% glycerol at −20°, it is stable for at least 6 months. Through these procedures, 3 mg of isovaleryl-CoA dehydrogenase is obtained. The final preparation shows a single band on sodium dodecyl sulfate-polyacrylamide gel electrophoresis (SDS-PAGE) both in the presence and absence of 2-mercaptoethanol (Fig. 5A and B). It also gives a single band on 7.0% and 5% gels in polyacrylamide gel electrophoresis without SDS (Fig. 5D and E).

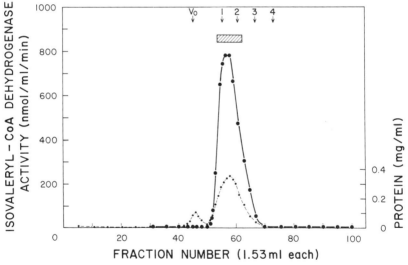

FIG. 4. The final step of isovaleryl-CoA dehydrogenase purification by BioGel A-0.5m chromatography. The sample (8.7 mg of protein in 1.7 ml) obtained from Matrex Gel Blue A chromatography was applied to a BioGel A-0.5m column (1.5 × 100 cm) equilibrated with 10 mM KPO$_4$ buffer, pH 8.0, containing 0.2 M NaCl. The arrows indicate the locations of the peaks of pure proteins used as standards: 1, catalase (M_r = 240,000); 2, aldolase (M_r = 158,000); 3, bovine serum albumin (M_r = 68,000); 4, ovalbumin (M_r = 45,000). V_0 indicates the void volume. The hatched rectangle indicates the fractions combined as the final preparation. From Ikeda and Tanaka.[3]

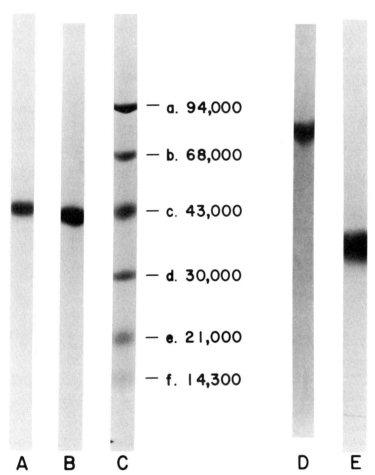

FIG. 5. Polyacrylamide gel electrophoresis with and without SDS of the purified isovaleryl-CoA dehydrogenase. The purified isovaleryl-CoA dehydrogenase (12 μg of protein each) was applied to SDS-PAGE without (A) and with (B) 2-mercaptoethanol. The authentic proteins were applied to SDS-PAGE with 2-mercaptoethanol (C) as calibration standards. The purified enzyme protein (10 μg of protein each) was also applied to 7.0% (D) and 5.0% (E) polyacrylamide gels without SDS. The migration is from the top to the bottom. From Ikeda and Tanaka.[3]

Properties of the Enzyme

Properties of isovaleryl-CoA dehydrogenase are summarized in Table II in comparison to those of short-chain[19] and 2-methyl branched-chain acyl-CoA dehydrogenase[13] purified from rat liver mitochondria.

Molecular Weight and Subunit Structure

The molecular weight of the native enzyme is estimated to be 175,000

TABLE II
PROPERTIES OF ISOVALERYL-CoA DEHYDROGENASE IN COMPARISON WITH 2-METHYL BRANCHED-CHAIN ACYL-CoA AND SHORT-CHAIN ACYL-CoA DEHYDROGENASES[a]

	Value		
Property	Isovaleryl-CoA dehydrogenase (IVD)	2-Methyl branched-chain acyl-CoA dehydrogenase[b] (2-meBCAD)	Short-chain acyl-CoA dehydrogenase (SCAD)
Native M_r	175,000	170,000	160,000
Subunit M_r	43,000	41,500	41,000
Isoelectric point	5.6 ± 0.2	5.5 ± 0.2	4.8 ± 0.2
$A^{1\%}_{280\,nm}$	12.3	12.5	12.8
Ratios of spectral	275:325:375:445	275:340:435	275:365:450
maxima (nm)	12.5:1.5:0.9:1.0	10.3:1.3:1.0	6.3:0.8:1.0
FAD content/subunit			
(mol/sub)	1	1	1
Fluorescence[b]			
Emission peak (nm)	530	530	530
Excitation peak (nm)	370 and 465	380 and 465	370 and 465
Intensity (% of FAD)	24	28	2
Kinetic parameters for the			
respective or best substrate[c]			
V_{max} (μmol min^{-1} mg^{-1})			
2-meC$_4$CoA	0	2.2	0
iC$_4$CoA	0	2.0	0
iC$_5$CoA	2.6	0	0
n-C$_4$CoA	0	0	7.5
K_m (μM)			
2-meC$_4$CoA	—	20	—
iC$_4$CoA	—	89	—
iC$_5$CoA	33	—	—
n-C$_4$CoA	—	—	18
Antibody to IVD	Positive reaction	No cross-reaction	No cross-reaction
Antibody to SCAD	No cross-reaction	No cross-reaction	Positive reaction

[a] From Ikeda and Tanaka.[13]
[b] See chapter [46], this volume.
[c] Substrate abbreviations: 2-meC$_4$CoA, 2-methylbutyryl-CoA; iC$_4$CoA, isobutyryl-CoA; iC$_5$CoA, isovaleryl-CoA; n-C$_4$CoA, n-butyryl-CoA.

by gel filtration on BioGel A-0.5m. Analysis of the subunit by SDS-PAGE (Fig. 5A and B) indicates that isovaleryl-CoA dehydrogenase is composed of four subunits of equal size with an approximate molecular weight of 43,000. It is slightly smaller than medium-chain and long-chain acyl-CoA dehydrogenases (both 45,000),[20] but larger than short-chain (41,000)[20] and 2-methyl branched-chain acyl-CoA dehydrogenases (41,500).[13] These differences in molecular size are clearly visible when they are analyzed together on a slab gel.[20]

Isoelectric Point, Amino Acid Composition, and Absorbance Coefficient

The enzyme shows a relatively broad peak in isoelectric focusing with an isoelectric point of 5.65 ± 0.2. Amino acid composition of isovaleryl-CoA dehydrogenase is described in number of residues per subunit as follows: Asx, 42; Thr, 17; Ser, 20; Glx, 45; Pro, 15; Gly, 37; Ala, 34; ½ Cys, 7; Val, 27; Met, 9; Ile, 19; Leu, 34; Tyr, 13; Phe, 19; His, 10; Lys, 25; Arg, 21; Trp, 4. The values are based on a subunit molecular weight of 43,000. This amino acid composition gives the subunit molecular weight of 44,160. The specific volume of the enzyme is 0.73. The absorbance coefficient is estimated to be 12.3 in 10 mM phosphate buffer, pH 7.8, with the protein concentration determined by the microbiuret method.

Absorption and Fluorescence Spectra, and Prosthetic Group

The visible and ultraviolet spectra of the purified isovaleryl-CoA dehydrogenase are shown in Fig. 6. The apparent major absorption maxima are found at 275 and 445 nm. Weak shoulders are observed at 325 and 375 nm. The ratios of absorbance at 275, 325, 375, and 445 nm are 12.0 : 1.5 : 0.9 : 1.0. The fluorescence emission spectrum of the enzyme excited at 450 nm shows a peak at 530 nm, as in the case of authentic FAD, and its fluorescence intensity is 24% of that of authentic FAD.[3] The excitation spectrum of the enzyme with fluorescence emitted at 530 nm shows three maxima at 370, 445, and 460 nm, as in the case of authentic FAD.[3] Its fluorescence intensity is 25% of that of authentic FAD, indicating quenching due to FAD–protein interaction. These spectroscopic studies indicate that isovaleryl-CoA dehydrogenase contains FAD as a prosthetic group. However, its absorption spectrum is peculiar, having a small shoulder at 325 nm. This shoulder is not observed in the absorption spectra of other acyl-CoA dehydrogenases containing FAD as a prosthetic group.[13,21] The nature of the 325 nm band remains to be elucidated. It is interesting to note that human liver isovaleryl-CoA dehydrogenase also exhibits the 325 nm band.[14]

[20] Y. Ikeda, K. O. Ikeda, and K. Tanaka, *J. Biol. Chem.* **260**, 1311 (1985).
[21] Y. Ikeda, K. O. Ikeda, and K. Tanaka, *Biochemistry* **24**, 7192 (1985).

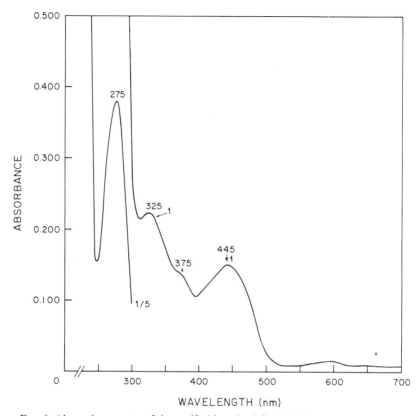

FIG. 6. Absorption spectra of the purified isovaleryl-CoA dehydrogenase. The enzyme solution, 1.5 mg of protein/ml in 0.01 M potassium phosphate buffer, pH 7.8, was used. The absorbance was scanned from 750 to 300 nm on the 0 to 0.5 scale and from 300 to 220 nm on the 0 to 2.0 scale. From Ikeda and Tanaka.[3]

The FAD content of the purified preparation was 0.6 mol of FAD per mole of subunit as calculated from these observed values. The absorption coefficient of FAD ($E_{450} = 11.3$ mM^{-1}) and the estimate of 43,000 for the subunit molecular weight are used for this calculation. The FAD content was lower than 1.0 mol per subunit, probably due to a partial loss of FAD during the purification process. The activity of the purified preparation was enhanced 1.5-fold by the addition of 0.1 mM FAD. Thus the FAD content of isovaleryl-CoA dehydrogenase is estimated to be originally 1 mol per mole of subunit in the native form. The relatively labile nature of FAD binding in this enzyme is indicated by the observation that the activity in the mitochondrial sonic supernatant was also enhanced 1.5 to 2.5-fold by the addition of 0.1 mM FAD.

Substrate Specificity, Reaction Product, and Kinetic Analysis

The optimum pH is 8.0 using isovaleryl-CoA as substrate. The enzyme exhibits its highest activity with isovaleryl-CoA as substrate. Its V_{max} value is 2.68 μmol per milligram per minute and its K_m value is 33 μM. The only other substrate which is dehydrogenated at a significant rate by this enzyme is n-valeryl-CoA (Table III). The V_{max} and K_m values for n-valeryl-CoA are 0.8 μmol per milligram per minute and 400 μM, respectively. The activity is extremely low or not detectable when the following acyl-CoAs are used as substrate: (S)-2-methylbutyryl-CoA,[22] isobutyryl-CoA, propionyl-CoA, n-butyryl-CoA, n-hexanoyl-CoA, n-octanoyl-CoA, palmitoyl-CoA, crotonyl-CoA, and glutaryl-CoA (Table III). The substrate specificity of human isovaleryl-CoA dehydrogenase is basically similar, but it exhibits low activities with n-butyryl-CoA and n-hexanoyl-CoA (18 and 22%, respectively, of that with isovaleryl-CoA).[14]

The product of the reaction of this enzyme with isovaleryl-CoA as substrate was identified as 3-methylcrotonyl-CoA by detection of its hydrolysis product, 3-methylcrotonic acid, using gas chromatographic analysis.[3] The enzyme activity is competitively inhibited by 3-methylcrotonyl-CoA with a K_i value of 20 μM. This product inhibition is highly specific, and the enzyme activity is not significantly inhibited by other 2,3-unsaturated acyl-CoAs such as crotonyl-CoA and tiglyl-CoA.

Requirement for ETF as Natural Electron Acceptor

ETF serves as a natural electron acceptor for isovaleryl-CoA dehydrogenase with isovaleryl-CoA as substrate. When ETF (3.3 nmol of ETF-bound FAD), purified from rat liver mitochondria,[2,3,13] is added to the assay media replacing PMS, a specific activity of 1.38 μmol per milligram per minute is observed with isovaleryl-CoA as substrate. Its value is approximately one-half of that observed with PMS.

Inhibitors

Rat isovaleryl-CoA dehydrogenase activity is completely inhibited by several organic sulfhydryl reagents such as 2 mM N-ethylmaleimide, 0.01 mM p-hydroxymercuribenzoate, and 0.1 mM methylmercury iodide. Iodoacetamide (2 mM) does not significantly inhibit the activity. The enzyme activity is also completely inhibited by heavy metals, such as Hg^{2+} (0.1 mM), Cu^{2+} (0.01 mM), and Ag^+ (0.01 mM) that are known to affect the thiol groups in proteins. These observations suggest that isovaleryl-CoA

---

[22] S-2-methylbutyryl-CoA was synthesized from CoA and S-2-methylbutyric acid that is prepared from L-isoleucine using L-amino acid oxidase.

TABLE III
SUBSTRATE SPECIFICITY OF PURIFIED ISOVALERYL-CoA
DEHYDROGENASE[a]

Substrate	Specific activity (nmol DCIP[a] reduced mg/min)	K_m (μM)
Isovaleryl-CoA	2680[b]	33
(S)-2-Methylbutyryl-CoA	0	—
Isobutyrl-CoA	0	—
Propionyl-CoA	0	—
n-Butyryl-CoA	86[c]	—
n-Valeryl-CoA	832[b]	400
n-Hexanoyl-CoA	0	—
n-Octanoyl-CoA	0	—
Palmitoyl-CoA	0	—
Crotonyl-CoA	0	—
Glutaryl-CoA	0	—
Sarcosine	0	—

[a] The dehydrogenating activity was determined by the dye-reduction assay in the presence of 20 μM FAD. The substrate concentration was 100 μM. Details of the method are described in the text. DCIP, Dichloroindophenol. From Ikeda and Tanaka.[3]
[b] Represents V_{max}.
[c] $\Delta 600$ nm was 0.005. Data expressed as zero were less than 0.005 in $\Delta 600$ nm. From Ikeda and Tanaka.[3]

dehydrogenase contains an essential cysteine residue that may be involved in the C-2 proton abstraction, as in the case of three straight-chain acyl-CoA dehydrogenases.[23] The activity of the purified enzyme is completely inhibited by 10 μM (methylenecyclopropyl)acetyl-CoA, the toxic metabolite of hypoglycin. Diisopropyl fluorophosphate and phenylmethylsulfonyl fluoride do not affect the activity of this enzyme at all.

Immunochemical Properties

In Ouchterlony double diffusion, the antibody raised against the purified isovaleryl-CoA dehydrogenase forms a single precipitin line with the pure enzyme preparation in serial dilutions of either the antibody or the enzyme.[3] From the immunotitration experiment, it is estimated that anti-isovaleryl-CoA dehydrogenase antibody (100 μl of original serum) precipitates approximately 15 μg of the pure isovaleryl-CoA dehydrogenase. The anti-isovaleryl-CoA dehydrogenase antibody does not cross-react with other acyl-CoA dehydrogenases such as 2-methyl branched-chain acyl-

[23] K. O. Ikeda, Y. Ikeda, and K. Tanaka, *J. Biol. Chem.* **260**, 1338 (1985).

CoA,[13] short-chain acyl-CoA,[20] medium-chain acyl-CoA,[20] and long-chain acyl-CoA dehydrogenases.[20] Individual antibodies, each raised against the pure short-chain acyl-CoA, medium-chain acyl-CoA, or long-chain acyl-CoA dehydrogenase do not cross-react with the pure isovaleryl-CoA dehydrogenase either.[3,20] The content of the isovaleryl-CoA dehydrogenase in rat liver mitochondria is estimated to be 0.1% of the whole protein. The anti-rat isovaleryl-CoA dehydrogenase antibody cross-reacts with human isovaleryl-CoA dehydrogenase in fibroblasts. The molecular size of the human enzyme is essentially identical to that of the rat counterpart.[24,25]

[24] Y. Ikeda and K. Tanaka, *Biochem. Med.,* in press (1986).
[25] Y. Ikeda, S. K. Keese, and K. Tanaka, *Proc. Natl. Acad. Sci. U.S.A.* **82,** 7081 (1985).

[48] Purification of Methylmalonate-Semialdehyde Dehydrogenase from *Pseudomonas aeruginosa* PAO

By KENNETH HATTER and JOHN R. SOKATCH

Methylmalonate-semialdehyde dehydrogenase (acylating, EC 1.2.1.27) catalyzes the final reaction unique to the catabolism of valine in pseudomonads.[1]

Methylmalonate semialdehyde + CoA + NAD$^+$ → propionyl-CoA + carbon dioxide + NADH + H$^+$

Carbon dioxide originates from the carboxyl carbon of methylmalonate semialdehyde. Methylmalonyl-CoA is probably not an intermediate in the reaction. Methylmalonate-semialdehyde dehydrogenase is an inducible enzyme, being formed during growth on valine, isobutyrate, 3-hydroxyiso-butyrate, and propionate as carbon sources.[2] Growth on isobutyrate and 3-hydroxyisobutyrate also results in induction of 3-hydroxyisobutyrate dehydrogenase, but not of earlier enzymes of the pathway such as branched-chain keto acid dehydrogenase. Methylmalonate-semialdehyde dehydrogenase has been purified from bacteria[1] but not from higher organisms. However, there is isotopic evidence for the existence of methylmalonate-semialdehyde dehydrogenase in man.[3] Propionyl-CoA is metabo-

[1] D. Bannerjee, L. E. Sanders, and J. R. Sokatch, *J. Biol. Chem.* **245,** 1828 (1970).
[2] V. P. Marshall and J. R. Sokatch, *J. Bacteriol.* **110,** 1073 (1972).
[3] K. Tanaka, I. M. Armitage, H. S. Ramsdell, Y. E. Hsia, S. R. Lipsky, and L. E. Rosenberg, *Proc. Natl. Acad. Sci. U.S.A.* **72,** 3692 (1975).

lized in pseudomonads by carboxylation to methylmalonyl-CoA, isomerization to succinyl-CoA, and then oxidization through the tricarboxylic acid cycle. Methylmalonate-semialdehyde dehydrogenase is an example of a rather small class of CoA acylating enzymes that are not multienzyme complexes, malonate-semialdehyde dehydrogenase being another example.[4]

Assay

Principle

The assay of methylmalonate-semialdehyde dehydrogenase is based on measurement of the reduction of NAD^+ in the presence of the substrate, either methylmalonate semialdehyde or propionaldehyde. Propionaldehyde is probably not a physiological substrate since the affinity of methylmalonate-semialdehyde dehydrogenase for this substrate is low. High concentrations of 2-mercaptoethanol can replace CoA, since methylmalonate-semialdehyde dehydrogenase catalyzes transacylation with 2-mercaptoethanol as the acyl acceptor. Since methylmalonate semialdehyde is not available commercially, propionaldehyde is useful for measurement of enzyme fractions during purification. For this reason two assays are provided, one with propionaldehyde and 2-mercaptoethanol for enzyme purification, and a second assay with methylmalonate semialdehyde and CoA for kinetic studies.

Reagents for Propionaldehyde Assay

1 M Tris, pH 9.2
1 M 2-mercaptoethanol
0.1 M NAD^+
0.5 M propionaldehyde

Procedure

The reaction mixture contains 100 μl Tris buffer, 100 μl 2-mercaptoethanol, 10 μl NAD^+, and enough enzyme to give an initial absorbance change at 340 nm of 0.03 to 0.2 against a water blank. The volume is made to 0.95 ml and the reaction is started with 50 μl of propionaldehyde. One unit is 1 μmol NADH formed per minute per milligram protein and specific activity is units per milligram protein.

---

[4] E. W. Yamada and W. B. Jakoby, J. Biol. Chem. 235, 589 (1960).

Reagents for Methylmalonate Semialdehyde Assay

1 M Tris, pH 9.2
0.01 1 M coenzyme A in 0.02 M dithiothreitol
0.1 M NAD$^+$
0.1 M methylmalonate semialdehyde, pH 7.0
Methylmalonate semialdehyde is prepared by the method of Kupiecki and Coon[5] and standardized either by titration with standard base or reduction by 3-hydroxyisobutyrate dehydrogenase with NADH. The purification of 3-hydroxyisobutyrate dehydrogenase free of methylmalonate-semialdehyde dehydrogenase was reported by Bannerjee *et al.*[1] The assay for this purpose contains 100 μmol Tris buffer, pH 9.2, 0.02 μmol NADH, and 0.5 unit of 3-hydroxyisobutyrate dehydrogenase in a 1-ml cuvette. The initial absorbance at 340 nm is read and then up to 0.1 μmol methylmalonate semialdehyde is added and the final absorbance is read every minute until the absorbance stabilizes. One micromole of NADH oxidized is equivalent to 1 μmol of methylmalonate semialdehyde.

Procedure

The reaction mixture for the methylmalonate-semialdehyde dehydrogenase assay contains 100 μl Tris buffer, 10 μl NAD$^+$, 10 μl coenzyme A, and enzyme. Water is added, leaving enough room to add 5 μmol of methylmalonate semialdehyde, which is used to start the reaction. One unit of enzyme is 1 μmol of NADH produced per minute per milligram protein.

Enzyme Purification

Buffers

A: 50 mM potassium phosphate, pH 7.0, 20 mM 2-mercaptoethanol
B: 300 mM potassium phosphate buffer, pH 7.0, 20 mM 2-mercaptoethanol
C: 20 mM potassium phosphate, pH 9.0, 20 mM 2-mercaptoethanol
D: 50 mM potassium phosphate, pH 7.5, 20 mM 2-mercaptoethanol

Procedure

The enzyme is purified from *Pseudomonas aeruginosa* PAO which is grown in valine/isoleucine medium as described in chapter [44] in this volume. The preparation described in the following paragraphs is scaled for

[5] F. P. Kupiecki and M. J. Coon, *Biochem. Prep.* **7,** 69 (1960).

20–50 g of cells grown. The cells are resuspended in 2 ml buffer A per gram of cell paste. The cell suspension is chilled in an ice–salt bath, then treated with sonic oscillation in a Heat Systems model W225R oscillator at a setting of 7 using a flat tip. The suspension is treated for a total of 1.2 min per gram of cells in 3-min cycles, not allowing the temperature to rise above 10°. The broken cell suspension is centrifuged at 48,000 g for 30 min at 4°.

The supernatant is adjusted to pH 6.5 with 2 M acetic acid prior to heat treatment. One-milliliter aliquots are treated at 65° for 3, 6, and 9 min and assayed. The data are interpolated to determine the correct time for heat treatment to ensure that no less than 75% of the activity remains. If cell wall fragments remain after the preceding centrifugation, the activity may actually increase after heat treatment due to destruction of NADH oxidizing enzymes. The heat-treated solution is chilled quickly in an ice bucket and coagulated protein is removed by centrifugation at 41,000 g for 40 min at 4°.

The heat-treated enzyme is added to a column of DEAE-Sepharose CL-6B, 2.5 × 40 cm, equilibrated in buffer A. Methylmalonate-semialdehyde dehydrogenase is eluted with a linear gradient of 50 to 300 mM phosphate ion. The mixing chamber contains 200 ml of buffer A and the reservoir contains 200 ml of buffer B. Methylmalonate-semialdehyde dehydrogenase begins to appear at about 200 mM phosphate ion.

The active fractions from the preceding column are collected and dialyzed against buffer C. The dialyzed sample is then added to a column of DEAE-Sepharose CL-6B, 1.5 × 20 cm, equilibrated in buffer C. Methylmalonate-semialdehyde dehydrogenase is eluted in this case by a pH gradient, from 9.0 to 7.5. The mixing flask contains 200 ml buffer C and the reservoir flask contains 200 ml buffer D. Methylmalonate-semialdehyde dehydrogenase elutes at about pH 7.8.

TABLE I
MICHAELIS CONSTANTS FOR PROPIONALDEHYDE AND
METHYLMALONATE SEMIALDEHYDE IN THE PRESENCE OF
2-MERCAPTOETHANOL AND COENZYME A

	Acyl acceptor (M)	
Substrate	Mercaptoethanol	Coenzyme A
Propionaldehyde	59.3×10^{-3}	6.4×10^{-3}
Methylmalonate semialdehyde	3.5×10^{-5}	2.4×10^{-5}

Properties

Methylmalonate-semialdehyde dehydrogenase is a dimer with subunit molecular weight of about 60,000. Mercaptoethanol can substitute for CoA, but the K_m for mercaptoethanol is $4.4 \times 10^{-2} M$ compared with $2.1 \times 10^{-5} M$ for CoA. It is not clear if propionaldehyde is a physiological substrate. The affinity of methylmalonate-semialdehyde dehydrogenase for propionaldehyde is increased by a factor of 10 in the presence of CoA to $6.4 \times 10^{-3} M$, which is still high for a catabolic enzyme (Table I). The K_m for methylmalonate semialdehyde is unaffected by the nature of the thiol acceptor (Table I). The K_m for NAD$^+$ is $8.7 \times 10^{-5} M$ and does not change when mercaptoethanol is the thiol provided.

[49] D-Methylmalonyl-CoA Hydrolase

By ROBIN J. KOVACHY, SALLY P. STABLER, and ROBERT H. ALLEN

Introduction

$$\text{HOOCCH (CH}_3\text{)CO-SCoA} \rightarrow \text{HOOCH(CH}_3\text{)COOH} + \text{CoA}$$

D-Methylmalonyl-CoA   methylmalonic acid   coenzyme A

As shown below in Fig. 1, degradation products of certain amino acids, odd-chain fatty acids, cholesterol, and thymine are metabolized to propionyl-CoA which is then carboxylated to form the D-isomer of methylmalonyl-CoA.[1,2] A highly specific hydrolase, D-methylmalonyl-CoA hydrolase, exists which converts D-methylmalonyl-CoA to methylmalonic acid and coenzyme A.[3] D-Methylmalonyl-CoA can also be metabolized to the L-isomer of methylmalonyl-CoA by DL-methylmalonyl-CoA racemase (DL-methylmalonyl-CoA epimerase, EC 5.1.99.1).[2,4] L-Methylmalonyl-CoA is then converted to succinyl-CoA by the adenosylcobalamin-dependent enzyme, L-methylmalonyl-CoA mutase (EC 5.4.99.2).[5,6] This chapter describes the assay, purification, and properties of the hydrolase. Subse-

[1] Y. Kaziro and S. Ochoa, *Adv. Enzymol.* **26**, 283 (1964).
[2] S. P. Stabler, P. D. Marcell, and R. H. Allen, *Arch. Biochem. Biophys.* **241**, 252 (1985).
[3] R. J. Kovachy, S. D. Copley, and R. H. Allen, *J. Biol. Chem.* **258**, 11415 (1983).
[4] R. Mazumder, T. Sasakawa, Y. Kaziro, and S. Ochoa, *J. Biol. Chem.* **237**, 3065 (1962).
[5] R. Mazumder, T. Sasakawa, and S. Ochoa, *J. Biol. Chem.* **238**, 50 (1963).
[6] J. F. Kolhouse, C. Utley, and R. H. Allen, *J. Biol. Chem.* **255**, 2708 (1980).

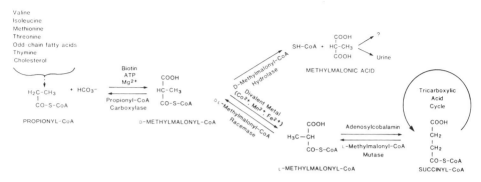

FIG. 1. Biochemical pathway of methylmalonyl-CoA metabolism in mammalian cells. Adapted from Refs. 1–3.

quent chapters describe the assay, purification, and properties of the race-mase[7] and the mutase.[8]

Assay Method[3]

Principle

D-Methylmalonyl-CoA hydrolase activity is followed by measuring the rate of formation of [methyl-^{14}C]methylmalonic acid from a racemic mixture of DL-[*methyl*-^{14}C]methylmalonyl-CoA. [*methyl*-^{14}C]methylmalonic acid is separated from unhydrolyzed DL-[*methyl*-^{14}C]methylmalonyl-CoA by taking advantage of the fact that the former is extracted from aqueous samples into ethyl acetate, whereas the latter is not. The amount of radioactive methylmalonic acid present in the ethyl acetate can then be determined directly in a liquid scintillation counter using an appropriate scintillation counting fluid.

     When applied to crude tissues, the assay is not specific for D-methylmalonyl-CoA hydrolase because it also measures the activity of L-methylmalonyl-CoA mutase. This problem arises from the fact that L-methylmalonyl-CoA mutase converts L-[*methyl*-^{14}C]methylmalonyl-CoA to [2(3)-^{14}C]succinyl-CoA, which then is hydrolyzed to [2(3)-^{14}C]succinic acid either spontaneously[9] or by the action of succinate-CoA ligase which simultaneously converts GDP to GTP. Radioactive succinic acid is also

[7] S. P. Stabler and R. H. Allen, this volume [50].

[8] J. F. Kolhouse, S. P. Stabler, and R. H. Allen, this volume [51].

[9] E. R. Stadtman, P. Overath, H. Eggerer, and F. Lynen, *Biochem. Biophys. Res. Commun.* **2**, 1 (1960).

extracted from aqueous samples into ethyl acetate and thus is measured together with radioactive methylmalonic acid.

However, the assay can be made largely or even entirely specific for D-methylmalonyl-CoA hydrolase, or L-methylmalonyl-CoA mutase by taking advantage of the fact that the pH activity curves of the two enzymes are very different, as shown in Fig. 2. Assays performed at pH 6.0 reflect mainly hydrolase activity and assays performed at pH 5.0 are specific for the hydrolase. Assays performed at pH 8.0, on the other hand, reflect mainly mutase activity and assays performed at pH 9.0 are specific for the mutase.

Reagents

DL-Methylmalonyl-CoA (containing 150 μCi of DL-[*methyl-*¹⁴C]methylmalonyl-CoA per millimole), 1 mM in 3 mM HCl
Potassium phosphate buffer, pH 6.0 (or pH 5.0), 1.0 M
HCl, 3 N
Ethyl acetate

Procedure

Pipet into 4.0-ml glass tubes in ice, 95 μl of DL-methylmalonyl-CoA, 45 μl of potassium phosphate buffer, pH 6.0 (or pH 5.0), sufficient H₂O to make the volume 450 μl after enzyme is added, and, finally, enzyme. The tubes are incubated in a 30° water bath for 30 min and the reactions are stopped by placing the tubes in an ice-water bath and adding 50 μl of HCl. Ethyl acetate, 1.3 ml, is added, followed by vigorous vortexing for 15 sec and centrifugation at 500 g for 4 min at 4°. A portion, 1.0 ml, of the upper

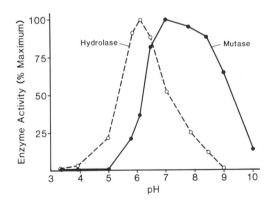

FIG. 2. pH activity profiles of D-methylmalonyl-CoA hydrolase and L-methylmalonyl-CoA mutase in rat liver.

ethyl acetate layer is removed and added to glass counting vials containing 10 ml of Research Products International scintillation cocktail 3a70. Under these conditions, 60% of the methylmalonic acid formed is recovered in the 1.0 ml of ethyl acetate extract.

Control tubes, containing the identical reaction mixture but lacking enzyme have less than 4% of the total radioactivity present in methylmalonic acid. Assays with rat liver supernatant, which contains large amounts of DL-methylmalonyl-CoA racemase, are linear with time and protein concentration up to the point at which approximately 40% of the total radioactivity is present in methylmalonic acid. Assays with pure hydrolase, that is, free of racemase, are linear with time and protein concentration up to the point at which approximately 20% of the total radioactivity is present in methylmalonic acid.

One unit of activity is defined as the amount of enzyme that forms 1 μmol of methylmalonic acid per minute. Specific activity is defined as units per milligram of protein. Protein concentration is determined by the method of Lowry using bovine serum albumin as the standard.[10]

Enzyme Purification Procedure[3]

All centrifugations and procedures are performed at 4° or in an ice-water bath. Enzyme activity is assayed at pH 6.0 using the standard hydrolase assay described above. Fresh and frozen rat livers give comparable activity and the latter are used for convenience.

Step 1. Preparation of Rat Liver Homogenate

Frozen rat livers, 200, weighing 1600 g are thawed, minced, and homogenized in 4000 ml of 28 mM sodium phosphate, pH 7.0, in a Waring blender. The homogenate is centrifuged at 30,000 g for 1 hr and the supernatant, 4600 ml, is decanted.

Step 2. Ammonium Sulfate Precipitation

Solid ammonium sulfate, 1228 g, is added slowly to the supernatant over 20 min with constant stirring followed by centrifugation at 30,000 g for 1 hr. The supernatant is decanted and 966 g of ammonium sulfate is added with stirring over 20 min, followed by centrifugation at 30,000 g for 30 min. The pellet is resuspended in 400 ml of 5 mM citric acid–NaOH, pH 6.0, and dialyzed against 20 liters of this same buffer, which is changed

[10] O. H. Lowry, N. J. Rosebrough, A. L. Farr, and R. J. Randall, *J. Biol. Chem.* **193**, 265 (1951).

four times over a 24-hr period. The sample is centrifuged at 30,000 g for 1 hr and the supernatant, 680 ml, is decanted.

Step 3. CM-Cellulose Chromatography

Two 90 × 2.5 cm cclumns are poured with CM-cellulose (Whatman CM-52) which is prepared by removing the fines, adjusting the pH to 6.0 with 1 N HCl, and equilibrating with three washes of 5 mM citric acid–NaOH buffer, pH 6.0. The columns are washed with 800 ml of the same buffer. One-half of the dialyzed sample from the preceding step, 340 ml, is applied to each column and the columns are washed with 400 ml of equilibrating buffer at a flow rate of 70 ml per hour. The hydrolase is eluted with a linear salt gradient consisting of 500 ml of the equilibrating buffer in the mixing chamber and 500 ml of equilibrating buffer containing 100 mM NaCl in the reservoir chamber. Fractions of 18 ml are collected starting at the time the sample is applied. The hydrolase elutes as a single peak in fractions 55 to 75. Those fractions from both columns containing greater than 0.1 enzyme units per milliliter are pooled and dialyzed for 24 hr against 4 liters of 10 mM Tris-HCl, pH 8.6, with dialysate changes at 2 and 16 hr.

Step 4. DEAE-Cellulose Chromatography

DEAE-cellulose (Whatman DE-52) is prepared by removing the fines, adjusting the pH to 8.6 with 1 N HCl, equilibrating with 10 mM Tris-HCl, pH 8.6, and then packing into a single 90 × 2.5 cm column. The column is washed with 800 ml of equilibrating buffer and the dialyzed sample, 290 ml, from the preceding step is applied at a flow rate of 70 ml per hr. After washing with 400 ml of equilibrating buffer, the hydrolase is eluted with a linear salt gradient in which 1 liter of equilibrating buffer is present in the mixing chamber and 1 liter of equilibrating buffer containing 100 mM NaCl is present in the reservoir chamber. Fractions of 18 ml are collected, starting at the time the sample is applied. The hydrolase elutes as a single peak in fractions 113 to 124 which are pooled, concentrated to 6.0 ml using an Amicon concentrator with a UM10 membrane, and dialyzed against 4 liters of 5 mM citric acid–NaOH, pH 6.0, for 24 hr with dialysate changes at 2 and 16 hr.

Step 5. Coenzyme A-Agarose Affinity Chromatography

A 0.9 × 8 cm column of coenzyme A-agarose (linked as a thiol ester via the SH moiety of CoA, P-L Biochemicals, type 5) is poured and washed with 10 ml of 5 mM citric acid–NaOH, pH 6.0, containing 1 N NaCl

followed by 50 ml of 5 mM citric acid–NaOH, pH 6.0. The sample from the preceding step is applied at a flow rate of 0.2 ml per minute. The column is washed successively with 24 ml of 5 mM citric acid–NaOH, pH 6.0; 18 ml of 5 μM DL-methylmalonyl-CoA–NaOH, pH 6.0; 10 ml of 5 mM citric acid–NaOH, pH 6.0; and 10 ml of 5 mM citric acid–NaOH containing 1 M NaCl. The eluates obtained with each wash solution are collected separately in their entirety, except for the eluates obtained with the DL-methylmalonyl-CoA solution. During this elution, successive individual fractions of 2, 6, 6, and 4 ml are collected and dialyzed for 24 hr against 4 liters of 5 mM citric acid–NaOH, pH 6.0, with dialysate changes at 2 and 16 hr to remove the DL-methylmalonyl-CoA which interferes with the hydrolase assay. Assay of these dialyzed fractions and the other undialyzed eluates show that hydrolase activity is present only in the first of the two 6-ml fractions obtained with 5 mM DL-methylmalonyl-CoA–NaOH, pH 6.0.

Step 6. Sephadex Chromatography

A 2 × 90 cm column of Sephadex G-100 is poured and equilibrated with 200 ml of 5 mM citric acid–NaOH, pH 6.0, containing 25 mM NaCl. The sample, 6 ml, from the preceding step is applied at a flow rate of 10 ml per hour and 3-ml fractions are collected. The hydrolase elutes as a single, symmetrical peak with peak activity present in tube 62. This gives a value for V_e/V_0 of 1.85 based on the determination of fraction 33 as V_0 when the column is rerun in an identical manner using a 6-ml sample of equilibrating buffer containing Blue dextran. Samples of 100 μl of individual fractions are subjected to polyacrylamide gel electrophoresis and stained for protein with Coomassie Blue. A single band is observed in fractions 60 through 64 and these fractions are pooled and utilized as the final enzyme preparation. Contaminating protein bands, representing approximately 5% of the total protein stained, are noted on the gels from the fractions before and after fractions 60 through 64. The pooled fractions are concentrated to 2.0 ml using an Amicon concentrator with a UM10 membrane. The concentrated sample is stored at $-20°$ where the hydrolase is stable for at least 1 month.

A sample purification is summarized in Table I. The hydrolase was purified 3100-fold with a net yield of 2.1%. The low yield is due to the many steps required to purify the hydrolase, each of which gives a yield of approximately 50%.

Comments on Enzyme Purification Procedure[3]

A single protein band is obtained on polyacrylamide gel electrophoresis at pH 8.6, and the position of the band corresponds with the position of a

TABLE I
PURIFICATION OF D-METHYLMALONYL-CoA HYDROLASE

Purification step	Volume (ml)	Total protein (mg)	Total activity (units)	Specific activity (units/mg)
Rat liver supernatant	4,600	516,000	236[a]	0.000046
Ammonium sulfate				
40–70% pellet	680	92,000	98.6	0.0011
CM-cellulose	290	1,500	45.8	0.031
DEAE-cellulose	198	148	14.3	0.096
Coenzyme A-agarose	6.0	7.6	9.4	1.24
Sephadex G-100	13.5	3.6	5.1	1.41

[a] This value represents 80% of the actual enzyme activity assayed and is based on the fact that approximately 20% of the enzyme activity in crude rat liver supernatant, as assayed by the standard hydrolase assay at pH 6.0, is L-methylmalonyl-CoA mutase activity (see text for additional details).

single peak of hydrolase activity that is observed when a duplicate unstained gel is sliced into 2-mm sections followed by elution and assay for hydrolase activity. SDS-polyacrylamide gel electrophoresis also gives a single protein band.

The capacity of the coenzyme A-agarose column used in step 4 is markedly decreased when either step 3 or step 4 of the purification scheme is omitted. The use of both step 3 and step 4 appears to eliminate more contaminating proteins that compete for sites on the coenzyme A-agarose column. It is necessary to concentrate the pooled fractions from step 4 to prevent leakage of the hydrolase during application of the sample and washing of the coenzyme A-agarose column. DL-methylmalonyl-CoA is used to elute the hydrolase from the coenzyme A-agarose column. Whether coenzyme A alone, or other compounds, would work as well has not been tested.

Properties of the Enzyme[3]

The molecular weight of the purified hydrolase is 35,000 based on gel filtration. A molecular weight of 35,000 is also obtained using SDS-polyacrylamide gel electrophoresis in the presence and absence of 1% 2-mercaptoethanol, indicating that the hydrolase contains a single polypeptide chain. The hydrolase is active on the D isomer, but not on the L isomer of methylmalonyl-CoA, based on experiments employing CD spectropolarimetry. The K_m for D-methylmalonyl-CoA is 0.7 mM and the molar activity is 1400 molecules of substrate per minute per molecule of enzyme. At substrate concentrations of 0.5 mM, the relative rate of hydrolysis of CoA esters is as follows: D-methylmalonyl-CoA (100%), malonyl-CoA (16%),

propionyl-CoA (3%), acetyl-CoA (1%), succinyl-CoA (<1%), and palmitoyl-CoA (<1%).

No cofactors have been identified for the hydrolase. Its functional significance is unknown. It may serve to limit the intracellular concentration of DL-methylmalonyl-CoA in cobalamin deficiency or to provide free coenzyme A under certain circumstances. It could also serve to provide a source of methylmalonic acid, although no biochemical function for this compound has been described. The hydrolase appears to account for the markedly increased amounts of methylmalonic acid that are excreted in the urine in cobalamin deficiency and by patients with genetic disorders involving adenosylcobalamin synthesis or L-methylmalonyl-CoA mutase.

[50] DL-Methylmalonyl-CoA Racemase from Rat Liver

By SALLY P. STABLER and ROBERT H. ALLEN

$$\text{HOOCH(CH}_3)\text{CO-SCoA} + \text{H}^+ \underset{}{\overset{\text{Divalent metal}\atop (\text{Co}^{2+}, \text{Mn}^{2+}, \text{Fe}^{2+})}{\rightleftharpoons}} \text{HOOCC(CH}_3)\text{HCO-SCoA} + \text{H}^+$$
D-Methylmalonyl-CoA   L-methylmalonyl-CoA

DL-Methylmalonyl-CoA racemase (EC 5.1.99.1, DL-methylmalonyl-CoA epimerase) catalyzes the interconversion of the D isomer and the L isomer of methylmalonyl-CoA. The enzyme has been partially purified from sheep kidney[1] and to homogeneity from *Propionibacterium shermannii*[2-4] and rat liver.[5]

As shown in Fig. 1 in chapter [49] in this volume,[6] propionyl-CoA is carboxylated to form the D isomer of methylmalonyl-CoA, which is then either hydrolyzed to methylmalonic acid and coenzyme A by D-methylmalonyl-CoA hydrolase or is converted to the L isomer of methylmalonyl-CoA by DL-methylmalonyl-CoA racemase. L-methylmalonyl-CoA is converted to succinyl-CoA by the adenosylcobalamin-dependent enzyme, L-methylmalonyl-CoA mutase.[7] Because the hydrolase acts only on the D

[1] R. Mazumder, T. Sasakawa, Y. Kaziro, and S. Ochoa, *J. Biol. Chem.* **237**, 3065 (1962).
[2] S. H. G. Allen, R. Kellermeyer, R. Stjernholm, B. Jacobson, and H. G. Wood, *J. Biol. Chem.* **238**, 1637 (1963).
[3] P. F. Leadlay, *Biochem. J.* **197**, 413 (1981).
[4] P. F. Leadlay and J. Q. Fuller, *Biochem. J.* **213**, 635 (1983).
[5] S. P. Stabler, P. D. Marcell, and R. H. Allen, *Arch. Biochem. Biophys.* **241**, 252 (1985).
[6] R. J. Kovachy, S. P. Stabler, and R. H. Allen, this volume [49].
[7] J. F. Kolhouse, S. P. Stabler, and R. H. Allen, this volume [51].

isomer of methylmalonyl-CoA whereas the mutase acts only on the L isomer of methylmalonyl-CoA, the racemase plays an important role in determining the fate of D-methylmalonyl-CoA after it is formed from propionyl-CoA. This chapter describes a method for assay of the racemase and its purification to homogeneity from rat liver.[5]

Assay Method[5]

Principle

The assay method described is based on the observation[1,8] that the proton on C-2 of DL-methylmalonyl-CoA is replaced by a proton from the media during both the spontaneous and the enzymatic racemization of DL-methylmalonyl-CoA. DL-methylmalonyl-CoA racemase activity is followed by measuring the rate of release of tritium from a racemic mixture of DL-[2-^{3}H]methylmalonyl-CoA. The released tritium is present in H_2O and is separated from that still present in DL-[2-^{3}H]methylmalonyl-CoA by adsorbing the latter onto an anion-exchange resin. The amount of radioactivity present in the H_2O can then be determined directly in a liquid scintillation counter using an appropriate scintillation counting fluid.

Preparation of DL-[2-^{3}H]Methylmalonyl-CoA

DL-methylmalonyl-CoA, 100 mg, is dissolved in 200 μl of H_2O containing 5 mCi of tritium and incubated at 100° for 10 min. After cooling to 4°, the sample is applied to a 1.5 × 1.7 cm column of DEAE-cellulose (Whatman DE-52) which has been adjusted to pH 6.5 with HCl and washed with 450 ml of H_2O. After standing for 16 hr, the column is washed with another 50 ml of H_2O and the DL-[2-^{3}H]methylmalonyl-CoA is eluted with 5 ml of 1.0 M NaCl and stored at −70°. This material has a specific activity of approximately 100 μCi per millimole.

When the DL-[2-^{3}H]methylmalonyl-CoA is treated with 10-fold excess of racemase at pH 6.0, about 60% of the tritium is exchanged with H_2O as determined by high-performance liquid chromatography. Similar amounts of tritium are released into H_2O when the DL-[2-^{3}H]methylmalonyl-CoA is heated at pH 6.0 at 100° for 10 min. These results indicate that about 60% of the radioactivity is located at C-2 in the methylmalonic acid portion of the molecule, since previous studies have shown that there is complete racemization of methylmalonyl-CoA after heating at near neu-

[8] P. Overath, E. R. Sladtman, G. M. Kellerman, F. Lynen, H. P. Fritz, and H. S. Keller, *Biochem. J.* **335**, 500 (1962).

tral pH. The remaining 40% of the tritium, whose location is unknown, can be released into H_2O by heating with base.

Reagents

DL-Methylmalonyl-CoA (containing 100 μCi of DL-[2-³H]methylmalonyl-CoA per millimole), 1 mM
Potassium phosphate buffer, pH 7.5, 1 M
NaCl, 1 M
$CoCl_2$, 10 mM
Bovine serum albumin in H_2O, 10 mg/ml
AG1-X8 resin, 200–400 mesh, formate form (Bio-Rad Laboratories), 50% slurry in H_2O

Procedure

Pipet into 4.0-ml glass tubes in ice, 100 μl of DL-methylmalonyl-CoA, 5 μl of potassium phosphate buffer, pH 7.5, 75 μl of NaCl, 5 μl of $CoCl_2$, 5 μl of bovine serum albumin, sufficient H_2O to make the volume 500 μl after enzyme is added, and, finally, enzyme. The tubes are incubated in a 30° water bath for 20 min and reactions are stopped by placing the tubes in an ice-water bath and adding 1.0 ml of AG1-X8 resin. After vortexing and allowing the resin to settle twice, the tubes are centrifuged at 1000 g at 4° for 10 min. A portion of the supernatant, 750 μl, is removed and added to glass counting vials containing 250 μl of H_2O and 10 ml of Research Products International Corp. scintillation cocktail 3a70.

Control tubes, containing the identical reaction mixture but lacking enzyme, have less than 2% of the total radioactivity released from the DL-[2-³H]methylmalonyl-CoA. Assays with rat liver supernatant and purified racemase are linear with time and protein concentration up to the point at which approximately 40% of the total radioactivity is released into H_2O.

One unit of activity is defined as the amount of enzyme that releases 1 μmol of proton from DL-methylmalonyl-CoA into H_2O per minute. Specific activity is defined as units per milligram of protein. Protein concentration is determined by the Bio-Rad protein assay using bovine serum albumin as the standard.

Alternative Methods

DL-Methylmalonyl-CoA racemase can also be assayed using relatively complex coupled enzyme assays that involve spectrophotometric techniques[2,3] or radioactive compounds.[1]

Enzyme Purification Procedure[5]

All centrifugations and procedures are performed at 4° or in an ice-water bath. Enzyme activity is assayed using the standard racemase assay described above. Fresh and frozen rat livers give comparable activity and the latter are used for convenience.

Step 1. Preparation of Rat Liver Homogenate

Frozen rat livers, 400, weighing 2.2 kg are partially thawed, minced, and homogenized in positions in a Waring blender for 2 min each in a total volume of 5500 ml of 28 mM sodium phosphate, pH 7.0, containing 5 mM EDTA. The homogenate is centrifuged at 30,000 g for 1 hr and the supernatant, 5230 ml, is decanted.

Step 2. Ammonium Sulfate Precipitation

Solid ammonium sulfate, 1098 g, is added slowly to the supernatant from the preceding step followed by centrifugation at 30,000 g for 30 min. Then, an additional 1098 g of ammonium sulfate is added to the supernatant followed by centrifugation as above. The pellet is resuspended in 1050 ml of 10 mM potassium phosphate, pH 7.5, containing 5 mM EDTA and 150 mM sodium chloride and dialyzed for 24 hr against 50 liters of this same buffer with three changes of dialysate.

Step 3. Acid Treatment

While stirring vigorously, 1 N HCl (approximately 25 ml) is added to 1960 ml of the dialyzed pellet from the preceding step to bring the pH to 5.0. The entire sample, which contains grossly precipitated protein, is immediately dialyzed against 100 liters of 5 mM citric acid–NaOH, pH 5.5, containing 1 mM EDTA for 24 hr with three changes of dialysate. The turbid sample is then centrifuged at 30,000 g for 30 min and 1730 ml of supernatant is decanted.

Step 4. CM-Cellulose Chromatography

The supernatant from the preceding step is applied to a 60 × 2.5 cm column of CM-cellulose (Whatman CM-52) that has been adjusted to pH 5.5 with 1 N HCl and equilibrated with 5 mM citric acid–NaOH, pH 5.5, containing 1 mM EDTA. After washing with 500 ml of the equilibrating buffer, the racemase is eluted with a linear salt gradient from 0 to 500 mM NaCl, 3 liters total volume. The racemase elutes as a single peak

between 1275 and 1650 ml of the gradient, and is then dialyzed for 24 hr against 4000 ml of 5 mM sodium phosphate, pH 7.5, containing 1 mM EDTA. The sample is then centrifuged at 30,000 g for 30 min and the supernatant decanted.

Step 5. Reactive Red Agarose Chromatography

The pooled dialyzed sample from the previous step, 275 ml, is applied to a 30 × 2.5 cm column of reactive red 120-agarose (Sigma Chemical Corp.) that has been washed with 400 ml of 5 mM sodium phosphate, pH 7.5, containing 1 mM EDTA and 2.0 M NaCl, followed by 500 ml of 5 mM sodium phosphate, pH 7.5, containing 1 mM EDTA. After applying the sample, the column is washed with 200 ml of the final equilibrating buffer, and the racemase is eluted with a linear salt gradient from 0 to 1.5 M NaCl, 2 liters total volume. The racemase elutes as a single peak between 900 and 1245 ml of the gradient and is then dialyzed for 24 hr against 4 liters of 5 mM sodium phosphate, pH 7.5, containing 1 mM EDTA.

Step 6. Coenzyme A-Agarose Affinity Chromatography

Step 6A. The pooled dialyzed sample from the preceding step, 295 ml, is applied to a 1.5 × 3 cm column of coenzyme A-agarose (linked as a thiol ester via the SH moiety of CoA, P-L Biochemicals, type 5) that has been previously washed with 10 ml of 5 mM sodium phosphate, pH 7.5, containing 1 mM EDTA and 2.0 M NaCl, followed by 100 ml of 5 mM sodium phosphate, pH 7.5, containing 1 mM EDTA. After applying the sample and washing with 50 ml of equilibrating buffer, the racemase is eluted with a linear salt gradient from 0 to 75 mM NaCl, 1 liter total volume. The racemase elutes in a single peak between 90 and 150 ml of the gradient and is then dialyzed for 24 hr against 4 liters of 5 mM sodium phosphate, pH 7.5, containing 1 mM EDTA.

Step 6B. A second column, 0.4 × 2 cm, of coenzyme A-agarose is prewashed with 5 ml of 5 mM sodium phosphate, pH 7.5, containing 1 mM EDTA and 2.0 M NaCl, followed by 3 ml of 1 mM DL-methylmalonyl-CoA, and a wash of 50 ml of 5 mM sodium phosphate, pH 7.5, containing 1 mM EDTA. The dialyzed sample from the preceding step, 67 ml, is applied to the column at a flow rate of 1 ml per minute. The racemase is eluted with five successive washes of 0.4, 2, 2, 2, and 1.6 ml of the final equilibrating buffer containing 1 mM DL-methylmalonyl-CoA. Epimerase activity, 65%, elutes in the first 2-ml fraction.

Step 7. Sephadex Chromatography

The sample, 2 ml, from the preceding step is applied to a 1.5×90 cm column of Sephadex G-100 which has been equilibrated with 5 mM sodium phosphate, pH 7.5, containing 1 mM EDTA, and fractions of 3 ml are collected. The racemase elutes as a single peak with peak activity in fraction 29. This gives a value for V_e/V_0 of 1.61, based on the determination of fraction 18 as V_0 when the column is rerun in an identical manner with Blue dextran. Samples of 150 μl of individual fractions are subjected to polyacrylamide disc. gel electrophoresis and stained for protein with Coomassie blue. A single band is observed in fractions 27 through 33 and these fractions are pooled. Contaminating protein bands, representing approximately 5% of the total protein stained, are noted on the gels run on fractions immediately before fraction 27. The pooled fractions are concentrated to 1.8 ml using an Amicon concentrator with a UM10 membrane. The concentrated sample is stored at $-20°$ where the racemase is stable for at least 3 months.

A sample purification is summarized in Table I. The epimerase was purified 23,000-fold with a net yield of 1.8%. The low yield was due to the many steps required to purify the racemase, each of which gives a yield of approximately 50%.

TABLE I
PURIFICATION OF DL-METHYLMALONYL-CoA RACEMASE

Purification step	Volume (ml)	Total protein (mg)	Total activity (units)	Specific activity (units/mg)
Rat liver supernatant	5,200	840,000	310,000[a]	0.36
Ammonium sulfate 30–60% pellet	2,000	110,000	240,000	2.2
Acid treatment	1,700	32,000	75,000	2.3
CM-cellulose	280	1,900	20,000	10
Reactive red-agarose	300	120	24,000	200
Coenzyme A-agarose				
A	67	22	17,000	770
B	2.0	1.3	11,000	8,500
Sephadex G-100[b]	1.8	0.67	5,600	8,400

[a] When rat liver supernatant is prepared without EDTA and assayed without adding Co^{2+} or any other metal to the assay tubes, only 5% as much racemase activity is observed.

[b] This step is performed to remove several minor (5%) contaminating proteins that are observed on polyacrylamide disc. gel electrophoresis.

Comments on Enzyme Purification Procedure[5]

A single protein band is obtained on polyacrylamide gel electrophoresis at pH 8.9, and the band corresponds with the position of a single peak of racemase activity that is observed when a duplicate unstained gel is sliced into 2-mm sections, followed by elution and assay for racemase activity.

EDTA at a concentration of 1 mM causes a slow and eventually complete loss ($>$99.7% after 144 hr) of racemase activity in crude rat liver supernatant with a 25% loss at 3 hr and a 90% at 24 hr. When these inactivated samples are assayed with the addition of 100 μM Co^{2+} in the standard racemase assay, recovery of 100 to 200% of the original maximal racemase activity is observed. Irreversible loss of activity can be prevented by storing samples of crude or purified racemase in the presence of 1 mM EDTA at 4°. Samples stored in buffer alone, or buffer containing various sulfhydryl compounds, or Co^{2+} are much less stable. Because of this, the entire enzyme purification is performed in the presence of 1 mM EDTA.

Properties of the Enzyme[5]

DL-Methylmalonyl-CoA racemase from rat liver has a molecular weight of 32,000 based on gel filtration. It gives a single band on SDS-polyacrylamide gel electrophoresis with a molecular weight of 16,000 in the presence and absence of 2-mercaptoethanol. This indicates that the racemase contains two 16,000-molecular-weight subunits that are not connected by disulfide bonds. The pH optimum is at 7.0, with 50% of maximal activity observed at pH 5.0 and 9.0. Racemase activity is completely inactivated by EDTA and can be reactivated by the addition of Co^{2+}, with 50% activation occurring at a concentration of 0.2 μM. Lower levels of maximal activation are obtained at higher concentrations of Co^{3+}, Fe^{2+}, and Mn^{2+}. Other metals such as Zn^{2+}, Cu^{2+}, Cu^{+}, and Cd^{2+} completely inhibit racemase activity even in the presence of equal concentrations of Co^{2+}. Purified racemase binds 1 mol of Co per mole of subunit. The kind of metal bound to the racemase *in vivo* is not known. The K_m for DL-methylmalonyl-CoA is 0.1 mM and the molar activity is 250,000 molecules of substrate per minute per molecule of enzyme.

[51] L-Methylmalonyl-CoA Mutase from Human Placenta

By J. Fred Kolhouse, Sally P. Stabler, and Robert H. Allen

Introduction

$$\text{HOOC(CH}_3\text{)HCO-SCoA} \xrightleftharpoons{\text{adenosylcobalamin}} \text{HOOCCH}_2\text{CH}_2\text{CO-SCoA}$$

L-Methylmalonyl-CoA   succinyl-CoA

L-Methylmalonyl-CoA mutase (EC 5.4.99.2) catalyzes the interconversion of L-methylmalonyl-CoA and succinyl-CoA. Adenosylcobalamin binds tightly to L-methylmalonyl-CoA mutase and is required for enzyme activity. Preceding chapters [49, 50] deal with D-methylmalonyl-CoA hydrolase[1] and D L-methylmalonyl-CoA racemase,[2] which are also involved in methylmalonyl-CoA metabolism. D L-Methylmalonyl-CoA mutase has been isolated in homogeneous form from *Propionibacterium shermanii*,[3,4] the intestinal worm *Ascaris*,[5] sheep liver,[6] human placenta,[7] and human liver.[8] This chapter describes two methods for assay of the mutase and its purification to homogeneity from human placenta.

Assay Methods[1,9,10]

We have utilized two different methods for the assay of L-methylmalonyl-CoA mutase. The first method is a modification of the method of Cannata *et al.*[6] It is more complex and time consuming than the second method, but is more specific in the sense that it does not measure the

[1] R. J. Kovachy, S. P. Stabler, and R. H. Allen, this volume [49].
[2] S. P. Stabler and R. H. Allen, this volume [50].
[3] R. W. Kellermeyer and H. G. Wood, this series, Vol. 13, p. 207.
[4] F. Francalanci, N. K. Davis, J. Q. Fuller, D. Murfitt, and P. F. Leadlay, *Biochem. J.* **235**, 489 (1986).
[5] Y. S. Han, J. M. Bratt, and H. P. C. Hogenkamp, *Comp. Biochem. Physiol. B* **78**, 41 (1984).
[6] J. J. B. Cannata, A. Focesi, R. Mazumder, R. C. Warner, and S. Ochoa, *J. Biol. Chem.* **24**, 3249 (1965).
[7] J. F. Kolhouse, C. Utley, and R. H. Allen, *J. Biol. Chem.* **255**, 2708 (1980).
[8] W. A. Fenton, A. M. Hack, F. W. Huntington, A. Gertler, and L. E. Rosenberg, *Arch. Biochem. Biophys.* **214**, 815 (1981).
[9] J. F. Kolhouse and R. H. Allen, *Proc. Natl. Acad. Sci. U.S.A.* **74**, 921 (1977).
[10] R. J. Kovachy, S. D. Copley, and R. H. Allen, *J. Biol. Chem.* **258**, 11415 (1983).

activity of D-methylmalonyl-CoA hydrolase at any pH. The second method is simpler and faster than the first method, although it also measures the activity of D-methylmalonyl-CoA hydrolase at acid and neutral pH. By utilizing a higher pH, however, the activity of the hydrolase is abolished and this assay is then also specific for L-methylmalonyl-CoA mutase. Both methods are described below.

Method I[9]

Principle. L-Methylmalonyl-CoA mutase activity is followed by measuring the rate of formation of [2-(3)-^{14}C]succinyl-CoA from a racemic mixture of D L-[*methyl*-^{14}C]methylmalonyl-CoA. At the end of the reaction, samples are treated with acid, heat, and permanganate ion, which hydrolyzes unreacted DL-[*methyl*-^{14}C]methylmalonyl-CoA to [*methyl*-^{14}C] methylmalonic acid and coenzyme A. This treatment also decarboxylates [*methyl*-^{14}C]methylmalonic acid to [3-^{14}C]propionic acid and CO_2. The [3-^{14}C]propionic acid is volatile at acid pH and is removed by drying.

The sample treatment also hydrolyzes the formed [2-(3)-^{14}C]succinyl-CoA to [2(3)-^{14}C]succinic acid and coenzyme A. The [2(3)-^{14}C]succinic acid is not decarboxylated, however, and is still present at the end of the drying step, where it can be determined directly in a liquid scintillation counter using an appropriate scintillation counting fluid.

Reagents

Tris-HCl buffer, pH 8.0, 1.0 *M*
NaCl, 5 *M*
DL-Methylmalonyl-CoA (containing 600 µCi of DL-[*methyl*-^{14}C]methylmalonyl-CoA per millimole), 3.0 m*M* in 3 m*M* HCl
Perchloric acid, 2.0 *M* (prepared daily in a glass container)
Potassium permanganate, 4% (prepared daily)
Propionic acid
Adenosylcobalamin, 1.0 mg/ml (stored in the dark)

Procedure. Pipet into 4.0-ml glass tubes in ice, 50 µl of Tris-HCl, pH 8.0, 18 µl of NaCl, 50 µl of DL-methylmalonyl-CoA, sufficient H_2O to make the volume 500 µl after enzyme is added, and, finally, enzyme. The tubes are incubated in a 30° water bath for 11 min and the reactions are stopped by placing the tubes in an ice-water bath and adding 100 µl of perchloric acid, followed by boiling for 3 min. After cooling, the tubes are centrifuged at 1500 *g* for 15 min at 4°. A portion of the supernatant, 250 µl, is added to 4.0-ml glass tubes containing 50 µl of perchloric acid, followed by the addition of 500 µl of potassium permanganate. The tubes are incubated in a boiling water bath for 10 min, followed by centrifugation at 1500 *g* for 15 min at 4°. A portion of the supernatant, 500 µl, is

added to glass counting vials containing 250 μl of propionic acid. The vials are then dried in an oven at 100°, followed by the addition of 1.0 ml of H_2O and 10 ml of Research Products International Corp. scintillation cocktail 3a70.

The above procedure measures holo-L-methylmalonyl-CoA mutase activity. Total L-methylmalonyl-CoA activity can be measured using the same procedure with the following changes: (1) an excess of adenosylcobalamin, 10–50 μl, replaces 10–50 μl of H_2O in the reaction mixture, and (2) assay tubes are prepared and incubated in the dark because free adenosylcobalamin is light sensitive, whereas adenosylcobalamin once bound to the mutase is light stable. Apo-L-methylmalonyl-CoA mutase activity can be calculated by subtracting the amount of holomutase activity from the amount of total mutase activity.

One unit of activity is defined as the amount of enzyme that forms 1 μmol of succinyl-CoA per minute. Specific activity is defined as units per milligram of protein. Protein concentration is determined by the method of Lowry et al.[11] using bovine serum albumin as the standard.

Method II[1,10]

Principle. L-Methylmalonyl-CoA mutase activity is followed by measuring the rate of formation of [2-(3)-14C]succinyl-CoA from a racemic mixture of DL-[*methyl*-14C]-methylmalonyl-CoA. Succinic thiokinase, GDP, and Mg^{2+} are present in the assay mix and convert [2-(3)-14C]succinyl-CoA and GDP to [2-(3)-14C]succinic acid, coenzyme A, and GTP, but have no effect on DL-[*methyl*-14C]methylmalonyl-CoA. [2-(3)-14C]Succinic acid is then separated from unreacted DL-[*methyl*-14C]methylmalonyl-CoA by taking advantage of the fact that the former is extracted from aqueous samples into ethyl acetate while the latter is not. The amount of [2-(3)-14C]succinic acid present in the ethyl acetate can then be determined directly in a liquid scintillation counter using an appropriate scintillation counting fluid.

When applied to crude tissues at acid or neutral pH, the assay is not specific for L-methylmalonyl-CoA mutase because it also measures the activity of D-methylmalonyl-CoA hydrolase. This problem arises from the fact that D-methylmalonyl-CoA hydrolase converts D-[*methyl*-14C]-methylmalonyl-CoA to [*methyl*-14C]methylmalonic acid and coenzyme A. Radioactive methylmalonic acid is also extracted from aqueous samples into ethyl acetate and thus is measured together with radioactive succinic acid.

[11] O. H. Lowry, N. J. Rosebrough, A. L. Farr, and R. J. Randall, *J. Biol. Chem.* **193**, 265 (1951).

The assay can be made largely or even entirely specific for L-methylmalonyl-CoA mutase, however, by taking advantage of the fact that the pH activity curves of the two enzymes are very different as shown earlier in Fig. 2 of chapter [49],[1] which deals with D-methylmalonyl-CoA hydrolase. Assays performed at pH 8.0 reflect mainly mutase activity and assays performed at pH 9.0 are specific for the mutase. Assays performed at pH 6.0, on the other hand, reflect mainly hydrolase activity, and assays performed at pH 5.0 are specific for the hydrolase.

Reagents.

Tris-HCl buffer, pH 9.0, 0.5 M
Succinate-CoA ligase (Sigma Chemical Corp, succinic thiokinase), 5 enzyme units/ml
GDP, 20 mM
MgCl, 0.1 M
DL-methylmalonyl-CoA (containing 150 μCi of DL-[*methyl*-^{14}C]methylmalonyl-CoA per millimole), 1 mM in 3 mM HCl
HCl, 3 N
Ethyl acetate
Adenosylcobalamin, 1.0 mg/ml

Procedure. Pipet into 4.0-ml glass tubes in ice, 95 μl of DL-[*methyl*-^{14}C]methylmalonyl-CoA, 45 μl of Tris-HCl buffer, pH 9.0, 20 μl of succinate-CoA ligase, 45 μl of GDP, 45 μl of MgCl$_2$, sufficient H$_2$O to make the volume 450 μl after eznyme is added, and, finally, enzyme. The tubes are incubated in a 30° water bath for 30 min and reactions are stopped by placing the tubes in an ice-water bath and adding 50 μl of HCl. Ethyl acetate, 1.3 ml, is added followed by vigorous vortexing for 15 sec and centrifugation at 500 g for 4 min at 4°. A portion, 1.0 ml, of the upper ethyl acetate layer is then added to glass counting vials containing 10 ml of Research Products International Corp. cocktail 3a70. Under these conditions, 45% of the succinic acid formed is recovered in the 1.0 ml of ethyl acetate extract.

The above procedure measures holomethylmalonyl-CoA mutase. Total and apomutase can be measured and calculated following the procedures described at the end of Method I above. Enzyme activity and specific activity are defined as above in Method I. Values for enzyme activity obtained with Method II are approximately 3 times higher than the values obtained with Method I.

Enzyme Purification Procedure[7]

All centrifugations and procedures are performed at 4° or in an ice-water bath. In the particular purification described below, the assay of

Method I (see above) was employed. Assay Method II has been utilized, however, in other purifications without any problems.

Preparation of Placental Homogenates

Human full-term placentas are obtained from healthy patients following vaginal delivery. The placentas are immediately immersed in an ice-water bath and kept there until assayed. The umbilical cord membranes are removed and four small random samples of each placenta are combined and frozen in dry ice–acetone for 30 min while the remainder of each placenta is frozen at −70°. Within 1–2 weeks of collection, the random samples from each placenta are thawed and homogenized in 2.5 volumes of 0.01 M Tris-HCl, pH 8.0, containing 2 mM GSH–NaOH, pH 8.0. Homogenization is achieved with five 1-min bursts in a Waring blender. The homogenates are centrifuged at 40,000 g for 30 min and the supernatants are decanted and assayed for L-methylmalonyl-CoA mutase activity. If the level of apomutase activity is greater than 0.001 unit per milliliter of supernatant, that placenta is suitable for inclusion in the purification scheme as outlined below. Approximately 25% of placentas screened in this way are suitable for use in purification. Eight placentas, 2400 g (250–400 g each), were utilized in the particular purification described below. Homogenization and the collection of 6460 ml of supernatant from placental homogenates was performed as just described for the small samples of placenta.

Affinity Chromatography on Adenosylcobalamin-Sepharose

A 1.5 × 3 cm column of adenosylcobalamin-Sepharose[7] (1 μmol adenosylcobalamin per milliliter Sepharose) is washed with 100 ml of 0.1 M glycine-NaOH, pH 10, containing 1 M NaCl, followed by 50 ml of 0.01 M Tris-HCl, pH 8.0, containing 2 mM GSH–NaOH, pH 8.0. An aliquot, 800 ml, of placental supernatant is then applied at a flow rate of 15 ml per hour. After sample application is completed, the column is washed at a flow rate of 50 ml per hour in the following sequence. One liter of 0.01 M Tris HCl, pH 8.0, containing 2 mM GSH–NaOH, pH 8.0, and 0.05 M NaCl; 2 liters of 0.01 M Tris-HCl, pH 8.0, containing 2 mM GSH–NaOH, pH 8.0, and 1 M NaCl; 50 ml of 0.01 M Tris-HCl, pH 8.0, containing 2 mM GSH–NaOH, pH 8.0, and 0.05 M NaCl; and 0.01 M KCN–HCl, ph 8.0. After incubating for 12 hr in the latter solution, the column is washed with 2 liters of 0.01 M Tris-HCl, pH 8.0, containing 2 mM GSH–NaOH, pH 8.0, and 1 M NaCl. The column is then equilibrated with 15 ml of 0.01 M Tris-HCl, pH 8.0, containing 2 mM GSH–NaOH, pH 8.0, 1 M NaCl, and 3 mM adenosylcobalamin. After incubating for

12 hr, the column is eluted with an additional 15 ml of 0.01 M Tris-HCl, pH 8.0, containing 2 mM GSH – NaOH, pH 8.0, and 1 M NaCl and 3 mM adenosylcobalamin. The entire 30 ml of the final eluate is then dialyzed against 3 liters of standard buffer for 16 hr.

DEAE-Cellulose Chromatography

A 0.7 × 1.25 cm column of DEAE-cellulose (Whatman DE-52) is prepared by removing the fines, adjusting the pH to 8.0 with 1.0 M Tris base, followed by washing with 50 ml of 0.01 M Tris-HCl, pH 8.0, containing 2 mM GSH – NaOH, pH 8.0, and 0.05 M NaCl. The dialyzed sample from the preceding step is applied to the column at a flow rate of 50 ml per hour and the column is then washed with 100 ml of 0.01 M Tris-HCl, pH 8.0, containing 2 mM GSH – NaOH, pH 8.0, and 0.05 M NaCl. After an additional wash with 100 ml of 0.01 M Tris-HCl, pH 8.0, containing 2 mM GSH – NaOH, pH 8.0, and 0.1 M NaCl, the mutase is eluted with 4 ml of 0.01 M Tris-HCl, pH 8.0, containing 2 mM GSH – NaOH, pH 8.0, and 0.125 M NaCl, followed by 2 ml of 0.01 M Tris-HCl, pH 8.0, containing 2 mM GSH – NaOH, pH 8.0, and 0.135 M NaCl. The latter two elutions are performed at a flow rate of 5 ml per hour. These two elutions are pooled to give a final volume of 6 ml. This preparation is stored at 4° and is stable for at least 2 weeks. Freezing leads to a marked decrease of activity.

A typical purification of L-methylmalonyl-CoA mutase is summarized in Table I. The mutase was purified 45,000-fold with a recovery of 10.4%.

Comments on Enzyme Purification[7]

A single protein band is obtained on polyacrylamide gel electrophoresis at pH 8.6. The single protein band corresponds to the single peak of

TABLE I
PURIFICATION OF L-METHYLMALONYL-CoA MUTASE

Purification step	Volume (ml)	Total protein (mg)	Total activity (units)	Specific activity (units/mg)
Placental homogenate	6,460	264,000	7.79	0.0000295
Eluate of affinity column	123	a	1.12	
Eluate of DEAE-cellulose	6	0.61	0.81	1.33

[a] Protein could not be determined due to interference by the large amount of free adenosylcobalamin.

mutase activity that is observed when a duplicate unstained gel, is sliced into 2-mm slices followed by elution and assay of mutase activity. SDS-polyacrylamide gel electrophoresis also gives a single protein band.

Greater than 95% of the apo-L-methylmalonyl-CoA mutase activity and less than 5% of the holo-L-methylmalonyl-CoA mutase activity bind to the adenosylcobalamin-Sepharose column under the conditions described above. A loss of 20 to 40% of apomutase activity is observed over the 16-hr period required to apply the sample to the adenosylcobalamin-Sepharose column. Loss of apomutase activity is greater when GSH was omitted from the application buffer. No mutase activity is eluted from the adenosylcobalamin-Sepharose column with any of the wash steps outlined in step 2 above, including the incubation with the solution that contained KCN and converted the excess adenosylcobalamin on the column to cyanocobalamin. The mutase was eluted with 0.01 M Tris-HCl, pH 8.0, 2 mM GSH–NaOH, pH 8.0, containing 1 M NaCl and 3 mM adenosylcobalamin. No elution of mutase occurred when identical concentrations of hydroxocobalamin or cyanocobalamin were substituted for the adenosylcobalamin. Mutase eluted slowly in the presence of adenosylcobalamin; elution was minimal after 1 hr and almost maximal after 12 hr. Strict adherence to the flow rates utilized with the adenosylcobalamin-Sepharose column is important, as is adherence to the concentrations of NaCl utilized in the various wash and elution procedures. The latter is also true with respect to the DEAE-cellulose column, since elution with NaCl concentrations above 0.135 M results in elution of a number of contaminating proteins.

Properties of the Enzyme[7]

L-Methylmalonyl-CoA mutase has a molecular weight of 144,000 based on gel filtration. A molecular weight of 72,000 is obtained using SDS-polyacrylamide gel electrophoresis in the presence of 2-mercaptoethanol. This indicates that the mutase contains two subunits of molecular weight 72,000. It is not clear if these subunits are joined by disulfide bonds, since SDS-polyacrylamide gel electrophoresis performed in the absence of 2-mercaptoethanol gives varying molecular weights ranging from 110,000 to 140,000. Purified mutase binds 2 mol of adenosylcobalamin per mole of mutase, i.e., 1 mol of adenosylcobalamin per mole of subunit. Both crude and purified mutase have a broad pH optimum that extends from pH 7 to pH 9. Kinetic parameters obtained with the purified mutase are difficult to interpret, since the purified enzyme appears to contain covalently bound cobalamin and differs in its behavior from that of crude or partially purified mutase that has not been exposed to the adenosylcobalamin-

Sepharose column or excess adenosylcobalamin. This phenomenon has also been observed with L-methylmalonyl-CoA mutase purified from sheep liver.[6] Based on studies performed with partially purified mutase, the K_m for DL-methylmalonyl-CoA is 0.2 mM and the K_m for adenosylcobalamin is 0.05 μM.

[52] α-Isopropylmalate Synthase from Yeast

By GUNTER B. KOHLHAW

α-Isopropylmalate synthase (EC 4.1.3.12) catalyzes the first reaction in leucine biosynthesis, the formation of α-isopropylmalate from acetyl-CoA and α-ketoisovalerate via an aldol condensation-type reaction. Recent genetic and structural analyses suggest that yeast cells contain three forms of α-isopropylmalate synthase.[1-3] α-Isopropylmalate synthase I, the major isoenzyme, is encoded by *LEU4*. It constitutes more than 80% of the α-isopropylmalate synthase activity in wild-type cells and consists of two subforms: synthase Ia (about 68 kDa) which is imported into the mito-chondrial matrix, and synthase Ib (about 65 kDa) which stays in the cytoplasm. The two forms differ in the N-terminal region, with Ib lacking the first 30 amino acid residues of Ia. The synthesis of synthase Ib is believed to be a consequence of differential transcription–translation and not of posttranslational modification.[3] (see section Cloning and Sequenc-ing below). α-Isopropylmalate synthase II, the minor isoenzyme, is poorly characterized. Its structural gene has not yet been identified. In the past, no distinction was made among the three forms of α-isopropylmalate synthase because their existence was not known. However, it is likely that the "wild-type enzyme" characterized in earlier work consisted for the most part of isoenzyme I, for the following reasons: first, the activity of isoen-zyme II (measured in strains carrying a total deletion of *LEU4*)[4] amounts to only 5–20% of the total α-isopropylmalate synthase activity of wild-type cells; second, when the *LEU4* gene is cloned on a multicopy vector in a

[1] V. R. Baichwal, T. S. Cunningham, P. R. Gatzek, and G. B. Kohlhaw, *Curr. Genet.* **7**, 369 (1983).
[2] D. M. Hampsey, A. S. Lewin, and G. B. Kohlhaw, *Proc. Natl. Acad. Sci. U.S.A.* **80**, 1270 (1983).
[3] J. P. Beltzer, L. L. Chang, A. E. Hinkkanen, and G. B. Kohlhaw, *J. Biol. Chem.* **261**, 5160 (1986).
[4] L. L. Chang, P. R. Gatzek, and G. B. Kohlhaw, *Gene* **33**, 333 (1985).

synthase II⁻ background, the properties of the α-isopropylmalate synthase of that strain resemble those established earlier for wild-type enzyme[5]; third, the response of isoenzyme II activity to increasing concentrations of leucine is very different from that of wild-type enzyme.[4]

Assay Method

α-Isopropylmalate synthase catalyzes the following reaction:

α-Ketoisovalerate + acetyl-CoA → α-isopropylmalate + CoA

In our hands, the most satisfactory way to measure synthase activity during purification or when determining kinetic parameters of partially or highly purified enzyme has been the determination with 5,5′-dithiobis(2-nitrobenzoic acid) (DTNB, Ellman's reagent) of free CoA formed within a timed incubation period.

Reagents

Tris-HCl buffer, pH 8.5, 0.625 M
KCl, 0.250 M
α-Ketoisovalerate, 6.25 mM
Acetyl-CoA, 6.25 mM
Ethanol
DTNB, 1 mM in Tris-HCl buffer, pH 8.0, 0.1 M

Procedure

The assay mixture contains in a final volume of 125 μl: 10 μl of Tris-HCl buffer, 20 μl of KCl, 10 μl each of α-ketoisovalerate and acetyl-CoA, and enough enzyme solution to give a final ΔA of no more than 0.2 (to assure proportionality). After a timed incubation period at 30° (usually < 15 min), the reaction is stopped by addition of 375 μl of ethanol. This is followed by 250 μl of the DTNB solution. After a 5-min centrifugation in a tabletop centrifuge, the absorption at 412 nm is measured against a blank reaction from which α-ketoisovalerate has been omitted. Linearity of the reaction with time must always be established. At pH 8.0, the molar extinction coefficient, ϵ_{412}, for 3-carboxylato-4-nitrothiophenolate (reduced Ellman's reagent) is 14,140 M^{-1} cm^{-1}.[6] One unit of activity is defined as the amount of enzyme catalyzing the formation of 1 μmol of CoA per minute. Specific activity is given as units per milligram of protein.

[5] L. L. Chang, T. S. Cunningham, P. R. Gatzek, W. Chen, and G. B. Kohlhaw, *Genetics* **108**, 91 (1984).
[6] H. B. Collier, *Anal. Biochem.* **56**, 310 (1973).

Protein concentration is determined by the biuret procedure or by the dye-binding method of Bradford.[7] To avoid interference from high concentrations of glycerol or ammonium sulfate, the protein is precipitated with trichloroacetic acid (final concentration 10%).

Other Assay Methods

The reproducibility of the procedure outlined above begins to wane when the specific activity drops below 0.02. However, we have found that wild-type cells, with specific activities around 0.01, can still be assayed by the DTNB method when an *in situ* assay is used. This involves permeabilization of cells according to one of the methods suggested by Miozzari *et al.*[8] Freshly harvested cells are washed once with deionized water. One part of cells (wet weight) is then suspended in 8 parts of 50 mM 4-morpholinopropanesulfonic acid (MOPS)–KOH buffer, pH 7.5, containing 0.01% of the nonionic detergent Triton X-100. The suspension is mixed well and kept frozen at temperatures of $-20°$ or below for at least 15 hr. After thawing, the cells are washed once with 50 mM MOPS-KOH buffer, pH 7.5, and resuspended in the same buffer [1 part of cells (wet weight) plus 8 parts of buffer]. This suspension is used in the assay mixture (above) in place of the enzyme solution. Specific activity may be expressed as micromoles of CoA formed per minute per milligram of cells (wet weight). Alternatively, whole-cell protein may be determined by a modified biuret method.[9]

A more sensitive assay procedure is based on the determination of a fluorescent umbelliferone derivate that forms when sulfuric acid-treated α-isopropylmalate is reacted with resorcinol.[10] Although this procedure is much more time-consuming, it is the method of choice when specific activities below 0.01 are encountered.

A continuous assay based on the decrease in absorption at 232 nm due to the conversion of acetyl-CoA to CoA plus the subsequent reaction of CoA with N-ethylmaleimide has been used in the study of α-isopropylmalate synthase of *Neurospora crassa*.[11] This assay has also been employed with yeast enzyme,[12] but it is not known whether N-ethylmaleimide is entirely without effect on the properties of this enzyme.

---

[7] M. M. Bradford, *Anal. Biochem.* **72**, 248 (1976).
[8] G. Miozzari, P. Niederberger, and R. Hütter, *Anal. Biochem.* **90**, 220 (1978).
[9] D. Herbert, P. J. Phipps, and R. E. Strange, *Methods Microbiol.* **5B**, 209 (1971).
[10] J. Calvo, J. Bartholomew, and B. Stieglitz, *Anal. Biochem.* **28**, 164 (1969).
[11] S. R. Gross, this series, Vol. 17, p. 777.
[12] E. H. Ulm, R. Böhme, and G. Kohlhaw, *J. Bacteriol.* **110**, 1118 (1972).

Purification Procedure for Yeast α-Isopropylmalate Synthase

The procedure described here is for α-isopropylmalate synthase from strain HB190/pLFC2. This strain contains multiple copies of an episomal plasmid carrying yeast 2 μm plasmid sequences, *E. coli* pBR322 sequences, and the yeast *LEU4* gene plus flanking sequences in a synthase II⁻ background.[5] The α-isopropylmalate synthase elaborated by strain HB190/pLFC2 therefore represents isoenzyme I. The amount of enzyme produced by this strain is approximately 20 times greater than that of wild-type strain S288C.[5] For purification purposes, cells are grown aerobically at 30° in Fink's minimal medium.[13] They are harvested in late log phase, washed once with deionized water, and stored as a paste at −20°. All subsequent operations are carried out at 0–5°.

Step 1. Preparation of Crude Extract

Fifty grams of cells (wet weight) is thawed in the presence of 75 ml of 50 mM potassium phosphate buffer, pH 7.5, containing 1.5 mM phenylmethylsulfonyl fluoride and 0.6% benzamidine-HCl. The suspension is passed through a French pressure cell twice and centrifuged at 18,000 g for 15 min. To the supernatant solution enough streptomycin sulfate is added to give a final concentration of 1.5%. The supernatant solution obtained after another centrifugation constitutes the crude extract.

Step 2. Ammonium Sulfate Fractionation

Protein precipitating between 48 and 68% of saturation at 0° is collected by centrifugation at 18,000 g for 15 min. The pellet obtained at this point is stable for extended periods of time when stored frozen. For further purification, it is dissolved in a small volume (< 10 ml) of 50 mM potassium phosphate buffer, pH 7.0 (buffer A), containing 1.2 M ammonium sulfate.

Step 3. Combined High Ionic Strength Hydrophobic and Affinity Chromatography on Leucine-Sepharose

The solution obtained in the preceding step is applied to a leucine-Sepharose column (2.3 cm × 30 cm) prepared by coupling L-leucine to cyanogen bromide-activated Sepharose 4B[14] and equilibrated with buffer A containing 1.2 M ammonium sulfate. Two column volumes of the same buffer are then used to elute unbound protein. Elution is continued with

[13] G. R. Fink, this series, Vol. 17, p. 59.
[14] P. Cuatrecasas, *J. Biol. Chem.* 245, 3059 (1970).

buffer A containing 0.8 M ammonium sulfate. When the protein concentration in the effluent has returned to near background levels (which requires about 1.5 column volumes), the eluant is changed to buffer A containing 0.8 M ammonium sulfate plus 10 mM leucine. This causes α-isopropylmalate synthase activity to be eluted as a sharp peak. The fractions containing most of the enzyme are combined and protein precipitating after raising the ammonium sulfate concentration to 70% saturation (0°) is collected by centrifugation.

Step 4. Chromatography on Blue Sepharose

The material obtained in the previous step is dissolved in a small volume (<5 ml) of 100 mM MOPS–KOH buffer, pH 7.5, containing 250 mM sucrose, 20 mM ammonium sulfate, and 0.25 mM dithiothreitol (buffer B). The solution is applied to a Blue Sepharose column (1.5 cm × 29 cm) equilibrated with the same buffer. Unbound protein is eluted by washing with 1.5 to 2 column volumes of buffer B. α-Isopropylmalate synthase is then eluted in essentially pure form by switching to buffer B containing 1 M KCl. The fractions containing enzyme activity are pooled, the ammonium sulfate concentration is raised to 70% saturation (0°), and the precipitate is collected by centrifugation. It is dissolved in a small volume (<5 ml) of 100 mM N-2-hydroxyethylpiperazine-N'-2-ethanesulfonic acid (HEPES)-KOH buffer, pH 7.5, containing 1.2 M ammonium sulfate, 20% (v/v) glycerol, and 1 mM dithiothreitol (buffer C), dialyzed against the same buffer, and reprecipitated with ammonium sulfate at 70% saturation. The precipitated material is dissolved in buffer C such that the final protein concentration is at least 20 mg/ml. When kept frozen in this form at −20° or less, the enzyme is usually stable for months.

The purification is summarized in Table I.

TABLE I
PURIFICATION OF α-ISOPROPYLMALATE SYNTHASE I

Step	Total activity (units[a])	Total protein (mg)	Specific activity (units/mg)	Purification (-fold)	Recovery (%)
Extract	165	1062	0.16	1.0	100
Ammonium sulfate	130	548	0.24	1.5	78.8
Leucine-Sepharose	107	51	2.10	13.1	64.8
Blue Sepharose	28.4	4	7.10	44.4	17.2

[a] Micromoles of CoA formed per minute.

Comments on Purification Procedure

The most efficient step of the procedure described above is chromatography on leucine-Sepharose. Since leucine is a natural ligand of α-isopropylmalate synthase I and its presence accelerates elution from the leucine-Sepharose column, the effectiveness of this step is believed to be, in part, the result of enzyme-ligand affinity. It is important that the degree of substitution of the Sepharose matrix with leucine be high, i.e., around 15 μmol of leucine per gram of Sepharose, dry weight.[15] Lesser degrees of substitution or the use of aged leucine-Sepharose will lead to less pure α-isopropylmalate synthase that is also considerably less stable because of protease contamination. Elution of the enzyme from Blue Sepharose, which in the present procedure is done by raising the salt concentration, may also be accomplished by adding 5 mM leucine. It is not known through which site(s) the enzyme interacts with Blue Sepharose; nucleotide binding sites are present, however (see below). Enzyme purified as shown in Table I is virtually homogeneous by the criterion of polyacrylamide gel electrophoresis in the presence of sodium dodecyl sulfate.

Properties

Some more recently established properties of yeast α-isopropylmalate synthase may be assigned unequivocally to either isoenzyme I or isoenzyme II. For the reasons given above (introductory statement), earlier work seems to have dealt chiefly with isoenzyme I. However, since the exact ratio of the α-isopropylmalate synthase forms in the preparations used in the early work is unknown, the following summary will show the source of the enzyme and its state of purity for each of the properties described.

Catalytic Properties

These were for the most part determined with a partially purified preparation from strain 60615 of the Lindegren collection.[12] The activity of the enzyme shows a broad optimum between pH 7.2 and 8.5; it drops off rapidly below pH 7.0 and above pH 9.0. At pH 7.2, the apparent K_m for acetyl-CoA was found to be $0.9 \times 10^{-5}\ M$, that for α-ketoisovalerate $1.6 \times 10^{-5}\ M$. (With highly purified enzyme from a strain related to wild-type S288C, the apparent K_m values for acetyl-CoA and α-ketoisovalerate have been determined as $4.3 \times 10^{-5}\ M$ and $4.8 \times 10^{-5}\ M$, respectively.[16]) Regarding α-keto acid specificity, α-ketobutyrate and pyruvate can serve

[15] R. Bigelis and H. E. Umbarger, *J. Biol. Chem.* **250,** 4315 (1975).
[16] P. R. Roeder and G. B. Kohlhaw, *Biochim. Biophys. Acta* **613,** 482 (1980).

as substrates, with apparent K_m values of $5.7 \times 10^{-4} M$ and $2 \times 10^{-4} M$, respectively, while α-ketoisocaproate and α-ketovalerate are competitive inhibitors.[12] The enzyme requires monovalent cations for full activity. K^+ is most effective, with an apparent K_a of $2 \times 10^{-3} M$.

Metal Content

Atomic absorption spectrometry of highly purified enzyme isolated from a derivative of wild-type S288C has shown the presence of two gram atoms of zinc per subunit.[16] Two types of zinc-binding sites can be distinguished based on the relative ease with which the metal is removed by ethylenediaminetetraacetic acid. Reduction of the zinc content of the enzyme to $<20\%$ ("apoenzyme") causes essentially total loss of catalytic activity. Activity can be restored to varying degrees by adding Zn^{2+}, Mn^{2+}, Fe^{2+}, Co^{2+}, or Cd^{2+} to the apoenzyme. In the case of Mn^{2+}-treated apoenzyme, V_{max} and the K_m values for both substrates increase significantly.

Feedback Inhibition by Leucine

The presence of leucine affects the K_m values for both substrates as well as V_{max} (partially purified enzyme from strain 60615 of the Lindegren collection).[12] The effect is strongly pH dependent: the same concentration of leucine that causes complete inhibition at pH 7.2 causes about 70% inhibition at pH 8.0 and about 15% inhibition at pH 8.8. At pH 7.2 and with saturating substrate concentrations, the leucine concentration required for half-maximal inhibition (apparent K_i) is about $2 \times 10^{-4} M$. From the leucine response curve, a Hill coefficient of 1.4 can be calculated. With cell-free extract prepared from HB190/pLFC2 cells, which contain only isoenzyme I activity, the apparent K_i for leucine is about $1 \times 10^{-4} M$.[5] By contrast, with cell-free extract from SK920 cells, which carry a total *LEU4* deletion and therefore exhibit only isoenzyme II activity, the apparent K_i is about $1.2 \times 10^{-3} M$.[4] With this enzyme, low concentrations of leucine (<0.3 mM) actually have a slight activating effect. Owing chiefly to the low level of isoenzyme II, no properties other than the leucine response have been studied.

Reversible Inactivation by CoA

Physiological concentrations of CoA cause a highly specific, time-dependent, reversible inactivation of α-isopropylmalate synthase when Zn^{2+} ions are also present. This effect has been observed with crude as well as highly purified preparations of enzyme from strain 60615 of the Lindegren

collection,[12] from bakers' yeast,[17] from a derivative of strain S288C,[18] and from strain HB190/pLFC2,[5] which shows that it is isoenzyme I that is so regulated. With purified enzyme, the presence of 10 μM each of CoA and $ZnCl_2$ will cause an inactivation of 80–90% within 10 min at 30° and pH 7.0–7.5. Five to 10 times as much CoA and Zn^{2+} is required with crude extract or permeabilized cells to achieve a similar effect. This CoA effect is distinct from product inhibition. In fact, there are two separate CoA binding sites on each monomer of α-isopropylmalate synthase.[17] The product site interacts with CoA and desulfo-CoA, both of which are competitive inhibitors with respect to acetyl-CoA. The apparent K_i for CoA interacting at the product site is $7 \times 10^{-5}\ M$. The "regulatory" CoA site, which becomes accessible to CoA in the presence of free Zn^{2+}, is highly specific for CoA, and there is no measurable competition by acetyl-CoA. The dissociation constant for the Zn^{2+}-dependent binding of CoA is $3.5 \times 10^{-5}\ M$. CoA-inactivated α-isopropylmalate synthase can be rapidly reactivated by millimolar concentrations of ATP or ADP, an effect which is not due to the chelating ability of these compounds.[18] Likewise, ATP can prevent CoA inactivation. Together with other results, these findings suggest that CoA inactivation functions to channel acetyl-CoA away from biosynthetic uses and toward catabolic reaction sequences under conditions of energy depletion.

Structural Properties

Molecular weight determinations carried out by sedimentation equilibrium centrifugation and polyacrylamide gel electrophoresis under nondenaturing and denaturing conditions with purified enzyme from baker's yeast[17] and from HB190/pLFC2 cells[5] indicate that the α-isopropylmalate synthase from these sources consists of identical subunits with a mass of 65–68 kDa that aggregate to form dimers. Ligands such as leucine, acetyl-CoA plus α-ketoisocaproate, or CoA plus $ZnCl_2$ have no demonstrable effect on the dimeric nature of the enzyme. Limited proteolysis of isoenzyme I generates two fragments of about 45 kDa and 23 kDa, respectively, long before the onset of secondary digestion.[5] It is noteworthy that the larger of these "domains" is similar in size to the entire subunit of the α-isopropylmalate synthase from *Salmonella typhimurium*[19] and from *Neurospora crassa*.[11]

[17] J. W. Tracy and G. B. Kohlhaw, *Proc. Natl. Acad. Sci. U.S.A.* **72**, 1802 (1976) and *J. Biol. Chem.* **252**, 4085 (1977).

[18] D. M. Hampsey and G. B. Kohlhaw, *J. Biol. Chem.* **256**, 3791 (1981).

[18a] J. P. Beltzer, S. R. Morris, and G. B. Kohlhaw *J. Biol. Chem.* **263**, 368 (1988).

[19] T. R. Leary and G. B. Kohlhaw, *J. Biol. Chem.* **247**, 1089 (1972).

Subcellular Localization

Most of the α-isopropylmalate synthase activity of wild-type yeast (strains D273-10B and S288C) and of strain HB190/pLFC2 (elaborating isoenzyme I only) is found associated with the mitochondria,[2,5] where it occupies the matrix space. Import into the matrix apparently occurs without cleavage of a presequence. *In vitro* translation of total RNA from galactose-grown wild-type cells gives rise to two anti-α-isopropylmalate synthase antibody-reactive proteins with molecular masses of 65–67 kDa and 63–64 kDa, respectively.[2] The same two species are observed when crude extract is treated with antibody. The smaller protein is very likely not a product of the larger protein. Only the larger protein is taken up by freshly isolated mitochondria.[2]

Cloning and Sequencing

At this writing, only the *LEU4* gene, which encodes isoenzyme I, has been cloned[5] and its nucleotide sequence determined.[3] An open reading frame specifies a protein of 619 amino acid residues with a calculated molecular weight of 68,416. The sequence between amino acid positions 32–47 and 415–438 has been confirmed by protein sequencing. Determination of the 5' ends of the *LEU4* message showed four major potential transcription starts (labeled α, β, γ, δ in Fig. 1). Two of these (γ and δ) are located downstream from the beginning of the long open reading frame. Productive translation of messages starting at γ or δ could begin at the nearest in-frame ATG (AUG) located at nucleotide position +91. The protein thus generated would have a calculated molecular weight of 65,169. It is likely that it was this short form of isoenzyme I that was seen in the *in vitro* translation experiments mentioned above. Both the long and

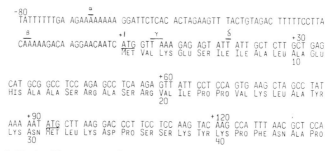

FIG. 1. Nucleotide sequence of yeast *LEU4* surrounding position +1, the beginning of the long open reading frame. Overlined positions labeled α, β, γ, δ signify major transcription starts. Two in-frame ATGs (at positions +1 and +91) are underlined. Sequence data from Beltzer *et al.*[3] with permission of the publisher).

short forms of isoenzyme I have been studied using α-isopropylmalate synthase (yeast)–β-galactosidase *(E. coli)* fusion proteins.[18a] Much of the long form of the fusion protein is imported into the mitochondria, the short form stays in the cytoplasm, suggesting that the first 30 amino acids of the long form contain all of the information necessary for mitochondrial import. The native short form of isoenzyme I is functional in leucine biosynthesis and is inhibited by leucine with an apparent K_i of about 4×10^{-4} M at pH 7.2.[18a]

α-Isopropylmalate Synthase from Other Sources

Purification procedures as well as structural and kinetic analyses have been published for α-isopropylmalate synthase from *Salmonella typhimurium*,[19-22] *Neurospora crassa*,[11,23,24] and *Alcaligenes eutrophus*.[25]

[20] G. B. Kohlhaw, T. R. Leary, and H. E. Umbarger, *J. Biol. Chem.* **244,** 2218 (1969).
[21] G. B. Kohlhaw and T. R. Leary, this series, Vol. 17, p. 771.
[22] J. C. Bartholomew and J. M. Calvo, *Biochim. Biophys. Acta* **250,** 577 (1971).
[23] R. E. Webster and S. R. Gross, *Biochemistry* **4,** 2309 (1965).
[24] R. E. Webster, C. A. Nelson, and S. R. Gross, *Biochemistry* **4,** 2319 (1965).
[25] J. Wiegel and H. Schlegel, *Arch. Microbiol.* **112,** 239 (1977), **112,** 247 (1977), and **114,** 203 (1977).

[53] Isopropylmalate Dehydratase from Yeast

By GUNTER B. KOHLHAW

Isopropylmalate dehydratase (EC 4.2.1.33, isopropylmalate isomerase) catalyzes the second pathway-specific reaction in leucine biosynthesis, the interconversion between α-isopropylmalate and β-isopropylmalate. Yeast isopropylmalate dehydratase is encoded by *LEU1*. The enzyme is located in the cytosol.[1] It is unstable when removed from the cells, but can be stabilized by a variety of conditions. The specific activity of isopropylmalate dehydratase decreases more than 10-fold when cells are grown in minimal medium supplemented with 2 mM leucine[2] or 0.5% tryptone.[3]

[1] E. D. Ryan, J. W. Tracy, and G. B. Kohlhaw, *J. Bacteriol.* **116,** 222 (1973).
[2] V. R. Baichwal, T. S. Cunningham, P. R. Gatzek, and G. B. Kohlhaw, *Curr. Genet.* **7,** 369 (1983).
[3] T. Satyanarayana, H. E. Umbarger, and G. Lindegren, *J. Bacteriol.* **96,** 2012 and 2018 (1968).

This regulation appears to take place, at least in part, at the transcriptional level.[4] *LEU1* is coregulated with *LEU2*, the gene encoding β-isopropylmalate dehydrogenase, even though the two genes reside on separate chromosomes.

Assay Method

Isopropylmalate dehydratase catalyzes the following reactions:

α-Isopropylmalate ⇌ (Dimethylcitraconate) ⇌ β-Isopropylmalate

The equilibrium of the reaction favors the formation of α-isopropylmalate.[5,6] The reaction mixture contains dimethylcitraconate when the reaction is at equilibrium,[5] but it is uncertain whether the compound exists as a free intermediate *in vivo*.

Principle of Assay

Use is made of the double bond character of dimethylcitraconate, which causes the compound to absorb in the ultraviolet range (190–250 nm; $\epsilon_{235} = 4530 \ M^{-1} \ cm^{-1}$). Measurements are routinely made at 235 nm. Since this wavelength lies on a steeply ascending branch of the absorption curve (λ_{max} is close to 200 nm), exact adjustment of the spectrophotometer is important.

Reagents

Potassium phosphate buffer, pH 7.0, 0.2 M
Dimethylcitraconate (DMC), 0.1 M
β-Isopropylmalate (β-IPM), 0.1 M
(Citraconate, 0.2 M)

Procedure

Either DMC or β-IPM can be used as substrate (see Comments on Assay Method below for the use of citraconate), recording either the decrease or the increase in absorption at 235 nm. A 0.5-ml reaction mixture typically contains 0.2 ml potassium phosphate buffer, 0.01 ml substrate solution, 0.27 ml H_2O, and 0.02 ml enzyme solution. To reduce total absorption, a cuvette with a 2 mm path length is used. Buffer and water are mixed and equilibrated at the desired temperature, usually 30°. Enzyme solution is added and the background rate recorded for about 30 sec. The

[4] Y.-P Hsu and P. Schimmel, *J. Biol. Chem.* **259**, 3714 (1984).
[5] S. R. Gross, this series, Vol. 17, p. 786.
[6] R. Bigelis, Ph.D. thesis. Purdue University, West Lafayette, Indiana, 1974.

reaction is then started by adding the substrate solution, and the rate is recorded for several minutes. With crude extract, the rate is stable for at least 2 min. Initial rate is calculated after subtracting the background rate. A unit of activity is defined as the amount of enzyme catalyzing the disappearance (formation) of 1 μmol of substrate (product) per minute. Specific activity is units per milligram of protein. Protein concentration is determined by the biuret procedure or by the dye-binding method of Bradford.[7] To avoid interference from high concentrations of glycerol or ammonium sulfate, the protein is precipitated with trichloroacetic acid (final concentration 10%).

Comments on Assay Method

Neither dimethylcitraconate nor β-isopropylmalate are commercially available. The natural isomer of β-isopropylmalate can be isolated from the culture medium of overproducing organisms[8,9]; dimethylcitraconate has to be commercially synthesized.[10] However, for routine determinations of isopropylmalate dehydratase activity, a commercially available substrate analog, citraconate, may be used:

$$
\begin{array}{cc}
\underset{\text{Dimethylcitraconate}}{
\begin{array}{l}
\mathrm{H_3C} \\
\qquad\diagdown \\
\qquad\qquad\mathrm{CH-C-CO_2^-} \\
\qquad\diagup \\
\mathrm{H_3C}
\end{array}
\begin{array}{l}
\mathrm{HC-CO_2^-} \\
\;\;\|
\end{array}
}
&
\underset{\text{Citraconate}}{
\begin{array}{l}
\mathrm{HC-CO_2^-} \\
\;\;\| \\
\mathrm{H_3C-C-CO_2^-}
\end{array}
}
\end{array}
$$

The UV absorption spectra of the two compounds are virtually superimposable, which means that the same extinction coefficient can be employed. The apparent K_m value for citraconate is approximately twice that for dimethylcitraconate[11] (see below, Properties). Since a cell-free extract from a *leu1* mutant fails to act on citraconate (under routine assay conditions; P. R. Brisco and G. B. Kohlhaw, unpublished observations), it is very likely that it is indeed isopropylmalate dehydratase and not another enzyme that catalyzes the conversion of the analog.

Purification Procedure for Yeast Isopropylmalate Dehydratase

Yeast isopropylmalate dehydratase can be purified from wild-type cells (strain S288c) if stabilizing agents such as β-isopropylmalate, glycerol,

[7] M. M. Bradford, *Anal. Biochem.* **72**, 248 (1976).
[8] J. M. Calvo and S. R. Gross, this series, Vol. 17, p. 791.
[9] P. N. Fultz, K. L. Choung, and J. Kemper, *J. Bacteriol.* **142**, 513 (1980).
[10] J. V. Schloss, R. Magolda, and M. Emptage, this volume [12].
[11] W. Chen, M.S. thesis. Purdue University, West Lafayette, Indiana, 1984.

and/or ammonium sulfate are utilized.[12] The cells are grown aerobically at 30° in a synthetic medium[3] containing salts, succinic acid (5.8 g/liter), trace elements, vitamins, and glucose (20 g/liter). They are harvested in late log phase, washed once with cold 0.05 M potassium phosphate buffer, pH 6.8 (buffer A), and stored at −20°.

Step 1. Preparation of Crude Extract

About 100 g of frozen cell paste is thawed and suspended in 100 ml of 0.1 M potassium phosphate buffer, pH 6.8, containing 1 mM β-isopropylmalate. Aliquots of 35 ml of the suspension are then disrupted for four consecutive 30-sec treatments with a Bronson S75 Sonifier set at 5 amp. This step, like all subsequent steps, is carried out at 0–5°.

Step 2. Ammonium Sulfate Fractionation

The suspension obtained after sonification is treated with solid ammonium sulfate to give 50% saturation. The pellet obtained after centrifugation at 27,000 g for 10 min is discarded. Enough solid ammonium sulfate is added to the supernatant solution to give 65% saturation. The precipitate is collected by centrifugation and dissolved in 2 ml of buffer A containing 1.24 M ammonium sulfate.

Step 3. High Ionic Strength Hydrophobic Chromatography on Valine-Sepharose

The solution obtained in the previous step is applied to a valine-Sepharose column (2 cm × 21.5 cm) prepared by coupling L-valine to cyanogen bromide-activated Sepharose 4B[13] and equilibrated with buffer A containing 1.24 M ammonium sulfate. The same buffer is used initially to elute excluded and loosely adsorbed proteins; a shift to buffer A containing 30% glycerol and no ammonium sulfate then elutes isopropylmalate dehydratase.

Step 4. High Ionic Strength Hydrophobic Chromatography on Leucine-Sepharose

Fractions from step 3 containing most of the isopropylmalate dehydratase activity are pooled and protein is precipitated by 70% saturation with ammonium sulfate. After standing for 1 hr, the precipitate is collected by

---

[12] R. Bigelis and H. E. Umbarger, *J. Biol. Chem.* **250,** 4315 (1975).
[13] P. Cuatrecasas, *J. Biol. Chem.* **245,** 3059 (1970).

TABLE I
PURIFICATION OF ISOPROPYLMALATE DEHYDRATASE

Step	Total activity (units)[a]	Total protein (mg)	Specific activity (units/mg)	Purification (-fold)	Recovery (%)
Extract	71	2285	0.031	1.0	100%
Ammonium sulfate	57	160	0.356	11.5	80.3
Valine-Sepharose	41	40	1.025	33.1	57.7
Leucine-Sepharose	21	3.4	6.176	199.2	29.6

[a] Micromoles of substrate utilized per minute.

centrifugation and dissolved in buffer A containing 1.24 M ammonium sulfate. The solution is applied to a leucine-Sepharose column (1.2 cm $\times$ 10 cm; prepared the same way as the valine-Sepharose column[13]) equilibrated with buffer A containing 1.24 M ammonium sulfate. Elution is accomplished by using a linear gradient of decreasing ammonium sulfate (1.24 $M \rightarrow 0$) and increasing glycerol concentrations (0 $\rightarrow$ 30%). The gradient (total volume 1 liter) is established in buffer A. Isopropylmalate dehydratase elutes around 1.1 M ammonium sulfate and 4% glycerol. The fractions containing most of the dehydratase activity are combined and treated with ammonium sulfate. Material precipitating between 54 and 62% saturation is collected. This precipitate is dissolved in buffer A containing 1.24 M ammonium sulfate and 30% (v/v) glycerol and stored at $-20°$. It should be homogeneous by the criterion of polyacrylamide disc. gel electrophoresis under denaturing conditions (0.1% sodium dodecyl sulfate).

A sample purification is shown in Table I.

Comments on Purification Procedure

Successful purification of isopropylmalate dehydratase depends to a large extent on the judicious use of the ammonium sulfate fractionations (step 2 and the concentration part of step 4). It is especially important that enough time be allowed for the precipitations to occur (no less than 1 hr).[12] The valine- and leucine-substituted Sepharoses are not interchangeable. Since neither valine nor leucine influences the activity of the dehydratase, the interaction between the substituents and the enzyme is believed to be strictly hydrophobic, resulting in different retentive abilities of the two types of Sepharose. Elevated concentrations of ammonium sulfate ($> 30\%$ saturation) increase retention, glycerol (30%) opposes this effect. At these

concentrations, both agents stabilize isopropylmalate dehydratase,[14] allowing for a wide variety of retention–elution programs.

Properties

Stability

Yeast isopropylmalate dehydratase, purified through step 2, has a half-life of 2–3 hr when kept at 0–5° and at a protein concentration of 0.35 mg/ml in 0.05 M potassium phosphate buffer, pH 6.8.[14] Dimethylcitraconate (1 mM) has no effect on stability, but β-isopropylmalate (1 mM) increases the half-life to 14 hr. Increasing concentrations of ammonium sulfate or glycerol increase stability; thus, the half-life of the enzyme is 27 hr in the presence of 1.24 M ammonium sulfate and 15 hr in the presence of 30% glycerol. A combination of these two agents at the indicated concentrations causes the half-life to increase to several months. In the absence of stabilizing agents, the enzyme is significantly less stable at pH 9.0 than at pH 7.0.

Catalytic Properties

The enzyme recognizes α-isopropylmalate, β-isopropylmalate, dimethylcitraconate, and also the analog citraconate, as indicated above; the *trans* isomer of citraconate, mesaconate, is not a substrate of the dehydratase.[11] The apparent K_m value of purified enzyme for dimethylcitraconate is 2.16×10^{-4} M under low-salt conditions and 2.37×10^{-4} M in the presence of 0.4 M ammonium sulfate.[14] With crude extract, an apparent K_m value of 4.8×10^{-4} M has been observed for dimethylcitraconate; under identical conditions, the value for citraconate is 8.3×10^{-4} M.[11]

The pH dependence of initial velocity, determined under standard (i.e., low ionic strength) assay conditions and with each of the three natural substrates, shows a sharp upturn at approximately pH 6 and a broad optimum between pH 7 and 9.

The presence of chelators such as ethylenediaminetetraacetic acid *o*-phenanthroline, or 8-hydroxyquinoline sulfonate in otherwise standard assay mixtures at pH 7.0 has only a very small effect on initial velocity.[14] There is, however, a time- and concentration-dependent inactivation by KCN that is almost instantaneous and complete in the presence of 10 mM KCN. This effect is held back by high concentrations of ammonium sulfate. No metal requirement has been established.

[14] R. Bigelis and H. E. Umbarger, *J. Biol. Chem.* **251**, 3545 (1976).

Structural Properties

Sucrose density gradient centrifugation of native, stabilized isomerase purified through step 2 yields an apparent molecular weight of approximately 90,000, according to the procedure of Martin and Ames.[14,15] The same value is obtained when highly purified, sodium dodecyl sulfate-denatured enzyme is subjected to polyacrylamide gel electrophoresis.[14] This suggests that yeast isopropylmalate dehydratase does not possess quaternary structure under the *in vitro* conditions tested; it does not necessarily reflect the *in vivo* situation.[16]

Cloning and Sequencing

By transformation of a *leu1* auxotroph, a 3.5-kb yeast genomic fragment has been identified that complements the *leu1* mutation.[4] The sequence of over 600 bp of the 5'-flanking region and 144 bp of what apparently is a long open reading frame has been established. A major transcription start is seen at position −79 (relative to the beginning of the open reading frame), and Northern blotting has revealed the presence of a transcript about 2.9 kilonucleotides in length that is large enough to accommodate an open reading frame for a 90-kDa protein.[4]

Isopropylmalate Dehydratase from Other Sources

Structure and regulation of isopropylmalate dehydratase from *Neurospora crassa* have been studied in considerable detail.[16,17] The enzyme from *Salmonella typhimurium* has been studied with respect to stability and several other properties,[6] but purification to homogeneity has apparently been elusive.

[15] R. G. Martin and B. N. Ames, *J. Biol. Chem.* **236**, 1372 (1961).
[16] V. E. Reichenbecher and S. R. Gross, *J. Bacteriol.* **133**, 802 (1978).
[17] V. E. Reichenbecher, M. Fischer, and S. R. Gross, *J. Bacteriol.* **133**, 794 (1978).

[54] β-Isopropylmalate Dehydrogenase from Yeast

By GUNTER B. KOHLHAW

β-Isopropylmalate dehydrogenase (EC 1.1.1.85) catalyzes the third pathway-specific reaction in the biosynthesis of leucine, an oxidative decarboxylation of β-isopropylmalate (IPM) to yield α-ketoisocaproate and

CO_2. In yeast, the enzyme is the product of the *LEU2* gene. Yeast β-isopropylmalate dehydrogenase is a cytosolic enzyme.[1] Outside its natural environment, it is inactivated by cold temperatures or by dilution unless protective measures are taken.[2] It is striking that no loss of activity is noted when yeast cells are kept frozen for several months, but that, when the yeast *LEU2* gene is expressed in *Escherichia coli,* the enzyme is very sensitive to cold temperatures, even in intact *E. coli* cells.[3] The specific activity of yeast β-isopropylmalate dehydrogenase decreases by more than 10-fold when cells are grown in minimal glucose medium that contains leucine (2 mM) or leucine plus threonine (at least 1 mM each).[4-6]

Assay Method

β-Isopropylmalate dehydrogenase catalyzes the following reaction:

$$\beta\text{-Isopropylmalate} + NAD^+ \longrightarrow \alpha\text{-ketoisocaproate} + NADH + H^+ + CO_2$$

Reagents

Potassium phosphate buffer, pH 8.0, 1.0 M
$MnCl_2$, 0.01 M
KCl, 1.0 M
NAD^+, 0.01 M
β-Isopropylmalate (β-IPM), 0.02 M
Pyrazole, 0.02 M

Procedure

The formation of NADH is followed in a recording spectrophotometer at 340 nm and 30°. A 0.2-ml reaction mixture typically contains 0.03 ml potassium phosphate buffer, 0.01 ml $MnCl_2$, 0.01 ml KCl, 0.02 ml NAD^+, 0.01 ml β-IPM, 0.02 ml pyrazole, 0.08 ml H_2O, and 0.02 ml enzyme solution. Pyrazole is an inhibitor of alcohol dehydrogenase,[7] but not of β-IPM dehydrogenase. With crude extract as the enzyme source, pyrazole reduces the background reduction of NAD^+ to very low levels. The background

[1] E. D. Ryan, J. W. Tracy, and G. B. Kohlhaw, *J. Bacteriol.* **116,** 222 (1973).
[2] Y.-P. Hsu and G. B. Kohlhaw, *J. Biol. Chem.* **255,** 7255 (1980).
[3] G. B. Kohlhaw, Y.-P. Hsu, R. D. Lemmon, and T. D. Petes, *J. Bacteriol.* **144,** 852 (1980).
[4] H. D. Brown, T. Satyanarayana, and H. E. Umbarger, *J. Bacteriol.* **121,** 959 (1975).
[5] V. R. Baichwal, T. S. Cunningham, P. R. Gatzek, and G. B. Kohlhaw, *Curr. Genet.* **7,** 369 (1983).
[6] Y.-P. Hsu, G. B. Kohlhaw, and P. Niederberger, *J Bacteriol.* **150,** 969 (1982).
[7] T. E. Singlevich and J. J. Barboriak, *Fed. Proc., Fed. Am. Soc. Exp. Biol.* **29,** 275 (1970).

rate is recorded for a maximum of 1 min with all ingredients of the reaction mixture except β-IPM present. Then β-IPM is added and the rate recorded for an additional period of 2–3 min. Initial rate is calculated after subtracting the background rate. A unit of activity is defined as the amount of enzyme catalyzing the formation of 1 μmol of product per minute. Specific activity is units per milligram of protein. Protein concentration is determined by the biuret procedure or by the dye-binding method of Bradford.[8] To avoid interference from high concentrations of glycerol or ammonium sulfate, the protein is precipitated with trichloroacetic acid (final concentration 10%).

Comments on the Assay Method

In our hands, the continuous assay described above is the method of choice, even with crude extracts. Pyrazole may be omitted when not working with crude preparations. A colorimetric assay that measures the formation of α-ketoisocaproate as the 2,4-dinitrophenylhydrazone within a timed incubation period has been described.[9] It should be used when NADH oxidation is a problem. Because of the limited stability of yeast β-IPM dehydrogenase, the use of a stabilizing buffer is recommended when preparing crude extracts (see step 1 of the purification procedure below). The substrate β-IPM is not commercially available, but can be isolated from the culture medium of overproducing organisms,[10,11] or a diastereomeric mixture can be prepared chemically (see chapters [56] and [57] in this volume).

Purification Procedure for Yeast β-Isopropylmalate Dehydrogenase

Yeast β-isopropylmalate dehydrogenase has been purified from a strain of *Saccharomyces cerevisiae* (21D/pYT14-*LEU2*)[12] that carries multiple copies of a *LEU2*-containing plasmid and produces about 30 times as much enzyme as wild-type strains, such that β-isopropylmalate dehydrogenase constitutes about 2% of the total extractable protein.[13] Cells are grown to late log phase in Fink's minimal medium[14] containing 2% glucose

[8] M. M. Bradford, *Anal. Biochem.* **72**, 248 (1976).
[9] S. J. Parsons and R. O. Burns, this series, Vol. 17, p. 793.
[10] J. M. Calvo and S. R. Gross, this series, Vol. 17, p. 791.
[11] P. N. Fultz, K. L. Choung, and J. Kemper, *J. Bacteriol.* **142**, 513 (1980).
[12] J. D. Cohen, T. R. Eccleshall, R. B. Needleman, H. Federoff, B. Buchferer, and J. Marmur, *Proc. Natl. Acad. Sci. U.S.A.* **77**, 1078 (1980).
[13] Y.-P. Hsu and G. B. Kohlhaw, *J. Biol. Chem.* **257**, 39 (1982).
[14] G. R. Fink, this series, Vol. 17, p. 59.

and appropriate amino acid supplements (0.1 mM histidine, 1 mM lysine). Harvested cells can be stored for at least 2 months at $-20°$. Throughout the purification, stabilizing conditions are used. All operations can therefore be carried out at between $0°$ and $4°$.

Step 1. Preparation of Crude Extract

Approximately 100 g of washed cells (wet weight) are suspended in 150 ml of 0.1 M potassium phosphate buffer, pH 6.9, (buffer A) containing 1.25 M ammonium sulfate, 20% (v/v) glycerol, 50 μM MnSO$_4$, 4 mM dithiothreitol, and 0.03% NaN$_3$. The suspension is passed through a French pressure cell twice. This is followed by centrifugation at 18,000 g for 1 hr.

Step 2. Ammonium Sulfate Fractionation

Material precipitating between 65 and 80% saturation of the crude extract with ammonium sulfate ($0°$) is collected by centrifugation at 18,000 g for 20 min and resuspended in buffer A containing 1.5 M ammonium sulfate, 20% (v/v) glycerol, and 0.03% NaN$_3$. During saturation with ammonium sulfate, the pH is maintained at neutrality with 2 N KOH.

Step 3. High Ionic Strength Hydrophobic Chromatography on Leucine-Sepharose

The solution obtained in step 2 is applied to a leucine-Sepharose column (2.6 cm × 37 cm) prepared by coupling L-leucine to cyanogen bromide-activated Sepharose 4B[15] and equilibrated with buffer A containing 1.5 M ammonium sulfate, 20% (v/v) glycerol, and 0.03% NaN$_3$. Unbound protein is removed by washing with the loading buffer (three column volumes), and β-isopropylmalate dehydrogenase is eluted with buffer A containing 1.25 M ammonium sulfate, 20% (v/v) glycerol, and 0.03% NaN$_3$. Active fractions are pooled, concentrated by ultrafiltration (Amicon PM10 membrane), and dialyzed against 50 mM Tris-HCl buffer, pH 8.2, containing 20% (v/v) glycerol, 1 mM dithiothreitol, 50 μM MnCl$_2$, and 0.03% NaN$_3$.

Step 4. DEAE-Sepharose Chromatography

The dialyzed solution from the preceding step is applied to a DEAE-Sepharose column (1.6 cm × 85 cm) equilibrated with dialysis buffer. The column is developed with a linear NaCl gradient (zero to 0.1 M, about 10

---

[15] P. Cuatrecasas, *J. Biol. Chem.* **245,** 3059 (1970).

TABLE I
PURIFICATION OF β-ISOPROPYLMALATE DEHYDROGENASE

Fraction	Total activity (units)[a]	Total protein (mg)	Specific activity (units/mg)	Purification (-fold)	Recovery (%)
Extract	775	2152	0.36	1.0	100
Ammonium sulfate	690	1326	0.52	1.5	89
Leucine-Sepharose	525	275	1.91	5.3	68
DEAE-Sepharose	147	8	18.40	51.1	19

[a] Micromoles of product formed per minute.

column volumes) established in the same buffer. β-Isopropylmalate dehydrogenase activity peaks at 0.072 M NaCl. Fractions with specific activities of > 15 are pooled and concentrated by ultrafiltration through an Amicon PM10 membrane. The purification procedure is summarized in Table I.

Comments on Purification Procedure

By the criterion of sodium dodecyl sulfate-polyacrylamide gel electrophoresis, the enzyme as purified in Table I is at least 95% pure. When cells are used that contain less starting material than strain 21D/pYT14-*LEU2,* an additional purification step such as chromatography on phenyl-Sepharose[2] must be included. Purified enzyme is stored at room temperature (20–23°) in 0.1 M potassium phosphate buffer, pH 6.9, containing 1.5 M ammonium sulfate, 20% (v/v) glycerol, and 0.03% NaN$_3$. Under these conditions, it is fully stable for 2 weeks. Thereafter, gradual loss of activity is observed.

Properties

Catalytic Properties

β-Isopropylmalate dehydrogenase is an NAD$^+$-dependent enzyme. Under standard assay conditions, the activity drops to < 5% when NADP$^+$ is substituted for NAD$^+$. The apparent K_m values for both NAD$^+$ and β-isopropylmalate are enzyme concentration dependent. The values range from 5.4×10^{-5} to 1.5×10^{-4} M for NAD$^+$ (enzyme concentration 0.008 and 0.160 mg/ml, respectively), and from 2.3×10^{-5} to 4.2×10^{-5} M for β-isopropylmalate (enzyme concentration 0.016 and 0.160 mg/ml, respec-

tively).[2] Both divalent and monovalent cations are required for maximal activity. Among the divalent cations, Mn^{2+} and Cd^{2+} are about equally effective (at a final concentration of 0.5 mM). Next in effectiveness are Co^{2+} and Mg^{2+}, which yield relative activities of 73 and 62%, respectively. Cu^{2+} and Ni^{2+} are inhibitory.

The pH optimum, determined under otherwise standard assay conditions, is seen as a plateau between pH 8.9 and 10.1. At pH 8.0, 85% of maximal activity is obtained. Routine assays are performed at pH 8.0 because of solubility problems for ions such as Mn^{2+} at higher pH values.

Quaternary Structure

Native yeast β-isopropylmalate dehydrogenase exists in a dynamic monomer–dimer equilibrium.[2] A 20-fold dilution of purified enzyme leads to a significant increase in the relative amount of monomers within a period of 1 hr. Based on data obtained from polyacrylamide gel electrophoresis in the presence of sodium dodecyl sulfate, the monomer has an estimated M_r of 43,000 to 45,000. A calculation based on amino acid composition derived from the nucleotide sequence of *LEU2* yields a value of 39,085.

Cloning and Sequencing

The *LEU2* gene has been cloned on several different hybrid yeast–*E. coli* vectors and expressed in both *E. coli*[16,17] and yeast.[17-20] It was the *LEU2* gene with which transformation of yeast was first performed successfully.[18] The amino acid sequence deduced from the nucleotide sequence of *LEU2* defines a protein of 364 amino acid residues.[21] Around position 295, there is a segment of 14 amino acid residues that shows substantial homology with a corresponding segment in yeast isopropylmalate dehydratase. It has been argued that this segment might be involved in binding of β-isopropylmalate, a substrate common to both enzymes.

The N-terminus is blocked in native β-isopropylmalate dehydrogenase.[22]

[16] B. Ratzkin and J. Carbon, *Proc. Natl. Acad. Sci. U.S.A.* **74,** 487 (1977).
[17] J. D. Beggs, *Nature (London)* **275,** 104 (1978).
[18] A. Hinnen, J. B. Hicks, and G. R. Fink, *Proc. Natl. Acad. Sci. U.S.A.* **75,** 1929 (1978).
[19] J. R. Broach, J. N. Strathern, and J. B. Hickes, *Gene* **8,** 121 (1979).
[20] R. K. Storms, J. B. McNeil, P. S. Khandekar, G. An, J. Parker, and J. D. Friesen, *J. Bacteriol.* **140,** 73 (1979).
[21] A. Andreadis, Y.-P. Hsu, M. Hermodson, G. Kohlhaw, and P. Schimmel, *J. Biol. Chem.* **259,** 8059 (1984).
[22] Y.-P. Hsu, Ph.D. thesis, p. 55. Purdue University, West Lafayette, Indiana, 1980.

β-Isopropylmalate Dehydrogenase from Other Sources

It appears that the only other source from which β-isopropylmalate dehydrogenase has been extensively purified and characterized is *Salmonella typhimurium*.[9]

[55] Purification and Assays of Acetolactate Synthase I from *Escherichia coli* K12

By LILLIAN EOYANG and PHILIP M. SILVERMAN

The acetolactate synthases (EC 4.1.3.18) catalyze the first pair of homologous reactions on the parallel valine/isoleucine biosynthetic pathway [reactions (1) and (2)].[1]

$$2CH_3CO—COOH \rightarrow CH_3COH—(COCH_3)—COOH + CO_2 \qquad (1)$$

Pyruvate   acetolactate

$$CH_3CO—COOH + CH_3CH_3CO—COOH \rightarrow$$

Pyruvate   2-ketobutyrate

$$CH_3CH_2COH—(COCH_3)—COOH + CO_2 \quad (2)$$

acetohydroxybutyrate

As a class, these enzymes are of interest for several reasons. First, it is not clear how or even whether they regulate the two reactions they catalyze, which at least have the potential to interfere with each other. Second, enteric bacteria contain several acetolactate synthase isozymes. Each of these isozymes have diverged from a common ancestor[2,3] to function under different cellular conditions.[4,5] These conditions and their relation to enzyme structure and function remain as important problems in microbial physiology.

The bacterial isozymes bear as much amino acid sequence homology to the yeast enzyme as they do to each other[6]; what physiological significance attaches to that homology is also not clear.

Here we describe the purification to homogeneity of acetolactate synthase I from *Escherichia coli* K12. The purification is somewhat simpler,

[1] H. E. Umbarger, *Annu. Rev. Biochem.* **47**, 533 (1978).

[2] R. Wek, C. Hauser, and G. W. Hatfield, *Nucleic Acids Res.* **13**, 3995 (1985).

[3] P. Friden, J. Donegan, J. Mullen, P. Tsui, M. Freundlich, L. Eoyang, R. Weber, and P. M. Silverman, *Nucleic Acids Res.* **13**, 3979 (1985).

[4] J. McEwen and P. M. Silverman, *J. Bacteriol.* **144**, 68 (1980).

[5] F. Dailey and J. Cronan, *J. Bacteriol.* **165**, 453 (1986).

[6] S. Falco, K. Dumas, and K. Livak, *Nucleic Acids Res.* **13**, 4011 (1985).

faster, and more efficient than our original method.[7] In addition, we describe a simple and accurate radiometric assay for acetohydroxybutyrate formation catalyzed by acetolactate synthase I (and the other acetolactate synthases) that requires no special or expensive equipment or reagents; our assay is a modification of one described originally by Shaw and Berg.[8]

Enzyme Assays

Flavin adenine dinucleotide (FAD), 2-ketobutyrate (Na⁺ salt), thiamin diphosphate, pyruvic acid (Na⁺ salt), β-nicotinamide adenine dinucleotide (reduced form) (NADH, grade III, disodium salt), pyridoxal phosphate, creatine·H_2O, and α-naphthol (grade III) were obtained from the Sigma Chemical Co., St. Louis, MO. L-Threonine (allo-free, A grade) was obtained from Calbiochem. Acetoin (3-hydroxy-2-butanone; practical grade), purchased from Kodak Laboratory Chemicals, was washed, with ether and air dried before use. Bovine serum albumin was from the Miles Chemical Co. and rabbit muscle lactate dehydrogenase (grade V, 920 U/mg), from the Sigma Chemical Co. L-[U-¹⁴C]Threonine (225 mCi/mmol), [1-¹⁴C]pyruvate (Na⁺ salt; 36 mCi/mmol), and sodium [¹⁴C]bicarbonate (0.1 mCi/mmol) were purchased from Amersham/Searle. Other materials were obtained from standard commercial sources.

Reaction mixtures (0.25 ml) contained 25 μmol of potassium phosphate buffer (pH 8.0), 2.5 μmol of $MgCl_2$, 20 μg of thiamin diphosphate, 0.25 μg of FAD, and substrates and enzyme as indicated.[4] Incubation was at 37° for 5 or 10 min.

Acetolactate formed in the reaction was measured colorimetrically as acetoin after acid-catalyzed decarboxylation.[9,10] Sulfuric acid (50% in H_2O, 25 μl/0.25 ml reaction mixture) was added and the reaction mixture incubated for 30 min at 37°. Thereafter, we add 0.5 ml of a solution composed of equal volumes of 0.5% creatine in H_2O and 5% α-naphthol in 4 N NaOH. The reaction mixtures were incubated for 30 min at 37°, after which the absorbance at 530 nm was determined.

Using acetoin as a calibration standard, we obtained a ratio of 35 nmol/absorbance unit over the range of 10–100 nmol/0.25 ml of reaction mixture. We obtained the same value by converting a measured amount of pyruvate quantitatively to acetolactate in the reaction catalyzed by aceto-

---

[7] L. Eoyang and P. M. Silverman, *J. Bacteriol.* **157**, 184 (1984).
[8] K. Shaw and C. Berg, *Anal. Biochem.* **105**, 101 (1980).
[9] H. E. Umbarger and B. Brown, *J. Biol. Chem.* **233**, 1156 (1953).
[10] R. Bauerle, M. Freundlich, F. Stormer, and H. E. Umbarger, *Biochim. Biophys. Acta* **92**, 142 (1964).

lactate synthase I. For this calibration, the pyruvate concentration of the solution used as substrate was measured enzymatically in the lactate dehydrogenase reaction in the presence of excess NADH. Alternatively, pyruvate consumption was measured as $^{14}CO_2$ released after acidification of acetolactate synthase I reaction mixtures in which [1-^{14}C]pyruvate was used as substrate (see below). With either measure, the ratio of pyruvate consumed (nmol)/2 to absorbance at 530 nm is 35. The agreement between the chemical and enzymatic calibrations establishes that the acid-catalyzed conversion of acetolactate to acetoin is complete within 30 min at 37° and that the stoichiometry of the reaction is as shown in reaction (1).

We measured acetohydroxybutyrate formation by a modification of the radiometric assay described by Shaw and Berg.[8,11] The assay measures $^{14}CO_2$ released from the 1 position of acetohydroxybutyric acid in acid solution [reaction (3)].[8]

$$CH_3CH_2COH(COCH_3)^{14}COOH \xrightarrow{H^+}$$
2-Aceto-2-hydroxybutyric acid

$$CH_3CH_2COHCOCH_3 + {}^{14}CO_2 \quad (3)$$
3-hydroxy-2-ketopentane

The $^{14}CO_2$ is captured in an alkaline trap and quantitated by liquid scintillation spectrometry.

As substrate we used 2-keto-[U-^{14}C]butyrate prepared from L-[U-^{14}C]theonine in the reaction catalyzed by theonine dehydratase (purified from *E. coli* K12 through DEAE-cellulose chromatography).[12] Reaction mixtures contained per milliliter (total volume usually 0.5 ml): 100 μmol of potassium phosphate buffer, pH 8.2; 10 μg of pyridoxal phosphate; 10 μmol of L-threonine; sufficient L-[U-^{14}C]threonine (225 mCi/mmol) to yield a specific radioactivity of 400–1000 counts/min/nmol carbon atom (1600–4000 counts/min/nmol threonine); and ~1 unit of threonine dehydratase. Incubation was at 37°. The conversion was followed routinely by removing a small portion of the reaction mixture and measuring 2-ketobutyrate as the phenylhydrazone.[12] An alternative is thin-layer chromatography on silica gel (Kieselgel 60 F_{254}, 0.25 mm thick, Merck) with acetonitrile/ethanol/glacial acetic acid (12:4:1, v/v) as solvent, followed by autoradiography (R_f relative to pyruvate: threonine, 0.44; 2-ketobutyrate, 1.07).[13] By either method, the conversion is essentially complete (≥0.9 mol of 2-ketobutyrate formed/mol of threonine added) within 30–40 min.

[11] L. Eoyang and P. Silverman, *J. Bacteriol.* **166**, 901 (1986).
[12] D. Calhoun, R. Rimerman, and W. Hatfield, *J. Biol. Chem.* **248**, 3511 (1978).
[13] P. Lowe and R. Perham, *Biochemistry* **23**, 91 (1984).

The radioactive 2-ketobutyrate was used as substrate without further purification. Reaction mixtures were exactly as described for acetolactate formation except for the inclusion of 2-ketobutyrate. After an appropriate incubation (see below), the reaction mixture or a portion of it was placed in a glass shell vial (Fisher Scientific; 15 mm × 45 mm) inside a standard glass vial for liquid scintillation spectrometry (Research Products International) containing 1 ml of 1 N NaOH as CO_2 trap. The outside vial was then sealed with a serum stopper (16 mm i.d.; Aldrich Chemical Co. #Z12,461-3 or Z10,076-5) and the reaction mixture inside the shell vial was acidified with 0.5 ml of 1 N HCl injected through the stopper. After 18 hr at ambient temperature, the shell vial was removed and radioactive carbonate in the NaOH trap determined by liquid scintillation spectrometry. Using sodium [^{14}C] bicarbonate in the shell vial, we could show that the above procedure quantitatively traps all of the CO_2 formed upon acidification. The sensitivity of the assay is about 0.5 nmol, measured as $^{14}CO_2$ released from an acetolactate synthase I reaction mixture incubated without enzyme.[11] In some conditions (Table II) the contents of the shell vial can be assayed for acetoin, so that both reaction products can be assayed in the same reaction mixture.[11]

As described, the assay for acetohydroxybutyrate is simple, requiring equipment routinely available in most laboratories, and about as sensitive as the colorimetric assay for acetolactate. The use of commercially available L-[U-^{14}C]threonine to prepare the substrate replaces the expensive custom synthesis of 2-keto-[1-^{14}C]butyrate. The assay can be carried out using crude extracts as source of enzyme.[11]

Enzyme Purification

DEAE-Sephacel, Blue Sepharose CL-6B, Sephacryl S-300, and poly-buffer exchanger PBE94 were purchased from Pharmacia Fine Chemicals, Inc., Piscataway, NJ. *Escherichia coli* K12 strain MF2348 containing the *ilvB*$^+$ plasmid pTCN12 was obtained from Dr. M. Freundlich.[14] Cells were grown in 80-liter batches in a New Brunswick fermentor at 37°. The medium contained, per liter, 10 g of Difco tryptone, 5 g of Difco yeast extract, 5 g of NaCl, and 50 mg of ampicillin. Cells were harvested by sedimentation when the culture reached an optical density (660 nm) of 1.25 and were stored at −80°.

All operations were carried out at 4°. About 20 g of frozen cells were thawed and suspended uniformly in 2.5 volumes of standard buffer

---

[14] T. Newman, P. Friden, A. Sutton, and M. Freundlich, *Mol. Gen. Genet.* **186**, 378 (1982).

[20 mM potassium phosphate, pH 7, 0.5 mM dithiothreitol, 0.1 mM thiamin diphosphate, 20% (v/v) glycerol] containing 10 μM FAD and 1 mM MgCl$_2$. Cells were broken by a single passage through a French pressure cell at 16,000 psi. Insoluble debris was removed by centrifugation for 20 min at 30,000 g (supernatant fluid = crude extract).

DEAE-Sephacel

After dilution with an equal volume of standard buffer containing 1 mM MgCl$_2$, the crude extract was precipitated by the addition of solid ammonium sulfate (0.36 g/ml). After the ammonium sulfate was dissolved, the suspension was subjected to centrifugation at 30,000 g for 20 min and the precipitate was dissolved in about 50 ml of standard buffer. The protein solution was dialyzed for 18 hr against 2 liters of standard buffer and then against 2 liters of standard buffer containing 1 mM MgCl$_2$. The dialysis membrane used was Spectraflor 3 which has a molecular weight cut-off of 3500. The dialyzed enzyme was applied to a 3.4 cm $\times$ 11 cm column of DEAE-Sephacel previously equilibrated with standard buffer containing 1 mM MgCl$_2$. A linear gradient of 1 liter standard buffer containing 1 mM MgCl$_2$ with concentration limits of 0.02–0.5 M NaCl was used. Enzyme eluted at 0.2 M NaCl. Peak fractions were pooled and precipitated with ammonium sulfate (0.44 g/ml). After centrifugation at 30,000 g for 20 min, the enzyme protein was resuspended in a final volume of 9.6 ml of standard buffer.

Sephacryl S-300

The DEAE-Sephacel fraction was applied to a 3.4 cm $\times$ 83 cm column of Sephacryl S-300 previously equilibrated with standard buffer containing 10 mM FAD and 1 mM MgCl$_2$. Enzyme activity was detected at approximately 1.2 times the void volume. Peak fractions were pooled. Protein was precipitated with ammonium sulfate (0.44 g/ml) and collected by centrifugation at 30,000 g for 20 min. The pellet was dissolved in a final volume of 6 ml of standard buffer. The fraction was dialyzed for 18 hr against 1 liter of standard buffer and then for 3 hr against 2 liters of standard buffer containing 10 mM MgCl$_2$. One-half of this fraction was used for the subsequent steps. The remainder was stored at 4° and purified identically as required.

Blue Sepharose

Blue Sepharose was precycled with 10 volumes of 6 M urea followed by exhaustive washes with H$_2$O. One-half of the gel filtration fraction was

applied to a 3.4 cm × 24 cm column of Blue Sepharose previously equilibrated with standard buffer containing 10 mM MgCl$_2$. The column was washed with column buffer until the protein concentration of the effluent was less than 0.3 mg/ml (about 500 ml). Enzyme was eluted with 500 ml of column buffer containing 0.2 mM FAD.

PBE94

Active fractions from the Blue Sepharose chromatography were pooled and applied directly to a 1 cm × 11 cm column of PBE94 polybuffer exchanger previously equilibrated with standard buffer containing 10 mM FAD and 1 mM MgCl$_2$. The column was developed with a 200-ml linear gradient with concentration limits of 0.02 to 0.4 M potassium phosphate. For long-term storage at − 80° glycerol was added to individual fractions at a final concentration of 40%. Enzyme activity was stable indefinitely under these conditions.

The purification is summarized in Table I. Final specific activities have ranged from 20 to 60 units/mg (10 purifications).

Purity and Subunit Composition

Active fractions from the PBE column contain only two polypeptide chains (Fig. 1).[7] The larger of these is the *ilvB* gene product (M_r 60,000) and the smaller is the *ilvN* gene product (M_r 11,000).[3] Genetic, enzymologic, and physiologic evidence show that both polypeptides are functionally important subunits of the enzyme.[3,7,11] Quantitative analyses of SDS-polyacrylamide gel electropherograms[7] and of chromatograms obtained by gel filtration in 4 M urea[15] indicate a 1 : 1 stoichiometry of the IlvB and IlvN

TABLE I
PURIFICATION OF ACETOLACTATE SYNTHASE I FROM
E. coli K12

Fraction	Specific activity (units/mg/protein)	Total activity (units)	Yield (%)
Crude	1.3	3674	100
DEAE-Sephacel	2.6	1759	48
Sephacryl S-300	6.1	1808	49
Blue-Sepharose	13.8	599	16
PBE exchanger	30	751	20

[a] Computed on the basis of 10 g of cells.

[15] P. Silverman and L. Eoyang. *J. Bacteriol.* **169**, 2494 (1987).

**M$_r$ × 10^{-3}
(kDa)**

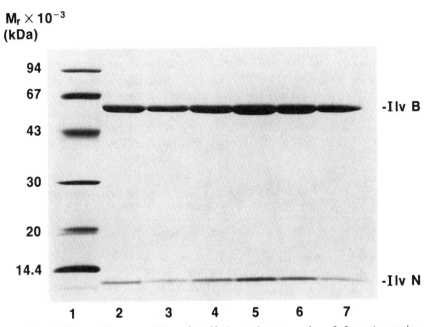

FIG. 1. Polypeptide composition of purified acetolactate synthase I. Lane 1 contains markers whose molecular weights are indicated to the left. Lanes 2–7 contain portions of successive fractions from the PBE column. The SDS-polyacrylamide slab gel (1.3 mm thick × 10 cm × 13 cm) contained a 10 to 20% linear acrylamide gradient and was otherwise run as described by U. K. Laemmli [*Nature (London)* **227**, 680 (1970)]. The gel was stained after electrophoresis with Coomassie Blue R250. IlvB and IlvN indicate the two acetolactate synthase I subunits as established by Friden *et al.*[3]

polypeptides. These data, along with the hydrodynamic properties of active enzyme[7,15] and cross-linking data,[7] indicate an overall enzyme composition of [(IlvB)$_3$ (IlvN)$_3$] and a molecular weight of about 200,000 for the largest active species.

Catalytic Properties of Purified Acetolactate Synthase I

While a detailed kinetic analysis is beyond the scope of this chapter, some basic catalytic features of the purified enzyme are appropriately presented here, especially as they pertain to the assay for acetohydroxybu-tyrate synthesis described above.

[16] H. Grimminger and H. E. Umbarger, *J. Bacteriol.* **137**, 846 (1979).

The requirements for the two reactions are compared in Table II. To remove components of the reaction mixture in which the enzyme is normally stored, enzyme (0.2 ml PBE fraction, 3.7 mg/ml) was dialyzed for 18 hr against 1 L of buffer (4°C) containing 20 mM potassium phosphate, pH 7/0.5 M KCl/1 mM EDTA/1 mM dithiothreitol/40% (v/v) glycerol, and then for 48 hr against a second liter of the same buffer also at 4°C. Even so, we were unable to demonstrate a complete requirement for added thiamin pyrophosphate (TPP), $MgCl_2$, or FAD, in apparent contrast to the report by Grimminger and Umbarger.[16] However, LaRossa and Schloss[17]

TABLE II
REQUIREMENTS FOR ACETOLACTATE SYNTHESIS
CATALYZED BY *E. coli* ACETOLACTATE SYNTHASE I

Reaction mixture[a]	Synthesis (nmol/10 min)	
	Acetolactate	Acetohydroxybutyrate
Complete	72.9	14.7
Pyruvate	<0.3	1.7[b]
2-Ketobutyrate	67.5	—
TPP	16.6	8.2
$MgCl_2$	21.7	6.6
FAD	35.6	8.9
BSA	68.4	12.2

[a] Reaction mixtures contained 1.2 μg of enzyme protein and, where appropriate, 1600 nmol of pyruvate (6.4 mM) and 265 nmol of 2-keto[U-14C]butyrate (1.1 mM). Acetohydroxybutyrate was measured as described in the text, and acetolactate was determined as acetoin in the shell vial after CO_2 evolution. Since the decarboxylation product of acetohydroxybutyrate contributes to color formation in the acetoin assay, albeit less than acetoin itself [R. Leavitt and H. E. Umbarger, *J. Biol. Chem.* **236**, 2486 (1961)], this method can be used most reliably in conditions where acetolactate synthesis exceeds acetohydroxybutyrate synthesis.

[b] CO_2 evolution in the absence of added pyruvate may occur as the result of condensation of two molecules of acetohydroxybutyrate [L. Abell, M. O'Leary, and J. Schloss, *Biochemistry* **24**, 3357 (1985)]; it is unlikely to reflect pyruvate contamination of the 2-ketobutyrate substrate, which was prepared from L-threonine.

[17] R. LaRossa and J. Schloss, *J. Biol. Chem.* **259**, 8753 (1984).

noted that complete resolution of acetolactate synthase II of *Salmonella typhimurium* in crude extracts from its cofactors failed to occur at 4°, but occurred readily at 37°. At any rate, both reactions catalyzed by purified acetolactate synthase I are affected by individual cofactor omissions to about the same extent. Finally, at the substrate levels used, 2-ketobutyrate had no apparent effect on the rate of acetolactate synthesis. The converse experiment cannot be done because pyruvate is a substrate for both reactions, but the rate of acetohydroxybutyrate synthesis is independent of pyruvate concentrations over significant ranges of substrate concentrations (see below).

As shown in Fig. 2, rates of acetolactate and acetohydroxybutyrate accumulation are both linear for about 10 min at initial substrate concentrations of 5 mM. Thereafter the rates of accumulation decay, owing in part to the reduction in substrate concentrations as the reactions proceed. The insets in Fig. 2 show that the amounts of product accumulated in 5 min are proportional to enzyme concentration at least up to 7 μg of enzyme/0.25 ml of reaction mixture. The different slopes indicate that at 5 mM substrate(s) the specific activity of acetolactate synthase I is severalfold higher for acetolactate formation than for acetohydroxybutyrate formation.

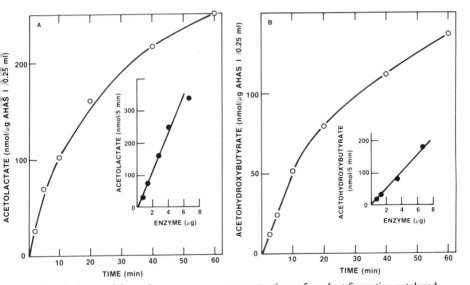

FIG. 2. Rate and dependence on enzyme concentrations of product formation catalyzed by purified acetolactate synthase I (AHAS I). Reaction conditions and assays were as described in the text. (A) Acetolactate formation (5 mM pyruvate substrate), (B) acetohydroxybutyrate formation (5 mM pyruvate and 5 mM 2-ketobutyrate substrates).

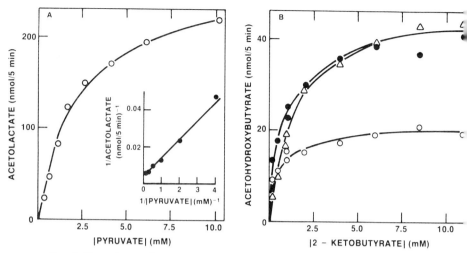

FIG. 3. Dependence of product formation catalyzed by acetolactate synthase I or substrate concentrations. Reaction conditions and assays were as described in the text. Each reaction mixture contained 1.2 μg of enzyme. In (B), acetohydroxybutyrate synthesis was measured as a function of 2-ketobutyrate concentration at three different levels of pyruvate: 0.25 mM (O), 1.5 mM (●), and 5 mM (△).

A substrate saturation curve for acetolactate formation as a function of pyruvate concentration is shown in Fig. 3A. The inset in Fig. 3A shows that in a double-reciprocal plot the data are linear over a 40-fold range of pyruvate concentration. The apparent K_m for pyruvate in this experiment was 3 mM, though we more commonly obtain a value closer to 1.5 mM.[16] In view of the kinetic mechanism proposed for acetolactate synthase II,[18] this K_m probably reflects binding of the aldehyde-acceptor pyruvate molecule. V_{max} for the reaction corresponds to about 40 μmol of acetolactate/min/mg of protein, comparable to that for purified acetolactate synthase II of *Salmonella typhimurium*.[17]

Figure 3B shows the rate of acetohydroxybutyrate formation as a function of both 2-ketobutyrate and pyruvate concentrations. At concentrations of 2-ketobutyrate in excess of ~2.5 mM, the rate of acetohydroxybutyrate formation was independent of the pyruvate concentration in the range of 1.5 to 5 mM; below 2.5 mM 2-ketobutyrate, the higher level of pyruvate somewhat inhibited the rate of the reaction. Double-reciprocal plots of these data were nonlinear, but in the limit as [2-ketobutyrate]$^{-1}$ approached 0, the apparent K_m for 2-ketobutyrate at 1.5 mM pyruvate was ~1 mM and, at 5 mM pyruvate, about 2 mM. The V_{max} for acetohydroxy-

[18] J. Schloss, D. Van Dyk, J. Vasta and R. Kutny, *Biochemistry* **24**, 4952 (1985).

butyrate formation was 5–6 μmol/min/mg at either pyruvate concentrations. At 0.25 mM pyruvate, the maximal rate of acetohydroxybutyrate formation apparently decreased by a factor of two, but this may have occurred because the combined synthesis of acetohydroxybutyrate and acetolactate (not measured here) exhausted the supply of pyruvate.

Acknowledgment

This work was supported by NIH grants CA-1330, GM-11301, and GM-30549. P.M.S. acknowledges support as a Cancer Scientist Awardee of the Irma T. Hirschl Trust.

[56] Acetolactate Synthase Isozyme II from *Salmonella typhimurium*

By JOHN V. SCHLOSS and DREW E. VAN DYK

$$
\begin{array}{cc}
CO_2^- & CO_2^- \\
| & | \\
C=O & +C=O \\
| & | \\
CH_2 & CH_2 \\
| & | \\
R & R'
\end{array}
\xrightarrow[\text{FAD}]{\text{TPP Mg}^{2+}}
\begin{array}{cc}
& O \quad CO_2^- \\
& \| \quad | \\
CH_2 & -C-C-OH \ +CO_2 \\
| & | \\
R & CH_2 \\
& | \\
& R'
\end{array}
$$

$$R = H \text{ or } CH_3; \ R' = H \text{ or } CH_3$$

The first common step in the biosynthesis of valine and leucine (where both R and R′=H) and of isoleucine (where R=H and R′=CH₃) in bacteria, yeast, and higher plants is catalyzed by acetolactate synthase.[1] In enteric bacteria there are multiple isozymes of acetolactate synthase,[2] which differ in their ability to handle α-ketobutyrate, affinity for flavin adenine dinucleotide (FAD),[3] allosteric inhibition by branched-chain amino acids,[4] pH optima,[3] and sensitivity to various active-site-directed

---

[1] EC 4.1.3.18.

[2] M. De Felice, C. T. Lago, C. H. Squires, and J. M. Calvo, *Ann. Microbiol. (Paris)* **133A**, 251 (1982).

[3] H. Grimminger and H. E. Umbarger, *J. Bacteriol.* **137**, 846 (1979).

[4] N. Gollop, D. M. Chipman, and Z. Barak, *Biochim. Biophys. Acta* **748** 34 (1983); M. De Felice, M. Levinthal, M. Iaccarino, and J. Guardiola, *Microbiol. Rev.* **43**, 42 (1979); M. De Felice, C. Squires, and M. Levinthal, *Biochim. Biophys. Acta* **541**, 9 (1978).

inhibitors (i.e., the sulfonylurea[5,6] and imidazolinone herbicides). In addition to the FAD-dependent isozymes of acetolactate synthase, which are unique to the biosynthetic pathway for branched-chain amino acids, there is also another form of the enzyme, which does not require FAD, involved in the biosynthesis of acetoin.[7] By contrast to the multiplicity of isozymes in prokaryotes, there is only a single acetolactate synthase involved in the biosynthesis of branch-chain amino acids in yeast,[8] so the significance of the diversity of this enzyme in enteric bacteria is not clear. Acetolactate synthase isozyme II (ALSII) is unique in bacterial biosynthesis of branched-chain amino acids in that it is the only isozyme not subject to feedback, allosteric inhibition by valine, leucine, or isoleucine.

Assay Method

Principle

The enzymatic reaction can readily be measured by two assays: (A) continuous spectrophotometric monitoring of the loss of absorbance of pyruvate and/or α-ketobutyrate at 333 nm; and (B) acid-catalyzed decarboxylation of acetolactate (1 mol produced for every 2 mol of pyruvate consumed) to acetoin, and the subsequent determination of this product by the method of Westerfeld.[9]

Reagents

Sodium pyruvate, 1.0 M
Tricine, 1.0 M in 0.5 N NaOH
MgCl$_2$, 1.0 M
FAD, 5 mM
Thiamin pyrophosphate, 5 mM
H$_2$SO$_4$, 12 N
α-Naphthol, 5% in 2.5 N NaOH
Creatine, 0.5%
Sodium hydroxide, 50%

Procedure

(A) The continuous assay is conducted in a standard 1-cm quartz cuvette at 37°. The assay mixture consists of 0.05 ml of sodium pyruvate,

[5] J. V. Schloss, *in* "Flavins and Flavoproteins" (R. C. Bray, P. C. Engel, and S. G. Mayhew, eds.), p. 737. de Gruyter, Berlin, Federal Republic of Germany, 1984.
[6] R. A. LaRossa and D. R. Smulski, *J. Bacteriol.* **160**, 391 (1984).
[7] F. C. Størmer, *J. Biol. Chem.* **242**, 1756 (1967).
[8] S. C. Falco and K. S. Dumas, *Genetics* **109**, 21 (1985).
[9] W. W. Westerfeld, *J. Biol. Chem.* **161**, 495 (1945).

0.1 ml of Tricine, 0.01 ml of $MgCl_2$, 0.02 ml of FAD, 0.02 ml of thiamin pyrophosphate (TPP), and 0.75 ml of water. After allowing sufficient time for thermal equilibration and a baseline at 333 nm to be established, 0.05 ml of enzyme solution is added to initiate the assay. The reaction is followed by monitoring the loss in absorbance at 333 nm (total change in absorbance at equilibrium ≈ 0.9) with a recording spectrophotometer. (B) The fixed time assay is carried out in an identical fashion, except that the assays are conducted in disposable 10×75 mm borosilicate glass tubes. After allowing sufficient time for the reaction to produce $0.01-0.2$ μmol of acetolactate, the reaction is quenched by the addition of 0.25 ml of H_2SO_4. Multiple samples can be stored at room temperature for several hours, or overnight at 4°, before being processed further with negligible effect. Samples are then incubated at 80° for 5 min to ensure decarboxylation, followed by cooling to 37°. In rapid succession, with mixing, are added 0.16 ml of NaOH, 0.3 ml of creatine, and 0.3 ml of naphthol. Color development is allowed to proceed for 1 hr at 37°, with efficient mixing at about 15-min intervals to ensure adequate aeration (color development involves air oxidation). Absorbance is read at 530 nm, after brief centrifugation (5 min at 2000 rpm) to clarify samples. Background absorbance is estimated from blanks containing equivalent amounts of enzyme added after acidification. Background absorbance can usually be lowered by using a good grade of commercial α-naphthol (e.g., EM Science) which has been recrystallized twice from chloroform prior to use.

Definition of Unit and Specific Activity

One unit is the amount of enzyme which catalyzes the formation of 1 μmol of acetolactate per minute from 2 μmol of pyruvate. In the continuous assay, the extinction of pyruvate at 333 nm is 17.5 M^{-1} cm^{-1}. For the fixed time assay, 0.1 μmol of acetolactate or acetoin results in 0.65 $A_{530\ nm}$. Either assay gives comparable estimates of enzymatic activity, except that the continuous assay is substantially less sensitive, but considerably more convenient, than the fixed-time assay. Specific activity is given in units per milligram of protein.

Protein Determinations

Protein is measured by the method of biuret.[10] The purified protein is measured by using an extinction of 0.90 $A_{280\ nm}$ mg^{-1} ml cm^{-1} for the enzyme–FAD complex, and 0.78 $A_{280\ nm}$ mg^{-1} ml cm^{-1} for the apoenzyme.[11]

[10] E. Lyane, this series, Vol. 3, p. 447.
[11] J. V. Schloss, D. E. Van Dyk, J. F. Vasta, and R. M. Kutny, *Biochemistry* **24**, 4952 (1985).

Purification

Cultivation of Organism and Preparation of Extracts

Plasmid pDU9 was obtained from the late Dr. R. O. Burns of Duke University.[12] This plasmid encodes the large and small subunits of the *Salmonella typhimurium* ALSII[11] and valine aminotransferase.[12] pDU9 was transformed into *Escherichia coli* HB101 by Dr. N. S. Yadav, E. I. du Pont de Nemours & Company. pDU9 transfers on *E. coli* K12 strains the ability to grow on minimal media in the presence of valine and resistance to ampicillin. *Escherichia coli* HB101/pDU9 is grown on a minimal salts medium[13] which contains per liter: 2.33 g of K_2HPO_4, 1 g of KH_2PO_4, 1 g of $(NH_4)_2SO_4$, 0.1 g of $MgSO_4$, 0.075 g of $CaCl_2 \cdot 2H_2O$, 0.02 g of ethylenediaminetetraacetic acid (EDTA), 0.012 g of $FeSO_4 \cdot 7H_2O$, 2.8 mg of H_3BO_3, 2.1 mg of $MnSO_4 \cdot 4H_2O$, 750 μg of $Na_2MoO_4 \cdot 2H_2O$, 240 μg of $ZnSO_4 \cdot 7H_2O$, 40 μg of $Cu(NO_3)_2 \cdot 3H_2O$, 0.16 g of L-leucine, 0.14 g of L-proline, 0.35 g of L-valine, 0.02 g of sodium ampicillin, and 0.1 g of thiamin-HCl. Initially, the culture contains 5 g/liter of dextrose, and three additions of dextrose (5 g/liter each of the first two additions and 10 g/liter for the third addition) are made during fermentation when the turbidity of the culture reaches 1,2, and 4 $OD_{650 \, nm}$, respectively. During growth, the pH of the culture is maintained at 7.1 by automated titration with anhydrous ammonia. The temperature of the culture is maintained at 35°, and the stirring and aeration rates are varied to achieve $\geq$ 30% of air saturation. Under these conditions, the turbidity doubles about every 1.5 hr. After inoculation of 1700 liters of medium with 300 liters of culture (8 $OD_{650 \, nm}$), the turbidity is allowed to increase to 6 $OD_{650 \, nm}$, at which time 240 kg of ice is added to rapidly reduce the culture temperature from 35 to 18°. It is important to prevent the culture from reaching stationary phase (which occurs at about 10–12 $OD_{650 \, nm}$) as the ALSII activity rapidly declines at this point. The chilled culture is harvested by use of a Sharples continuous-flow centrifuge. The cell paste is frozen as small lumps in liquid nitrogen, and then broken, under liquid nitrogen, into a finely divided gravel with a spatula. A yield of 16.5 kg of frozen cell paste is obtained, which represents 8.2×10^5 units of acetolactate synthase (32 g of enzyme). Cell paste can be stored indefinitely in liquid nitrogen (> 3 yr) with no noticeable loss of enzymatic activity.

Three stock reagents are prepared for the formulation of buffer solutions. The first is a 1 M solution of Tris buffer, composed of 0.5 M Tris

[12] D. L. Blazey, R. Kim, and R. O. Burns, *J. Bacteriol.* **147**, 452 (1981).
[13] J. G. Ormerod, K. S. Ormerod, and H. Gest, *Arch. Biochem. Biophys.* **94**, 449 (1961).

base and 0.5 M Tris-HCl. The second stock solution is a 0.1 M solution of EDTA, prepared by adjusting the pH of a solution of disodium EDTA to pH 8 at room temperature with NaOH. The third stock is a 4 M solution of ultrapure ammonium sulfate (Schwarz/Mann), which is further purified by passage through a column of Chelex 100 (Bio-Rad, about 200 ml of resin for 4 liters of ammonium sulfate solution). Buffer A consists of 0.1 M Tris buffer, 10 mM EDTA, and 1 mM dithiothreitol. Buffer B consists of 25 mM Tris buffer, 1 mM EDTA, and 1 mM dithiothreitol. All purification steps are carried out in the dark or in subdued light, at 4°, except the heat step. Frozen cell paste is suspended in 2 volumes (w/v) of buffer A containing 0.1 mM FAD, and allowed to thaw. After obtaining a uniform suspension, 350-ml aliquots of the thawed cells are subjected to full-power pulsed sonication four times for 5-min periods with a Model W-375 sonicator (Heat Systems-Ultrasonics, Inc.) equipped with a 1.3-cm tip at 70% duty cycle. The suspension is centrifuged in a Sorvall GSA rotor at 7000 rpm for a minimum of 5 hr (more uniform results are obtained if the centrifugation is carried out overnight, about 15 hr). The supernatant is decanted into 600-ml beakers, and incubated in a 50° water bath for sufficient time to raise the temperature of the solution to 50° with stirring (30 min), followed by incubation on ice with stirring until the temperature is lowered to 4°. Insoluble material is removed by centrifugation for 30 min. After discarding the pellet, 176 g of solid ammonium sulfate is added per liter of supernatant, and the solution is centrifuged for 30 min. The supernatant of the third centrifugation step is collected, and a second addition of 267 g of solid ammonium sulfate per liter is made. After collecting the precipitate by centrifugation for 3 hr, the pellet is suspended with a minimal volume (about 0.1 volume per gram of cells) of buffer A containing 0.1 mM FAD. The suspended pellet is further diluted with a total of 0.25 ml of buffer per gram of cell paste, and the suspension is stirred at 4° for 30 min. After collection of the precipitate by centrifugation for 30 min, the pellet is dissolved in 0.15 ml of buffer A, containing 0.1 mM FAD, per gram of cell paste. This solution can be stored frozen at −60° indefinitely.

Phenyl-Sepharose Chromatography

All columns are eluted at a flow rate of 80 ml/hr, and 22-ml fractions are collected. A solution of the second ammonium sulfate fraction (containing up to 10 g of protein) is mixed with 0.17 volume of 4 M Chelex-treated ammonium sulfate and applied to a 2.4 × 55 cm column of phenyl-Sepharose CL-4B. The column is developed with a 2-liter linear gradient, starting with buffer A (1 liter), containing 1 M ammonium sulfate

and 20 μM FAD, and ending with buffer B (1 liter), containing 20 μM FAD. Following the gradient, the column is eluted with 1 liter of the limit buffer. ALSII elutes after the completion of the gradient, with the majority of activity eluting in fractions 96–108 for a freshly prepared column. If the phenyl-Sepharose column is used repeatedly, however, ALSII begins to elute in an earlier position in subsequent usages of the same column. Fractions containing ALSII are concentrated to < 100 ml by pressure dialysis (Amicon PM30 membrane).

Sephacryl S-200 Chromatography

The concentrated solution of ALSII obtained from phenyl-Sepharose chromatography is applied to a 5 × 107 cm column of Sephacryl S-200. Prior to sample application, the column is equilibrated with 2 liters of buffer B, containing 20 μM FAD, and, following application of ALSII, the column is developed with an additional 2 liters of the same eluant. ALSII elutes between fractions 34 and 46. The same column has been used repetitively for > 2 years with no signs of deterioration.

DEAE-Fractogel Chromatography

A 5 × 108 cm column of DEAE-650M Fractogel TSK is prepared by suspending the column packing in buffer B and adjusting the pH of the suspension to 8.1 with HCl at room temperature. Prior to sample application, the column is equilibrated with about 6 liters of buffer B. Pooled ALSII from Sephacryl S-200 chromatography is applied to the DEAE-Fractogel column, and eluted with an 8-liter linear gradient, composed of 4 liters of buffer B (initial) and 4 liters of buffer A, containing 0.4 M KCl (puratronic grade, obtained from Alfa) and 0.1 mM FAD. The majority of ALSII elutes between fractions 134 and 145 (at about 0.15 M KCl). Fractions containing ALSII can be concentrated by pressure dialysis to a protein concentration > 200 mg/ml and stored frozen indefinitely at −60° without loss of enzymatic activity. Purification of the enzyme is summarized in Table I.

Preparation of FAD-Free ALSII

FAD can be removed from ALSII by extended dialysis or charcoal treatment in the presence of high salt (2 M KCl) with good retention of enzymatic activity upon reconstitution with FAD, thiamin pyrophosphate, and Mg^{2+}.[11] A somewhat more convenient procedure for removing FAD is treatment of ALSII with acidic ammonium sulfate in the presence of high salt, similar to the protocol described for glyoxylate carboligase (tartron-

TABLE I
PURIFICATION OF ACETOLACTATE SYNTHASE II

Step	Activity (units)	Protein (mg)	Specific activity (units/mg)
Extract (600 g of cells)	36,700	56,300	0.65
Heat	30,800	45,600	0.67
$(NH_4)_2SO_4$	26,100	7,300	3.57
Phenyl-Sepharose	26,400	2,440	10.8
Sephacryl S-200	23,400	1,770	13.2
DEAE-Fractogel	23,000	910	25.3

ate-semialdehyde synthase) by Gupta and Vennesland.[14] To 50 μl of ALSII (155 mg/ml) is added 150 μl of 4 M KCl (puratronic grade). The sample is maintained in an Eppendorf tube, and all steps are conducted at 4°, using prechilled reagents. After the addition of 100 μl of saturated $(NH_4)_2SO_4$ containing 30 mM H_2SO_4, the sample is briefly vortexed, followed by centrifugation in a minifuge for 5 min. The supernatant is decanted, and the interior of the tube blotted dry with a paper towel. The pellet is resuspended with 1 ml of 0.1 M Bicine (in 0.05 N NaOH, final pH equivalent to the pK of Bicine), 1 mM EDTA (at room temperature) divided into four 0.25-ml aliquots, and added to four 0.75-ml aliquots of 4 M KCl in Eppendorf tubes. After the addition of 0.5 ml of acidic ammonium sulfate to each sample, they are each vortexed, and centrifuged for 5 min. The pellets are suspended with four 1-ml aliquots of 0.1 M Bicine, 1 mM EDTA (at room temperature) and recentrifuged. The remaining pellets are further extracted with four additional 1-ml aliquots of Bicine buffer, which after pooling, filtration through glass wool to remove denatured protein, and concentration with a centrifugal membrane concentrator (Centricon PM10), give 7.4 mg of apo-ALSII at 9.7 mg/ml. Assay of the apoenzyme in the absence of FAD gives a specific activity of 0.51 units/mg, and in the presence of FAD gives 30.7 units/mg.

Properties

Stability

The enzyme–FAD complex (as purified above) is quite stable aerobically in the dark, or anaerobically in the light. When exposed to light for an

[14] N. K. Gupta and B. Vennesland, *J. Biol. Chem.* **239**, 3787 (1964).

extended period of time (aerobically), severe losses of enzymic activity can be encountered ($\approx 80\%$ loss overnight at $4°$ under conditions of normal fluorescent lighting). After removing FAD, the apoenzyme is no longer light sensitive. Enzymatic activity can be maintained for several days, even at room temperature. Both the apoenzyme and the enzyme–FAD complex can be stored indefinitely at $-20°$ (>2 years) or $-60°$ without loss of activity. Dilution of the apoenzyme at low ionic strength, particularly at low protein concentrations, can result in severe losses of activity, probably as a result of surface adsorption.[11] After reconstitution of ALSII–FAD with TPP and Mg^{2+} (or other metal), the enzyme exhibits oxygen lability, even in the dark. Although losses of enzymatic activity of the fully reconstituted enzyme (ALSII–FAD–TPP–Mg^{2+}) are rather slow at $4°$, with a half-time for irreversible loss of activity of about 2 days, the enzyme has a half-time at $37°$ of 1.5–3 hr under assay conditions.[11,15]

Physical Constants and Structure

ALSII is composed of two dissimilar subunits[11] of molecular weight 59,300 and 9,700, which are encoded by the genes *ilvG* and *ilvM*, respectively, and are present in equimolar amounts in the enzyme.[11,16] Native ALSII–FAD has a molecular weight of about 140,000, as determined by gel filtration,[11] thus having a native $\alpha_2\beta_2$ structure. Each $\alpha\beta$ protomer contains an FAD-binding site, as there are two molecules of FAD bound per ALSII dimer.

Activators and Cofactors

ALSII has an absolute requirement for FAD. Several "partial analogs" of FAD, e.g., FMN, NAD, NADH, NADP, NADPH, FMN in combination with AMP, and ATP, fail to activate the apo-ALSII.[11] Despite this seemingly stringent requirement for FAD, the homologous condensation of two pyruvates to acetolactate is virtually unaffected by reduction of the enzyme-bound FAD, or by substitution of FAD with FAD analogs which are more easy (8-chloro-FAD) or more difficult (5-deaza-FAD) to reduce.[17] Since there is no "hidden" or internal redox change, and no net oxidation or reduction, the requirement of this enzyme for FAD remains one of its most puzzling features. By contrast, the enzyme's requirement for TPP is expected, given the reactions catalyzed. One of the expected intermediates

[15] R. A. LaRossa and J. V. Schloss, *J. Biol. Chem.* **259,** 8753 (1984).
[16] R. P. Lawther, D. H. Calhoun, C. W. Adams, C. A. Hauser, J. Gray, and G. W. Hatfield, *Proc. Natl. Acad. Sci. U.S.A.* **78,** 922 (1981).
[17] J. V. Schloss and C. Thorpe, *Proc. Int. Congr. Biochem., 13th* p. 25 (1985).

of the enzymatic reaction, hydroxyethylthiamin pyrophosphate, has been trapped by chemical quench techniques.[18] Although identified in its protonated form, this intermediate is almost certainly never protonated in the normal catalytic sequence, as the enzyme has neither the ability to abstract the proton from the hydroxyl-bearing carbon to generate the catalytically competent ene-amine form, nor to convert the protonated form of this intermediate to acetaldehyde and TPP (as does pyruvate decarboxylase).[19] The protonated form of hydroxyethylthiamin pyrophosphate and tetrahydrothiamin pyrophosphate (an analog in which the thiazole ring is doubly reduced) are simple dead-end inhibitors of ALSII, with affinities for the enzyme comparable to TPP.[19] By contrast, thiamin-thiazolone pyrophosphate, which is a much better analog of the ene-amine form of hydroxyethylthiamin pyrophosphate, is essentially an irreversible inhibitor of ALSII.[19] Tight binding of TPP, and its various analogs mentioned above, requires the presence of a metal (L. M. Ciskanik and J. V. Schloss, unpublished observations), a role best fulfilled by Mn^{2+} or Mg^{2+}. Given the promiscuity of ALSII in its metal requirement, it would seem that the metal plays little role in catalysis other than assisting TPP binding. The following metals exhibit some activity with ALSII, in order of decreasing activity: Mn^{2+}, Mg^{2+}, Co^{2+}, Ca^{2+}, Ni^{2+}, Cd^{2+}, Zn^{2+}, Ba^{2+}, Al^{3+}, and Cu^{2+}.[11]

Kinetic Constants

Despite the fact that two molecules of pyruvate are condensed to form acetolactate, sigmoidal saturation kinetics are not observed. Presumably this is a consequence of the first pyruvate being added in a highly committed fashion, as deduced from rather small ^{13}C- isotope effects for ALSII,[20] and/or the release of product (CO_2) before addition of the second pyruvate. In either case, the apparent Michaelis constant for saturation with pyruvate as the sole substrate is 11 mM.[11] This constant presumably reflects occupancy of the second pyruvate site, the first being silent. In addition to the physiological reactions catalyzed by ALSII, it catalyzes the homologous condensation of α-ketobutyrate with itself to form α-propio-α-hydroxybutyrate and CO_2.[20] This reaction proceeds at about 20% of the rate of the homologous condensation of pyruvate, with a Michaelis constant for α-ketobutyrate of about 10 mM. When presented with an equimolar mixture of pyruvate and α-ketobutyrate, ALSII carries out the heterologous reaction (pyruvate and α-ketobutyrate to α-aceto-α-hydroxybutyrate) almost exclusively, at the expense of the two homologous condensation reactions (phys-

---

[18] L. M. Ciskanik and J. V. Schloss, *Biochemistry* **24**, 3357 (1985).

[19] L. M. Ciskanik and J. V. Schloss, *Fed Proc., Fed. Am. Soc. Exp. Biol.* **45**, 1607 (1986).

[20] L. M. Abell, M. H. O'Leary, and J. V. Schloss, *Biochemistry* **24**, 3357 (1985).

iological and nonphysiological), as measured by examining the isotopic composition of CO_2 derived from a mixture in which the carboxyl-carbons of pyruvate and α-ketobutyrate have different $^{13}C/^{12}C$ ratios.[20] This observation regarding substrate preference has recently been corroborated by the use of an assay method which can follow both of the physiological reactions simultaneously.[21] The saturation constants for activation of the apo-ALSII by TPP, FAD, and Mg^{2+} are 1.5, 0.8, and 22 μM, respectively.[11]

Inhibitors

Inhibition of acetolactate synthase in higher plants is responsible for the herbicidal activities of two new broad classes of commercial herbicides, the sulfonylurea herbicides, exemplified by sulfometuron methyl, and the imidazolinone herbicides, exemplified by Scepter.[22-24] Sulfometuron methyl[15] and Scepter both inhibit ALSII. Both compounds are slow-binding,[25] time-dependent inhibitors. Sulfometuron methyl has an initial K_i of 1.7 μM, a final steady-state K_i of 82 nM, and a maximal rate constant for transition between initial and final inhibition of 0.15 min^{-1}.[5] Scepter has an initial K_i of 0.8 mM, a final steady-state K_i of 20 μM, and a maximal rate constant for transition between initial and final inhibition of 0.6 min^{-1} (J. V. Schloss, unpublished observations). Formation of the slowly reversing complex with sulfometuron methyl requires the presence of pyruvate; however, both the initial and final levels of inhibition are competitive with pyruvate at high pyruvate concentrations.[5,15] The binding site for this inhibitor overlaps with the second pyruvate binding site and, to form a slowly reversible complex, requires the addition of the first pyruvate. ALSII is not inhibited by valine, leucine, or isoleucine at concentrations up to 10 mM, a feature that distinguishes this isozyme from isozymes I (*ilvBN* encoded) or III (*ilvIH* encoded) in enteric bacteria,[2] or the corresponding enzyme from yeast[8] or higher plants.[23]

[21] N. Gollop, Z. Barak, and D. M. Chipman, this volume [29].
[22] D. L. Shaner, P.C. Anderson, and M. A. Stidham, *Plant Physiol.* **76**, 545 (1984).
[23] T. B. Ray, *Plant Physiol.* **75**, 827 (1984).
[24] R. S. Chaleff and C. J. Mauvais, *Science* **224**, 1443 (1984).
[25] J. F. Morrison, *Trends Biochem. Sci.* **7**, 102 (1982).

[57] Acetolactate Synthase Isozyme III from *Escherichia coli*

By ZE'EV BARAK, JOSEPH M. CALVO, and JOHN V. SCHLOSS

Acetolactate synthase (ALS, EC 4.1.3.18), also known as acetohydrox-yacid synthase, is an enzyme which catalyzes the condensation of pyruvate either with another pyruvate molecule or with α-ketobutyrate to form α-acetolactate or α-aceto-α-hydroxybutyrate, respectively.[1-3] ALSIII is one of the two isozymes functioning in *Escherichia coli* K12.[1-3] It is encoded by the *ilvIH* operon, which is regulated by leucine alone,[2-5] and is closely linked to the *leuABCD* operon.[4] Although the gene is transcribed, this isozyme is cryptic in *Salmonella typhimurium*.[6] The lability of ALSIII like that of the other isozymes, has hampered its isolation, despite the availability of bacterial clones capable of its overproduction.[3,7,8] Hence, all characterizations of ALSIII have been based on crude or partially purified preparations from mutants which synthesize this single isozyme only.[2,9] This article describes the procedure for purification of ALSIII and the properties of the isolated enzyme.

Assay Method

Activity is determined at 37° in an assay mixture containing 50 mM sodium pyruvate, 0.1 M Tricine–NaOH (pH 7.8), 0.1 mM flavin adenine dinucleotide (FAD), and 0.1 mM thiamin pyrophosphate (TPP). The reaction is followed by monitoring the loss in absorbance at 333 nm as pyruvate is converted to α-acetolactate.[10] One unit of activity is that amount of enzyme which converts 2 μmol of pyruvate to 1 μmol of α-acetolactate per minute. Enzymatic activity is also determined with a gas–liquid chromato-

[1] H. E. Umbarger, *Annu. Rev. Biochem.* **47**, 533 (1978).

[2] M. De Felice, C. T. Lago, C. H. Squires, and J. M. Calvo, *Ann. Microbiol. (Paris)* **133A**, 251 (1982).

[3] H. E. Umbarger, *In* "Amino Acid Biosynthesis and Genetic Regulation" (K. M. Hermann and R. L. Somerville, eds.), p. 245. Addison-Wesley, Reading (1983).

[4] M. De Felice, J. Guardiola, B. Esposito, and M. Iaccarino, *J. Bacteriol.* **120**, 1068 (1974).

[5] M. De Felice and M. Levinthal, *Biochem. Biophys. Res. Commun.* **79**, 82 (1977).

[6] C. H. Squires, M. De Felice, C. T. Lago, and J. M. Calvo, *J. Bacteriol.* **154**, 1054 (1983).

[7] H. Grimminger and H. E. Umbarger, *J. Bacteriol.* **137**, 846 (1979).

[8] C. T. Lago, G. Sannia, G. Marino, C. H. Squires, J. M. Calvo, and M. De Felice, *Biochim. Biophys. Acta* **824**, 74 (1985).

[9] M. De Felice, C. H. Squires, and M. Levinthal, *Biochim. Biophys. Acta* **541**, 9 (1978).

[10] J. V. Schloss and D. E. Van Dyk, this volume [56].

graphic (glc) method,[11] which allows the formation of α-acetolactate and α-aceto-α-hydroxybutyrate to be monitored simultaneously. Protein is determined by the method of biuret.[12]

Purification

Cultivation of Organism

A strain is used which overproduces isozyme III, and does not produce the other isozymes, constructed from an $ilvB^-$ E. coli K12 strain, carrying a multicopy plasmid with the entire $ilvIH$ operon. It is also advantageous that it bear a Leu$^+$ phenotype to prevent repression (by leucine) of ALSIII production. Strain E. coli K12 CV942 (HfrC, thi-1, trpR, rbs-115, ilvB ::Mu1/pCV88[8]) gives good yields of bacterial culture with more than 100-fold increase in the activity of ALSIII (compared to the haploid clone). Escherichia coli CV942 is grown at a 200-liter scale on a minimal salts medium as described for E. coli HB101/pDU9,[10] with the exception that glycerol is substituted for glucose. The final bacterial yields with glucose as sole carbon source are threefold lower than those obtained with glycerol. ALSIII activity does not decline markedly after the cessation of exponential growth, as occurs for ALSII[10] in E. coli HB101/pDU9, and rapid cooling of cells prior to the end of their exponential growth phase is unnecessary.

Preparation of Extracts and Purification of ALSIII

The protocol used is identical to the procedure described for purifying isozyme II from E. coli HB101/pDU9.[10] From 246 g of cell paste is obtained an initial extract containing 13,600 units of ALS and 23.4 g of protein. After heat treatment, ammonium sulfate fractionation, phenyl-Sepharose chromatography, Sephacryl S-200 chromatography, and DEAE-Fractogel chromatography,[10] 2.11 g of nearly homogeneous ALSIII is obtained. The purified ALSIII has a specific activity of 7.3 units/mg and a total of 15,400 units of activity is recovered. Based on the initial and final specific activities, ALSIII comprises about 8% of the extractable protein in E. coli CV942.

Properties

Stability

When purified (ALSIII–FAD complex), the enzyme survives repetitive freezing and thawing, and can be stored frozen indefinitely. In the absence

[11] N. Gollop, Z. Barak, and D. M. Chipman, this volume [29].
[12] E. Lyane, this series, Vol. 3, p. 447.

of light, the enzyme can survive for several hours at room temperature without loss of activity.

Structure

The purified enzyme is composed of two dissimilar subunits, which by SDS-gel electrophoresis appear to have molecular weights close to those predicted from the gene sequences of *ilvI* and *ilvH*, 61,800 and 17,500, respectively.[13] Quantitation of the resolved subunits by amino acid analysis after reductive carboxymethylation and gel filtration[14] reveals that they are present in equimolar amounts. Polyclonal antibodies raised against ALSII do not cross-react with ALSIII, despite the approximately 40% amino acid sequence homology calculated in the large subunits of these isozymes.[13,15] Analysis of the first 25 amino acids at the amino terminal end of the purified small subunit of ALSIII conforms to the nucleotide sequence at the beginning of *ilvH*.[13] The small subunit of the enzyme does not undergo reductive carboxymethylation, as would be expected from the lack of any codons for cysteine in the corresponding gene.[13]

Activators and Cofactors

In general, the results obtained with the purified ALSIII confirm those previously recorded for the crude preparations.[7,9] The purified enzyme requires thiamin pyrophosphate and Mg^{2+} for activity. However, in contrast to previous notions,[1-3,6,9] ALSIII does require FAD for its activity. FAD is bound tightly by the enzyme, and is not released upon dilution of the enzyme into an assay medium lacking this cofactor. After treatment with acidic ammonium sulfate in the presence of high salt (conditions which reversibly remove the FAD from ALSII[10]), ALSIII retains less than 2.6% of its initial activity. Addition of FAD back to the enzyme restores approximately a third of the activity present prior to treatment.

Substrates

The Michaelis constant for pyruvate obtained with the purified enzyme is 12 mM for the homologous condensation of two pyruvates to form α-acetolactate and CO_2. The K_m for pyruvate measured in crude extracts is 7.6 mM.[9] The products formed by ALSIII at equimolar pyruvate and

[13] C. H. Squires, M. De Felice, J. Devereux, and J. M. Calvo, *Nucleic Acids Res.* 11, 5299 (1983).
[14] J. V. Schloss, D. E. Van Dyk, J. F. Vasta, and R. M. Kutny, *Biochemistry* 24, 4952 (1985); we would like to thank Dr. F. C. Hartman, Biology Division, Oak Ridge National Laboratory, Oak Ridge, TN for the amino acid analyses.
[15] R. C. Wek, C. A. Hauser, and G. W. Hatfield, *Nucleic Acids Res.* 13, 3995 (1985).

α-ketobutyrate concentrations (monitoring both reactions simultaneously with the glc method[12,16]) is approximately 50-fold greater toward α-aceto-α-hydroxybutyrate versus α-acetolactate.[17] However, at the physiological concentrations of pyruvate and α-ketobutyrate in *E. coli* grown on glucose, calculated to be 0.3 and 0.003 mM, respectively,[18,19] ALSIII produces comparable amounts of α-acetolactate and α-aceto-α-hydroxybutyrate. ALSIII does not carry out the nonphysiological homologous condensation of two molecules of α-ketobutyrate to α-propio-α-hydroxybutyrate as efficiently as ALSII.[10] For ALSIII, the rate of this reaction is at most 0.75% of the rate of the homologous condensation of pyruvate, whereas for ALSII it is 20%.

Inhibitors

By contrast to ALSII, purified ALSIII is, as expected,[1-3,9] sensitive to inhibition by valine. In the presence of 4 mM valine, the standard assay is inhibited by 85%. The reaction is also inhibited by isoleucine (70% at 8 mM) and leucine (35% at 25 mM), similar to the observations in crude extracts.[9,20] ALSIII is 100-fold less sensitive to inhibition by sulfometuron methyl (a sulfonylurea herbicide) than ALSII.[21,22] Inhibition of ALSIII by sulfometuron methyl is time dependent, as is it for ALSII, although the onset of more potent inhibition is somewhat slower.

[16] N. Gollop, Z. Barak, and D. M. Chipman, *Anal. Biochem.*, **160**, 323 (1987).
[17] Z. Barak, D. M. Chipman, and N. Gollop, *J. Bacteriol.* **169**, 3750 (1987).
[18] O. H. Lowry, J. Carter, J. B. Ward, and L. Glaser, *J. Biol. Chem.* **246**, 6511 (1971).
[19] J. Daniel, L. Dondon, and A. Danchin, *Mol. Gen. Genet.* **190**, 452 (1983).
[20] N. Gollop, D. M. Chipman, and Z. Barak, *Biochim. Biophys. Acta* **748**, 34 (1983).
[21] R. A. LaRossa and J. V. Schloss, *J. Biol. Chem.* **259**, 8753 (1984).
[22] J. V. Schloss, L. M. Ciskanik, and D. E. Van Dyk, *Nature (London)*, **331**, 360 (1988).

Section IV

Use of Animals and Animal Organs in Study of Branched-Chain Amino Acid Metabolism

[58] Amino Acid Control of Intracellular Protein Degradation

By Glenn E. Mortimore and A. Reeta Pösö

In most vertebrates an extra supply of amino acids, above that required for protein synthesis, is needed for gluconeogenesis and other oxidative and biosynthetic processes that utilize amino acids irreversibly. With food intake, this need is met by the breakdown of ingested protein, but when absorption ceases, endogenous sources must be called into play. Early studies in the rat have shown that decreases in cellular protein after short periods of starvation vary widely among different tissues.[1] Losses in liver, kidney, and intestine are among the largest whereas decreases in skeletal muscle are comparatively small, even when the total quantity of tissue is taken into account. In 48-hr starved rats and mice, for example, as much as 25 to 40% of liver protein can be lost without apparent alteration in the protein content of muscle.[1-4] During prolonged starvation, however, degradation accelerates, and protein metabolism becomes strongly catabolic.[3,4]

It follows from this that tissues such as liver are capable of serving as an important source of free amino acids in periods between food intake. Because losses of hepatic protein are achieved primarily by regulatory effects on degradation rather than synthesis,[5-7] our attention will be focused on the former. Accelerated rates of protein degradation comparable to those *in vivo* can be induced in the isolated liver,[2-8] and it is largely for this reason that the perfused organ has become a useful model for investigating the mechanism of control by amino acids. Isolated hepatocytes have also been employed to advantage, but their responsiveness has been rather small and variable.[9-11] Hence, their value in studies aimed at understanding regulation under physiological conditions must be qualified.

[1] T. Addis, L. J. Poo, and W. Lew, *J. Biol. Chem.* **115**, 111 (1936).
[2] N. J. Hutson and G. E. Mortimore, *J. Biol. Chem.* **257**, 9548 (1982).
[3] D. J. Millward and J. C. Waterlow, *Fed. Proc., Fed. Am. Soc. Exp. Biol.* **37**, 2283 (1978).
[4] D. J. Millward, *in* "Comprehensive Biochemistry" (M. Florkin and E. K. Stotz, eds.), p. 153. Elsevier/North-Holland, Amsterdam, The Netherlands, 1980.
[5] P. J. Garlick, D. J. Millward, and W. P. T. James, *Biochem. J.* **136**, 935 (1973).
[6] P. J. Garlick, D. J. Millward, W. P. T. James, and J. Waterlow, *Biochim. Biophys. Acta* **414**, 71 (1975).
[7] R. D. Conde and O. A. Scornik, *Biochem. J.* **158**, 385 (1976).
[8] A. R. Pösö, J. J. Wert, Jr., and G. E. Mortimore, *J. Biol. Chem.* **257**, 12114 (1982).
[9] M. F. Hopgood, M. G. Clark, and F. J. Ballard, *Biochem. J.* **164**, 399 (1977).
[10] P. O. Seglen, P. B. Gordon, and A. Poli, *Biochim. Biophys. Acta* **630**, 103 (1980).
[11] J. M. Sommercorn and R. W. Swick, *J. Biol. Chem.* **256**, 4816 (1981).

The aims of this chapter are threefold: (1) to outline some of the principles in the measurement of protein degradation. (2) to provide procedural details for its determination in the perfused rat liver, and (3) to describe some major regulatory effects of amino aids.

Principal Features of General Protein Turnover in the Hepatocyte

Classes of Protein Degradation

Despite the fact that the rates of breakdown of individual proteins are variable,[12] only two kinetic classes of turnover have been consistently observed in isolated cells or tissues when protein degradation is determined from the release of amino acids.[2,13-15] The first is a rapidly turning over or short-lived fraction; the second comprises the remainder and is termed long-lived or resident protein degradation. The distinction between these two classes is especially well illustrated in perfusion experiments with the mouse liver (Fig. 1) where the release of [^{14}C]valine that had been incorporated into liver protein during a 10-min pulse could be divided into a short-lived fraction with a half-life of about 10 min and a slow, linear release, representing the breakdown of resident proteins. Kinetically, the two components were readily separable, and no intermediate fractions were found.[2] Similar results have been obtained in cultured fibroblasts and HTC cells.[13,14]

Although the nature of the short-lived pool is not known, there is considerable evidence, summarized elsewhere,[2] that it is not extracted with liver protein and could exist largely in the form of acid-soluble peptides.[16] If true, the peptides might have arisen from the breakdown of fractions destined for rapid removal, possibly signal peptides or mistakes in synthesis. This possibility is consistent with the observation that short-lived turnover is resistant to physiological regulation (Fig. 1) and remains constant despite wide alterations in the rate of resident protein degradation.[2,13,14]

Quantitation of Short-Lived and Resident Protein Turnover

Because the average turnover time of resident proteins is at least two orders of magnitude greater than that of the short-lived pool, a reasonable degree of selectivity of labeling of the two classes can be achieved by

[12] A. L. Goldberg and J. F. Dice, Annu. Rev. Biochem. 43, 835 (1974).
[13] B. Poole and M. Wibo, J. Biol. Chem. 248, 6221 (1973).
[14] D. Epstein, S. Elias-Bishko, and A. Hershko, Biochemistry 14, 5199 (1975).
[15] N. T. Neff, G. N. DeMartino, and A. L. Goldberg, J. Cell. Physiol. 101, 439 (1979).
[16] A. E. Solheim and P. O. Seglen, Eur. J. Biochem. 107, 587 (1980).

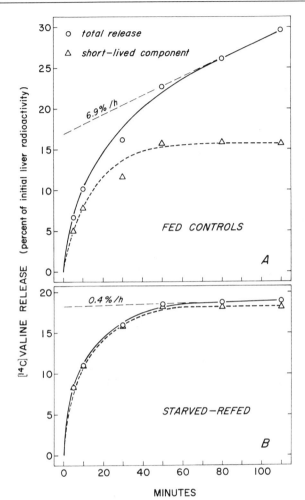

FIG. 1. Release of [¹⁴C]valine from pulse-labeled livers. Livers from (A) nonfasted and (B) 48 hr starved–12 hr refed mice were perfused 10 min in the single-pass mode with a 1× complete mixture of plasma amino acids and [¹⁴C]valine. Perfusion was then switched to a medium containing 15 mM unlabeled valine; after a 10-min washout, the liver outflow was reconnected to the perfusion flask and the remaining medium recirculated for 110 min. Labeled valine was determined at intervals, and the cumulative release expressed as a percentage of the total label in the liver at the beginning of the recirculation. O, Total label; Δ, short-lived release calculated as the difference between total release and the linear release of valine from resident protein breakdown. Note the similarity of short-lived release in the presence of a large, 17-fold difference in the average rate of resident protein degradation.

Inset in A, degradation of resident proteins in livers of nonfasted mice previously labeled with [¹⁴C]valine. Mice were labeled for 18 hr, and the livers cyclically perfused in the presence of 15 mM valine. The rates were the same as those determined after pulse-labeling. From Hutson and Mortimore.[2]

varying the duration of incorporation; short-lived components are preferentially labeled by pulses of 1 hr or less whereas 18-hr periods are frequently employed to obtain adequate labeling of resident proteins *in vivo*. In rat livers that were labeled with [^{14}C]valine over a 3-day period,[17] only 0.6% of the incorporated label was released by short-lived turnover during the initial 60 min of perfusion; 99.4% remained in resident proteins.[18] It should be pointed out that because of the very rapid turnover of the short-lived fraction, its contribution to protein synthesis is appreciably greater than its representation in cellular protein.

For reasons mentioned earlier, the perfused liver has proved useful in the investigation of proteolytic regulation. However, since the tissue is not uniform in cellular composition, it is appropriate to ask to what extent overall measurements of protein turnover reflect hepatocyte activity. Virtually all the nonhepatocytic cells are found within or along the sinusoidal border. Although numerous, they are smaller than hepatocytes and appear to contain lower concentrations of protein.[19] From these data we have calculated that they comprise no more than 1% of the total quantity of cellular protein in the whole liver.[20] Thus, from the point of view of general protein turnover and its regulation, the perfused liver may be regarded as a preparation of isolated hepatocytes. Such a conclusion, however, would not preclude regulatory influences generating from zonal differences in hepatocytes along the sinusoids.[21] Although such differences are not known to play a role in proteolytic regulation by amino acids, this possibility should be kept in mind in future investigations.

Measurement of Intracellular Protein Degradation

General Principles

At present, the most feasible approach for determining intracellular protein degradation over short intervals is one that employs the release of amino acids (labeled or unlabeled) as an end point. Measurements based on the decrease of intracellular protein would be too small for the degree of accuracy required, and estimates derived from the release of posttranscriptionally modified amino acids, such as 3-methylhistidine in muscle,[22] are

[17] G. E. Mortimore and C. E. Mondon, *J. Biol. Chem.* **245**, 2375 (1970).
[18] C. M. Schworer, K. A. Shiffer, and G. E. Mortimore, *J. Biol. Chem.* **256**, 7652 (1981).
[19] A. Blouin, R. P. Bolender, and E. W. Weibel, *J. Cell. Biol.* **72**, 441 (1977).
[20] G. E. Mortimore, N. J. Hutson, and C. A. Surmacz, *Proc. Natl. Acad. Sci. U.S.A.* **80**, 2179 (1983).
[21] D. Häussinger, *Eur. J. Biochem.* **133**, 269 (1983).
[22] V. R. Young and H. N. Munro, *Fed. Proc., Fed. Am. Soc. Exp. Biol.* **37**, 2291 (1978).

useful only for specific proteins or protein groups. But regardless of the particular method used, all share two strict requirements: (1) protein synthesis or the reutilization of released marker must be decreased to negligible values and (2) the mode of inhibition must not perturb the underlying proteolytic process.

The choice of amino acid marker depends heavily on the type of cell to be studied. Ideally, the amino acid should be widely distributed among proteins and metabolically stable. In liver[17] and the isolated hepatocyte,[23] valine has fulfilled these requirements; in addition to its overall abundance, it is neither synthesized nor appreciably oxidized.[17] Moreover, it is devoid of any regulatory activity, even at high concentrations.[8-10] In the perfused rat liver, its rate of oxidation has been calculated to be 1.075 nmol min^{-1} per gram at an average concentration of 253 μmol per milliliter of perfusate.[8] This is the equivalent of 0.3% of the total quantity of valine entering the liver of a nonstarved rat.

Past investigations have shown that the use of livers previously labeled with L-[^{14}C] valine *in vivo* provides a convenient and direct way to determine rates of resident protein degradation during perfusion in the cyclic or recirculating mode.[17,24-27] Owing to the rapid release of label from the short-lived pool, the initial specific radioactivity of free valine is high.[23] It decreases quickly, however, and by 60 min equals that of valine newly released from protein.[25] Thereafter the specific activity remains constant for at least 2 hr of perfusion.[24-26] By flooding precursor sites of protein synthesis with 10–15 mM additions of unlabeled valine, it is possible to prevent virtually all reincorporation of label. The subsequent accumulation of labeled valine thus would represent the total amount of label generated proteolytically. Resident protein breakdown is then determined from the rate of label accumulation divided by the specific radioactivity of valine obtained before the addition of the unlabeled valine load.

While the foregoing method is applicable in a large variety of experimental situations, the inability to control the perfusate concentrations of amino acids during the course of perfusion has been a major obstacle in examining the regulation of protein degradation by amino acids. The problem was solved by devising a two-stage perfusion.[18] In the first or experimental phase, livers are perfused in the single-pass (nonrecirculating) mode with a medium of defined amino acid composition. Perfusion is then

[23] P. O. Seglen and A. E. Solheim, *Eur. J. Biochem.* **85,** 15 (1978).
[24] K. H. Woodside and G. E. Mortimore, *J. Biol. Chem.* **247,** 6474 (1972).
[25] G. E. Mortimore, K. H. Woodside, and J. E. Henry, *J. Biol. Chem.* **247,** 2776 (1972).
[26] K. H. Woodside, W. F. Ward, and G. E. Mortimore, *J. Biol. Chem.* **249,** 5458 (1974).
[27] E. A. Khairallah and G. E. Mortimore, *J. Biol. Chem.* **251,** 1375 (1976).

rapidly switched to an adjacent, second perfusion and continued in the cyclic mode. The second flask contains cycloheximide to block protein synthesis. Rates of protein degradation are subsequently determined from the accumulation of perfusate valine.

Because resident protein degradation cannot be altered instantaneously by regulatory agents, but instead remains unchanged for 15–20 min after the addition of the agent,[2,18,22,24,26,27] the initial rate of valine accumulation in the second flask will closely reflect the activity of protein degradation in the experimental phase. Since 5 min are required before cycloheximide inhibits protein synthesis maximally and 15–20 min before it inhibits degradation,[27] valine accumulation must be determined in the comparatively brief interval between 5 and 15 min. Despite this limitation, extensive experience with this technique has shown that measured rates are highly linear and reproducible.

Experimental Procedures

Operative Technique for Perfusing Rat Livers in Situ. The operative procedure for perfusing the rat liver *in situ* has remained essentially the same since its original description.[28] Male rats, weighing from 100 to 140 g, are anesthetized with sodium pentobarbital. The viscera are then fully exposed through a wide abdominal incision, and the portal vein cannulated by direct puncture with a section of short-beveled, 20-gauge hypodermic tubing connected by vinyl tubing to the pump. The latter is started immediately, ensuring no significant interruption in the supply of oxygen to the liver. The flow of perfusate is maintained at 8–10 ml per minute for livers weighing 4–6 g.

The isolated liver remains *in situ*, and the flow of medium is returned to the oxygenating flask by a larger cannula (14-gauge) inserted through the right atrium and into the inferior vena cava at the level of the diaphragm. Both cannulae are ligated immediately. Loss of perfusate by retrograde flow down the inferior vena cava is prevented by ligation of the vena cava between the liver and the right renal vein. Although the liver remains in its normal position, it is completely isolated in a functional sense, and there is no detectable interaction with adjacent tissues.

Perfusion Apparatus. A single perfusion apparatus of the type used by one of us (G.E.M.) consists of six 500-ml spherical flasks (Kontes) or cylindrical Lucite flasks attached to horizontal shafts set 15 cm apart at the rear of a temperature-regulated (37°) plywood box[25-29]; flask rotation is

---

[28] G. E. Mortimore, F. Tietze, and D. Stetten, *Diabetes* **8,** 307 (1959).
[29] G. E. Mortimore and F. Tietze, *Ann. N.Y. Acad. Sci.* **82,** 329 (1959).

150 rpm. The liver donor animals, on operating boards, are placed on a rack within the box, each above its respective flask. The perfusate is pumped directly into the liver by hydraulically operated diaphragm pumps or commercial peristaltic pumps. The outflow from the liver is returned to the perfusion flask in cyclic perfusions and to waste during single-pass runs. Openings in each flask for perfusate and gas inflow and outflow as well as for the sampling of perfusate are contained in a fixed cylindrical block that was designed to seal the main opening of the rotating flask, which lies opposite the flask's attachment to the horizontal shaft. Oxygenation of the perfusate within the rotating flasks takes place by exchange with fully humidified 95% O_2-5% CO_2 flowing through each flask at 850 ml per minute.

Perfusion Medium.[18] The basic perfusion medium is composed of Krebs–Ringer bicarbonate buffer, 4% bovine plasma albumin (fraction V, Pentex, Miles Laboratories), and a suspension of red cells (0.27, v/v) obtained from fresh, heparinized bovine blood; for single-pass perfusions, the albumin is decreased to 3% and 10 mM glucose is added. The albumin–buffer solution is passed through 3.0- and 0.3-μm Millipore filters to remove fine particulate matter before subsequent additions are made. The red cells are washed four times with equal volumes of 0.85% NaCl and twice with similar volumes of the buffer before the final addition of the albumin–buffer.

For single-pass perfusion experiments, "normal" plasma amino acid concentrations (1 ×), or multiples/fractions thereof, are obtained by adding appropriate amounts of a concentrated solution of 20 amino acids in 0.85% NaCl, pH 7.4, to the basic perfusion medium.[18] Amino acid analyses of seven samples of perfusate plasma after 1× additions had equilibrated with red cell water gave the following average micromolar values[8,18] (numbers in parentheses represent the percentage molar composition of the stock solution): Ala, 484 (10.3); Arg, 219 (4.4); Asn, 78 (1.8); Asp, 63 (1.0); Cys (1.5); Gln, 738 (15.3); Glu, 169 (2.8); Gly, 380 (7.0); His, 104 (1.8); Ile, 116 (2.2); Leu, 212 (3.8); Lys, 442 (8.7); Met, 56 (1.1); Phe, 96 (1.9); Pro, 416 (8.2); Ser, 632 (13.4); Thr, 354 (6.3); Trp (1.7); Tyr, 98 (1.8); and Val, 258 (4.7).

Determination of Protein Degradation after Perfusion in the Single-Pass Mode. In a typical experiment, the liver is perfused in the single-pass mode for 40 min with a medium that either lacks added amino acids or contains one or more of the above plasma amino acids at multiples or fractions of the 1× complete mixture. In this type of experiment, we use cylindrical perfusion flasks that are capable of holding 250 ml of medium, more than three times the effective capacity of the spherical flasks. As described earlier, the flow of perfusate is then switched to an adjacent spherical flask

containing approximately 50 ml of the basic cyclic perfusion medium to which 18 μM cycloheximide had been added. After a 45-sec washout, the outflow tubing from the liver is then connected to the second reservoir and perfusion continued in the cyclic mode for 15 min. Five perfusate samples are taken at 2.5-min intervals between 5 and 15 min for the determination of plasma valine by one of the following procedures: quantitative paper chromatography,[17] tRNA binding,[30] or HPLC.

Total free valine accumulation is computed from the increase in plasma valine concentration multiplied by its volume of distribution in perfusate and liver. Since valine equilibrates with liver but not with bovine red cell water,[2,25] the space of valine distribution will equal the perfusate plasma volume plus liver water (0.72 × liver weight) minus the volume of red cells within the sinusoids.[20] For the method described here (packed red cell volume = 0.27), the sinusoidal red cell fraction is approximately 0.054 × liver weight. The total rate of valine release is then calculated by least squares regression and expressed as micromoles or nanomoles of valine per minute per liver (100-g rat); the content of valine in liver protein has been found to average 465 μmol per 100-g rat.[18]

Because the foregoing rates of valine release include contributions from short-lived turnover and the breakdown of plasma albumin acquired by endocytosis, they will be consistently larger than the actual rates of resident protein degradation. The difference, nominally 53 nmol of valine per minute per liver (100-g rat), has been determined in cyclically perfused livers under both basal and accelerated states in direct comparison with authentic rates of resident protein breakdown established in experiments with previously labeled livers.[18] In practice, this difference (53 nmol min^{-1}) is subtracted from observed rates of total valine accumulation to obtain rates of resident protein degradation.

Amino Acid Control of Intracellular Protein Degradation

Autophagic and Proteolytic Responses to Complete Amino Acid Mixtures

Amino acids may be regarded as the primary regulators of resident protein degradation in the hepatocyte since they are capable of evoking responses over the full range of acute regulation without assistance from their agents (Fig. 2). Thus, for example, accelerated rates obtained in the absence of added amino acids are not increased further by glucagon, a

---

[30] I. B. Rubin and G. Goldstein, *Anal. Biochem.* **33**, 244 (1970).

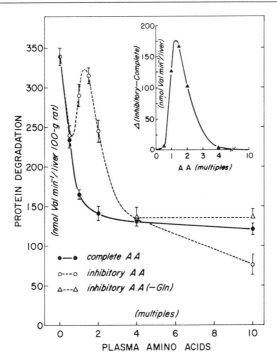

FIG. 2. Rates of resident protein degradation in the single-pass perfused rat liver at various concentrations of inhibitory and complete amino acid mixtures. The inhibitory mixture consisted of Leu, Tyr, Gln, Met, Trp, and His; Phe was added in some experiments, but no effects attributable to its addition were observed (see text). The lysosomotropic effect of Gln in the 10× inhibitory mixture is discussed in the text. The inset depicts differences between the complete and inhibitory mixtures from 0 to 4×. Each plotted value is the mean ± 1 SE of 5 to 28 observations. From Pösö and Mortimore.[45]

known inducer of autophagy and protein degradation.[26,31-33] At the other extreme, the lowest rate that is regularly achieved with 10× normal plasma amino acid concentrations (basal) is not appreciably affected by the addition of insulin, an established inhibitor of accelerated degradation.[9,17,34] In the majority of studies with perfused livers or hepatocyte suspensions from nonfasted rats, the amino acid deprivation response covers approximately two-thirds of the total range of resident protein breakdown, from about 4.5

[31] A. N. Neely, J. R. Cox, J. A. Fortney, C. M. Schworer, and G. E. Mortimore, *J. Biol. Chem.* **252**, 6948 (1977); see also L. L. Miller, *Nature (London)* **185**, 248 (1960).
[32] T. P. Ashford and K. R. Porter, *J. Cell Biol.* **12**, 198 (1962).
[33] M. F. Hopgood, M. G. Clark, and F. J. Ballard, *Biochem. J.* **186**, 71 (1980).
[34] A. N. Neely, J. R. Cox, J. A. Fortney, C. M. Schworer, and G. E. Mortimore, *J. Biol. Chem.* **252**, 6948 (1977).

to 1.6% per hour of intracellular protein.[8,10,18] Since rates of resident protein synthesis are nearly the same as rates of degradation in the presence of $10\times$ plasma amino acids, liver protein may be regarded as being in balance under basal conditions.[8]

There is now general agreement that autophagy is the cellular mechanism responsible for the acceleration of proteolysis during amino acid deprivation in the hepatocyte.[35,36] The term macroautophagy will be used in this chapter to distinguish deprivation-induced autophagy from other types believed to play a role in basal turnover (microautophagy).[20] It is known that macroautophagic vacuoles (AVi) are induced rapidly on amino acid withdrawal and, after a lag of 7–8 min, are transformed into digestive macroautophagic vacuoles or AVd.[18] The sequestration and digestion of cytoplasm by this means appears to be a rapid ongoing process in which a steady state between the formation of AVi and the disappearance of AVd is established within 20 min. Because AVi formation can be switched off almost instantaneously by amino acids,[18] the turnover of macroautophagic vacuoles can be indirectly determined from the exponential regression of the aggregate volume of AVd, measured stereologically.[18] Estimates of the half-life have ranged from 8 to 9 min[18,37,38]; we have taken 0.087 min^{-1} as an approximation of the first-order rate constant for macroautophagic turnover.[18]

The aggregate volume of cytoplasm that is sequestered by AVi at any level of plasma amino acids between 0 and $10\times$ has been shown to correlate directly with the increase in resident protein degradation above the basal value.[18] As would be expected, macroautophagy virtually disappears at the highest amino acid concentrations.[18] Calculations of cytoplasmic turnover, based on steady-state volumes of AVi and the turnover constant 0.087 min^{-1}, have agreed closely with corresponding fractional turnover rates of resident proteins, suggesting that the volumes sequestered are sufficient to account for the observed proteolytic responses.[18] The quantities of internalized cytoplasmic protein predicted by these calculations have been confirmed by direct measurements of degradable protein entrapped within lysosomes.[2,39] Thus there seems to be little doubt that the autophagic compartment is the source of the valine released during the accelerated phase of protein degradation.

[35] J. S. Amenta and S. C. Brocher, *Life Sci.* **28**, 1195 (1981).
[36] G. E. Mortimore, *Nutr. Rev.* **40**, (1982).
[37] A. N. Neely, P. B. Nelson, and G. E. Mortimore, *Biochim. Biophys. Acta* **338**, 458 (1974).
[38] U. Pfeifer, *J. Cell Biol.* **78**, 152 (1978).
[39] G. E. Mortimore and W. F. Ward, *J. Biol. Chem.* **256**, 7659 (1981).

Inhibitory and Noninhibitory Amino Acids

Several past studies with the perfused liver and isolated hepatocytes have indicated that the number of amino acids contributing to the overall suppression of proteolysis is comparatively small,[9-11,24] although larger than in skeletal muscle and heart where only leucine is inhibitory.[40-42] More recently, experiments were undertaken by the authors to define the inhibitory group under more physiological conditions, utilizing the single-pass perfusion technique as a means of controlling the concentration of external amino acids.[8] Results were evaluated largely on the basis of effects obtained at $4\times$ normal plasma levels because they represent the lowest concentrations that give consistently maximal or near-maximal suppression, and also correspond to the upper limit of amino acid concentrations in portal vein plasma. Aiming first at establishing the largest number of ineffective amino acids, we found that 12 of the 20 amino acids that normally turn over in liver protein were devoid of direct inhibitory activity when tested as a group at $1\times$, $4\times$, and $10\times$ normal plasma concentrations. The complementary group of 8 amino acids (leucine, phenylalanine, tyrosine, glutamine, proline, methionine, tryptophan, and histidine) thus appeared to contain all the active inhibitors.

But whether all of the above 8 amino acids are actually inhibitory was not easily answered, since in some instances individual activity was too low to measure.[8] In the case of phenylalanine and tyrosine, though, a different problem arose. Because phenylalanine is rapidly hydroxylated to form tyrosine,[43] it is possible that the former, the latter, or perhaps both are the active regulators. Each of these amino acids suppressed the deprivation response by more than 30%, and a similar inhibition was observed when the two were combined.[8] It is also known that phenylalanine activates phenylalanine hydroxylase strongly under the same conditions used for testing proteolytic regulation.[44] Taken together, these and related findings suggested that tyrosine is the actual inhibitor, and it was sufficient for our purposes to exclude phenylalanine from the inhibitory category.

When tested alone, leucine was by far the most effective inhibitor at the $4\times$ the normal level, suppressing the full deprivation response by more than 60%.[8] The other two branched-chain amino acids, valine and isoleu-

[40] R. M. Fulks, J. B. Li, and A. L. Goldberg, *J. Biol. Chem.* **250,** 290 (1975).
[41] M. G. Buse and S. S. Reid, *J. Clin. Invest.* **56,** 1250 (1975).
[42] B. Chua, D. L. Siehl, and H. E. Morgan, *J. Biol. Chem.* **254,** 8358 (1979).
[43] R. Shiman, G. E. Mortimore, C. M. Schworer, and D. W. Gray, *J. Biol. Chem.* **257,** 11213 (1982).
[44] R. Shiman and D. W. Gray, *J. Biol. Chem.* **255,** 4793 (1980).

cine, are devoid of inhibitory activity.[8-10] Tyrosine, glutamine, and proline also evoked significant effects individually, but responses to histidine, tryptophan, and methionine were too small to evaluate separately.[8] When various combinations of the inhibitory amino acids were examined, it became clear that leucine and proline were required for maximal inhibition. On the other hand, glutamine and tyrosine could be deleted from the inhibitory group without affecting the overall response. However, despite the fact that glutamine is dispensible, its effect was unmistakenly additive to that of leucine when all combinations of leucine, tyrosine, and glutamine were tried. Proline, histidine, tryptophan, and methionine as a group also appeared to act additively with leucine. Our results[8] are generally in accord with those of Seglen et al.[10] in suggesting that leucine plays a dominant role in providing numerous regulatory combinations with other inhibitory amino acids.

Proteolytic Responses to Graded Levels of Inhibitory Amino Acids

When the above inhibitory amino acid mixture was evaluated systematically from 0 to 10× normal plasma amino acid concentrations,[45] it deviated from the complete mixture in two significant ways (Fig. 2). First, while the seven amino acids as a group duplicated inhibitory effects of the complete mixture at 0.5× and 4×, they lost nearly all their suppressive effectiveness within a narrow zone centered between 1× and 1.5×. The second difference was the fact that the inhibitory mixture suppressed degradation below the basal level at 10×. This was attributed to excess ammonia production from the hydrolysis of glutamine at unphysiologically high concentrations.[46] As depicted in Fig. 2, the effect was abolished by the deletion of glutamine from the inhibitory mixture. It is also clear that inhibitory effects of glutamine at the 4× level (3 mM) and lower are not lysosomotropic, but rather the result of a direct suppression of AVi formation.[8,46,47] Despite the high concentrations of glutamine in the complete amino acid mixture at 10×, no lysosomotropic suppression has been observed. It is possible that glutamine metabolism is modulated in one or more ways by members of the noninhibitory group. This notion is supported by the ability of the latter to potentiate the inhibitory effectiveness of glutamine at 1× and 4× plasma concentrations.[47]

[45] A. R. Pösö and G. E. Mortimore, *Proc. Natl. Acad. Sci. U.S.A.* **81,** 4270 (1984).
[46] A. R. Pösö, C. M. Schworer, and G. E. Mortimore, *Biochem. Biophys. Res. Commun.* **107,** 1433 (1982).
[47] G. E. Mortimore and A. R. Pösö, *in* "Glutamine Metabolism in Mammalian Tissues" (D. Häussinger and H. Sies, eds.) pp. 138–157. Springer-Verlag, Berlin, Federal Republic of Germany, 1984.

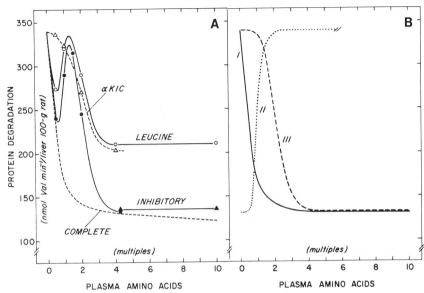

FIG. 3. (A) Rates of resident protein degradation in rat livers perfused as in Fig. 2. Effects of leucine and equimolar concentrations of α-ketoisocaproate (αKIC) were taken from Pösö et al.[8]; the inhibitory and complete amino acid dose responses were redrawn from Fig. 2. (B) Hypothetical components of the dose response of the inhibitory mixture; see text for explanation. From Pösö and Mortimore.[55]

Fig. 3A shows that leucine elicits the same unique dose response that was found in the inhibitory group. Although the overall inhibition is smaller with leucine, both curves exhibit a sharply defined zone of ineffectiveness that peaks between 1× and 1.5×. The similarity in pattern suggests that dose responses of the other inhibitory amino acids would follow the same course. The significance of this zonal loss will be discussed below.

Permissive Role of Alanine in the Expression of Proteolytic Inhibition

In attempting to explain the zonal loss, the possibility was considered that 1 or more of the 12 noninhibitory amino acids is necessary for expression of inhibitory activity by the complementary group at the 1–2× level. Because glucagon induces significant decreases of intracellular alanine, glutamate, glutamine, and glycine,[48,49] we had proposed that its catabolic action is mediated by a fall in glucogenic amino acids.[49] Although glutamine was considered a likely candidate, most of its inhibitory effec-

[48] L. E. Mallette, J. H. Exton, and C. R. Park, *J. Biol. Chem.* **244**, 5724 (1969).
[49] C. M. Schworer and G. E. Mortimore, *Proc. Natl. Acad. Sci. U.S.A.* **76**, 3169 (1979).

tiveness at the high concentrations used can be attributed to increased ammonia formation (see above).

Nevertheless, deletion of the above four glucogenic amino acids from the complete mixture at 1× accelerated proteolysis to near-maximum, an effect similar to that induced by glucagon.[46] Although the deletion of glutamine evoked a moderate accumulation of degradation, the omission of alanine alone fully accounted for the foregoing effect.[45] Later experiments showed that it is the only amino acid of the noninhibitory group that is required for effectiveness of the direct inhibitors at 1×.[45] Pyruvate and lactate could be substituted for 1× alanine (0.5 mM) in these experiments, but the concentrations required were 20 times greater. Similar effects with leucine and alanine in perfused hepatocytes have recently been reported by Leverve et al.[50]

Studies over the last 2 years have revealed that the dose response to complete mixtures of plasma amino acids can shift spontaneously from the form shown in Fig. 2 to one having a small, but distinct peak at 1.25×, corresponding in position to the sharp rise of the inhibitory amino acid curve.[51] The only differences that were observed between the two dose responses were found at 1× and 2×; responses at all other concentrations (0×, 0.5×, 4×, and 10×) were the same. Although the peak was suppressable by alanine, the amount required (5 mM) was 10 times greater than that needed to restore the effectiveness of the inhibitory amino acid mixture at 1× amino acid concentrations in the experiments of Fig. 2. Interestingly, the addition of insulin (2.4 μg hr^{-1}) at all amino acid levels eliminated the peak and yielded a response curve that was virtually identical in shape to that in Fig. 2.

The switchover between the high and low forms of alanine sensitivity has been abrupt though, happily, each phase has been sustained for a sufficiently long period of time that it has been possible to identify and partially characterize the two states.[51] The reason for the alteration is not known, but it could relate to changes in the pattern of food intake in ad libitum fed animals (light off, 1900–0700 hr). The low sensitivity form has appeared in livers perfused early in the postabsorptive period (1100–1200 hr), but has not been seen at 1500 hr or after 24 hr of starvation. The elucidation of this phenomenon is clearly an important question for future investigation.

[50] X. M. Leverve, L. H. P. Caro, P. J. A. M. Plomp, and A. J. Meijer, Biochem. Soc. Trans. 14, 1062 (1986).
[51] G. E. Mortimore, J. J. Wert, Jr., and C. E. Adams, J. Biol. Chem., in press (1988).

Mechanism of Proteolytic Regulation by the Inhibitory Amino Acids

Virtually nothing is known of the cellular locus or nature of the amino acid recognition that mediates the regulatory effects depicted in Fig. 1. Because the macroautophagic response to amino acid deprivation is almost instantaneous,[18] it is tempting to suggest that recognition may occur on the plasma membrane or close to it within the cell. The unique shape of the dose–response curve, which is observed with leucine as well as the inhibitory group (Fig. 3A), indicates that regulation of proteolysis by amino acids is complex and involves both positive and negative modulation from independent sites (Fig. 3B).

In the interpretation of the primary effect, a comparison of responses to α-ketoisocaproate and leucine (Fig. 3A and B, curve I) is of particular interest. The keto acid failed to exert any effect at the $0.5 \times$ level, although responses were identical to those of leucine at higher concentrations.[8] Because rates of leucine transamination in liver are extremely low,[8,52] the response to leucine at $0.5 \times$ indicates that the amino acid itself is required in the initial inhibition. Although the mechanism is unknown, it could involve binding of leucine as well as other inhibitory amino acids to their cognate tRNAs in a manner analogous to that described by Scornik *et al.*[53] in histidine-dependent mutants of CHO cells.

Why the inhibitory amino acids lose their effectiveness as their concentrations rise above $0.5 \times$ is equally unknown (Figs. 2 and 3), but the fact that the maximal loss occurs at amino acid concentrations approximating those in portal vein plasma[54] strongly suggests that it has physiological significance. This effect (Fig. 3B,[55] curve II) is not a feature of proteolytic regulation by leucine in muscle.[42,56] It could prove to be unique for liver, providing a way to accelerate proteolysis in response to a specific deficiency of glucogenic substrate.

A second inhibitory site is required to explain the proteolytic suppression that occurs at concentrations of inhibitory amino acids greater than $1.25 \times$ (Fig. 3B, curve III). Such a mechanism would be important as a safeguard against excessive rates of proteolysis mediated by glucagon or alanine deprivation. In suggesting an explanation for this second inhibition, one should recall the striking similarity in the inhibitory effects of

[52] D. W. Crabb and R. A. Harris, *J. Biol. Chem.* **253**, 1481 (1978).

[53] O. A. Scornik, M. L. S. Ledbetter, and J. S. Malter, *J. Biol. Chem.* **255**, 6322 (1980).

[54] K. M. Ojamaa, Ph.D. Dissertation, Pennsylvania State University, Hershey, Pennsylvania, 1985.

[55] A. R. Pösö and G. E. Mortimore, *in* "Intracellular Protein Catabolism" (E. A. Khairallah, J. S. Bond, and J. W. C. Bird, eds.), pp. 427–436. Liss, New York, 1985.

[56] M. E. Tischler, M. Desautels, and A. L. Goldberg, *J. Biol. Chem.* **257**, 1613 (1982).

leucine and α-ketoisocaproate at concentrations ranging from $1\times$ to $4\times$ (Fig. 3A). Because leucine is not transaminated at sufficiently rapid rates in liver to explain the inhibition by the keto acid, the latter must be recognized directly. This could involve specific structural features of the inhibitory amino acid side chains.

Acknowledgments

We thank Maxine L. Gerberich for her help in preparing the manuscript. The work was supported by USPHS Grant AM-21624. A.R.P. is a recipient of a Research Fellowship from the Research Council for Natural Services of the Academy of Finland.

[59] Use of Rat Hindquarter Preparations in Studies of Branched-Chain Amino Acid Metabolism

By E. Jack Davis, Sung-Hee Cho Lee, Øystein Spydevold, and Jon Bremer

The rat hindquarter, perfused *in situ* following ligation of all of the major arteries feeding blood to the viscera and other tissues, is almost exclusively a mixed muscle preparation. It can be maintained viable, and is hormone-responsive for at least 2 hr. It is generally accepted that muscle is the major source of carbon precursors for hepatic gluconeogenesis in starvation.[1] This carbon is presumed to take the form of amino acids produced via proteolysis. Glutamine and alanine together comprise more than one-half of the amino acids released by muscle during starvation, despite the fact that they account for less than 15% of the component amino acids found in muscle proteins.[2] Until a few years ago the number of amino acids, particularly the nondispensable ones, that contribute their carbon to muscle alanine, glutamine, and partial degradation products was largely unknown. In recent years, however, work in several laboratories has emphasized the branched-chain amino acids in this interorgan (mainly muscle, gut and liver) disposition of amino acid carbon (reviewed in Refs. 1 and 3). Progress in this area is due in large part to the use of the *in situ* rat

---

[1] T. N. Palmer, M. A. Caldecourt, K. Snell, and M. C. Sugden, *Biosci. Rep.* **5**, 1015 (1985).

[2] N. B. Ruderman, *Annu. Rev. Med.* **26**, 245 (1975).

[3] A. E. Harper and C. Zapalowski, *in* "Metabolism and Clinical Implication of Branched-Chain Amino and Ketoacids" (M. Walser and J. R. Williamson, eds.), p. 195. Elsevier/North-Holland, New York, 1981.

hindquarter as a viable muscle preparation. It has also found use in studies of endocrine control and of exercise on amino acid metabolism in muscle. Experiments on the hindquarter of rats have been reported in many early investigations,[4-7] but no systematic evaluation of the metabolic activities of this preparation in these early studies was attempted. In 1971, Ruderman et al.[8] described in detail (1) the surgical preparation, (2) the perfusion system, and (3) metabolic, functional, and morphologic characteristics of the preparation, compared to hindquarter muscle in vivo. A similar preparation, the rat hemicorpus, which includes also the psoas muscles, has been described by Jefferson and co-workers.[9,10] Since the surgical procedure for both preparations is detailed in the references cited, description of this procedure will be abbreviated for simplicity. It should be borne in mind that the intent is to remove all nonmuscle tissues from the circulation. Thus, the descending aorta carries perfusate to the hindquarter, and it passes through the muscle and exits via the vena cava. The perfusate is either collected at intervals or can be recirculated to allow accurate measurement of substrate consumption and product formation. Preparation of the hindquarter, described briefly below, is essentially that of Ruderman et al.[8] except that the hind legs were skinned at the end of the surgery in order to remove as much nonmuscle tissue from the systems as possible, and for ease in rapid sampling of tissue.

Preparation of the Isolated Hindquarter

Female rats, usually fasted for 48 hr and weighing 180–200 g are anesthetized by intraperitoneal injection of aqueous pentabarbital (50 mg/kg body wt). The animal's legs are secured to the operating platform and a midline incision of the abdominal wall is made. Then the uterine, ovarian, and inferior mesenteric arteries are ligated, and the upper half of the uterus, part of the descending colon, and adhering adipose tissue are excised. Ligatures are then placed around the neck of the bladder and the remaining portions of the uterus and descending colon. Adipose tissue in the perineal and retroperitoneal regions are cut away. Next, two loose

[4] D. S. Robinson and P. M. Harris, Q. J. Exp. Physiol. 44, 80 (1959).
[5] G. E. Mortimore, F. Tietze, and D. Stetten, Diabetes 3, 307 (1959).
[6] R. J. Mahler, O. Szabo, and J. C. Penhos, Diabetes 17, 1 (1968).
[7] A. J. Szabo, R. J. Mahler, and O. Szabo, Horm. Metab. Res. 1, 156 (1969).
[8] N. B. Ruderman, C. R. S. Houghton, and R. Hems, Biochem. J. 124, 639 (1971).
[9] L. S. Jefferson, J. O. Koehler, and H. E. Morgan, Proc. Natl. Acad. Sci. U.S.A. 69, 816 (1972).
[10] L. S. Jefferson, this series, Vol. 39, p. 72.

ligatures are placed around the aorta and vena cava, the first above the origin of the right iliolumbar vessels, and the other above the origin of the renal vessels. It is between these two ligatures that the aortic and vena cava cannulae will later be secured. The inferior epigastric, iliolumbar, and renal arteries and veins, the coeliac axis, and portal vein are then ligated, and a ligature placed around the tail.

The loose ligatures previously placed around the aorta and vena cava above the renal vessels are then tied. (Reinstitution of circulation must be done quickly at this point to prevent anaeorbic damage to tissues. With practice, this is accomplished in 60–90 sec.) The aorta is then incised below the tied ligature using iridectomy scissors, and a No. 18 polyethylene catheter filled with 0.9% sodium chloride and 200 units/ml of heparin is introduced until the tip is midway between the incision and the point of bilateral branching of the descending aorta feeding blood to the hind legs. A small amount of heparin-HCl is injected from a syringe, and the catheter is tied into place. The vena cava is then quickly cannulated with a No. 16 needle which is injected through the vessel wall, is positioned so that its tip is at the same level as the aortic cannula, and is secured. The cannula is connected through an adaptor to a piece of clear vinyl tubing. The preparation is then transferred to the perfusion apparatus, where the aorta is attached to tubing containing the previously oxygenated perfusion medium, and perfusion is begun. The perfusion reservoir will normally contain 150 ml of perfusion medium. Fifty milliliters of effluvium from the vena cava is discarded, after which the tube attached to the vena cava is returned through a hole in the operating platform to the reservoir for recirculation. The surgical procedure can be mastered by a novice in a few trial perfusions after careful referral to a rat anatomy book containing illustrations and anatomical nomenclature.[11]

Preparation of Perfusion Medium

The standard perfusion medium used by Ruderman et al.,[8] as well as most previous and subsequent workers perfusing organs, is composed of Krebs–Henseleit bicarbonate buffer[12] containing 4 g% of bovine serum albumin and aged erythrocytes (7–8 g% hemoglobin[13]). The perfusate is also usually fortified with substrates (5–8 mM glucose). In addition, lac-

---

[11] E. C. Greene, "Anatomy of the Rat." Hefner, New York, 1959.
[12] H. A. Krebs and K. Henseleit, *Hoppe-Seyler's Z. Physiol. Chem.* **210**, 33 (1932).
[13] R. P. Geyer, *Fed. Proc., Fed. Am. Soc. Exp. Biol.* **34**, 1499 (1975).

tate plus pyruvate is always present in the erythrocytes, giving a perfusate concentration of about 2–3 mM.

For our purposes in studying amino acid balance and their metabolism, it is advantageous to simplify the perfusion system for balance studies of amino acid metabolism: (1) omission of red cells avoids coaddition of interfering substrates (principally lactate and citrate) and of metabolites of the erythrocytes; and (2) omission of serum albumin avoids coaddition of fatty acids, acetate, and citrate and perhaps bound hormones, which are difficult to remove completely from the albumin. Perfluorocarbon emulsions can be effectively used in place of erythrocytes for oxygen-carrying capacity, and pluronic polyol can substitute for serum albumin as an oncotic agent, as well as an aid in stabilizing the perfluorocarbon emulsion. The Krebs–Henseleit buffer used contains (in mM) NaCl, 118; KCl, 5; $MgCl_2$, 1.2; KH_2PO_4, 1.2; and $NaHCO_3$, 25. ($CaCl_2$, to 2.5 mM, is added after the perfluorocarbon, pluronic polyol, and gas equilibration.) For convenience, 0.5 M stock solutions of salts are prepared and stored at 4°. All water used is freshly glass distilled and deionized. The perfusion medium is composed of the above buffer containing 10% (v/v) perfluorocarbon FC-47 (perfluorotributylamine, now numbered FC-43, 3 M Co., St. Paul, MI) and 3% (w/v) pluronic polyol (F108, BASF Wyandotte Corp., Wyandotte, MI). Ten grams of the pluronic polyol is dissolved in 260 ml Krebs–Henseleit buffer lacking calcium and bicarbonate ions. In a glass container chilled with ice and water, the pluronic polyol solution in buffer is gassed with CO_2 for 5–10 min, after which the CO_2 entrance tube is raised to about 0.5 cm above the surface to maintain a CO_2 atmosphere, and the 1 cm diameter probe of the sonicator is placed in the above solution about 3 cm below the surface. With this probe activated (100 W), 30 ml of perfluorocarbon is slowly introduced to the solution through a long slightly bent hypodermic needle, the tip of which is positioned slightly below the end of the probe.[13] Sonication is performed for 10 min, and stopped for 10 to 15 min to cool the solution. This procedure is repeated twice to obtain a stable perfluorocarbon emulsion in buffer. Care should be taken to keep the temperature below 5° during sonication. Immediately before use, the emulsion is gassed with an O_2/CO_2 (19:1) mixture for 30 min, $CaCl_2$ is added, and the pH carefully adjusted to pH 7.4 with $NaHCO_3$. Finally, the medium is filtered through a Millipore filter 1 PKG SSWPO 4700, 3 μm. (This amount of perfusate, enough for two perfusions, can be scaled up to 600 ml, but no greater, owing to limitation of the sonication step. The emulsion as prepared is stable for several days in the refrigerator, but as a precaution we always filter the emulsion immediately before use.) After a washout of 50 ml of the medium, time is set at zero and hindquarters are cyclically perfused (100 ml) for various periods of time.

Perfusion Apparatus

The rat is mounted on a Plexiglas platform which is placed on a glass container, the perfusate reservoir, which has two or three side arms for ease of sampling. The animal and reservoir are contained in a closed (with folding front door) Plexiglas chamber which is gassed with humidified and heated (37°) air to maintain temperature and to prevent dehydration of the skinned hindquarter. The perfusate circuit is as follows: reservoir → roller pump (or any source having multiple stepwise or continuously variable outputs) → plastic mesh filter (taken from a disposable blood-transfusion set) → top entrance of a multiple-bulb glass oxygenator. Near the bottom of the oxygenator there is an overflow circuit which returns perfusate directly back to the reservoir. Flow through the first roller-pump is set at 250–300 ml/min in order to assure adequate oxygen exchange. From the bottom of the oxygenator the perfusate is pumped by a second roller pump (or second channel) back to the aortic inflow to the hindquarter. The perfusate leaving the animal is then returned to the reservoir from which it is recycled. Arterial pressure (70–90 mm Hg in the tubing leading to the arterial cannula) is maintained by a screw clamp or needle valve inserted between the overflow and hindquarter circuits. This provides a constant flow rate through the hindquarter of 12–15 ml/min. The oxygenator is water-jacketed and temperature-controlled to 37°. The gas circuit, pre-heated and humidified (O_2:CO_2, 19:1), enters the oxygenator just above the perfusate overflow bypass, and leaves the top of the oxygenator to the atmosphere. Gas flow is 300 ml/min. If it is desired to measure CO_2 production from ^{14}C-labeled substrates, bifurcations in the tubing leading to and from the hindquarter can be installed for sampling. Since the flow rate is known, the difference in [14]CO_2 in the perfusate entering and leaving the hindquarter gives a measure of the rate of [14]CO_2 production over any sampling interval. The perfusate samples are quickly added to closed vials with a hanging well containing triethanolamine or other alkaline material. Perchloric acid is added by injection and, after an appropriate period of shaking for complete release and capture of CO_2, the alkaline material in the hanging well is transferred to a scintillation vial and counted for radioactivity. Alternatively, an indirect measurement of [14]CO_2 production over an entire perfusion period has proved to be fairly accurate.[14] This simply entails measurement of total radioactivity added at the beginning of a perfusion, and measurement of total acid-soluble radioactivity recovered in tissue plus perfusate at the end of a perfusion.

[14] S.-H. C. Lee and E. J. Davis, *Biochem. J.* **233**, 621 (1986).

Sampling of Perfusate and Tissue

A few milliliters of medium are taken from the reservoir at various time intervals. Perchloric acid is added to a concentration of 5% and the samples are centrifuged at 17,500 g for 30 min to remove perfluorocarbon. The cold supernatant fluids are neutralized with 4 M K_2CO_3, or with 4 M KOH containing 0.5 M Tris base. At the end of a given perfusion period, portions of muscle (mainly quadriceps and biceps) from hind legs are rapidly freeze-clamped with Wollenberger forceps which have been precooled in liquid nitrogen. Between 7 and 8 g of tissue from each rat is frozen and extracted. If there is delay (one or more days) between freeze-clamping of tissue and assay for metabolites, it is best to store the intact frozen tissue at $-70°$ until immediately before metabolite assays are to be carried out. To pulverized frozen tissues, three volumes of 2 M $HClO_4$ is added, followed by about 1 min of ultrasonic homogenization. After centrifugation of the acid precipitates, the fluids are neutralized with K_2CO_3. The temperature is maintained near $0°$ throughout extraction and neutralization.

Weight of Muscle Perfused

Ruderman et al.[8] have carefully estimated the total muscle (plus adipose) in the hindquarter, based on total body weight. Since adipose tissue comprises only a small portion of the soft tissue mass (especially for the fasted animal), total soft tissue is equated with muscle mass. The ratio of total body weight to grams of muscle perfused which they found was 6.0.

Metabolic Integrity of the Tissue; Examples of Perfusion Conditions

When hindquarters are perfused with glucose plus insulin, the rates of glucose consumption and lactate production, and the lactate/pyruvate ratio are near the normal expected range.[15] A battery of critical metabolite concentrations and ratios indicative of the functional integrity of each preparation is routinely carried out. These include phosphocreatine, ATP, ADP, AMP, citrate cycle intermediates, glycogen, and several amino acids.[14-16] Ordinarily, these are all well maintained. In our experience, only occasionally (about 5 % of animals) did the perfusion rate drop. If the flow rate dropped more than 50% the phosphocreatine content declined almost immediately, followed within minutes by glycogen shedding with attendant accumulation of lactate, lowering of the ATP content and the ATP/

[15] S.-H. C. Lee and E. J. Davis, J. Biol. Chem. 254, 420 (1979).
[16] E. J. Davis and S.-H. C. Lee, Biochem. J. 229, 19 (1985).

ADP ratio, and changes in the proportions of muscle glutamate, aspartate, and alanine. Data from such perfusions were discarded.

In virtually all previous studies that have been done on the metabolism of intact muscle preparations (diaphragms, perifused and perfused muscles), glucose is present in plasma — or supraphysiological — concentrations. However, in order to study the metabolism of individual amino acids to greater advantage, it would be useful to study metabolism without other interfering substrates. The use of an artificial perfusion medium makes this possible. In fact, this hindquarter preparation is stable for at least 2 hr without any added substrate whatsoever.[16] Table I summa-

TABLE I
CRITICAL METABOLITES IN MUSCLE BEFORE AND AFTER
PERFUSION[a]

| Tissue | Concentration of metabolite (μmol/g tissue) | |
	After washout	120 min perfusion
ATP	9.2 ± 0.4	8.6 ± 0.7
ADP	0.93 ± 0.02	0.83 ± 0.04
AMP	0.22 ± 0.02	0.15 ± 0.03
ATP/ADP	9.9	10.4
Phosphocreatine	17.0 ± 1.1	17.7 ± 0.4
Lactate	7.6 ± 1.5	7.7 ± 1.0
Citrate	0.15 ± 0.02	0.13 ± 0.03
Malate + fumarate	0.25 ± 0.05	0.14 ± 0.04
Glucose	2.0 ± 0.4	0.84 ± 0.14
Glutamine	2.7 ± 0.3	2.0 ± 0.2
Alanine	2.1 ± 0.2	1.7 ± 0.3
Aspartate	0.40 ± 0.06	0.10 ± 0.02
Glycogen	11.9 ± 1.7	13.0 ± 3.3
Perfusate	60 min perfusion	120 min perfusion
Lactate	4.5 ± 0.6	6.5 ± 1.2
Pyruvate	0.45 ± 0.06	0.59 ± 0.12
Lactate/pyruvate	9.9	11.1
Citrate	0.17 ± 0.02	0.28 ± 0.02
Glucose	0.88 ± 0.12	0.57 ± 0.15

[a] Perfusions are from 48-hr fasted rats without added substrates. Tissues were freeze-clamped immediately after "washout" (50 ml) or after 2 hr perfusion. Samples are mean values from 5–7 animals ± SEM. From Davis and Lee.[16]

rizes some of the above criteria of substrate-free perfusions. Hence, these data demonstrate that endogenous substrates are adequate to keep the resting hindquarter viable for 2 hr. Although there was slow glycolysis of endogenous glucose, presumably the principle substrate was endogenous lipid, as well as some endogenous amino acids and those formed by proteolysis. With the knowledge that the aromatic amino acids are not metabolized, and of the frequency of occurrence of the individual amino acids in muscle proteins, it is possible to calculate the rate of catabolism of the individual indispensable (not interconvertible) amino acids. From such balance studies it can be concluded that, of the indispensable amino acids, only the branched-chain amino acids and methionine are significantly catabolized.

Muscle is in a protein-catabolic state after a fast. This is due, at least in part, to a decrease in the insulin/glucagon ratio. Even in substrate-free perfusions, the muscle is insulin-responsive, resulting in decreased net appearance of all of the indispensable amino acids.[16] In fact, if hindquarters from fasted rats are perfused with plasma levels of all amino acids, the muscle rapidly returns to protein balance (protein synthesis and degradation are equal) after insulin addition.[14] Therefore, under these conditions, any net consumption of a nondispensable amino acid is a measure of degradation.

Partial or Complete Degradation of Branched-Chain Amino Acids

The types of balance studies described above give an estimate of the rates of initial stages in the catabolism[1] of the branched-chain amino acids, but provide no direct information as to the fate of the amino acid carbon. A useful technique to resolve this question is to perfuse with plasma levels of all amino acids plus a single amino acid which is ^{14}C-labeled, followed by determination of radioactivity retained in metabolites. Numerous procedures have been, and are being developed for separation, quantitation, and identification of amino acid metabolites. The reader is referred to several chapters in this volume which specifically address this subject.

Acknowledgments

The authors' work is supported by grant to E.J.D., USPHS AM13939, and the Grace M. Showalter Trust.

[60] Measurement of Branched-Chain α-Keto Acid Dehydrogenase Flux Rates in Perfused Heart and Liver

By David S. Lapointe, Ellen Hildebrandt, Denis B. Buxton, Tarun B. Patel, Parvan P. Waymack, and Merle S. Olson

The use of perfused organs to investigate metabolic processes offers many advantages and insights not available to similar studies utilizing tissue homogenates and isolated cells, though such techniques have advantages of their own. The main advantage in the use of isolated perfused organs is in the facility to preserve intact the control systems and relationships found in the organ of the intact animal. The ability to control and to specify the input concentrations of metabolites to a perfused organ and to measure the outflow of the metabolic products from perfused organs allows the calculation of flux rates and transport parameters. By combining data obtained from measurements of pathway fluxes with data obtained from freeze-clamped tissue where determinations of metabolic intermediates and enzymatic activity parameters can be made, the importance of single steps in the control of metabolic pathways can be ascertained.

In certain metabolic situations, such as during starvation,[1] catabolism of amino acids derived from muscle protein is used to provide energy and maintain blood glucose levels. The initial step in metabolism of the branched-chain amino acids, transamination to the branched-chain α-keto acids, occurs primarily in the extrahepatic tissues, while the highest cellular activity for the branched-chain α-keto acid dehydrogenase is found in the liver. In the liver, the metabolic products of valine and isoleucine are available for conversion into blood glucose, whereas leucine catabolism results in ketone body production. In the heart, the branched-chain amino acids are able to be transaminated to their corresponding branched-chain α-keto acids and, albeit at a lower rate than in the liver, metabolized via the branched-chain α-keto acid dehydrogenase. Perfused organs, heart and liver, thus serve as model systems to study the control of branched-chain α-keto acid metabolism and the interactions of other metabolic pathways with the metabolism of the branched-chain amino acids in these organs.

Procedures

Perfusion Medium

Krebs–Henseleit bicarbonate buffer (pH 7.4; 118 mM NaCl, 25 mM NaHCO$_3$, 5.9 mM KCl, 1.2 mM NaH$_2$PO$_4$, 0.58 mM MgSO$_4$, 1.25 mM

---

[1] M. Kaser, R. Kaser, and H. Lestradet, *Metabolism* **9**, 926 (1960).

$CaCl_2$) was used for both the heart and the liver perfusions.[2] Perfusion buffer was brought to 37° and saturated with $O_2 : CO_2$ (95% : 5%) by pumping the buffer through a membrane oxygenator made from thin-walled silicone tubing (Silastic; i.d. 0.058 in., o.d. 0.077 in.,) 25 ft in length, wound around a milled aluminum cylinder which was thermostated to 37° and enclosed in a Plexiglas cylinder continually gassed with a mixture of O_2 and CO_2 (95% : 5%). At the flow rates used in the perfusions (less than 40 ml/min), complete temperature and gas saturation was obtained using this device.[3]

Perfused Heart

The simplest perfusion system for rat heart perfusion is the Langendorff[4] method where hearts are perfused retrograde through the aorta to the coronary vasculature. Alternatively, a working heart preparation which is more complicated to set up but yields information having greater physiological significance can be used.[5] The analysis of the metabolic flux of branched-chain α-keto acids in the heart is very similar to that for the perfused liver and is explained for both organs below.

Male Sprague-Dawley rats (180–250 g), either fed *ad libitum* or fasted for 24 to 48 hr depending on the experimental protocol, were anesthetized by an intraperitoneal injection of sodium pentabarbitol (10 mg/100 g body wt). The chest cavity was opened, the beating heart was excised and rinsed briefly by placing it in a small beaker containing cold perfusion buffer before being attached to the perfusion apparatus. Using forceps, the aorta was opened gently to allow the heart to be slipped over the cannula and tied securely in order to begin the retrograde perfusion.[4] Underneath the cannulated heart, a small funnel was positioned to collect the effluent perfusate and direct it to the fraction collector. Flow rates were maintained at 10 ml/min, which resulted in a stable aortic pressure of approximately 50 mm Hg throughout the perfusion. At a minimum, 5 min of substrate-free perfusion occurred prior to starting the experimental protocols. Infusion of the 1-^{14}C-labeled branched-chain α-keto acids as well as of unlabeled substrates was made into the perfusion curcuit immediately prior to the heart using syringe pumps to provide uniform delivery of infused substances. Oxygen consumption was measured using a Clark-type electrode placed in the effluent portion of the perfusion system. Effluent samples were collected at 30-sec intervals.

[2] H. A. Krebs and K. Henseleit, *Z. Physiol. Chem.* **210**, 33 (1932).
[3] R. Scholz, W. Hansen, and R. G. Thurman, *Eur. J. Biochem.* **38**, 64 (1973).
[4] O. Langendorff, *Pfleugers Arch. Gesamte Physiol. Menschen Tiere* **61**, 291 (1895).
[5] J. R. Neely and M. J. Rovento, this series, Vol. 39, p. 43.

Perfused Liver

The technique for perfusion of rat livers which we have chosen to use is based on the nonrecirculating perfusion technique of Scholz *et al.*[3] Other rat liver perfusion techniques and apparatuses have been described elsewhere.[6,7] The advantages of the nonrecirculating perfusion technique are several. The flexibility to alter rapidly the perfusate composition by either removal or addition of substances (substrates, inhibitors, hormones, metal ions) allows precise experimental manipulations of the perfused organ. Since perfusate passes through the organ once, there is neither accumulation nor depletion of metabolites in the perfusate which could alter the perfusion conditions.

Male Sprague-Dawley rats (180–250 g), either fed *ad libitum* or fasted for 24 to 48 hr depending on the experimental protocol, were anesthetized by an intraperitoneal injection of sodium pentabarbitol (10 mg/ 100 g body wt). Details of the surgical techniques for cannulation of the liver are presented elsewhere.[6,7] Flow rates were maintained at 35 ml/min throughout the perfusion. Prior to starting the experimental protocol, livers were perfused with substrate-free buffer for a period of not less than 15 min to ensure stabilization of the liver and washout of endogenous hormones. Infusion of the 1-^{14}C-labeled branched-chain α-keto acids as well as of unlabeled substrates was made into the perfusion circuit immediately prior to the liver. Oxygen consumption was measured using a Clark-type electrode placed in the effluent portion of the perfusion system. Typically, during the substrate-free perfusion period, livers consumed 40–55% of the supplied oxygen. Effluent samples were collected at 30-sec intervals.

For the analysis of the released $^{14}CO_2$ from the metabolism of branched-chain α-keto acids in the perfused heart or liver, 2.5-ml aliquots of the effluent perfusate were placed in 25-ml Erlenmeyer flasks equipped with stoppers and center wells. Addition of 0.5 ml of 2 M acetate buffer, pH 3.3, to the sample of perfusate via a syringe and needle into the stoppered flask initiated the release of the $^{14}CO_2$ from the perfusate. Into the center well of each flask assembly, 0.3 ml of phenylethylamine was added to absorb the released $^{14}CO_2$. Flasks were shaken gently for 60 min before removing the center wells to scintillation vials containing 10 ml scintillation fluid. The specific activity of the infused branched-chain α-keto acid was determined by removing an aliquot of the influent perfusate for determination of the radioactivity and concentration of the branched-

[6] B. D. Ross, "Perfusion Techniques in Biochemistry." Oxford University Press, Clarendon, England, 1972.

[7] J. H. Exton, this series, Vol. 39, p. 25.

chain α-keto acid. The concentration of aliquots of branched-chain α-keto acids was determined enzymatically using purified branched-chain α-keto acid dehydrogenase and dihydrolipoyl dehydrogenase that were generously supplied by Dr. Lester J. Reed (University of Texas, Austin), by measuring the stoichiometric formation of NADH produced in a spectrophotometric assay using the procedure of Pettit et al.[8] Metabolic flux through the branched-chain α-keto acid dehydrogenase in the perfused heart or liver was calculated using the specific activity of the infused branched-chain α-keto acid, the flow rate, and the weight of the organ to correct the $^{14}CO_2$ disintegrations per minute released from the perfusate. Metabolic fluxes are expressed as micromoles/hour/gram liver weight or nanomoles/minute/gram heart weight.

Ketogenesis, expressed as the production of acetoacetate and β-hydroxybutyrate from α-ketoisocaproate, was measured in the effluent perfusate samples using the procedures of Mellenby and Williamson[9] and Williamson and Corkey,[10] respectively. Radiolabeled branched-chain α-keto acids (α-ketoisocaproate and α-ketoisovalerate) were produced from their corresponding branched-chain amino acids (leucine and valine, respectively) using the enzymatic preparation method and purification described by Rüdiger et al.[11]

Freeze-Clamp Tissue Analysis of Branched-Chain α-Keto Acid Dehydrogenase

For the analysis of the tissue activity of the branched-chain α-keto acid dehydrogenase from perfused heart or liver, perfused tissues were freeze-clamped at selected intervals during perfusions using aluminum tongs cooled in liquid nitrogen. Frozen tissue was powdered with a porcelain mortar and pestle that was cooled to $-70°$ in a dry ice:acetone bath. During the grinding to a powder, the mortar was kept on powdered dry ice. Frozen tissue powder (0.2 g) was weighed in a liquid nitrogen-cooled Potter homogenizer to which 2 ml of 50 mM potassium phosphate, pH 7.4, with 2.5 mM cysteine, was added subsequently and the contents were homogenized (8–10 strokes) using a motor-driven Teflon pestle. A 1-ml aliquot of the homogenate was removed and mixed with 0.25 ml rabbit serum and stored on ice until needed (no longer than 30 min). The activity of the

[8] F. H. Pettit, S. J. Yeaman, and L. J. Reed, *Proc. Natl. Acad. Sci. U.S.A.* **75**, 4881 (1978).
[9] J. Mellenby and D. H. Williamson, *in* "Methods of Enzymatic Analysis" (H. V. Bergmeyer, ed.), Vol. 4, p. 1840. Academic Press, New York, 1974.
[10] J. R. Williamson and B. E. Corkey, this series, Vol. 13, p. 434.
[11] H. W. Rüdiger, U. Langenbeck, and H. W. Goedde, *Biochem. J.* **126**, 445 (1972).

branched-chain α-keto acid dehydrogenase present in homogenized samples was determined in duplicate by adding 0.5 ml of the homogenate/rabbit serum mixture to 0.5 ml of assay solution contained in the bottom of a 25-ml Erlenmeyer flask equipped with rubber stopper and center well and incubating for 5 min at 37°. The final assay mixture (1 ml) contained 1 mM α-keto[1-^{14}C]isocaproate (100 dpm/nmol), 5 mM NAD$^+$, 1 mM cysteine, 0.3 mM CoASH, 0.4 mM thiamin pyrophosphate, 0.04% (w/v) Lubrol WX, 4 mM magnesium chloride, and 20 mM potassium phosphate buffer, pH 7.4. The assay was terminated by the addition of 0.5 ml of 2 M acetate buffer (pH 3.3). Collection and determination of ^{14}CO$_2$ was performed as described above for perfusate samples.

Washout Kinetics of Branched-Chain α-Keto Acids in Perfused Rat Liver

Prior to performing any measurement of branched-chain α-keto acid metabolic flux rates, or the metabolic flux rates of any other metabolite for that matter, it is necessary to ascertain whether more than one process is responsible for the production of the monitor metabolite, in this case ^{14}CO$_2$. In the case of the branched-chain α-keto acids, transamination back to the homologous amino acids could create extra pools of radiolabeled species which might lead to erroneous interpretations. The measurement of the washout kinetics for the branched-chain α-keto acids is simple to perform, but must be interpreted correctly. Two conditions must be met for proper interpretation. First, the tracer species must be present in trace quantities, that is, it must have an insignificant concentration relative to the unlabeled species. The second condition requires that the system be operating under steady-state conditions. This means that the metabolic fluxes must be time-invariant during the period of measurement and that the specific activity of the pools be time-invariant at the start of the washout period.

To assess the nature of the kinetics of ^{14}CO$_2$ production from the branched-chain α-keto acids, a liver from a fed rat was perfused with 1 mM α-ketoisocaproate for 15 min prior to label infusion, when a tracer amount of α-keto[1-^{14}C]isocaproate was infused for a 10-min period. During the tracer infusion period, sample collection occurred at 30-sec intervals to quantitate the rate of ^{14}CO$_2$ production following introduction of the tracer. At the beginning of the washout period, after the 10 min of label infusion had ended, samples were collected at 12-sec intervals for the first 3 min and at 30-sec intervals thereafter. The results of this experiment are shown in Fig. 1A, where the data, corrected for the specific activity of α-ketoisocaproate during the tracer infusion period and the weight of the liver, are plotted in a semilog plot to illustrate the two distinct kinetic

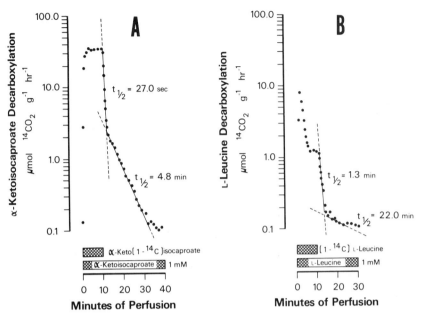

FIG. 1. Washout kinetics of $^{14}CO_2$ produced from α-keto[1-^{14}C]isocaproate (A) and [1-^{14}C]leucine (B) in the perfused rat liver. Livers were perfused with unlabeled substrates for 15 min prior to infusion of 1-^{14}C-labeled substrates. Samples were collected at 12-sec intervals during the first 3 min of the washout and at 30-sec intervals elsewhere. From Patel et al.[15]

components of the metabolism of α-ketoisocaproate. The data, which can be analyzed by fitting the data to a sum of exponentials using a computer program such as SAAM[12] or MLAB[13] or using a calculator to do curve peeling of the data,[14] indicate a rapid kinetic component having a $t_{1/2}$ of 27 sec which accounts for 94% of the $^{14}CO_2$ washed out from the liver. The other component has a $t_{1/2}$ of 4.8 min. Similar results are obtained with the other branched-chain α-keto acids[15] and in the perfused heart.[16] Since α-ketoisocaproate can be transaminated to leucine using glutamate as an amino group donor, it was necessary also to examine the kinetics of $^{14}CO_2$

[12] M. Berman and M. F. Weiss, "SAAM Manual," NIH Publ. No. 78-1168. U.S. Dept. of Health, Education and Welfare, Washington, D.C., 1978.

[13] G. D. Knott, Comp. Prog. Biomed. 10, 271 (1979).

[14] J. A. Jacquez, "Compartmental Analysis in Biology and Medicine." University of Michigan Press, Ann Arbor, Michigan, 1985.

[15] T. B. Patel, M. S. DeBuysere, L. L. Barron, and M. S. Olson, J. Biol. Chem. 256, 90009 (1981).

[16] C. K. Buffington, M. S. DeBuysere, and M. S. Olson, J. Biol. Chem. 254, 10453 (1979).

production from [1-^{14}C]leucine. The washout of leucine from the perfused rat liver was performed in the same manner as the washout of α-ketoisocaproate. These results, shown in Fig. 1B, indicate that the kinetic components of the leucine washout have different half-lives than in the experiment with α-ketoisocaproate. However, there are still two major kinetic components to the leucine washout. The increased $t_{1/2}$ values associated with leucine most likely reflect the rate-limiting nature of the transamination of leucine to α-ketoisocaproate in the liver. It should be mentioned, though a detailed analysis is outside of the scope of this article, that the measured $t_{1/2}$ could represent many processes associated with the metabolism of α-ketoisocaproate or leucine, such as transfer of α-ketoisocaproate or leucine between different compartments (for example, cytosolic or extracellular compartments) and different metabolite pools (for example, protein or transaminations between the branched-chain α-keto acid and the branched-chain amino acid). The magnitude and the short $t_{1/2}$ associated with the production of $^{14}CO_2$ from α-ketoisocaproate indicates that it is possible to measure rapid changes in the metabolism of the branched-chain α-keto acids.

Dose–Response Curve for Metabolism of Branched-Chain α-Keto Acids

In a manner similar to that used to characterize enzymes, the metabolism of the branched-chain α-keto acids can be described by determining the maximal metabolic rates and the half-maximal concentrations for the utilization of the branched-chain α-keto acids. It should be appreciated, however, that utilization of the branched-chain α-keto acids by the heart or liver is dependent on the transport of the branched-chain α-keto acids into the organ as well as the metabolic processes occurring within the organ. In the case of the branched-chain α-keto acids, intermediates along the metabolic pathway, for example, acyl-CoAs and NADH, can exert inhibitory feedback effects which can affect the steady-state flux of branched-chain α-keto acid decarboxylation. Thus, differences among the various branched-chain α-keto acids in their steady-state flux through the branched-chain α-keto acid dehydrogenase reflect a number of processes important to the metabolism of the branched-chain α-keto acids.

Using sequentially increasing concentrations of infused branched-chain α-keto acids, the relationship between perfusate concentration of the branched-chain α-keto acids and the resultant flux rate through the branched-chain α-keto acid dehydrogenase complex can be determined. Figure 2 shows this relationship for α-ketoisocaproate in the perfused liver.[15] Perfusate concentrations of α-ketoisocaproate ranging from 0.05 to 5 mM were obtained using stock solutions at several different concentrations of α-keto[1-^{14}C]isocaproate at known specific activity. By varying the

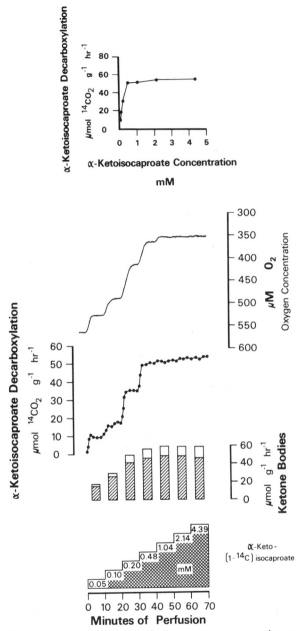

FIG. 2. Effects of varying the α-keto[1-^{14}C]isocaproate concentration on the rates of α-ketoisocaproate decarboxylation and ketogenesis in a perfused rat liver. Infused α-ketoiso-caproate concentrations were increased in stepwise fashion indicated at the bottom of the figure. Oxygen consumption, $^{14}CO_2$ production, and ketone bodies, β-hydroxybutyrate and acetoacetate, were measured as described in Procedures. At the top of the figure, the graph illustrates the relationship between perfusate α-ketoisocaproate concentration and the rate of decarboxylation of the substrate. From Patel et al.[15]

infusion rate of the stock solutions, by using different settings on the syringe pumps, a range of concentrations was obtained as indicated in the bars in Fig. 2. Each separate concentration was infused for a 10-min period to ensure that steady-state conditions prevailed. As the concentration of α-keto[1-^{14}C]isocaproate increased, the rates of production of $^{14}CO_2$, oxygen consumption, and ketogenesis increased until the perfusate concentration of α-ketoisocaproate exceeded 1 mM. The metabolic flux through the branched-chain α-keto acid dehydrogenase complex demonstrated saturation kinetics when plotted versus the perfusate α-ketoisocaproate concentration (see graph of Fig. 2). The rates of ketogenesis increased with increasing concentrations of α-ketoisocaproate until, at saturation ($>$ 1 mM α-ketoisocaproate), the combined rates of β-hydroxybutyrate and acetoacetate production (60 μmol g^{-1} hr^{-1}) were approximately 75% of the rate obtained in a perfused liver from a fed rat using octanoate (0.5 mM) as the ketogenic substrate. The ratio of β-hydroxybutyrate to acetoacetate, an indication of the mitochondrial redox level, doubled in value (0.11 to 0.22) as the concentration of α-ketoisocaproate went from 0.05 m to 5 mM. There was close correspondence between the rates of ketogenesis and the rates of production of $^{14}CO_2$ from α-keto[1-^{14}C]isocaproate at all perfusate concentrations of α-ketoisocaproate. Similarly, the rates of α-ketoisovalerate decarboxylation were shown to saturate at perfusate concentrations in excess of 1 mM.[15] The maximal decarboxylation flux rates obtained for α-ketoisovalerate were 2-fold less than the rates for α-ketoisocaproate infusion (30 versus 54 μmol g^{-1} hr^{-1}). In the perfused rat heart similar determinations demonstrated that α-ketoisovalerate had a higher rate of decarboxylation than α-ketoisocaproate for the same perfusate concentration.[16] For α-ketoisovalerate, which is glucogenic in the liver, the rates of glucose production have been shown to be 10% of the rates measured for $^{14}CO_2$ production, suggesting that loss of carbon mass was occurring at some point in the pathway.[17] Comparison of these results with similar studies on the metabolism of α-ketobutyrate,[18] a homolog of pyruvate which is metabolized directly to propionyl-CoA, suggests that metabolic products derived from the decarboxylation of α-ketoisovalerate do not completely enter the propionyl-CoA pool.

Effect of Fatty Acids on Metabolism of Branched-Chain α-Keto Acids

The branched-chain α-keto acid dehydrogenase complex, similar to the pyruvate dehydrogenase complex, is regulated by feedback inhibition by

[17] T. B. Patel, D. S. Lapointe, and M. S. Olson, *Arch. Biochem. Biophys.* **233**, 362 (1984).
[18] D. S. Lapointe and M. S. Olson, *Arch. Biochem. Biophys.* **242**, 417 (1985).

the products of its reaction, the branched-chain acyl-CoAs and NADH. Regulation by phosphorylation/dephosphorylation of the branched-chain α-keto acid dehydrogenase complex has been demonstrated in mitochondria,[19] purified enzyme,[20] and perfused heart.[21] It is of interest to determine whether other pathways, such as fatty acid oxidation, which produce similar metabolic products would have effects on the *in vivo* activity of the branched-chain α-keto acid dehydrogenase complex. By subjecting the perfused organ to metabolic transitions, such as the coinfusion of various fatty acids[22] or other agents during the metabolism of branched-chain α-keto acids, the effects of metabolic perturbations on the measured flux rates and the extractable activity of the branched-chain α-keto acid dehydrogenase can be measured.

In the rat heart, the extractable levels of branched-chain α-keto acid dehydrogenase are normally low when the heart is freeze-clamped immediately after removal from the rat. This low level of branched-chain α-keto acid dehydrogenase activity increases 10-fold during the first 15 to 20 min of substrate-free perfusion. If α-ketoisocaproate is included in the perfusion fluid, at levels as low as 0.05 mM there is a dramatic increase in the extractable levels of the tissue branched-chain α-keto acid dehydrogenase: 5- to 6-fold in the first 3 min of perfusion with 2 mM α-ketoisovalerate or α-ketoisocaproate.[23] Stable rates of production of $^{14}CO_2$ from α-keto[1-^{14}C]isocaproate or α-keto[1-^{14}C]isovalerate occur within 5 min after perfusion with these branched-chain α-keto acids.

Figure 3 illustrates the correlation between extractable branched-chain α-keto acid dehydrogenase activity and the metabolic flux through the complex in perfused rat hearts. Hearts from rats fed *ad libitum* or fasted for 48 hr were perfused using the Langendorff method described above for 10, 20, or 30 min with [1-^{14}C]leucine. In the rat heart leucine is readily transaminated to α-ketoisocaproate. After the time period for [1-^{14}C]leucine infusion had ended, hearts were immediately freeze-clamped and assayed for active branched-chain α-keto acid dehydrogenase levels. Whether hearts were obtained from fed or fasted rats, the extractable activity of the branched-chain α-keto acid dehydrogenase complex was directly proportional to metabolic flux measured in the perfusate just prior to freeze-clamping the perfused heart. Hearts from fasted rats showed 3- to 6-fold less activity of branched-chain α-keto acid dehydrogenase, measured

[19] K. S. Lau, H. R. Fatania, and P. J. Randle, *FEBS Lett.* **126**, 66 (1981).

[20] R. Paxton and R. A. Harris, *J. Biol. Chem.* **257**, 14433 (1981).

[21] D. B. Buxton and M. S. Olson, *J. Biol. Chem.* **257**, 15026 (1981).

[22] M. Walzer, P. Lund, N. B. Ruderman, and A. W. Coulter, *J. Clin. Invest.* **52**, 2865 (1973).

[23] P. P. Waymack, M. S. DeBuysere, and M. S. Olson, *J. Biol. Chem.* **255**, 9773 (1980).

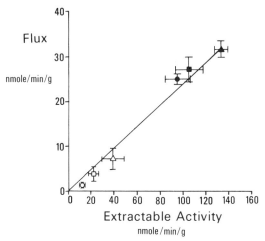

FIG. 3. Correlation between metabolic flux and extractable branched-chain α-keto acid dehydrogenase activity. Hearts from rats fasted for 48 hr (open symbols) or fed *ad libitum* (closed symbols) were perfused for 10 (O, ●), 20 (□, ■), or 30 min (△, ▲) with 0.1 mM [1-^{14}C]leucine. Hearts were freeze-clamped immediately following sample collection for the determination of metabolic flux through the branched-chain α-keto acid dehydrogenase complex. The mean ± standard error of the mean for each treatment group (n = 3 to 8) has been plotted.

as metabolic flux rates in the perfused organ or as extractable activity. Other factors can alter the relationship between extractable activity and metabolic flux through the branched-chain α-keto acid dehydrogenase complex.

Figure 4 shows an experiment where an infusion of octanoate alters the rate of decarboxylation of α-keto[1-^{14}C]isocaproate in the perfused rat heart. As described above, after cannulation of the heart using the Langendorff procedure, a steady infusion of 0.5 mM α-keto[1-^{14}C]isocaproate was begun and samples were taken for determination of the metabolic flux through the branched-chain α-keto acid dehydrogenase complex. After 7 min of α-keto[1-^{14}C]isocaproate infusion, the addition of 0.5 mM octanoate was begun and continued for an additional 7 min. In parallel perfusions, hearts were freeze-clamped at points prior to the beginning and near to the end of the octanoate infusion. Freeze-clamped samples were assayed using the procedure described above for tissue branched-chain α-keto acid dehydrogenase activity. Infusion of octanoate decreased the metabolic flux through the branched-chain α-keto acid dehydrogenase complex by almost 95%. Complete reversal of this inhibition followed within minutes of the termination of the octanoate infusion. The extractable branched-chain α-keto acid dehydrogenase activity showed only a modest decrease in measurable activity (15%). Additional experiments with freeze-clamped

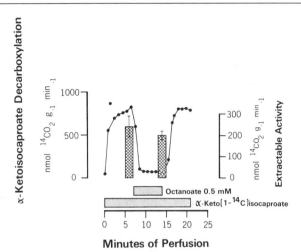

Fig. 4. Comparison of the effects of octanoate on the decarboxylation of α-keto[1-^{14}C]isocaproate and the extractable branched-chain α-keto acid dehydrogenase activity in the isolated perfused rat heart. Hearts were perfused with α-keto[1-^{14}C]isocaproate (0.5 mM) and octanoate (0.5 mM) as indicated in the figure. Hatched bars indicate the extractable branched-chain α-keto acid dehydrogenase activity determined in rat hearts freeze-clamped at the indicated times ($n = 4$) using the assay procedure described in the text. From Waymack et al.[23]

perfused hearts where the infusion of α-ketoisocaproate was replaced by an infusion of 0.5 mM octanoate alone demonstrated that α-ketoisocaproate did not exert a protective effect over octanoate inhibition of tissue branched-chain α-keto acid dehydrogenase activity. Thus the inhibitory effects of octanoate seem to be mediated solely through direct effects on the branched-chain α-keto acid dehydrogenase complex and not through inactivation of the complex. Similar experiments using an infusion of 1 mM pyruvate during the decarboxylation of 0.5 mM α-keto[1-^{14}C]isovalerate demonstrated that reduction of metabolic flux through the branched-chain α-keto acid dehydrogenase complex was accompanied by a reduction in the extractable tissue branched-chain α-keto acid dehydrogenase activity.[23] Both the metabolic flux through the enzyme complex and the extractable activity returned to previous levels once the pyruvate infusion had ended.

Effect of Fatty Acids on Metabolism of Leucine to Branched-Chain α-Keto Acids

In the heart, branched-chain amino acids are able to be transaminated to their corresponding branched-chain α-keto acids, using the specific cytosolic aminotransferase (EC 2.6.1.42, branched-chain-amino-acid aminotransferase) with α-ketoglutarate as the amino acceptor, prior to

being metabolized by the branched-chain α-keto acid dehydrogenase in the mitochondria. It is of interest to know what effect various metabolic states have on the metabolism of the branched-chain amino acids. Several laboratories report stimulation of branched-chain amino acid metabolism by short-chain fatty acids in muscle and heart preparations;[24,25] however, branched-chain α-keto acid metabolism has been shown to be inhibited by octanoate.[16,23]

Using a perfused rat heart, the metabolism of [1-^{14}C]leucine to ^{14}CO$_2$ and α-keto[1-^{14}C]isocaproate was examined in the absence and presence of short- and long-chain fatty acids during an infusion of [1-^{14}C]leucine.[26] Metabolic flux rates through the branched-chain α-keto acid dehydrogenase were measured as the output of ^{14}CO$_2$ from the perfused heart, while the rate of transamination of [1-^{14}C]leucine to α-keto[1-^{14}C]isocaproate was measured by quantitating the amount of ^{14}CO$_2$ produced from the α-keto[1-^{14}C]isocaproate released into the perfusate from the heart using the method of Odessey and Goldberg[27] to oxidize the α-keto acid. Perfusate samples were first acidified to release the ^{14}CO$_2$ resulting from the decarboxylation of α-keto[1-^{14}C]isocaproate produced from [1-^{14}C]leucine in the heart. Following the collection of released ^{14}CO$_2$ for the quantitation of the metabolic flux through the branched-chain α-keto acid dehydrogenase, the same perfusate samples were oxidized with H$_2$O$_2$ to decarboxylate chemically the α-keto[1-^{14}C]isocaproate released into the perfusate from the heart in order to estimate the flux of leucine into α-ketoisocaproate.

Hearts were perfused with 0.1 mM leucine for 30 min prior to infusion of [1-^{14}C]leucine. After 10 min of [1-^{14}C]leucine perfusion, a coinfusion of 0.1 mM palmitate was begun for 15 min and then terminated. Palmitate was bound to defatted bovine serum albumin prior to infusion into the heart. The results of this experiment are shown in Fig. 5. Infusion of 0.1 mM palmitate led to a gradual and substantial decrease in the decarboxylation of [1-^{14}C]leucine, which was readily reversible upon removal of the palmitate. The output of α-keto[1-^{14}C]isocaproate, which was low normally, rose dramatically during the infusion of palmitate but returned to its previous low value after the palmitate infusion had ended. This behavior was contrasted by a similar experiment using the infusion of octanoate (0.5 mM) where a stimulation of [1-^{14}C]leucine decarboxylation was observed.[26] The infusion of pyruvate caused similar decreases in the decarboxylation of [1-^{14}C]leucine as were observed for palmitate infusion.

[24] M. G. Buse, J. F. Biggers, K. H. Friderici, and J. F. Buse, *J. Biol. Chem.* **247**, 8085 (1972).
[25] O. Spydevold and B. Hokland, *Biochim. Biophys. Acta* **676**, 279 (1981).
[26] D. B. Buxton, L. L. Barron, M. K. Taylor, and M. S. Olson, *Biochem. J.* **221**, 593 (1984).
[27] R. Odessey and A. L. Goldberg, *Biochem. J.* **178**, 475 (1979).

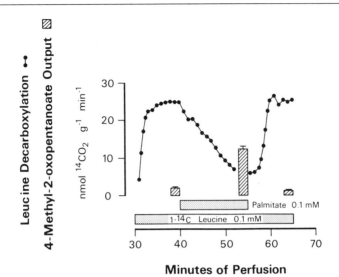

Minutes of Perfusion

FIG. 5. Effect of palmitate on [1-¹⁴C]leucine decarboxylation and α-keto[1-¹⁴C]isocaproate output in the perfused rat heart. Hearts were perfused for 30 min with 0.1 mM leucine before the infusion of [1-¹⁴C]leucine. For infusion, palmitate is bound to bovine serum albumin. Reprinted by permission from Buxton et al.[26] Copyright © 1984 by The Biochemical Society, London.

Summary

The methods described here represent a flexible set of procedures for investigating the metabolism of the branched-chain α-keto acids and other substances in perfused organs, notably the rat heart and liver. These procedures have been used to investigate many aspects of the metabolism of the branched-chain α-keto acids not discussed here, such as the effects on branched-chain α-keto acid metabolism by exposure to α-adrenergic agents,[28] by inhibition of the monocarboxylate translocator,[29] and by the coinfusion of other metabolites.[30,31]

Acknowledgment

The work reported here was supported by grants from the National Institutes of Health, AM-19473 and HL-24654, and the Robert A. Welch Foundation AQ-728.

[28] D. Buxton, L. L. Barron, and M. S. Olson, J. Biol. Chem. **257**, 14318 (1982).
[29] T. B. Patel, P. P. Waymack, and M. S. Olson, Arch. Biochem. Biophys. **201**, 629 (1980).
[30] M. S. Olson, R. Scholz, C. Buffington, S. C. Dennis, A. Padma, T. B. Patel, P. Waymack, and M. S. DeBuysere, in "The Regulation of Carbohydrate Formation and Utilization in Mammals" (C. M. Venezia, ed.), p. 153. University Park Press, Baltimore, MD (1981).
[31] T. B. Patel, L. L. Barron, and M. S. Olson, Arch. Biochem. Biophys. **212**, 452 (1981).

Author Index

Numbers in parentheses are footnote reference numbers and indicate that an author's work is referred to although the name is not cited in the text.

A

Abell, L. M., 237, 453
Abell, L., 442
Aberhart, D. J., 14, 17, 18, 131, 133, 360, 371(3), 374
Adams, C. E., 474
Adams, C. W., 452
Addis, T., 461
Adelberg, E. A., 254
Adibi, S. A., 210, 211
Aftring, R. P., 176, 189, 202, 205, 206, 207(19a), 210, 211(8), 212(24), 213
Aguis, L., 79, 88(3)
Aitken, A., 11, 12, 117, 121(23), 166, 169(3), 202, 208(5), 315
Akaboshi, I., 113
Aki, K., 269, 270
Allen, R. H., 393, 394, 396(3), 398(3), 399(3), 400, 401(5), 403(5), 406(5), 407, 408(9), 409(1, 10), 410(1, 7), 411(7), 412(7), 413(7)
Allen, S. H. G., 400
Allred, J. B., 223, 224
Amédéé-Manesme, O., 136
Amen, R. J., 39
Amenta, J. S., 470
Ames, B. N., 429
An, G., 434
Anderson, J. J., 246, 249
Anderson, P. C., 97, 454
Andreadis, A., 434
Armitage, I. M., 389
Armstrong, N., 110, 136, 146(11)
Ashford, T. P., 469
Atsusaka, T., 282, 287(4), 288(4)
Azzi, A., 253

B

Bachmann, B. J., 105
Bagdasarian, M. M., 350

Bagdasarian, M., 350, 357, 359(12)
Baichwal, V. R., 414, 423, 430
Bakay, B., 221
Baker, F. C., 56, 71, 85
Baker, J. E., 140
Baker, J. J., 134
Baker, K. M., 40
Baker, L., 223
Ballard, F. J., 461, 465(9), 469, 471(9), 472(9)
Bannerjee, D., 389, 391(1)
Barak, Z., 235, 237, 240, 445, 454, 456, 458
Barber, E. F., 254
Barboriak, J. J., 430
Bardawill, C. J., 183, 310, 332
Barron, L. L., 489, 490(15), 491(15), 492(15), 496, 497
Bartholomew, J. C., 423
Bartholomew, J., 416
Bartholomous, R. C., 352
Bartlett, K., 70, 78(2), 79, 81(4), 83, 85, 88(3), 90(4), 92(4), 215, 289
Baudhuin, P., 378
Bauerle, R. H., 234, 235(3), 243
Bauerle, R., 436
Baumgarter, R., 47
Baverel, G., 6, 286
Bayer, E., 115
Bazil, H., 182
Becker, J. E., 259
Becker, K., 28
Beggs, J. D., 434
Beggs, M., 118, 167, 169, 170(15), 175(9), 189, 198(2), 199(2), 201
Beinert, H., 360, 374
Beltzer, J. P., 414, 422(3)
Benedetti, E., 262
Benson, P. F., 140
Berg, C. M., 234, 235(5)
Berg, C., 436, 437(8)
Berg, D. E., 103
Berman, M., 489
Bernard, S., 97, 104(5)

Subject Index

A